Biomechanics of the Female Reproductive System: Breast and Pelvic Organs

From Model to Patient

Biomechanics of Living Organs Series

SERIES EDITORS
Jacques OHAYON, PhD and Yohan PAYAN, PhD

Volumes published in this series

Biomechanics of Living Organs – Hyperelastic Constitutive Laws for Finite Element Modeling, edited by Yohan Payan, PhD and Jacques Ohayon, PhD, 2017.

Biomechanics of Coronary Atherosclerotic Plaque – From Model to Patient, edited by Jacques Ohayon, PhD, Gerard Finet, MD, PhD and Roderic Pettigrew, PhD, MD, May 2020.

Biomechanics of the Female Reproductive System: Breast and Pelvic Organs

From Model to Patient

Edited by

Mathias Brieu

Michel Cosson

Poul Nielsen

ELSEVIER

ACADEMIC PRESS

An imprint of Elsevier

Academic Press is an imprint of Elsevier
125 London Wall, London EC2Y 5AS, United Kingdom
525 B Street, Suite 1650, San Diego, CA 92101, United States
50 Hampshire Street, 5th Floor, Cambridge, MA 02139, United States
The Boulevard, Langford Lane, Kidlington, Oxford OX5 1GB, United Kingdom

Notices

Knowledge and best practice in this field are constantly changing. As new research and experience broaden our understanding, changes in research methods, professional practices, or medical treatment may become necessary.

Practitioners and researchers must always rely on their own experience and knowledge in evaluating and using any information, methods, compounds, or experiments described herein. In using such information or methods they should be mindful of their own safety and the safety of others, including parties for whom they have a professional responsibility.

To the fullest extent of the law, neither the Publisher nor the authors, contributors, or editors, assume any liability for any injury and/or damage to persons or property as a matter of products liability, negligence or otherwise, or from any use or operation of any methods, products, instructions, or ideas contained in the material herein.

ISBN: 978-0-323-90319-6

For information on all Academic Press publications
visit our website at https://www.elsevier.com/books-and-journals

Publisher: Mara E. Conner
Acquisitions Editor: Carrie L. Bolger
Editorial Project Manager: Sara Valentino
Production Project Manager: Surya Narayanan Jayachandran
Cover Designer: Vicky Pearson Esser

Typeset by VTeX

Printed in the United States of America

Last digit is the print number:
9 8 7 6 5 4 3 2 1

Contents

Contributors

Steven Abramowitch
Translational Research Laboratory in Urogynecology, Pittsburgh, PA, United States

Laure Astruc
Univ. Lille, ENSAIT, ULR2461-GEMTEX-Génie et Matériaux Textiles, Lille, France
Junia, Equipe robotique-mécanique, Lille, France

Stéphane Avril
Mines Saint-Étienne, Université Jean Monnet, INSERM, U 1059 SAINBIOSE, Saint-Étienne, France

Thiranja Prasad Babarenda Gamage
Auckland Bioengineering Institute, University of Auckland, Auckland, New Zealand

O. Bou Matar
Centrale Lille, Institute of Electronics, Microelectronics and Nanotechnology, IEMN UMR 8520, Lille, France

Shaniel Bowen
Department of Bioengineering, University of Pittsburgh, Pittsburgh, PA, United States

Mathias Brieu
College of Engineering, Computer Science and Technology, Dept. of Mechanical Engineering, California State University, Los Angeles, CA, United States

Michel Cosson
Centre Hospitalier Universitaire Lille, Université Lille, Lille, France
CHU Lille, Service de Chirurgie Gynécologique, Lille, France
Univ. Lille, CNRS, Centrale Lille, UMR 9013-LaMcube – Laboratoire de Mécanique, Multiphysique, Multiéchelle, Lille, France

Hadrien Courtecuisse
AVR/MLMS ICube, CNRS, University of Strasbourg, Strasbourg, France

John O.L. DeLancey
Norman F. Miller Professor of Gynecology, University of Michigan, L4000 Women's Hospital, Ann Arbor, MI, United States

Sergio Dempsey
School of Biomedical Engineering, Western University, London, ON, Canada

Jérémie Dequidt
Univ. Lille, Inria, CNRS, Centrale Lille, UMR 9189 CRIStAL, Lille, France

Anthony Doyle
Department of Anatomy and Medical Imaging, University of Auckland, Auckland, New Zealand

Bjoern Eiben
Joint Department of Physics, The Institute of Cancer Research and The Royal Marsden NHS Foundation Trust, London, United Kingdom
Centre for Medical Image Computing, Department of Medical Physics and Biomedical Engineering, University College London, London, United Kingdom

Matthew French
Auckland Bioengineering Institute, University of Auckland, Auckland, New Zealand

Xavier Fritel
Poitiers University Hospital, Department of Obstetric and Gynecology, Poitiers, France
Poitiers University, INSERM CIC 1402, Poitiers University Hospital, Poitiers, France

Bertrand Gachon
Poitiers University Hospital, Department of Obstetric and Gynecology, Poitiers, France
Poitiers University, INSERM CIC 1402, Poitiers University Hospital, Poitiers, France
Nantes University, Movement Interaction Performance Lab, MIP EA4334, Nantes, France

Celine Giordano
Centre Hospitalier Universitaire Lille, Université Lille, Lille, France

Geraldine Giraudet
Centre Hospitalier Universitaire Lille, Université Lille, Lille, France

Madeline Hackett
Translational Research Laboratory in Urogynecology, Pittsburgh, PA, United States

David Hawkes
Medical Physics and Biomedical Engineering, Centre for Medical Image Computing (CMIC) and Wellcome/EPSRC Centre for Surgical and Interventional Science (WEISS), University College London, London, United Kingdom

John Hipwell
Toshiba Medical Visualization Systems, Edinburgh, United Kingdom
Centre for Medical Image Computing, Department of Medical Physics and Biomedical Engineering, University College London, London, United Kingdom

Khaled M. Ismail
Department of Obstetrics and Gynecology, Faculty of Medicine in Pilsen, Pilsen, Czech Republic
Biomedical center, Faculty of Medicine in Pilsen, Pilsen, Czech Republic

Zhifan Jiang
Sheikh Zayed Institute for Pediatric Surgical Innovation, Children's National Hospital, Washington, DC, United States

Vladimir Kalis
Department of Obstetrics and Gynecology, Faculty of Medicine in Pilsen, Pilsen, Czech Republic
Biomedical center, Faculty of Medicine in Pilsen, Pilsen, Czech Republic

Kejia Khoo
Auckland Bioengineering Institute, University of Auckland, Auckland, New Zealand

Pauline Lecomte-Grosbras
Univ. Lille, CNRS, Centrale Lille, UMR 9013-LaMcube – Laboratoire de Mécanique, Multiphysique, Multiéchelle, Lille, France

Liam Martin
Translational Research Laboratory in Urogynecology, Pittsburgh, PA, United States

Gonzalo Daniel Maso Talou
Auckland Bioengineering Institute, University of Auckland, Auckland, New Zealand

Olivier Mayeur
Univ. Lille, CNRS, Centrale Lille, UMR 9013-LaMcube – Laboratoire de Mécanique, Multiphysique, Multiéchelle, Lille, France

Kristin Miller
University of Texas at Dallas, Dallas, TX, United States

Pamela Moalli

Department of Obstetrics, Gynecology & Reproductive Sciences, University of Pittsburgh, Pittsburgh, PA, United Stated

Division of Urogynecology & Pelvic Reconstructive Surgery, UPMC Magee-Womens Hospital, Pittsburgh, PA, United States

Department of Bioengineering, University of Pittsburgh, Pittsburgh, PA, United States

Annie Morch

Univ. Lille, CNRS, Centrale Lille, UMR 9013-LaMcube – Laboratoire de Mécanique, Multiphysique, Multiéchelle, Lille, France

Martyn Peter Nash

Auckland Bioengineering Institute, University of Auckland, Auckland, New Zealand

Department of Engineering Science, University of Auckland, Auckland, New Zealand

Poul Michael Følnss Nielsen

Auckland Bioengineering Institute, University of Auckland, Auckland, New Zealand

Department of Engineering Science, University of Auckland, Auckland, New Zealand

John Pan

Auckland Bioengineering Institute, University of Auckland, Auckland, New Zealand

P. Pernod

Centrale Lille, Institute of Electronics, Microelectronics and Nanotechnology, IEMN UMR 8520, Lille, France

Fabrice Pierre

Poitiers University Hospital, Department of Obstetric and Gynecology, Poitiers, France

Barbara Röhrnbauer

Zurich University of Applied Sciences, Institute of Mechanical Systems, Biomechanical Engineering, Winterthur, Switzerland

Chrystèle Rubod

Univ. Lille, CNRS, Centrale Lille, UMR 9013-LaMcube – Laboratoire de Mécanique, Multiphysique, Multiéchelle, Lille, France

CHU Lille, Service de Chirurgie Gynécologique, Lille, France

Zdenek Rusavy
Department of Obstetrics and Gynecology, Faculty of Medicine in Pilsen, Pilsen, Czech Republic
Biomedical center, Faculty of Medicine in Pilsen, Pilsen, Czech Republic

Abbas Samani
School of Biomedical Engineering, Western University, London, ON, Canada
Department of Medical Biophysics, Western University, London, ON, Canada
Department of Electrical and Computer Engineering, Western University, London, ON, Canada
Imaging Research, Robarts Research Institute, Western University, London, ON, Canada

Marianne Sanchez
Auckland Bioengineering Institute, University of Auckland, Auckland, New Zealand

Freda Song
Department of Anatomy and Medical Imaging, University of Auckland, Auckland, New Zealand

Rhonda-Joy Sweeney
Department of Anatomy and Medical Imaging, University of Auckland, Auckland, New Zealand

Vasileios Vavourakis
Centre for Medical Image Computing, Department of Medical Physics and Biomedical Engineering, University College London, London, United Kingdom
Department of Mechanical and Manufacturing Engineering, University of Cyprus, Nicosia, Cyprus

Shelby White
Tulane University, New Orleans, LA, United States

Jean-François Witz
Univ. Lille, CNRS, Centrale Lille, UMR 9013-LaMcube – Laboratoire de Mécanique, Multiphysique, Multiéchelle, Lille, France

Florence Zara
Univ Lyon, UCBL, CNRS, INSA Lyon, LIRIS, UMR5205, Villeurbanne, France
Domaine scientifique de la Doua, Bâtiment Nautibus, Villeurbanne, France

Editor's biography of 2022

Mathias Brieu is a professor of mechanical engineering at California State University in Los Angeles - CSULA, United States, and is affiliated as a professor of mechanics at Centrale Lille, France. He received his PhD in mechanical engineering in 1999 at Ecole Normale Supérieure Paris-Saclay, France. From 1999 to 2005, he was Assistant Professor and then Associate Professor at the University of Lille. In 2005 he was appointed Full Professor of Mechanical Engineering at Centrale Lille and left this institution for CSULA in 2019. After fifteen years of research in the field, he received the 2020 Senior Prize from the French Biomechanics Society for the research he developed. His main research interests concern the biomechanics of the women's pelvic system with a specific focus on the mechanical behavior of soft tissue, their evolution with age and pathology, the modeling of the pelvic system, and the interaction between organs and the integration of these models into numerical models to study the physiology and physiopathology of the women's pelvic organs to assist surgeons for evaluation of pathology and planning (pre-operative surgical evaluation). Over the last two decades, he has co-supervised almost 25 PhD students and written more than 100 papers. He has continuously developed a research activity answering and focusing on industrial problems and is the inventor of 5 international patents. Mathias Brieu is also co-founder and Chief Scientific Officer of Digyne company, awarded in 2021 for an Innovative industrial project (Biomechanics of women's pelvic system for surgery planning).

Poul Nielsen is a professor in the Auckland Bioengineering Institute at the University of Auckland. His research focuses on using instrumentation, computational models, and quantitative descriptions of physical processes to gain a better understanding of human physiology. Many of his projects couple mathematical modeling with novel instrumentation to improve our ability to understand and interpret measurements of complex biological systems, subject to the constraints of well-understood physical conservation and balance laws.

Poul's research interests include: The Physiome Project (CellML and FieldML); Breast mechanics; Muscle microcalorimetry; Soft tissue mechanics; Pelvic floor mechanics; Modeling abusive head trauma; Soft sensors and actuator arrays; Stereoscopic tracking of soft body deformation; Bioinstrumentation and medical devices.

Poul received an OCCAM Visiting Fellowship at the University of Oxford (2010), and a Royal Society of New Zealand James Cook Research Fellowship (2011). He is currently chair of the Physiome Incorporated Executive Committee and a member of the VPH Institute Board of Directors.

Michel Cosson is currently a full Professor of gynecology and obstetrics at Lille University (Lille, France) and associated member of the French Medical Academy. He is the president of the group for Pelviperineology of the French medical society of gynecology and obstetrics - CNGOF (college gynecologues et obstetriciens Francais). He is the former director of the pedagogic medical council of Lille University and the past Medical director for French medical numerical university - UNF3S. His clinical activities are focused on pelvic floor dysfunction and benign pelvic surgery. After his medical doctorate in gynecology and obstetrics, he completed PhD in Science at Lille University - France. His domain of research deals with pelvic floor disorders. His main research interests concern the characterization of the mechanical properties of pelvic tissues and implants for pelvic surgery, the biomechanics modeling of the pelvic cavity for prolapse and vaginal delivery and the development of medical devices. Michel Cosson is the author of more than 300 papers in peer-reviewed international journals and more than 600 contributions to conferences. He is the editor of surgical gynecological books for Elsevier France and the author of several books on surgical techniques in gynecology.

Foreward

Biomechanics, usually considered a sub-discipline of Mechanical Engineering, is concerned with the structure, function, and motion of the mechanical aspects of biological systems at every level, from whole organisms to organs, cells, and cell organelles. The primary methodology of biomechanics is Theoretical and Applied Mechanics.

The history of biomechanics goes back to antiquity and Aristotle's book "On the Movement of Animals." During the Renaissance, Leonardo da Vinci studied anatomy in the context of mechanics. The beginning of modern biomechanics is attributed to Etienne-Jules Marey who pioneered the use of cinematography to record and study locomotion.

Biomechanics has seen great advances during the last two decades, driven by progress in many supporting technologies, especially imaging, sensors, and computing. The progress is evident on a full spectrum of scales, from the macroscopic all the way down to the molecular. Greatly expanded computer simulation capabilities have led to new insights and applications in medicine that directly contribute to the improvement of patient outcomes.

The field of biomechanics is now in an exciting period of transition from 20th century "theoretical" science to 21st century, enabling technology for precision, patient-specific medicine. The volume "Biomechanics of the female reproductive system: breast and pelvic organs. From model to patient" clearly and forcefully contributes to this evolution.

"Biomechanics of the female reproductive system: breast and pelvic organs. From model to patient" gives a comprehensive account of the current state of the art in the field. It contains nineteen chapters contributed by experts from academic and medical institutions in France, USA, United Kingdom, Canada, Switzerland, New Zealand, and Czechia. It covers anatomy, epidemiology, physiology, and pathophysiology, imaging, surgical treatment options, mechanical properties of tissues (as well as their evolution with age and pathology, and their inverse characterization), mechanical properties of implants and their interaction with living tissue, and computer modeling. It will serve as a classical reference on these subjects for years to come and will be useful for established researchers as well as the new generation of scientists entering the field.

Karol Miller, PhD, DSc
Intelligent Systems for Medicine Laboratory, The University of Western Australia, Perth
Harvard Medical School, Boston, MA

Preface

Since the seminal work of Y.C. Fung,[1] biomechanics has evolved significantly. In the early 1990s, biomechanical models of organs using the finite element (FE) method were proposed. At that time, the biomechanical models were more challenges addressed to mechanical engineers than predictive models, and their clinical applications, if existing, were minimal. The current century sees an incredible increase in computation power, enabling the development of complex biomechanics models, including those dealing with large deformations and nonlinear materials, such as soft organs. Such extraordinary developments open new opportunities for realistic biomechanical simulation of organs, allowing us to improve our understanding of pathologies and the quality of diagnosis, promote patient-specific treatments, and increase the quality of the treatment. Today, first initiated in orthopedy, where materials' behaviors are linearly elastic and deformations are small, several research teams and startup companies are developing biomechanical models and numerical tools for different pathologies and showing promising results.

Every year in the US, about four million pelvic and half a million breast surgeries are performed. Breasts and pelvic organs are part of the women's reproductive system. Each of a woman's reproductive organs is composed of soft tissues experiencing large stresses, strains, and displacements. In addition, women's reproductive organs cannot be considered in isolation, and contact between organs must be considered. Taking each of these biomechanical difficulties into consideration helps us to understand the complexity of the biomechanical models and the numerical simulation for modeling patient-specific physiological, pathological, or surgical cases of the women's reproductive system.

Nonetheless, from the methodological point of view, the steps in constructing FE models remain unchanged: (1) collecting accurate medical images and using the 3D image analysis method to reconstruct the patient-specific anatomy; (2) defining accurate constitutive models that will represent the mechanical behavior of the different tissues involved and identify the parameters of such constitutive models; (3) modeling and characterizing the boundary conditions applied to the system of organs considered; (4) developing and/or using FE software able to handle large deformation, displacements, and nonlinear elastic problems, taking into consideration contact to estimate stresses, strains, and displacements of organs; and (5) analyzing the results to propose additional information to the clinician. However, though this process seems straightforward, such a numerical goal is not as easy to achieve as expected. Several questions are raised or remain open: "How can we develop biomechanical models of the physiology or physio-pathology and the medical procedures, such as delivery or surgery, of such a multi-organ system?"; "How can we define and model

[1] Fung Y.C., Biomechanics: mechanical properties of living tissue, 1981. Springer, New York.

the geometry of observable and non-observable anatomical elements from medical images?"; "Which kinematics of transformation and constitutive models should we consider and how should we identify them?"; "Do such biomechanical model choices depend on the nature of the medical situation considered?"

"Biomechanics of the Female Reproductive System: Breast and Pelvic Organs. From Model to Patient" is the first book to cover biomechanical studies of the woman's reproductive system. Clinicians and scientists gathered to propose a book presenting the main results and studies completed over the last decade to increase the understanding of women's reproductive physiology, diseases, and treatment and improve the design of medical devices and clinical interventions, including surgical procedures. This book is composed of five complementary parts:

Part 1 introduces, with biomechanical perspectives, the anatomy and physiology of women's pelvic organs, physiopathology of delivery, pelvic organ prolapse, urinary incontinence, and the current surgical treatment of women's pelvic floor.

Part 2 moves forward with biomechanical modeling, focusing on the mechanical properties of women's reproductive tissues, their evolution with age or pathology, as well as the mechanical properties of implants. This part ends with a discussion of the constitutive models and a methodology to select a suitable or realistic constitutive model.

Part 3 presents promising results about medical imaging analysis to reconstruct the patient-specific anatomy, providing an objective evaluation of the pathology for the breast or women's pelvic floor, or identifying the mechanical properties of soft tissue.

Part 4 presents FE models for different women's reproductive medical applications, such as delivery, breast or prolapse surgery plan, and real-time assistance during surgery.

International key opinion leaders in medical, mechanical engineering, biomedical engineering, physics, and computer sciences have described, discussed, and presented existing approaches, methodologies, and results. We hope this book will provide the reader with a comprehensive overview of the state-of-the-art research, development, and perspectives in biomechanical modeling and numerical simulation of the women's reproductive system. From a medical perspective, spreading this culture of soft tissue biomechanics and their evolution with aging and damage could improve the knowledge and understanding of pathologies such as genital prolapse, pelvic pain, and pelvic organ dysfunction. One step beyond the modeling and prediction value of the models that could be developed could solve medical issues and improve patient care regarding these pathologies in an area lacking technologies, medical software, and devices. The primary motivation of every researcher in biomedical engineering is to improve the quality of the treatment and the patient's life. We hope our efforts will contribute to this.

Finally, we would like to thank all the authors for their contributions and enthusiasm during the writing of this book. This project was initiated a few months before the COVID-19 pandemic, but fortunately, it has not altered our motivation to bring this project to maturity.

Note of the series editors

The biomechanics of human soft tissues and organs has been an emerging research field since the publication of the Y.C. Fung's book series in the 1990s. We are very pleased to present this new volume, part of our *Biomechanics of Living Organs* book series, on the biomechanics of female reproductive system, breast, and pelvic floor. The publication of such volumes, entirely dedicated to a specific biomechanical subject, is necessary to advance scientific research in the field of biomechanics and to transfer important knowledge to future generations. Therefore we decided to create and edit a series of volumes on the *Biomechanics of Living Organs*. This series began in 2017 with the publication of a first volume on the fundamentals of *Hyperelastic Constitutive Laws for Finite Element Modeling of Living Organs*, and a second one in 2021 on the *Biomechanics of Coronary Atherosclerotic Plaque*. Other volumes on *Reduced Order Models for the Biomechanics of Living Organs*, *Biomechanics of the Aorta, and Biofluids for the Modeling of Physiological Microflows at Cellular Scale and Flows at Organ Scale*, are under preparation. The current new volume of the *Biomechanics of Living Organs series* is entitled *Biomechanics of the Female Reproductive System: Breast and Pelvic Organs. From Model to Patient*. It is the first book to cover women biomechanical studies conducted in this last decade, to increase our understanding of reproductive diseases and treatment as well as improving the design of medical devices and clinical interventions, including surgical procedures. This volume also allows students, researchers, clinicians, and engineers to study the main topics related to biomechanics of the reproductive system in one book written by several experts in the field. We would like to thank the volume editors and all the authors for their valuable contributions to the writing of this book.

Jacques Ohayon, PhD and Yohan Payan, PhD
Editors of the Biomechanics of Living Organs Series, Elsevier
September 2nd, 2022

PART

1

Backgrounds: anatomy, physiology, and physio-pathology

CHAPTER

Pelvic floor functional anatomy☆

1

John O.L. DeLancey

Norman F. Miller Professor of Gynecology, University of Michigan, L4000 Women's Hospital, Ann Arbor, MI, United States

1.1 Overview

There are two compartments in the pelvis: anterior and posterior. They are separated by the vaginal walls, which are attached to the pelvic sidewalls [Fig. 1.1a]. The anterior compartment is occupied by the bladder and urethra. The pubic bones and pelvic sidewalls form inflexible anterior and lateral boundaries to the anterior compartment, so only the anterior vaginal wall is a flexible component. The posterior compartment contains the rectum and anus. In contrast to the anterior compartment, where the front and sides are inflexible, the only inflexible portion of the posterior compartment is the sacrum. The levator ani muscles, the perineal body, and the posterior vaginal wall can be deformed.

The external openings of the bladder and rectum are closed by muscles: the urethral sphincter and the anal sphincter. The vagina and the entire pelvic cavity are closed by the actions of the levator ani muscles. At rest, the levator ani muscles squeeze to hold the openings in the pelvic organs closed and support the weight placed from above by the abdominal and pelvic organs. They also resist the forces created by differences between abdominal and atmospheric pressure both at rest and during pressure increases that occur with activities such as coughing, straining, and jumping.

Pelvic floor anatomy should be understood in its functional context. The muscles, connective tissue support, and organs comprising the pelvic floor are interdependent and consist of uterus and vagina, their connections to the pelvis, and a neuromuscular system that adjusts to changing forces placed on the pelvic floor by pressure differentials, which can exist between abdominal and atmospheric pressure. The weight of the organs is trivial compared to pressure-induced forces. The century-old argument about whether it is failure of the muscles or failure of the connective tissue that causes prolapse has fortunately been replaced by discussion of how these two types

☆ Portions of this chapter are taken from an earlier publication by the same author with permission of the publisher: Hoyte L, Damser M. Biomechanics of the Female Pelvic Floor. Chapter 2: Pelvic floor Anatomy and Pathology. Academic Press. 2016.

Biomechanics of the Female Reproductive System: Breast and Pelvic Organs. https://doi.org/10.1016/B978-0-12-823403-7.00012-9

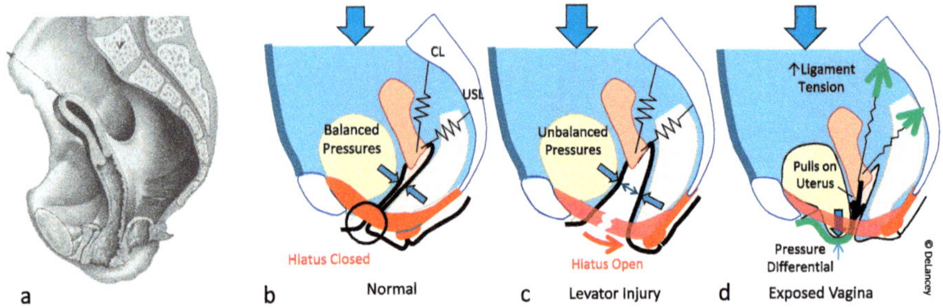

FIGURE 1.1

Sagittal view of the pelvic organs after removal of the bladder and rectum (a) showing the anterior and posterior compartments separated by the uterus and vagina and diagrammatic representation of interactions between levator ani muscle, anterior vaginal wall prolapse, and cardinal/uterosacral ligament suspension. With normal levator function (b) the vaginal walls are in apposition and anterior and posterior pressures generated by abdominal pressure are balanced (opposing blue arrows). Levator damage (c) results in hiatal opening and the vagina becoming exposed to a pressure differential between abdominal and atmospheric pressures. This pressure differential (d) creates a traction force on the cardinal ligament (*CL*) and uterosacral ligament (*USL*). ©DeLancey (Modified from DeLancey, 2012).

of tissues interact to provide pelvic organ support and where individual failure sites occur (Chen et al., 2006). The several structural failures associated with prolapse can occur singly, or in combination, to cause prolapse.

Let us begin by considering the normal situation and how it is altered in prolapse. Under normal conditions, the actions of the levator ani muscles hold the pelvic floor closed and provide a lifting force to prevent pelvic floor descent [Fig. 1.1b] (DeLancey, 2012). In this situation, the pressures in the anterior and posterior compartments are equal and balanced, canceling one another out, making the pelvis, in essence, an isobaric chamber. When the muscles are damaged or weakened, the hiatus in the levator can be pushed open and allow the pelvic organs to descend [Fig. 1.1c]. In this situation, one or both vaginal walls will descend. The resultant misalignment creates a pressure differential between abdominal pressure and atmospheric pressure [Fig. 1.1d]. This pressure, acting on the vaginal walls, places abnormal stresses on the tissues that attach the uterus and vagina to the pelvic walls. Therefore it is this combination of muscle activity and connective tissue attachments that determines the state of pelvic organ support. The sections that follow will provide additional detail to these general observations.

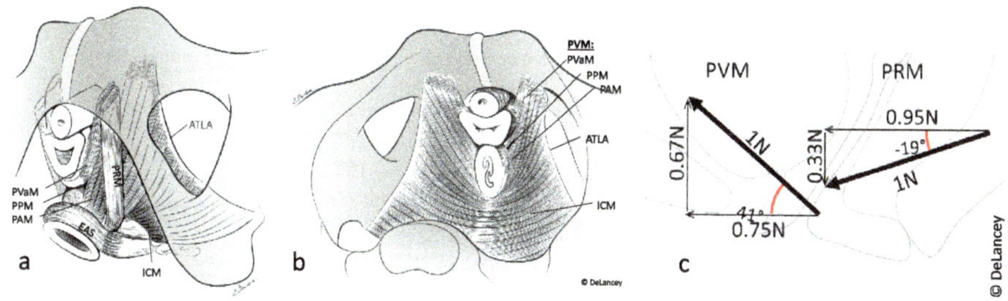

FIGURE 1.2

Schematic view of the levator ani muscles from below (a) after the vulvar structures and perineal membrane have been removed showing the arcus tendineus levator ani (*ATLA*); external anal sphincter (*EAS*); puboanal muscle (*PAM*); puboperineal muscle (*PPM*); iliococcygeal muscle (*ICM*); and puborectal muscle (*PRM*). Note that the urethra and vagina have been transected just above the hymenal ring. Panel b shows the levator ani muscle from above, looking over the sacral promontory (*SAC*) and showing the pubovaginal muscle (*PVaM*). The urethra, vagina, and rectum have been transected just above the pelvic floor. *PAM*, puboanal muscle; *ATLA*, arcus tendineus levator ani; *ICM*, iliococcygeal muscle. (The internal obturator muscles have been removed to clarify levator muscle origins.) (Kearney, 2004) Panel c shows the lines of action for the pubococcygeal and puborectal muscle. (Betschart et al., 2014) ©DeLancey.

1.2 Levator ani

1.2.1 Levator ani muscle and its three subdivisions

1.2.1.1 Overall anatomy of the levator ani muscle

The levator ani muscle consists of three primary subdivisions: pubococcygeal (or pubovisceral), iliococcygeal, and puborectal muscles [Fig. 1.2, 1.3]. Both pubococcygeal and its synonym have been used in illustrations from our published works, so will both occur interchangeably in this chapter. We favor the pubovisceral term coined by Lawson (1974), because it more accurately describes the origin and insertion of this muscle, whereas the older pubococcygeal term is based on evolutionary considerations, rather than human anatomy.

The pubovisceral muscle, in turn, has three components: pubovaginal, puboperineal, and puboanal. Although these origin-insertion pairs are simple and described consistently by all authors who have personally studied the muscle, a profusion of conflicting terms has been applied to this region, making it somewhat difficult to interpret the literature (Kearney et al., 2004). The muscle arises from the pubic bone, passes beside the pelvic organs, and has three areas of insertion that 1) attach to the perineal body (puboperineal), 2) insert into the anal canal and skin (puboanal), and 3) attach the vaginal wall to the medial portion of the pubovisceral muscle (pubovaginal). These are not distinct muscles, but rather three aspects of one muscle.

The iliococcygeal muscle (ICM) begins at the lateral margin of the pubovisceral muscle, with the transition usually occurring at approximately the level of the obtu-

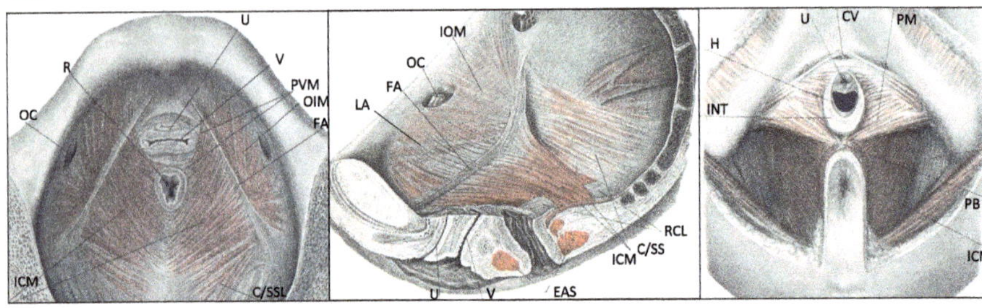

Halban and Tandler, 1907

FIGURE 1.3

Three views of the pelvic floor showing the urogenital hiatus, levator ani, and perineal membrane shown after removal of the pelvic organs, in sagittal view and from below, including the perineal membrane. *C/SS*, coccygeus/sacrospinous; *CV*, clitoral vessels; *EAS*, external anal sphincter; *FA*, fascial arch; *H*, hymen; *ICM*, iliococcygeal muscle; *INT*, introitus; LA, levator arch; *OC*, obturator canal; OIM, obturator internus muscle; *PB*, perineal body; *PM*, perineal membrane; *PVM*, pubovisceral muscle; *RCL*, recto-coccygeal ligament; *R*, rectum; *U*, urethra; *V*, vagina. (From Halban and Tandler, 1907).

rator canal. It is a thin sheet of muscle that spans the pelvic cavity from the arcus tendineus levator ani ("levator arch") to the midline iliococcygeal raphe, where it interdigitates with the muscle of the other side and connects with the superior surface of the sacrum and coccyx. Details of this anatomy will be provided in the next sections.

The puborectal muscle (PRM) arises near the perineal membrane and courses lateral to the other parts of the levator ani muscles. It forms a sling behind the rectum and is distinct from the pubovisceral muscle. The puborectal muscle creates an angulation in the rectum and the pubovisceral muscle elevates the anus, perineal body, and vagina. In the literature, the terms pubovisceral and puborectal are often confused or used interchangeably, which is an incorrect practice, because they are two distinct muscles with different origins, insertions, and lines of action, as evidenced by their separate names in *Terminologia Anatomica* (Terminology, 1998). The latter text is the gold standard for anatomical terminology based on over a century of international anatomical collaboration.

1.2.2 Levator ani lines of action

The action of the pelvic floor muscles is determined by the direction of their muscle fiber bundles (fascicles), the muscle shape, and the points of attachment. Within the levator ani, each muscle element has an origin and insertion that determines its unique mechanical action. Therefore damage to one component may have different mechanical effects than damage to another. The angle of each subdivision has been established in living women (Betschart et al., 2014) and forms the basis for understanding the mechanical effect of muscle contraction and tone. As shown in Fig. 1.2c, the pubovisceral muscle fibers course, 41° from the horizontal in the standing posture. By contrast, the puborectal muscle direction is 60° different, acting −19° from the horizontal. The anal sphincter is also below the horizontal in its line of action. The subcutaneous portion of the external anal sphincter constricts the anal verge, while the superficial portion pulls dorsally, opposing the puborectal muscle.

The large difference (60°) in mean fiber angle between the pubovisceral and the puborectal shows that they have two different mechanical actions. These can be understood within the context of the standing position. The diagonal vectors of the muscles can be resolved into two physiological vectors. The first, a vertically oriented "lifting" vector, acts against gravity by lifting the perineal structures and can be attributed to the pubovisceral muscle. Clinically, descent of the perineal structures is found in women with prolapse. This descent is associated with injury to the pubovisceral portion of the levator, either with our without prolapse (Clark et al., 2010; Lewicky-Gaupp et al., 2010). The second vector is oriented horizontally toward the pubic bone. Both the puborectal and pubovisceral muscles have a horizontal component. This vector helps to develop a "closing" force, acting in a horizontal direction to close the levator hiatus. The horizontal component of the pubococcygeal and puborectal muscles creates a vaginal high pressure zone (Jung et al., 2007; Guaderrama et al., 2005). As shown in Fig. 1.2c, the small vertical puborectal component acts in a downward (caudal) direction, so it has no "lifting" action.

1.2.2.1 Hiatal closure and the perineal complex

The levator ani is an important—but not the only—part of the perineal complex that maintains hiatal closure. The other aspects of the perineal complex are the perineal membrane and perineal body, which work in concert with the muscles to keep the hiatus from opening [Fig. 1.3]. In fact, levator injury severity only predicts about a quarter of variation in hiatal size (Nandikanti et al., 2018). Acting together, the perineal complex and its neuromuscular control system surround the lower third of the vagina to create a high-pressure zone (Raizada et al., 2010) that holds the urogenital hiatus closed. If the hiatus is closed, it protects the connective tissues from being subjected to forces that would injure them. The advent of clinical assessment and modern imaging, both ultrasound and MRI, has proven the central role that hiatal closure failure plays in causing prolapse and prolapse recurrence after surgery (Vaughan et al., 2018; Andrew et al., 2013; DeLancey and Hurd, 1998; Handa et al., 2019).

1.2.2.2 Perineal membrane

The perineal membrane (formerly known as the urogenital diaphragm) is a dense triangular membrane with a central opening, through which the vagina and urethra pass (Stein and DeLancey, 2008; Brandon et al., 2009). In its ventral aspect, the urethra, anterior vaginal wall, and perineal membrane are fused together as a single solid mass that is attached to the pubic bone laterally. This is quite different than the dorsal aspect, where the membrane is an actual sheet of tissue, to which surrounding structures are attached. The levator ani muscles attach to the cranial surface of the perineal membrane and to the perineal body, forming the perineal complex responsible for hiatal closure. Failures in the perineal membrane and perineal body are poorly understood at this time. Their structure clearly indicates their importance, yet the exact role these structures play in maintaining hiatal closure is unclear.

1.2.3 Levator ani muscle injury

1.2.3.1 Levator ani muscle injury associated with prolapse

The concept that injury to the levator ani muscles is important to pelvic organ prolapse has been discussed for over a century. The meticulous illustrations of dissections made of over 50 cadavers with prolapse presented in Halban and Tandler's landmark 1907 work (Halban and Tandler, 1907) on the etiology of genital prolapse clearly make this point. These remarkable illustrations can be viewed in Google Books.[1] Since that time, debate has been polarized between those who believe that pelvic organ support is determined by connective tissue ligaments and fascia and those who believe in the importance of pelvic muscle support (Paramore, 1908). Not surprisingly, both factors are important, and they interact with one another. It is this interaction that makes the pelvic floor such an interesting biomechanical structure. Our understanding of muscle-ligament interactions is improving with the ability to independently assess the relative contributions of both components (Chen et al., 2006).

There is strong evidence that levator failure is causal in pelvic organ prolapse. Injury to the pubovisceral muscle is found in 55% of women with prolapse but only 16% of women with normal pelvic organ support, thus proving the association between levator ani muscle injury and pelvic organ prolapse (DeLancey et al., 2007). It does not involve the puborectal portion (DeLancey et al., 2012), a point made somewhat confusing, because some authors have used the term puborectal and pubovisceral muscle interchangeably. It is interesting to note that levator ani muscle injury plays different roles in different pelvic floor disorders. It does not appear to be associated with stress incontinence in middle-aged women in the absence of pelvic organ prolapse (DeLancey et al., 2008).

As one might expect with a muscle injury, the amount of injured muscle matters. It would be easy for the body to compensate for the loss of a few muscle fascicles by strengthening adjacent muscle, and this would not be expected to cause problems. When more than half of the muscle is injured, referred to as a major loss, the remaining muscle is no longer able to compensate. In addition, if the muscle origin tears away from its normal origin, loss of muscle function can be expected to occur. When major loss is present, prolapse is found more often than not. Analysis of a large dataset involving 503 subjects, evenly divided between those with and without prolapse, has supported the idea that prolapse is only associated with damage of this

[1] https://books.google.com/books?id=5wgTAwAAQBAJ&printsec=frontcover&source=gbs_book_other_versions_r&cad=4#v=onepage&q&f=false.

extent (Berger et al., 2014). This finding has been independently confirmed in studies using ultrasound (Dietz and Simpson, 2008), establishing the certainty that this injury is clinically associated with prolapse. Levator injury involving more than half the muscle bulk is associated with a 40% reduction in the force that the levator ani muscles can generate during a pelvic muscle contraction (DeLancey et al., 2007).

1.2.3.2 How does childbirth cause injury?

Many competing theories have been proposed to explain the cause of birth-related pelvic floor injury. Muscle damage from compression or tearing and neuropathy from compression or stretching have all been proposed. There is clear neurophysiological evidence of alterations in neural function (e.g., Weidner et al., 2006), as well as evidence from blood samples suggesting ischemia/reperfusion (Conner et al., 2006) from compression. Computational simulations of birth revealed remarkable elongation of the levator ani muscles, where portions of the muscle more than tripled their lengths (Lien et al., 2004). These observations, however, do not establish the mechanisms responsible for the levator defect that is associated with prolapse. Since efforts at prevention must be based on a proper understanding of why the injury occurs, it is necessary to decide between these hypotheses. For example, if compression is the proposed mechanism of the injury, lessening the duration during which the tissues are compressed is logical. If tearing of the muscle is the mechanism involved, then allowing a slow and gradual delivery, which may take somewhat longer, would be the right solution. Because the clinical actions suggested by these two theories are opposite, knowing the specific mechanism of injury matters.

It is possible to study these injury mechanisms by imaging women with risk factors for pelvic floor trauma using fluid-sensitive MRI scans, which show edema in injured areas, as has been done in both the early (1–2 months) and late (7–8 months) postpartum period (Miller et al., 2015). Among 19 women in this mechanistic study, three women had unilateral high-grade tears (more than 50% of the muscle involved), three had unilateral low-grade tears, and one had bilateral high- and low-grade tears of the levator ani muscles. All tears were located at the pubic portion of the levator ani muscles. Regarding the compression hypothesis, levator edema was present in all women, but the internal obturator muscle was never involved. Edema resolved at the late postpartum period. None of the women showed a pattern of an initially normal appearing muscle becoming atrophic that would support atrophy due to neural injury. In a larger subsequent study, these findings were confirmed (Miller et al., 2010): 90% (53/59) had levator ani muscle edema, and 41% (28/68) had low-grade or greater LA tear at 7 weeks' postpartum. The magnitude of levator ani muscle tear did not substantially change by 8 months' postpartum, but levator ani muscle edema and bone injuries showed total or near total resolution at 8 months' postpartum ($P < .05$).

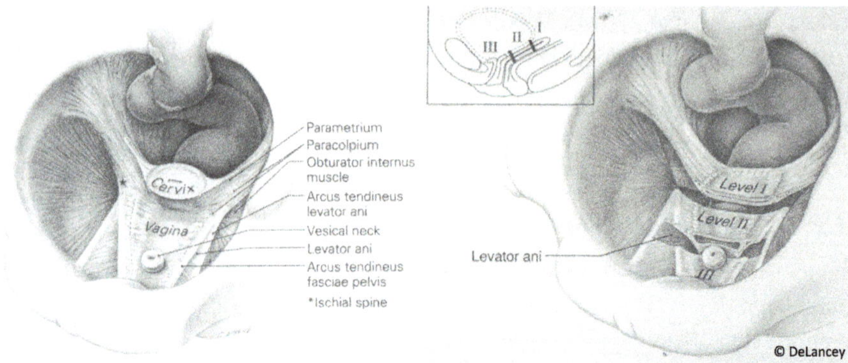

FIGURE 1.4

The left panel is a view of the pelvic organs from above, looking over the pubic symphysis, showing structures of the pelvic sidewall in relationship to the vagina (*outlined by dotted line*) after removal of the bladder and uterine corpus. The right panel shows different levels of support in a post-hysterectomy cadaver. The parametrium and paracolpium are two portions of the cardinal ligament. Arcus tendineus fasciae pelvis and arcus tendineus levator ani refer to the fascial and levator arches, respectively. (DeLancey, 1992.)

1.3 Connective tissue supports of the pelvic organs

1.3.1 Overview

The connective tissues of the pelvis connect the vagina and uterus to the pelvic walls. These can be divided into three levels (I, II, and III, numbered respectively from cephalad to caudad), reflecting the multifaceted nature of support [Fig. 1.4] (DeLancey, 1992). In Level I, the upper third of the vagina and the cervix are attached to the pelvic walls by mesenteric structures that suspend these organs. Next is Level II, wherein the middle third of the vagina is attached laterally to fascial structures (arcus tendineus fascia pelvis [ATFP] or fascial arch). The most caudad portion is Level III, where the distal third of the vagina is fused with the surrounding structures, namely the levator ani muscles and perineal membrane laterally and the perineal body dorsally. The paragraphs below discuss the connective tissue supports at each level in detail.

1.3.2 Apical supports: Level I

Downward descent of the cervix and vagina are important aspects of pelvic organ prolapse. Understanding how the tissues that connect the uterus and upper vagina to the pelvic sidewalls work and defining how they fail is relevant to understanding the pathophysiology of apical prolapse. In addition, there are important interactions between the apical supports and those of the anterior and posterior vaginal wall that, when dysfunctional, are involved in the development of cystoceles and rectoceles.

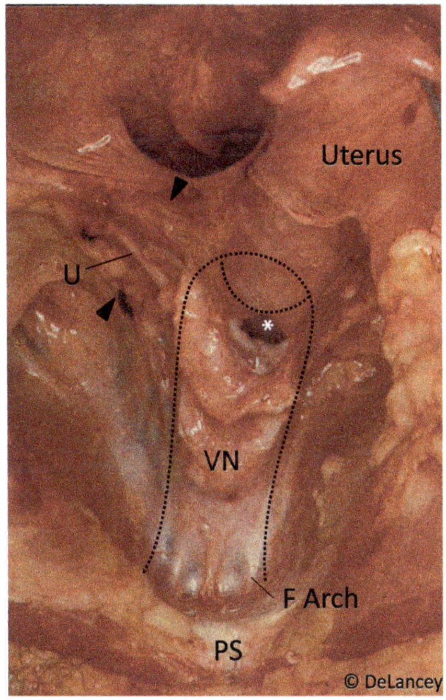

FIGURE 1.5

View of the uterus and vagina from the front, looking over the pubic symphysis (*PS*) after removal of the bladder above the vesical neck (*VN*) in a normal 17-year-old nulliparous cadaver. The cardinal ligament (*between the arrow heads*) can be seen, in addition to the ureter (*U*). The round ligament and upper peritoneal portion of the broad ligament have been excised to the level of the uterine artery. An incision has been made into the vagina at the cervico-vaginal junction (∗). The lumen of the vagina and the location of the cervix are shown in dotted outline. Note that the cardinal ligament goes to both the cervix and upper vagina. The distal attachment of the fascial arch (F Arch) can be seen attached to the pubic bone. ©DeLancey.

1.3.2.1 Cardinal and uterosacral ligaments

In Level I, there are two aspects to the suspending tissues: a) the cardinal ligaments and b) the uterosacral ligaments. These are unlike knee ligaments, which are uniformly dense regular connective tissue that connect two bony structures. Instead, the cardinal/uterosacral ligaments are bilateral mesenteries of the genital tract that consist of blood vessels, lymphatics, and nerves embedded in adipose (fatty) tissue with intervening septa of connective tissue [Fig. 1.5] (Kieserman-Shmokler et al., 2020).

These mesenteries have mechanical properties that resist downward displacement of the vagina and uterus, while allowing a great range of normal movement. But they have quite different mechanical characteristics when compared to classical "ligaments" in the body. Thus considering them to be analogous to, for example, knee ligaments is fundamentally incorrect and decidedly unhelpful in aiding an understanding of pelvic floor function. By way of comparison, the stiffness of the cardinal-uterosacral complex, tested in vivo, is 0.49 N/mm (Luo et al., 2014a), which is *three orders of magnitude* less than that of the anterior cruciate ligament of the knee (200 N/mm) (Woo et al., 1991). It is also important to recognize that the cardinal/uterosacral tissues have a heterogeneous, non-uniform composition. This contrasts with a piece of a rubber band, for example, which has a homogeneous, uniform composition. This fundamental difference has important implications for biomechanical modeling of the pelvic tissues. The high stiffness implied by the term "ligament" may lead to development of surgical strategies that are not normally compliant. Mechanical testing results from a small piece of rubber would be representative of a larger piece of the same material. However, results obtained from testing a single small, excised piece of cardinal/uterosacral tissue would not be representative of the whole structure in the same way.

The structural effect of the cardinal and uterosacral ligaments is most evident when the uterine cervix is pulled downward with a tenaculum, applying traction on the cervix. After several centimeters of descent within the elastic range of the "ligaments" (usually about 3–5 cm with traction in the operating room), the cardinal-uterosacral complex becomes tight and arrests the further downward cervical movement. Similarly, these same tissues resist downward descent of the vaginal apex after hysterectomy. The fact that these ligaments do not determine the position of the uterus in normal healthy women is attested to by the observation that the cervix may be easily pulled down to the level of the hymen (Bartscht and DeLancey, 1988) or pushed up to the level of the sacrum.

1.3.2.2 Tissue composition of cardinal and uterosacral ligaments

One of the major misconceptions in gynecology is the thought that there is a "ligament" attaching the cervix to the pelvic wall that is made of dense regular connective tissue like skeletal ligaments (Fritsch et al., 2012). A recent review of the literature indicates the following consensus concerning the structure of the cardinal-uterosacral complex (Ramanah et al., 2012). The cardinal ligament consists mainly of vascular, neural, and areolar tissue, containing fat and a delicate meshwork of connective tissue between different elements. The cardinal ligament tends to be more vascular in the cranial portion, with more neural components caudally. The vascular section is an extension of the perivascular sheath of visceral branches of the internal iliac vessel going to the genital tract, whereas the neural section contains parts of the inferior hypogastric plexus.

The origin of the cardinal ligament is at the upper border of the greater sciatic foramen, where the vascular aspect of this structure is fixed to the pelvic sidewall. This observation has previously been made on structural grounds (Range and Woodburne, 1964), and we have confirmed it by cutting a pelvis in half and moving the cervix in all directions so that the center of its movement can be determined and this location confirmed (unpublished data).

The uterosacral ligament originates from tissues in the region of S2–S4 sacral vertebrae, with no direct insertion to the bone. Insertion into the genital tract occurs at the dorsal margin of the uterine cervix and to the upper third of the posterior vaginal wall (Campbell, 1950; Umek et al., 2004). The uterosacral ligament is positioned lateral to the rectum and medial to the ureter. It is a multifaceted, mesentery-like structure containing loose connective tissue, smooth muscle, vessels, and autonomic nerve fibers from the inferior hypogastric plexus, with contributions from sacral nerves. The superficial portion typically seen during abdominal surgery is composed primarily of smooth muscle and attaches at the cervico-vaginal junction. The deep uterosacral ligament lies in the perirectal fat below and attaches to the cervix and upper vagina. It has a substantial neural content, explaining its origin from S2–S4. This anatomy will be illustrated below in considering the posterior compartment.

1.3.2.3 Ligament geometry

It is important to view these ligaments in the standing posture, because it is in this orientation that they have their mechanical effect during normal activities of daily living. Most clinical and anatomical drawings show the cardinal and uterosacral ligaments with the body supine, as they are usually encountered in the outpatient clinic, operating room, and dissecting lab. In addition, these drawings are often made while the uterus is being elevated, rather than drawn downward. Both of these conditions give rise to a false impression of their structural mechanics. To illustrate, consider a picture of a hammock shown perpendicular to its usual orientation. This would be out of context and would misrepresent the hammock's function.

Modern imaging allows the appearance of the cardinal and uterosacral ligaments (Umek et al., 2004; Kim et al., 2014) and their structural orientation (Chen et al., 2013) to be directly evaluated and displayed as if the individual is standing. Although it would be ideal to make the images of individuals in the upright posture, the low magnet strengths of open MRI units and the difficulties of keeping the subject motionless while standing preclude routine use of this posture. Fortunately, there are not significant shifts in position between upright and supine postures (Fielding et al., 1996), presumably because the levator ani adjusts its activity to the loads placed on it so as to maintain a constant position.

The cardinal ligaments are in a relatively vertical orientation in the upright posture, a logical direction for them to resist downward forces (DeLancey, 1999). The uterosacral is more dorsally directed in an orientation where it can prevent the uterus and upper vagina from sliding down the inclined plane of the levator plate toward the opening in the levator muscles, through which prolapse occurs.

The lines of action for the cardinal and uterosacral vary among different women as measured in MR images [Fig. 1.6] (Chen et al., 2013). In connecting the origin and insertion points, cardinal ligaments are essentially vertical in the standing posture being $18.1° \pm 6.8°$ from the cephalic-caudal body axis with a slight dorsal inclination, and their curved-line length from origin to the middle of their insertion in the genital tract is 5.7 cm $\pm$ 1 cm. The deep uterosacral ligaments are dorsally directed and $92.5° \pm 13.5°$ from the body axis and 2.7 cm $\pm$ 9.5 cm in length, about half the length

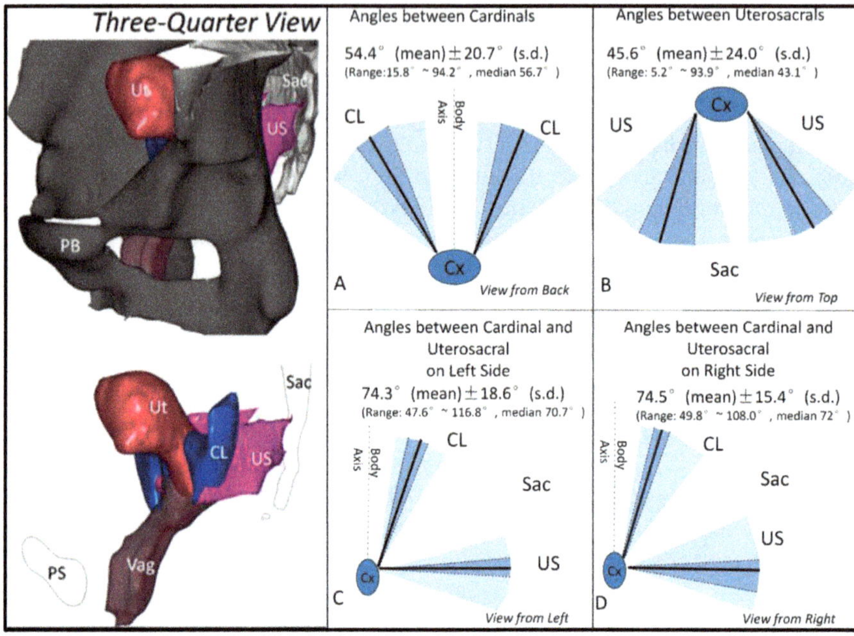

FIGURE 1.6

On the left are a 3D models of the pubic bone (*PB*), sacrum (*Sac*), uterus (*Ut*), vaginal (*Vag*), cardinal ligaments (*CL*), and deep uterosacral ligament (*US*) in three-quarter view. The bottom left shows the same structures without the bones. The four panels at right show the mean angle for the lines of action for the cardinal ligament seen from the top (A) and uterosacral ligament seen from the bottom (B). Lateral views and data for the left (C) and right (D) ligaments are also shown. (From Chen et al., 2013) ©DeLancey.

of the cardinal ligament. Using the lines of action for the cardinal and uterosacral ligaments and the inclination of the levator plate (the portion of the levator ani muscles in the midline behind the rectum), it is possible to make a theoretical calculation of the tensions on the two ligaments for a given 1N unit load. Analysis shows a cardinal ligament load that is 52% larger than the load on the uterosacral ligament when computer simulations are used to assess the relative stresses on these two different structures.

1.3.2.4 Ligament changes with prolapse

It has been possible to examine, *in vivo*, the resting length and force-displacement behavior of the cervix and supporting ligaments in women undergoing surgery and compare it to the clinical assessment of apical support that is measured as the location of the cervix (Point C on the clinical pelvic organ prolapse quantification [POP-Q] examination) during a maximal Valsalva maneuver. In a study of 17 women with varying degrees of uterine support from normal to prolapse, intraoperative measurements were made using a tripod-mounted computer-controlled linear servo actuator (Smith et al., 2013). The device produced a caudally directed force on a tenaculum that was attached to the cervix at 4 mm/sec velocity until the traction force reached 17.8 N (4 lbs). Cervix location at rest was assessed to serve as a proxy for ligament length. POP-Q point C correlated most strongly with cervix location at maximum force ($r = +0.68$; $P = .003$) and at rest ($r = +0.62$; $P = .009$). There was a trend between cervix location and ligament stiffness ($r = -0.44$; $P = .079$), but this was not statistically significant at this sample size. Cervix location at rest and at maximum traction force indicates that ligament length predicts about 38%–46% of cervix location during maximal Valsalva. By contrast, only 19% of the variation in POP-Q point C location was explained by ligament stiffness. The viscoelastic properties of these tissues have also been assessed *in vivo* and quantified (Luo et al., 2014a).

The data presented in the previous paragraph demonstrate that the cardinal and uterosacral ligaments exhibit visco-hyperelastic behavior and provide data in living women. These findings have implications for the repetitive loading that would occur during daily activities and may help to explain changes over time.

The cardinal and uterosacral ligaments change their length and angle in response to the forces applied to the pelvic organs during increases in abdominal pressure. Using 3D reconstructions from MR images of women with and without pelvic organ prolapse at rest and during maximal straining, ligament length and orientation have been measured (Luo et al., 2014b). The cardinal ligament measured at rest is longer in women with prolapse compared to women with normal support (71 mm ± 16 mm vs. 59 mm ± 9 mm, $P = 0.051$). In contrast, the deep uterosacral ligament at rest has a similar length in women with prolapse and those with normal support: 38 mm ± 16 mm vs. 36 mm ± 11 mm, $P = 0.8$, respectively. During a maximal Valsalva maneuver, the cardinal ligament lengthens by 30 mm ± 16 mm in women with prolapse vs. 15 mm ± 9 mm in women without prolapse ($P = 0.03$). During Valsalva, deep uterosacral ligament elongation was 15 mm ± 12 mm vs. 7 mm ± 4 mm ($P = 0.09$). In contrast, the cardinal ligament angle remains the same from rest to strain, while the deep uterosacral ligament angle was significantly different from rest to strain ($P < 0.001$). To summarize, cardinal ligament elongation was greater with Valsalva in women with prolapse than in those with normal support, whereas under the same conditions, the deep uterosacral ligament exhibited greater changes in ligament inclination angle.

1.3.3 Anterior compartment: Level II

In Level II, in the mid-vagina, the vaginal wall is loosely attached to the pelvic sidewalls at the anterior fascial arch [Fig. 1.4, 1.7]. The structural layer that supports the bladder is composed of a) the anterior vaginal wall and b) its attachment to the anterior fascial arch by the endopelvic fascia. What surgeons dissect and call the "fascia" is the fibromuscular layer of the vagina, which consists of the smooth muscle layer of the vaginal wall that contains a great deal of interstitial connective tissue. Similarly, the posterior vaginal wall and endopelvic fascia ("rectovaginal fascia") form the restraining layer that prevents the rectum from protruding forward, blocking formation

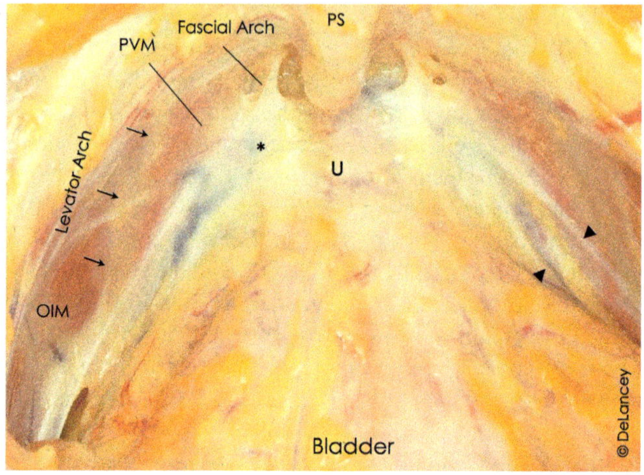

FIGURE 1.7

View into space of Retzius looking into the pelvis from above with the pubic symphysis (*PS*) in the midline. On the left side is the fascial arch, as well as the inner surface of the of the pubovisceral muscle (*PVM*) and obturator internus muscle (OIM). The bladder and urethra (*U*) have been pulled away from the inside of the pubic bones for exposure. The edge of the levator arch is marked with arrows as it lies on the inner surface of the obturator internus muscle, approximately 1 cm from its origin. The fascial arch enters a "node" marked with an asterisk (∗), where there is fusion with the vaginal wall periurethrally and an attachment to the medial portion of the levator ani. On the right side, a small paravaginal defect can be seen, with its margins marked with arrowheads. ©DeLancey.

of a rectocele, and there is a posterior fascial arch to which the tissues are connected. The concept that there is a separate structurally important layer between the vagina and bladder or rectum that is responsible for holding the organs in place is still an issue of active disagreement among experts.

1.3.3.1 Fascial and levator arches

The fascial arch and the levator arch can be seen in the anterior compartment lying on the medial surface of the pelvic sidewall [Fig. 1.7]. The fascial arch has a fine tendinous attachment to the inner surface of the pubic bone 1 cm from the midline and 1 cm above the bottom of the pubic symphysis. There is a short tendon that is about 1 cm long that broadens into a node that consists of the vaginal wall, the attachment of the pubovaginal portion of the pubovisceral muscle, and the endopelvic fascia. Extending dorsally from this "node," the fascial arch continues to its dorsal attachment 1–2 cm above the ischial spine. The arch is just the visible margin of this connective tissue and not actually a cord-like band as often described. The connective tissue fibers in this area run parallel to the edge of the fascial arch and not transversely from the vagina to the arch. Therefore dorsal to the "node," there is not a firm attachment between the vagina and the fascial arch. The levator "arch" is formed by the tendinous ends of the levator fibers.

These connective tissue structures can be affected by injury that occurs to the levator ani muscle at vaginal birth. And it is important to consider the different structural implications of muscle injury, where there is a loss of a portion of the muscle tissue, but the levator arch remains intact, and injuries where the levator arch and pubovisceral muscle become detached from its origin. Both the fascial and levator arches lie in this region, and since the fascial arch lies on the medial portion of the pubovisceral muscle, its location could be affected by pubovisceral detachment from the pubic bone.

1.3.3.2 Alteration in pelvic sidewall anatomy due to injury

The levator ani structural failure involves displacement in both fascial and levator arches (Larson et al., 2012a). When the pubovisceral muscle tears away from its origin on the pubis, both the levator arch and the fascial arch move downward. The most pronounced differences in location are seen in the ventral region near the pubic bone. These changes in location have two effects: 1) they affect the ability of the muscle to close the urogenital hiatus due to muscle detachment from its origin, and 2) they redirect the vectors of the fascial connections to the tendineus arch of the pelvic fascia. The levator arch (*arcus tendineus levator ani*) is lower on the defect side than the non-defect side in the region nearest the pubic bone. A similar, though smaller, change is seen with the fascial arch. In looking at the medial to lateral displacements, the levator arch is more lateral on the normal side than the side with architectural distortion.

1.3.4 Anterior compartment: Level III

In Level III, the vagina is attached directly to its surrounding tissues. It is connected to the levator ani muscles on either side just above the hymenal ring. The portion of the muscle involved in this attachment is referred to as the pubovaginal muscle. It is one of three portions of the pubovisceral muscles. This anatomy has been covered earlier in the section on the levator ani muscles. The role of other aspects of Level III, namely its connections to the perineal membrane and perineal body, as well as the consequences of disruption, will be considered under the description of posterior compartment Level III failure. These perineal tissues are important in anterior vaginal wall support, because they close the urogenital hiatus. When effective, they hold the posterior wall against the anterior wall, preventing its backward movements during increases in abdominal pressure.

1.3.5 Structural failure sites in anterior vaginal wall prolapse

The anterior vaginal wall and its suspension to the cardinal and uterosacral ligaments in Level I and its lateral attachment in Level II play a critical role in anterior compartment support. Understanding the impact of changes in these structures and relationships is a key to understanding anterior vaginal wall prolapse [Fig. 1.8]. 3D

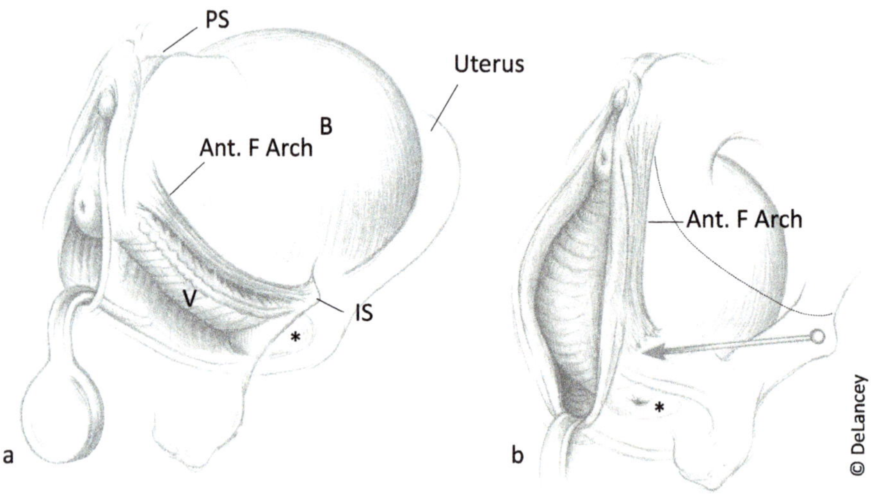

FIGURE 1.8

Diagram showing (a) the normal location of the anterior vaginal wall (*V*) and descent and appearance through the vaginal opening (b) in anterior vaginal prolapse or cystocele. Structures seen from the side with a woman in the lithotomy position and a weighted speculum in place include the bladder (*B*) uterus, cervix (*), ischial spine (*IS*), and pubic symphysis (*PS*). The mechanical effect of apical descent (arrow) and concomitant downward movement of the anterior fascial arch (*Ant. F Arch*), whose normal position is indicated with a dashed line, can be appreciated. Note that it is not changes in the vaginal wall that allow the vaginal wall to protrude, but rather loss of its normal attachments. (From DeLancey, 2002) ©DeLancey.

stress MRI, where the structures are imaged in living women during a maximal Valsalva maneuver, has provided the first opportunity to study the spatial changes in anterior vaginal wall prolapse and compare them to women with normal support.

The separation between the vagina and the fascial arch seen during surgery always starts at the end near the ischial spine and proceeds toward the pubic end, much as a zipper separates its two elements from one end to the other (DeLancey, 2002). This can be seen in Fig. 1.7, where a small right-sided paravaginal defect is visible and compared with the left side that is intact. Descent of the anterior vaginal wall involves this paravaginal detachment, which is usually (but not always) associated with descent of cervix (Rooney et al., 2006; Summers et al., 2006).

3D stress MRI in women shows that the normal configuration and deformation of the anterior vaginal wall at rest and under load is different than previously thought (Larson et al., 2010). In women with normal support, the upper vagina is usually above the fascial arch, and the anterior wall is not parallel to the fascial arch (Larson et al., 2009). Although they are attached, as described in the earlier sections, elevation of the vaginal wall by the levator ani muscles places it in a position oblique to the line of the fascial arch.

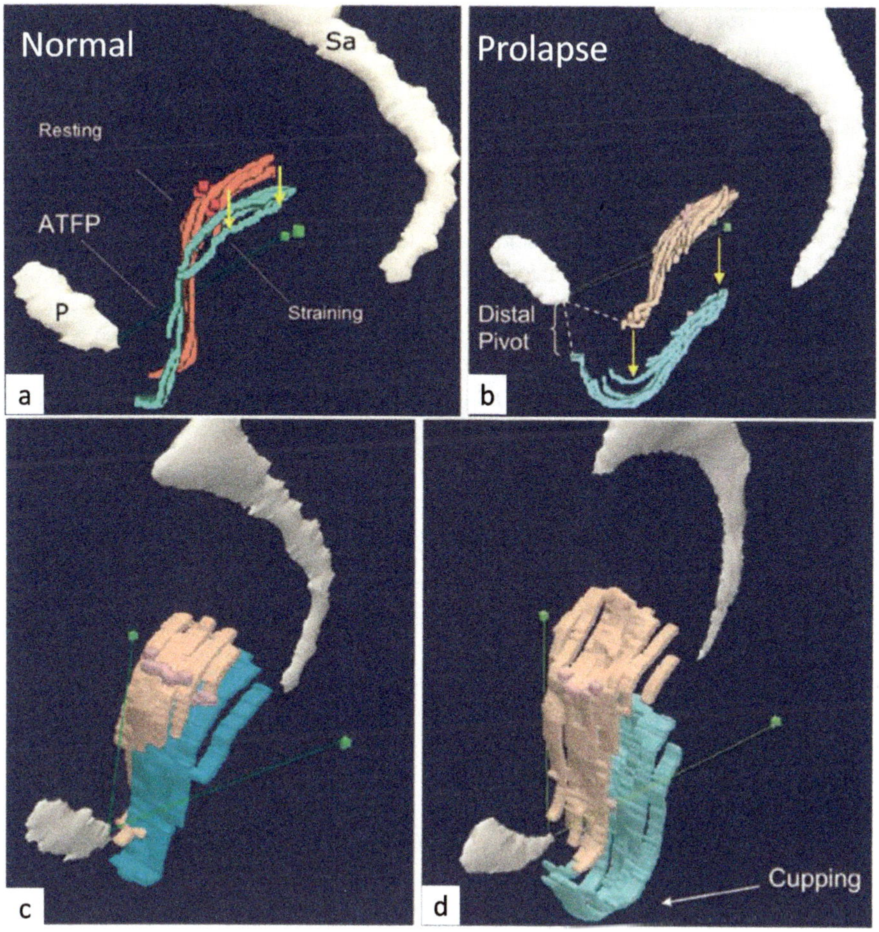

FIGURE 1.9

3D model showing the lateral view (a, b) and three-quarter view (c, d) of the anterior vaginal wall, in a normal individual (a, c) and one with cystocele (b, d). Resting (*tan*) and straining (*turquoise*) positions are shown. The ATFP position is shown (*green*). Descent along the length of the vagina with Valsalva, as well as the distal pivot seen in cystocele compared to a representative normal support. *P*, pubic symphysis; *Sa*, sacrum. An oblique view of the same models is shown in (c) and (d). Note cupping of the vaginal wall in (d) (*Cupping*). ©DeLancey (From Larson et al., 2010).

Quantitative assessment of vaginal wall support that compares women with and without anterior vaginal wall prolapse has allowed structural hypotheses to be tested concerning the changes associated with prolapse [Fig. 1.9] (Larson et al., 2012b), and these original findings have been reproduced (Chen et al., 2016). The difference between individuals with anterior wall prolapse and normal support can be quantified by the effect size statistic that determines the magnitude of difference between groups independent of the type of measurement made. The largest difference is in the descent of the vaginal wall's lateral margin that measures separation in the paravaginal location. It is 4.5 cm lower in the mid-vagina, with a large effect size of 2. This was abnormal (outside the normal range established in the control subjects) in 80% of women with anterior wall prolapse. Apical descent was also strongly associated with anterior prolapse. It was 3.9 cm lower with an effect size of 1.8, and 56.7% of cases were outside normal range. These factors—paravaginal location and apical location—are strongly related ($r = 0.84$). In addition, and consistent with the relationship between vaginal wall support and hiatal closure, the urogenital hiatus, through which the prolapse descends, is 55% larger in women with prolapse with an effect size of 1.9, with 57% outside the normal range. It is strongly correlated with apical and paravaginal location ($r = 0.84$ and 0.89, respectively). Vaginal wall length is also abnormal in 23% of individuals with anterior wall prolapse, being on average 14.4 mm longer, with an effect size of 1.1.

1.3.6 Anatomy and structural failure sites in posterior compartment prolapse

1.3.6.1 Anatomy of the posterior vaginal wall support as it applies to rectocele

Level I: Support of the posterior compartment can be seen in Fig. 1.10. The support of the upper third of the vagina has previously been described as suspended by the cardinal-uterosacral complex and generally similar for the posterior and anterior compartments. However, there is one important difference. The uterosacral ligaments, especially the deep component below what is visible during abdominal surgery, is of particular significance to the posterior compartment. The vaginal portion of the uterosacral ligaments attaches to the upper posterior vaginal wall and holds it dorsally. This part of the uterosacral ligament lies deep below the peritoneum and is easily seen in MRI (Umek et al., 2004), but is not visible from above. It can, however, be palpated when the apex is displaced ventrally (toward the pubic symphysis) during vaginal surgery. This portion of the suspensory system holds the upper posterior vaginal wall dorsally over the levator plate, a vector that cannot be operational for the more vertically oriented cardinal ligament. This is a critical element of the mechanism that prevents development of enterocele, where the space between the vagina and the rectum becomes distended (Hsu et al., 2008).

Level II: The mid-portion of the posterior vaginal wall is connected laterally to the posterior fascial arch by the endopelvic fascia (Leffler et al., 2001) in a way that is analogous to the anterior vaginal wall. The posterior fascial arch is stretched between an attachment near the ischial spine and the perineal body. The distal attachment to the moveable perineal body differs from the anterior fascial arch, whose distal end has an immovable attachment to the pubic bone. These lateral connections of the vaginal wall to the posterior fascial arch prevent ventral movement of the vagina during increases in abdominal pressure.

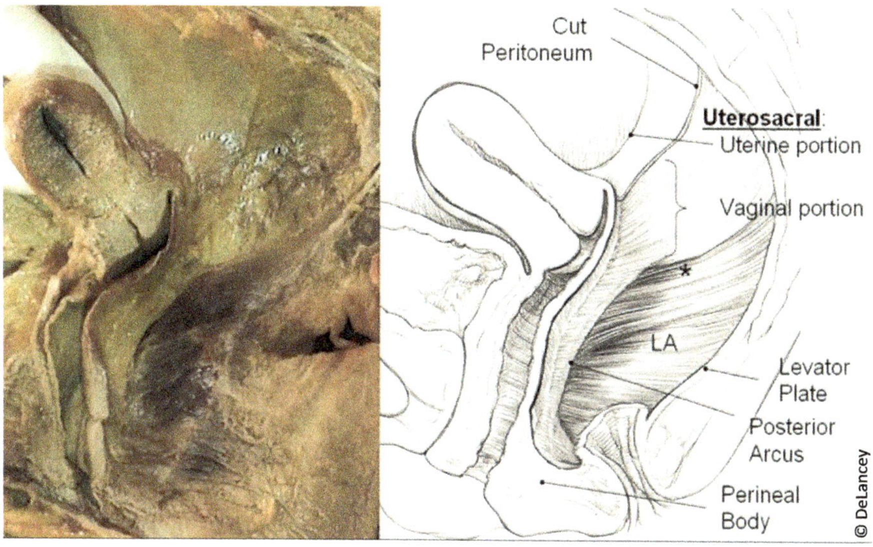

FIGURE 1.10

Cadaver dissection (left) and illustration (right) of the posterior compartment of a 56-year-old multiparous female showing structural relationships after the rectum has been removed. Note the apical connections of the upper posterior vagina to the inside of the pelvic wall in a retroperitoneal position. These lie below the peritoneum and are dorsal and caudal to what is traditionally referred to as the uterosacral ligament seen from above. These structures are continuous with the posterior arcus tendineus fascia pelvis. At the distal end of the vagina, the wall merges into the top of the perineal body. The lateral and dorsal margins of the compartment are formed by the levator ani muscles (*LA*) and the levator plate. The asterisk denotes the region of the sacrospinous ligament overlain by the coccygeus muscle. ©DeLancey (From Hsu et al., 2008).

A fascial layer has been frequently described in articles concerning posterior vaginal wall support as lying between the rectum and vagina and called the rectovaginal fascia, or Denonvilliers's fascia. Its existence as a separate layer has been documented as the presence of a visible layer in women with Müllerian agenesis (Huebner et al., 2008). The visible presence of this layer, however, does not mean that it has sufficient mechanical strength to be used during surgery.

Level III: The lower third of the vagina is fused laterally with the perineal membrane and medial aspect of the levator ani muscles and distally to the perineal body [Fig. 1.11].

The perineal membrane is critical to support, because its connection to the levator ani muscles and perineal body controls the location of the muscles. Separation at the perineal body can result in a diastasis of the levator ani, enlarging the opening through which prolapse occurs.

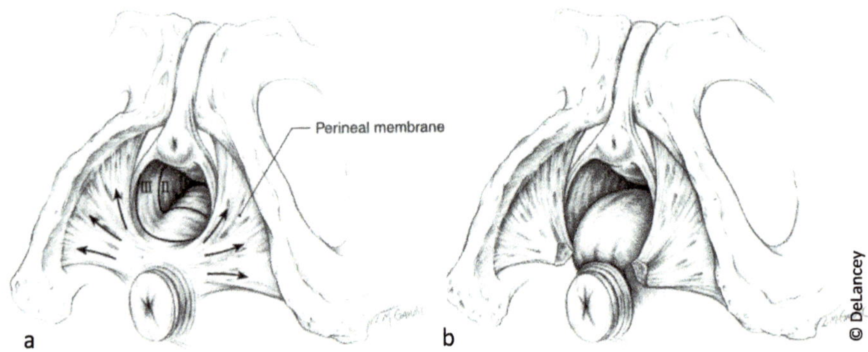

FIGURE 1.11

Peripheral attachments of perineal membrane to ischiopubic rami (a) and direction of tension on fibers uniting through perineal body (arrows). The U-shaped contour of the posterior vaginal wall in Level III and W-shaped contour in Level II are demonstrated. Loss of connection (b) shown as would occur due to obstetrical trauma that could lead to distal rectocele. ©DeLancey (From DeLancey, 1999).

The perineal membrane is composed of two regions: one dorsal and the other ventral (Stein and DeLancey, 2008). The dorsal portion consists of bilateral transverse fibrous sheets that attach the lateral wall of the vagina and perineal body to the ischiopubic ramus. This portion is devoid of striated muscle. The ventral portion is part of a solid 3D tissue mass, in which several structures are embedded. It is intimately associated with the compressor urethrae and the urethrovaginal sphincter muscle of the distal urethra. In this region, the perineal membrane is continuous with the insertion of the anterior fascial arch. The levator ani muscles connect to the cranial surface of the perineal membrane. The vestibular bulb and clitoral crus are fused with the membrane's caudal surface.

1.3.6.2 How does posterior vaginal wall support fail?

There has been a century of conjecture concerning the actual structural defects that are involved in rectocele formation, and many competing hypotheses have been put forward. The rectal shape and distortion have long been studied using defecography. More recently, MRI and 3D stress MRI have allowed the objective documentation of posterior vaginal wall support itself to be described [Fig. 1.12]. Presently, the scientific foundation for structural changes that are associated with posterior vaginal wall prolapse are not as well-developed as they are for anterior compartment support, but this is rapidly changing with new techniques that allow changes in the anatomical structures associated with posterior prolapse to be objectively documented and quantified. The strongest factors in rectocele formation are paravaginal descent and urogenital hiatus size during straining, with effect sizes of 2.4 and 72% abnormal (outside the normal range established in control women) and perineal body height (effect size 2.1 and 60% abnormal). Apical descent is not as strong a factor as it is

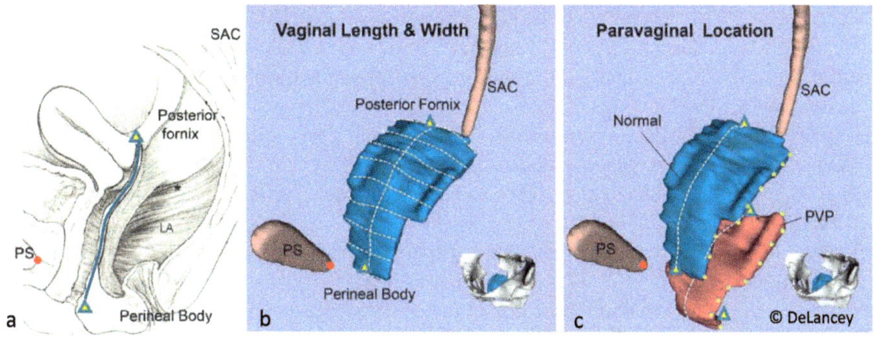

FIGURE 1.12

Panel (a) shows the posterior compartment after removal of the rectum, indicating the boundaries of the posterior compartment showing the levator ani (*LA*) and location of the ischial spine (∗). The red dot at the pubic symphysis (*PS*) and triangles at the top and bottom of the posterior wall correspond to locations shown in (b) and (c). Panel (b) shows the length and width measurement locations of a normally supported posterior vaginal wall, and panel (c) shows normal and abnormal locations (*tan*) during maximal Valsalva in posterior vaginal prolapse (*PVP*).

in anterior prolapse, with an effect size of only 0.9 and 24% abnormal (Chen et al., 2021).

References

Andrew, B.P., et al., 2013. Enlargement of the levator hiatus in female pelvic organ prolapse: cause or effect? Aust. N. Z. J. Obstet. Gynaecol. 53 (1), 74–78.

Bartscht, K.D., DeLancey, J.O., 1988. A technique to study the passive supports of the uterus. Obstet. Gynecol. 72 (6), 940–943.

Berger, M.B., Morgan, D.M., DeLancey, J.O., 2014. Levator ani defect scores and pelvic organ prolapse: is there a threshold effect? Int. Urogynecol. J. 25 (10), 1375–1379.

Betschart, C., et al., 2014. Comparison of muscle fiber directions between different levator ani muscle subdivisions: in vivo MRI measurements in women. Int. Urogynecol. J. 25 (9), 1263–1268.

Brandon, C.J., et al., 2009. Anatomy of the perineal membrane as seen in magnetic resonance images of nulliparous women. Am. J. Obstet. Gynecol. 200 (5), 583e1–583e6.

Campbell, R.M., 1950. The anatomy and histology of the sacrouterine ligaments. Am. J. Obstet. Gynecol. 59 (1), 1–12. illust.

Chen, L., et al., 2006. Interaction among apical support, levator ani impairment, and anterior vaginal wall prolapse. Obstet. Gynecol. 108 (2), 324–332.

Chen, L., et al., 2013. Cardinal and deep uterosacral ligament lines of action: MRI based 3D technique development and preliminary findings in normal women. Int. Urogynecol. J. 24 (1), 37–45.

Chen, L., et al., 2016. Structural failure sites in anterior vaginal wall prolapse: identification of a collinear triad. Obstet. Gynecol. 128 (4), 853–862.

Chen, L., et al., 2021. Structural failure sites in posterior vaginal wall prolapse: stress 3D MRI-based analysis. Int. Urogynecol. J.

Clark, N.A., et al., 2010. Levator defects affect perineal position independently of prolapse status. Am. J. Obstet. Gynecol. 203 (6), 595e17–595e22.

Conner, E., et al., 2006. Vaginal delivery and serum markers of ischemia/reperfusion injury. BJOG 94 (2), 96–102.

DeLancey, J.O., 1992. Anatomic aspects of vaginal eversion after hysterectomy. Am. J. Obstet. Gynecol. 166 (6 Pt 1), 1717–1724. discussion 1724-8.

DeLancey, J.O., 1999. Structural anatomy of the posterior pelvic compartment as it relates to rectocele. Am. J. Obstet. Gynecol. 180 (4), 815–823.

DeLancey, J.O., 2002. Fascial and muscular abnormalities in women with urethral hypermobility and anterior vaginal wall prolapse. Am. J. Obstet. Gynecol. 187 (1), 93–98.

DeLancey, J.O., 2012. Surgery for cystocele III: do all cystoceles involve apical descent?: observations on cause and effect. Int. Urogynecol. J. 23 (6), 665–667.

DeLancey, J.O., Hurd, W.W., 1998. Size of the urogenital hiatus in the levator ani muscles in normal women and women with pelvic organ prolapse. Obstet. Gynecol. 91 (3), 364–368.

DeLancey, J.O., et al., 2007. Comparison of levator ani muscle defects and function in women with and without pelvic organ prolapse. Obstet. Gynecol. 109 (2 Pt 1), 295–302.

DeLancey, J.O., et al., 2008. Stress urinary incontinence: relative importance of urethral support and urethral closure pressure. J. Urol. 179 (6), 2286–2290. discussion 2290.

DeLancey, J.O., et al., 2012. Comparison of the puborectal muscle on MRI in women with POP and levator ani defects with those with normal support and no defect. Int. Urogynecol. J. 23 (1), 73–77.

Dietz, H.P., Simpson, J.M., 2008. Levator trauma is associated with pelvic organ prolapse. BJOG 115 (8), 979–984.

Fielding, J.R., et al., 1996. MR imaging of the female pelvic floor in the supine and upright positions. J. Magn. Reson. Imaging 6 (6), 961–963.

Fritsch, H., Zwierzina, M., Riss, P., 2012. Accuracy of concepts in female pelvic floor anatomy: facts and myths! World J. Urol. 30 (4), 429–435.

Guaderrama, N.M., et al., 2005. The vaginal pressure profile. Neurourol. Urodyn. 24 (3), 243–247.

Halban, J., Tandler, J., 1907. Anatomie und Atiologie der Genital prolapse biem Weibe. Wilhelm Braumüller, Vienna and Leipzig.

Handa, V.L., et al., 2019. Longitudinal changes in the genital hiatus preceding the development of pelvic organ prolapse. Am. J. Epidemiol. 188 (12), 2196–2201.

Hsu, Y., Lewicky-Gaupp, C., DeLancey, J.O., 2008. Posterior compartment anatomy as seen in magnetic resonance imaging and 3-dimensional reconstruction from asymptomatic nulliparas. Am. J. Obstet. Gynecol. 198 (6), 651e1–651e7.

Huebner, M., Margulies, R.U., DeLancey, J.O., 2008. Pelvic architectural distortion is associated with pelvic organ prolapse. Int. Urogynecol. J. Pelvic Floor Dysfunct. 19 (6), 863–867.

Jung, S.A., et al., 2007. Vaginal high-pressure zone assessed by dynamic 3-dimensional ultrasound images of the pelvic floor. Am. J. Obstet. Gynecol. 197 (1), 52e1–52e7.

Kearney, R., Sawhney, R., DeLancey, J.O., 2004. Levator ani muscle anatomy evaluated by origin-insertion pairs. Obstet. Gynecol. 104 (1), 168–173.

Kieserman-Shmokler, C., et al., 2020. From molecular to macro: the key role of the apical ligaments in uterovaginal support. Am. J. Obstet. Gynecol. 222 (5), 427–436.

Kim, J., et al., 2014. Anatomy of the pubovisceral muscle origin: macroscopic and microscopic findings within the injury zone. Neurourol. Urodyn.

Larson, K.A., Hsu, Y., DeLancey, J.O., 2009. The relationship between superior attachment points for anterior wall mesh operations and the upper vagina using a 3-dimensional magnetic resonance model in women with normal support. Am. J. Obstet. Gynecol. 200 (5), 554e1–554e6.

Larson, K.A., et al., 2010. Magnetic resonance imaging-based three-dimensional model of anterior vaginal wall position at rest and maximal strain in women with and without prolapse. Int. Urogynecol. J. 21 (9), 1103–1109.

Larson, K.A., et al., 2012a. Measurement of the 3D geometry of the fascial arches in women with a unilateral levator defect and "architectural distortion". Int. Urogynecol. J. 23 (1), 57–63.

Larson, K.A., et al., 2012b. 3D analysis of cystoceles using magnetic resonance imaging assessing midline, paravaginal, and apical defects. Int. Urogynecol. J. 23 (3), 285–293.

Lawson, J.O., 1974. Pelvic anatomy. I. Pelvic floor muscles. Ann. R. Coll. Surg. Engl. 54 (5), 244–252.

Leffler, K.S., et al., 2001. Attachment of the rectovaginal septum to the pelvic sidewall. Am. J. Obstet. Gynecol. 185 (1), 41–43.

Lewicky-Gaupp, C., et al., 2010. Structural position of the posterior vagina and pelvic floor in women with and without posterior vaginal prolapse. Am. J. Obstet. Gynecol. 202 (5), 497e1–497e6.

Lien, K.C., et al., 2004. Levator ani muscle stretch induced by simulated vaginal birth. Obstet. Gynecol. 103 (1), 31–40.

Luo, J., et al., 2014a. In vivo properties of uterine suspensory tissue in pelvic organ prolapse. J. Biomech. Eng. 136 (2), 021016.

Luo, J., et al., 2014b. Using stress MRI to analyze the 3D changes in apical ligament geometry from rest to maximal Valsalva: a pilot study. Int. Urogynecol. J. 25 (2), 197–203.

Miller, J.M., et al., 2010. MRI findings in patients considered high risk for pelvic floor injury studied serially after vaginal childbirth. Am. J. Roentgenol. 195 (3), 786–791.

Miller, J.M., et al., 2015. Evaluating maternal recovery from labor and delivery: bone and levator ani injuries. Am. J. Obstet. Gynecol. 213 (2), 188e1–188e11.

Nandikanti, L., et al., 2018. Levator ani defect severity and its association with enlarged hiatus size, levator bowl depth, and prolapse size. Am. J. Obstet. Gynecol. 218 (5), 537–539.

Paramore, R.H., 1908. The supports-in-chief of the female pelvic viscera. Proc. R. Soc. Med. 1 (Obstet Gynaecol Sect), 195–214.

Raizada, V., et al., 2010. Dynamic assessment of the vaginal high-pressure zone using high-definition manometery, 3-dimensional ultrasound, and magnetic resonance imaging of the pelvic floor muscles. Am. J. Obstet. Gynecol. 203 (2), 172e1–172e8.

Ramanah, R., et al., 2012. Anatomy and histology of apical support: a literature review concerning cardinal and uterosacral ligaments. Int. Urogynecol. J. 23 (11), 1483–1494.

Range, R.L., Woodburne, R.T., 1964. The Gross and microscopic anatomy of the transverse cervical ligament. Am. J. Obstet. Gynecol. 90, 460–467.

Rooney, K., et al., 2006. Advanced anterior vaginal wall prolapse is highly correlated with apical prolapse. Am. J. Obstet. Gynecol. 195 (6), 1837–1840.

Smith, T.M., et al., 2013. A novel technique to measure in vivo uterine suspensory ligament stiffness. Am. J. Obstet. Gynecol. 209 (5), 484e1–484e7.

Stein, T.A., DeLancey, J.O., 2008. Structure of the perineal membrane in females: gross and microscopic anatomy. Obstet. Gynecol. 111 (3), 686–693.

Summers, A., et al., 2006. The relationship between anterior and apical compartment support. Am. J. Obstet. Gynecol. 194 (5), 1438–1443.

Terminology, F.C.o.A., 1998. Terminologia Anatomica: International Anatomical Terminology, 1 ed. Thieme Stuttgart, Stuttgart, p. 292.

Umek, W.H., et al., 2004. Quantitative analysis of uterosacral ligament origin and insertion points by magnetic resonance imaging. Obstet. Gynecol. 103 (3), 447–451.

Vaughan, M.H., et al., 2018. Surgical alteration of genital hiatus size and anatomic failure after vaginal vault suspension. Obstet. Gynecol. 131 (6), 1137–1144.

Weidner, A.C., et al., 2006. Neuropathic injury to the levator ani occurs in 1 in 4 primiparous women. Am. J. Obstet. Gynecol. 195 (6), 1851–1856.

Woo, S.L., et al., 1991. Tensile properties of the human femur-anterior cruciate ligament-tibia complex. The effects of specimen age and orientation. Am. J. Sports Med. 19 (3), 217–225.

CHAPTER

2

Epidemiology & pathophysiology of pelvic organ prolapse & urinary incontinence in women

Pamela Moalli[a,b,c] **and Shaniel Bowen**[c]

[a]*Department of Obstetrics, Gynecology & Reproductive Sciences, University of Pittsburgh, Pittsburgh, PA, United Stated*

[b]*Division of Urogynecology & Pelvic Reconstructive Surgery, UPMC Magee-Womens Hospital, Pittsburgh, PA, United States*

[c]*Department of Bioengineering, University of Pittsburgh, Pittsburgh, PA, United States*

2.1 Epidemiology of pelvic organ prolapse

2.1.1 Prevalence & incidence

Pelvic organ prolapse (POP) is defined as the descent of one or more of the anterior vaginal wall, posterior vaginal wall, uterus (cervix), or vaginal apex (vault or cuff scar after hysterectomy) with herniation of the pelvic organs (e.g., bladder, rectum, uterus, colon) into the vagina (Haylen et al., 2010; Razzak, 2018). There are several types of POP: cystocele is prolapse of the anterior vaginal wall; uterine or vaginal vault prolapse is the decent of the vaginal apex or cervix into the vaginal canal towards or beyond the hymen; rectocele is characterized by prolapse of the posterior vaginal wall, and enterocele is prolapse of the intestines to the vagina (Salvatore et al., 2021).

The reported worldwide prevalence of POP varies widely; its symptom-based estimate ranges from 3%–7%, while its gynecological assessment-based approximation is between 42%–56% (Maher et al., 2011; Walker and Gunasekera, 2011). This discrepancy in POP prevalence estimates is due to the substantial presence of asymptomatic cases of POP (Barber and Maher, 2013). It is predicted that by 2050, the number of women with symptomatic POP will increase by at least 46%, from 3 to 5 million women (Wu et al., 2009). It is likely that the actual prevalence of POP is higher than what has been reported as only 10%–20% of women with POP seek medical treatment (Ghetti et al., 2007; Maher et al., 2011).

Each year, 300,000 surgical procedures for POP are performed in the United States (Olsen et al., 1997; Smith et al., 2010; Wu et al., 2014a). The rate of retreatment for recurrent POP is 11.9% for uterosacral ligament suspension patients and 8.1% for sacrospinous ligament fixation patients (Jelovsek et al., 2018). To date, the

Biomechanics of the Female Reproductive System: Breast and Pelvic Organs. https://doi.org/10.1016/B978-0-12-823403-7.00013-0

most common type of POP and site of recurrent POP is cystocele, which accounts for ~73% cases of prolapse recurrence (Morgan et al., 2011; Śliwa et al., 2018).

2.1.2 Risk factors

2.1.2.1 Age & obesity

Evidence suggests that the risk of POP increases with age, with an estimated increase in prevalence of 40% every 10 years (Swift et al., 2005). Particularly, the incidence of POP varies greatly by age among women with symptomatic prolapse (Weintraub et al., 2020). The rate of symptomatic POP is 6% among women aged 20–29 years, 31% for women aged 50–59 years, and 50% among women over 56 years old (Wu et al., 2014b). In addition, older women (60+ years) are 8–11 times more likely to seek medical care for POP and other pelvic floor disorders than younger women (30–39 years) (Luber et al., 2001). By age 80, the lifetime risk of women undergoing surgery for prolapse is 12.6% (Wu et al., 2014b).

Obesity is another well-known contributor to POP, presumably related to chronically increased intraabdominal pressure (i.e., strain/burden) placing stress on the pelvic floor and other supportive structures (Pandey and Bhattacharya, 2010; Washington et al., 2010). Alternatively, chronically increased inflammation that occurs with excess visceral fat accumulation can lead to reduced contraction, overactivity, and ultimately dysfunction of the muscles responsible for maintaining continence (Otsubo et al., 2021). Weight change has also been associated with both the progression and regression of POP (Braga and Caccia, 2018). In a 5-year study relating weight change and POP progression/regression, it was observed that among postmenopausal women with an intact uterus, the risk of POP regression was higher in obese women compared to women with healthy BMIs (Kudish et al., 2009; Salvatore et al., 2021). In particular, when compared to women with healthy BMIs, the risk of POP progression for obese women increased by 48% for cystocele, 69% for uterine prolapse, and 58% for rectocele (Kudish et al., 2009). It was also found that weight loss did not significantly correlate with POP regression, thus signifying that compromise of pelvic organ support attributed to weight gain may be permanent (Braga and Caccia, 2018; Kudish et al., 2009). Comorbidities, such as diabetes, further exacerbate impairment of pelvic floor tissue integrity likely due to chronic inflammation, leading to weakened tissues and neuropathy. Joint hypermobility associated with collagen and elastin disorders also increase risk for POP (de Sam Lazaro et al., 2016; Ramalingam and Monga, 2015; Weintraub et al., 2020).

2.1.2.2 Race & ethnicity

Some studies have shown that prevalence of POP is lower among Black women compared to other racioethnic groups in the United States, where the risk of POP for black women, as well as Asian women, was reported to be 4–5 times lower than Hispanic and White women in a prospective cohort study (Hendrix et al., 2002; Kim et al., 2005; Whitcomb et al., 2009). In a cadaveric study comparing pelvic tissue features between White and Asian women, it was found that the pubourethral ligaments, endopelvic attachment to the obturator fascia, and endopelvic fascia were weaker and thinner in White women (Zacharin, 1977). In a transperineal ultrasound study, it was demonstrated that Asian women had less pelvic organ mobility than White women (Dietz, 2003). To date, the reason for these racial differences observed are poorly understood (Salvatore et al., 2021).

2.1.2.3 Childbirth, parity & mode of delivery

Pelvic floor muscle and connective tissue trauma (e.g., stretching/ballooning, avulsion) during childbirth is a significant factor for postpartum POP (Braga and Caccia, 2018; Glazener et al., 2013; Gyhagen et al., 2013; Tegerstedt et al., 2006). In particular, the incidence of postpartum levator ani muscle trauma is stated to range from 15% to up to 40%, where levator ani avulsion has been reported to occur in one in three women who deliver vaginally (Dietz and Lanzarone, 2005; Garriga et al., 2011; Heilbrun et al., 2010). Moreover, women with postpartum levator ani avulsion are two times more likely to develop POP—specifically anterior vaginal wall and apical prolapse—than women without levator ani avulsion (Dietz and Simpson, 2008).

Parity is perhaps the strongest risk factor for developing POP. In comparison to nulliparous women, having one childbirth was associated with a four-fold increased risk for POP, while having two childbirths showed an 8.4 times likelihood of experiencing POP (Patel et al., 2006). In addition, it has been reported that each subsequent childbirth further increases the risk of POP by 10–20% (Hendrix et al., 2002). It is believed that caesarean delivery has a potential protective effect that decreases the risk for developing POP (Salvatore et al., 2021). In a study by Gyhagen et al., results demonstrated that vaginal delivery increased the likelihood of experiencing POP 20 years postpartum by 255% compared to caesarean section (Gyhagen et al., 2013). Instrumented deliveries, particularly forceps, substantially increases the risk for POP, where the adjusted odds ratio for forceps delivery in comparison to vacuum delivery was 1.72 (95% CI, 1.06–2.79, p =0.03) for POP $\geq$ Stage 2 (Handa et al., 2013, 2011; Volløyhaug et al., 2015b).

2.1.2.4 Hysterectomy

Women who have undergone hysterectomy are at an increased risk for POP later on in life compared to women with an intact uterus, particularly cystocele (Koelbl et al., 2009; Weintraub et al., 2020). According to a study by Altman et al., 3.2% of women who had a hysterectomy underwent POP repair, while 2.0% of non-hysterectomized women had undergone POP surgery (Altman et al., 2008). Furthermore, the hysterectomy group was at higher risk for reoperation following their index or previous POP procedure (Abrams et al., 2016; Altman et al., 2008). In a longitudinal study by Mant et al., the cumulative risk for POP increases from 1% at 3 years post-hysterectomy to 5% at 15 years post-hysterectomy (Mant et al., 1997; Salvatore et al., 2021).

2.2 Pathophysiology of pelvic organ prolapse

2.2.1 Anatomical structure, support & function

2.2.1.1 Pelvic floor anatomy & supports

2.2.1.1.1 Levator ani

The levator ani muscles are comprised of the iliococcygeus, pubococcygeus, and puborectalis (Corton, 2012). The iliococcygeus originates from the arcus tendineus to the ischial spine and inserts into the anococcygeal, also known as the levator plate. The pubococcygeus (i.e., pubovisceralis) originates from the anterior aspect of the arcus tendineus and posterior portion of the pubic symphysis, inserts at the levator plate, and forms the levator hiatus (Razzak, 2018). The pubovisceralis can be further partitioned into the puboperinealis, puboanalis, and pubovaginalis muscles (Kearney et al., 2004). The puboperinealis is attached to the perineal body and pulls it toward the pubic symphysis; the puboanalis is connected to the anus, and the pubovaginalis is attached to the lateral margins of the vagina (Razzak, 2018). The puborectalis muscle originates from the inferior aspect of the pubic symphysis, passes inferiorly and posteriorly by the lateral margins of the vagina, and wraps around the rectum, forming a sling at the level where the rectum and anus are joined. Together, the muscles of the levator ani tonically contract at rest to close the genital hiatus, providing the pelvic viscera a stable platform (Koelbl et al., 2009). Trauma or denervation of the levator ani reduces its muscle tone and/ volume, which can result in widening the urogenital hiatus, hyper-relaxation (i.e., straightening) of the levator plate, and distension (i.e., ballooning) of the pelvic floor muscle complex (DeLancey et al., 2003; Koelbl et al., 2009; Singh et al., 2003).

2.2.1.1.2 Arcus tendineus levator ani & fascia pelvis

The arcus tendineus levator ani, the dense connective tissue extending from the pubic tubercle anteriorly to the ischial spines posteriorly, traverses the medial aspect of the obturator internus muscles and serves as the origin point for the iliococcygeus muscle of the levator ani (Razzak, 2018).

The arcus tendineus fascia pelvis, a fibrous recess of pelvic fascia extending from the inner portion of the superior pubic rami to the ischial spine and anteriorly to the pubovesical ligament, lies medially to the obturator internus muscles and laterally to the peritoneum (Occelli et al., 2001; Razzak, 2018). It provides an attachment point for the pubocervical fascia and proximal rectovaginal septum (Razzak, 2018).

2.2.1.1.3 Perineal membrane (urogenital diaphragm)

The perineal membrane, also known as the urogenital diaphragm, is a dense sheet of connective tissue comprised of inferior and superior fascial layers separated by the deep transverse perineal muscles (Razzak, 2018). It is attached to the ischiopubic rami laterally and to the distal vagina and perineal body medially. The perineal membrane is horizontally oriented (in the standing position) and provides support to the urethra and vesicourethral junction by its attachment to the perineal body (Corton, 2012; Razzak, 2018).

2.2.1.1.4 Perineal body

The perineal body, a dense pyramidal fibromuscular structure located between the anus and posterior aspect of the genital hiatus, provides attachments to several pelvic floor muscles and connective tissues/septa that help balance the biomechanical forces and maintain the integrity of the pelvic floor muscles (Razzak, 2018; Siccardi and Bordoni, 2021). More specifically, the perineal body is connected to the following muscles: bulbospongiosus, deep transverse perinei, compressor urethrae, internal and external anal sphincter, external urethral sphincter, recto-urethralis, and longitudinal anis muscle (Siccardi and Bordoni, 2021). The connective tissue/septa structures that join to the perineal body include the superficial perineal fascia of the perineal membrane and the rectovaginal septum (Siccardi and Bordoni, 2021).

2.2.1.1.5 Uterosacral-cardinal ligament complex

The uterosacral-cardinal ligament complex, comprised of the uterosacral ligaments and cardinal ligaments, provides apical support to the vagina by suspending the uterus and upper vagina to the sacrum and pelvic sidewalls, while positioning them over the levator plate (Razzak, 2018; Shahryarinejad and Vardy, 2008). The uterosacral ligaments (USLs) extend from the posterior surface of the cervix and upper vagina to the second-to-fourth sacral vertebrae, forming the lateral boundaries of the rectouterine and rectovaginal spaces. They are composed mainly of connective tissue along with vessels and splanchnic nerve fibers (Vu et al., 2010).

2.2.1.2 Pelvic floor support mechanisms

Although essentially a continuous sheet, for conceptual purposes, DeLancey has subdivided the muscular and connective tissue supports of the vagina—and by extension, the other pelvic organs—into three separate levels. The apical, middle, and distal vaginal supports are represented by Levels I, II, and III, respectively (DeLancey, 1992). Level I (i.e., apical vaginal supports) consists of the uterosacral-cardinal ligament complex; Level II (i.e., midvaginal supports) is comprised of the paravaginal attachments into the arcus tendineus levator ani and fascia pelvis, and Level III support (i.e., distal vaginal supports) is comprised of the perineal membrane and body (Doshani et al., 2007).

2.2.2 Physiological mechanisms of POP

2.2.2.1 Pelvic floor defects

Damage to, or defects of, one or more levels of support causes different types of pelvic organ prolapse (POP) and other gynecologic issues to arise (DeLancey, 1994a; Koelbl et al., 2009). In particular, loss of Level I support contributes to uterine or vaginal vault prolapse; loss of Level II support can lead to cystocele or rectocele, and loss of Level III support may predispose to rectocele and a widened genital hiatus (Koelbl et al., 2009; Salvatore et al., 2021).

2.2.2.2 The integral theory

Petros popularized the integral theory, which states that the development and symptoms of POP stem from a weakening of the collagenous connective tissues and ligaments that support the vagina (Petros and Woodman, 2008; Weintraub et al., 2020). The ligamentous vaginal supports (and hence, those of the bladder and rectum) as suggested above are linked to the perineal body by the pelvic floor fascia. Through the multidirectional pull of the pelvic floor muscles on the pelvic fascia and organs, the vagina and other pelvic organs are provided constant support, where the amount and direction of the tension play a role in maintaining their position, shape, and function (Petros and Woodman, 2008). The degree of support given is dependent on the strength (i.e., integrity) and form of the fascia, ligaments, and muscles of the pelvic floor (Petros and Woodman, 2008; Weintraub et al., 2020).

Injury or weakening of the collagenous supports can result in prolapse of the pelvic organs near the site of the defect(s) and abnormal symptoms. There are three areas of connective tissue laxity and associated supportive structures for each zone that impact the form and function of the pelvic organs: (1) anterior zone, consisting of the pubourethral ligament, suburethral hammock, and external urethral ligament; (2) middle zone, comprised of the pubocervical fascia, cardinal ligament-cervical ring complex, and arcus tendineus fascia pelvis; and (3) posterior zone, including the uterosacral ligaments, rectovaginal fascia, and perineal body (Petros, 2011; Petros and Woodman, 2008). Impairment in the anterior zone is associated with a high prevalence (probability) of stress urinary and fecal incontinence symptoms; defects in the middle zone is linked with a high risk of cystocele and urgency urinary incontinence symptoms; weakness or injury in the posterior zone is correlated with enterocele, rectocele, and uterine or vaginal vault prolapse (Petros, 2011; Petros and Woodman, 2008). It is important to note that POP severity may not necessarily correlate with the severity of prolapse symptoms (Weintraub et al., 2020).

While some researchers have challenged the existence of the supportive structures of vagina (and uterus) described by the integral theory, there is a paucity of studies employing the scientific method to confirm the accuracy of the theory as it pertains to POP (Fritsch et al., 2012; Petros and Woodman, 2008).

2.2.2.3 Changes in collagen & smooth muscle of the vagina & supportive tissues

Abnormal production and degeneration of collagen of the vaginal wall and its supportive connective tissues are hypothesized to play a role in the pathogenesis of POP (Kerkhof et al., 2009; Strohbehn et al., 1997). The vaginal wall is comprised of four layers: the superficial layer, consisting of non-keratinized stratified squamous epithelium; the subepithelium layer, containing dense connective tissue composed mainly of elastin and collagen; the muscularis layer, consisting of smooth muscle; and the adventitia layer, made up of loose connective tissue (Koelbl et al., 2009; Razzak, 2018). The subepithelium and muscularis layers are the primary contributors to the tensile strength of the vaginal wall. Normally, the connective tissue supports pull the vagina up and back towards the sacral hollow, where the upper vagina lies over the levator ani muscles (i.e., levator plate). In turn, a normally supported vagina also supports the uterus, bladder, urethra, and rectum (Koelbl et al., 2009). Thus the vagina is central to pelvic organ support (Feola et al., 2010). Hence, modification (e.g., change in the amount or proportion of collagen types), disruption, or injury of the connective tissue supports or vaginal wall are believed to be causative mechanisms of POP and have been associated with the development of POP (Koelbl et al., 2009; Moalli et al., 2002; Razzak, 2018).

In particular, lysyl oxidase-like 1 (LOXL1) and fibulin-5 (Fbln5) deficiency have been demonstrated to be critical for pelvic organ support and associated with the development and presence of POP (Klutke et al., 2008; Takacs et al., 2009; Word et al., 2009). LOXL1 plays a role in (1) crosslinking collagen and elastin and (2) elastin fiber homeostasis and repair in the connective tissue supports and walls of the vagina (Alperin et al., 2008; Couri et al., 2012; Liu et al., 2004). Fbln5 is responsible for (1) binding elastic fibers to cell surface integrin to facilitate cellular connections to the extracellular matrix of the pelvic supports and (2) regulating the expression of matrix metalloproteinases (MMPs), that is, enzymes that degrade collagen and elastin (Couri et al., 2012; Word et al., 2009). In LOXL1 knockout mice, abnormal elastic fibers and genitourinary pathology were observed, particularly weakening of the vaginal wall and lower urinary tract dysfunction, as well as impaired elastic fiber repair postparturition that led to POP (Liu et al., 2006). Similarly, structural alteration of elastic fibers and upregulation of MMP9 activity have been exhibited in Fbln5 knockout mice which later led to the development of POP (Budatha et al., 2011). These trends were also demonstrated in the vaginal tissue and suspensory ligaments of the vagina in women with POP (Budatha et al., 2011; Jung et al., 2009; Zhao and Zhou, 2012).

2.2.2.4 Age & obesity

Mitochondrial dysfunction and oxidative stress are common elements of the aging process often accompanied by POP. Mitochondrial dysfunction is caused by the buildup of mitochondrial DNA and reactive oxygen species that results in cellular and tissue damage, and oxidative stress is caused by the imbalance between antioxidant defenses and reactive oxygen and nitrogen species (Kim et al., 2013; Liguori et al., 2018; Srivastava, 2017; Wang et al., 2019). More specifically, mitochondrial dysfunction, given by increased expression of mitofusin 2, which modulates mitochondrial activity, is believed to be related to the inhibition of collagen synthesis, which can lead to weakening of the ligamentous pelvic organ supports and predispose to POP (Wang et al., 2019). Increased oxidative stress and mitochondrial apoptosis was observed in the uterosacral ligaments of women with POP and women with higher stage POP, respectively (Kim et al., 2013).

Obesity, which involves increased adipose content in the intraabdominal organs and often linked with poor tissue integrity, is marked by chronically increased intraabdominal pressure and increased inflammation, and has been associated with an increased risk of POP (Bradley et al., 2007; Jones and Moalli, 2010).

2.2.2.5 Pregnancy, parity, & postpartum pelvic floor muscle trauma

The compliance of the vagina increases with pregnancy through adaptive remodeling that result in increased elastin and decreased collagen to prevent injury during vaginal delivery (da Cunha et al., 2021; Rynkevic et al., 2017; Ulrich et al., 2014). Of note, the first vaginal delivery—the time of peak risk of damage—has been associated with increased compliance and reduced tensile strength of the vagina that arise from changes in collagen and elastin content (Damron and Capeless, 2004; Knight et al., 2016; Urbankova et al., 2019). These microscopic and biomechanical changes may predispose to POP (Chaliha, 2009). Postpartum POP is more common in women following vaginal delivery versus cesarean section, where vaginal (forceps) delivery has a higher risk of pelvic floor muscle trauma (i.e., levator avulsion injury, increased levator hiatal area) (Chiaffarino et al., 1999; Kearney et al., 2006; Volløyhaug et al., 2015a, 2016). While levator avulsion is often associated with a larger levator hiatus (particularly in the anterior portion), an enlarged levator hiatus has been associated with POP with and without the presence of levator avulsion (Dietz and Simpson, 2008; Dietz et al., 2012). Women with POP are 7.3 times more likely to have major levator ani defects than women without POP (DeLancey et al., 2007).

2.2.3 Clinical diagnosis & evaluation

Physical examination for POP involves pelvic examination, starting with the inspection of the vagina and vulva for superficial defects (e.g., ulcers, erosions, lesions) (Weber and Richter, 2005). Next, the degree (i.e., stage) and site(s) of prolapse are assessed using a vaginal speculum or retractor, where the appearance of a vaginal bulge signifies the presence of POP. The pelvic organ prolapse quantification (POP-Q) system is the standardized tool used during physical examination to describe, quantify, and stage pelvic support and POP, where prolapse is evaluated via vaginal anatomic landmarks with respect to the plane of the hymen (Bump et al., 1996). In particular, there are six points used to obtain POP-Q measures: Aa and Ba on the anterior vaginal wall; C and D on the superior vagina; and Bp and Ap on the posterior vaginal wall. In addition, there are three other POP-Q measures: genital hiatus (GH), the distance between the middle of the external urethral meatus to the posterior margin of the hymen; perineal body (PB), the distance between the posterior portion of the GH to the middle of the anal opening; and total vaginal length (TVL), the length of the vagina from the hymen to the posterior fornix when Point C or D of the vagina is fully reduced to its normal position (Madhu et al., 2018; Persu et al., 2011).

These nine measures are taken with the patient straining in the dorsal lithotomy position to determine the stage of POP at the leading edge (i.e., most advanced site) of prolapse, where POP stage ranges from 0 (no prolapse) to IV (total eversion of the vagina) (Persu et al., 2011; Salvatore et al., 2021; Weber and Richter, 2005). The POP-Q system has been shown to have proven intra-rater and inter-rater reliability (Hall et al., 1996). It is important to note that the POP-Q is limited to anatomic findings of POP; it does not account for symptoms and bothers perceived by the patient (Persu et al., 2011).

In addition to physical examination, as well as questionnaires for assessing patient-reported symptoms (e.g., bothersome bulge, bothersome pressure, vaginal laxity, vaginal looseness), radiographic imaging is used to better describe support defects and exact nature of POP through various imaging modalities, such as ultrasound and magnetic resonance imaging (MRI) (Dietz, 2015). In particular, ultrasonography has been used to identify and assess paravaginal defects, while MRI has been used to more comprehensively describe the pelvic tissues and viscera (Aronson et al., 1995). Dynamic MRI, which involves the imaging of the pelvic organ with strain maneuvers, allows the depiction and evaluation of pelvic organ mobility, levator integrity (e.g., avulsion), and levator distensibility (i.e., ballooning) in patients with POP (Dietz, 2015) and other pelvic organ disorders. However, there are no universal standards or guidelines concerning radiographic criteria for diagnosing POP (Chen and Walters, 2008). Furthermore, the high cost of MRI and its limited use in clinical decision-making make its applicability and feasibility in routine clinical care debatable (Chen and Walters, 2008; Matsuoka et al., 2001).

2.2.4 Treatment & management

Nonsurgical management of POP includes pelvic floor muscle training, pessaries, or a combination of the two (Tso et al., 2018; Weber and Richter, 2005). Pelvic floor muscle training (i.e., Kegel exercises) are used help prevent excessive pelvic organ descent into the vaginal canal and alleviate bothersome POP symptoms by strengthening the levator ani muscles and ligaments of the pelvic floor (Culligan, 2012; Tso et al., 2018). Pessaries are plastic devices inserted into the vagina to provide vaginal support and a non-invasive alternative to pelvic reconstructive surgery. The combined use of a pelvic floor exercise regimen and vaginal pessary has been shown to improve prolapse symptoms and the quality of life of women with symptomatic Stage I to Stage III POP (Cheung et al., 2016; Tso et al., 2018).

The main goal of surgical management of POP is to alleviate or improve prolapse symptoms and other bothersome symptoms associated with other gynecologic issues, if present. This may involve obliterative procedures with or without concomitant hysterectomy or a repair that aims to restore normal vaginal anatomy (Weber and Richter, 2005). Prolapse surgeries can use an abdominal (laparoscopic, robotic, or open), or vaginal approach depending on the type and severity of POP being treated. Anterior (cystocele) and apical (uterine, vaginal vault) prolapse procedures typically use either the abdominal or vaginal route, while posterior (rectocele) prolapse surgeries are generally approached by transvaginal route (Lazarou and Grigorescu, 2021; Weber and Richter, 2005).

Cystocele is traditionally repaired via anterior colporrhaphy, which involves the separation of the vaginal epithelium from underling connective tissue and plication of the vaginal muscularis overlying the bladder (pubocervical fascia) to reduce the herniation of the bladder into the anterior vaginal wall (Lazarou and Grigorescu, 2021; Weber and Richter, 2005). Paravaginal repair, where the anterolateral vaginal sulcus is reattached to the obturator internus and endopelvic fascia at the level of the arcus tendineus fascia pelvis bilaterally, is typically performed via the retropubic or transvaginal route (either abdominally or laparoscopically) to restore Level II (i.e., midvaginal) support (DeLancey, 1992; Weber and Richter, 2005).

Apical prolapse, which can also include enterocele, is conventionally treated using the abdominal or vaginal route with or without hysterectomy. The primary abdominal procedures performed for apical prolapse are minimally invasive hysterectomy with sacrocolpopexy or a uterosacral ligament suspension (Lazarou and Grigorescu, 2021). These operations resuspend the vaginal apex to the sacrum using grafts and sutures through the anterior longitudinal ligament to restore Level I (i.e., apical) support (Weber and Richter, 2005). Vaginal approaches of apical prolapse surgeries are often preferred over abdominal repairs due to their shorter intraoperative and recovery time (Lazarou and Grigorescu, 2021). Common vaginal procedures for apical repair include sacrospinous ligament suspension, attachment of the vaginal apex to the sacrospinous ligament unilaterally; iliococcygeal vaginal suspension, attachment of the vaginal apex to the iliococcygeus muscle and fascia (usually bilaterally); and uterosacral ligament suspension, attachment of the vaginal apex to the uterosacral ligaments bilaterally (Weber and Richter, 2005).

Rectocele is commonly repaired using posterior colporrhaphy via a transvaginal approach and may also involve posterior colpoperineorrhaphy with levator ani muscle plication (Lazarou and Grigorescu, 2021; Weber and Richter, 2005). During this procedure, the vaginal epithelium is separated from the rectovaginal fascia and excess vaginal mucosa is excised and repaired with absorbable sutures to eliminate the protrusion of the rectum into the posterior vaginal wall (Lazarou and Grigorescu, 2021).

Pelvic reconstructive surgeries have also been augmented with grafts, such as synthetic (polypropylene) mesh to reduce the rate of prolapse recurrence. However, mesh-augmented POP repairs, particularly those performed transvaginally, are often associated with complications that require mesh removal, and there is little evidence on the long-term efficacy and safety of mesh over traditional native tissue procedures. This lack of substantial empirical evidence on the safety and effectiveness of transvaginal mesh devices in POP surgeries led the Food and Drug Administration to halt the sale of distribution of mesh products for transvaginal mesh repair of cystocele in April 2019 (Dyer, 2019). Evidence that sacrocolpopexy provides more durable support over native tissue repair is more robust. This procedure has lower rates of mesh complications and is still considered the "gold standard" apical prolapse procedure. The main limitation of sacrocolpopexy is the steep learning curve, especially when minimally invasive techniques are used, and the potential for serious complications related to bowel and vascular injury.

2.3 Epidemiology of urinary incontinence

2.3.1 Prevalence, incidence, & remission

Urinary incontinence (UI), the involuntary loss of urine, is a common condition that adversely affects the quality of life of millions of women worldwide (Haylen et al.,

2010; Pizzol et al., 2021). Typically, UI is classified into three subtypes: (1) stress UI (SUI, the involuntary loss of urine on effort or exertion), (2) urgency UI (UII, the leakage of urine associated with a strong urge or desire to urinate), and (3) mixed UI (MUI, leakage associated with both urgency and physical exertion) (Leach et al., 1997). While the true prevalence of UI is unknown, it is estimated that 20–30% of women will develop UI during their lifetime, with SUI being the most common subtype among older women (Landefeld et al., 2008; Minassian et al., 2020; Patel and Chapple, 2008). Other forms of UI (e.g., nocturnal enuresis, reflex, overflow, functional) also occur, albeit much less frequently (Bardsley, 2016; Patel and Chapple, 2008). Literature suggests that the prevalence of UI increases with age, where there is a gradual increase until age 50 when it reaches 30%, and there is stabilization or slight decrease until age 70, after which the prevalence rises again (Hannestad et al., 2000; Hunskaar et al., 2000). The weekly incidence rates of UI range from 1%–4%, while the annual incidence rates for UI are between 1%–19% across several studies, where the incidence of UI among women increases about 0.5%–1% per year and the remission rates fall between 2%–5% (Abrams et al., 2016).

2.3.2 Risk factors

2.3.2.1 Age & obesity

Large studies have reported an increase in UI prevalence with age, where the rate of any urinary incontinence was found to be higher among older women (35%) versus younger women (28%) (Thom, 1998). More specifically, older women were more likely to experience UUI and MUI (due to UUI), while younger women were more likely to experience SUI (Thom, 1998). In addition, UI prevalence—particularly UUI—is known to increase with age-related comorbidities, such as coronary artery disease, pulmonary diseases, and other proinflammatory conditions (McGrother et al., 2006).

Several studies have also demonstrated a positive correlation between BMI and UI, where obese women had a nearly two-fold increased risk of UI regardless of type and severity (Hunskaar, 2008; Subak et al., 2009a). This positive association is most pronounced for severe and MUI, in comparison to frequent/occasional UI and SUI/UUI, respectively (Danforth et al., 2006; Hannestad, 2003). Furthermore, there is a temporal aspect of the relationship between BMI and UI, in which an earlier onset of obesity is associated with greater risk of UI among middle-aged women, more so for MUI and SUI compared to UII (Dallosso et al., 2003; Mishra et al., 2010; Waetjen et al., 2007). Obesity is known to increase intraabdominal pressure and predispose to SUI, whereas metabolic syndrome, which is coincident with obesity, is shown to predispose to UUI likely due to increased inflammation (Kupelian et al., 2009; Tai et al., 2010; Uzun and Zorba, 2012). Moderate weight loss has been correlated with alleviation or cure of SUI and UUI, with an increased likelihood of resolution with greater weight loss (Abrams et al., 2016; Auwad et al., 2008; Bump et al., 1992; Subak et al., 2009b; Wing et al., 2010).

2.3.2.2 Race

Some data suggests that White women are more likely to have UI than Black women (Hunskaar et al., 2000). In a study by Howard et al., Black women were reported to have more vesical mobility, a larger urethral volume, and higher urethral closure pressure (Howard et al., 2000). Bump et al. found that the prevalence of SUI was 2.3 times greater among White women than Black women, while another study showed that this trend persisted even when controlling for age and other health issues (Brown et al., 1999; Jackson et al., 2004; Bump, 1993). However, the significant racial differences may be attributed to differences in medical care-seeking behavior between Black and White women (Hunskaar et al., 2000). Thus additional studies are needed to determine the presence of racial differences in UI prevalence and the manner by which UI uniquely impacts racial groups differently.

2.3.2.3 Pregnancy, parity & mode of delivery

Parity is one of the most critical risk factors of UI (particularly, SUI), where increasing parity has been associated with greater risk of UI, especially between the 30–40 years ages range (Connolly et al., 2007; Danforth et al., 2006; Grodstein et al., 2003; Rortveit et al., 2001; Waetjen et al., 2007). In particular, this positive correlation between parity and UI is more evident for SUI and MUI than for UUI (Hirsch et al., 2010; Lukacz et al., 2006).

Moreover, there is a significant difference in this effect between cesarean and vaginal deliveries (Press et al., 2007; Tähtinen et al., 2016). Findings suggest that vaginal delivery increases the risk of SUI by two-fold compared to cesarean section, while for UUI this effect is less pronounced (Abrams et al., 2016). Conversely, cesarean was demonstrated to have a protective effect, in which the prevalence for UI was found to be similar between women that exclusively delivered by caesarean and age-matched nulliparous women (Hannah et al., 2004). Despite this, the onset of UI for many women is during pregnancy and is strongly predictive of postpartum UI (Altman et al., 2006; Foldspang et al., 2004; Hvidman et al., 2003; Viktrup et al., 2006).

During pregnancy, the prevalence of UI increases rapidly during the second trimester compared to the first and third trimesters (Marshall et al., 1998; Mørkved and Bø, 1999). When comparing the prevalence of SUI before and during pregnancy, it increased from 9% to 31% among nulliparous women and from 24% to 42% among parous women, whereas for UUI, rates stayed relatively the same (Wesnes et al., 2009). It is believed that the temporary, pregnancy-related physiological changes (i.e., vaginal tissue protease activity, a measure of tissue remodeling) can predispose women to postpartum UI (Giugale et al., 2021).

2.3.2.4 Hysterectomy

For several women, the onset of UI is immediately after hysterectomy (Abrams et al., 2016). With concomitant oophorectomy, surgical menopause is induced, which may cause hormonal changes associated with increased incidence of UI (Heliövaara-Peippo et al., 2010). It is hypothesized that intraoperative nerve injury or disruption of the musculofascial supports (i.e., loss of pelvic floor support) may be potential causes of UI post-hysterectomy (Abrams et al., 2016; Brown et al., 2000).

2.4 Pathophysiology of urinary incontinence

2.4.1 Anatomical structure, support & function

2.4.1.1 Lower urinary tract anatomy & supports

2.4.1.1.1 Urinary bladder

The urinary bladder is a hollow, distensible, muscular organ that has two primary functions: (1) it acts as a lower-pressure reservoir of urine (up to 500 ml in the normal adult) and (2) it creates enough pressure to expel urine (Fry and Vahabi, 2016; Mangera et al., 2013). Its muscular wall confers good compliance and contractility to the bladder, permitting the storage of large volumes of urine and coordinated contraction for the expulsion of urine at an appropriate time (Mahadevan, 2016; Mangera et al., 2013). When empty, the bladder is solely extraperitoneal, lying behind the pubic symphysis and bones, and appears as a flattened tetrahedron. As the bladder fills with urine, it distends, becoming more ovoid in shape, and moves upwards towards the abdomen (Mahadevan, 2016).

In females, the superior surface of the bladder lies over the inferior portion of the uterus and upper cervix; the inferolateral aspects of the bladder are related to the obturator and levator ani muscles; and the bladder neck (i.e., the connection of the bladder to the urethra) and anterior aspect of the bladder are adjacent to the retropubic space; the posterior surface of the bladder is attached to the anterior aspect of the vagina, supravaginal cervix, and body of the uterus by connective tissue (Mahadevan, 2016).

The bladder wall is comprised of four layers: (1) *urothelium* (i.e., mucosa), the innermost lining of transitional cell epithelium that serves as an impervious barrier to urine and plays a role in sensing bladder filling and modulating bladder contractions; (2) *lamina propria* (i.e., submucosa); (3) *detrusor muscle* (i.e., muscularis propria); interlacing smooth muscle fibers; and (4) *serosa*, the outermost adventitial connective tissue layer made from peritoneum (Bolla and Jetti, 2019; Fry and Vahabi, 2016; Mangera et al., 2013).

2.4.1.1.2 Urethra

The urethra is a small muscular tube, whose functions are to (1) permit the passage of urine from the bladder during voiding and (2) maintain urinary continence during the storage phase (Patel and Chapple, 2008). In women, the urethra is about 4 cm long and 6 mm wide and embedded in the anterior vaginal wall (Mahadevan, 2016; Patel and Chapple, 2008). It begins at the bladder neck (i.e., internal urethral sphincter), passing behind the pubic symphysis through the urogenital diaphragm and external urethral sphincter, and opens through the external urethral meatus that lies between the clitoris and vaginal introitus (Sekhar et al., 2019).

Histologically, the urethra is composed of three layers: (1) *superficial mucosal layer*, that is continuous with the urothelium layer of the bladder; (2) *erectile submucosa layer*, a rich network of interconnected blood vessels and collagen, elastic, and muscle fibers that help keep the urethra closed at rest; and (3) *muscular layer*, smooth muscle that shortens the urethra during urination and gives tone (Sekhar et al., 2019).

2.4.1.1.3 Pelvic floor muscles & the external urethral sphincter

The pelvic floor is a group of muscles and connective tissues that has two main functions: (1) support the abdominal viscera (including the urethra) and (2) act to constrict or maintain continence of the urethral, vaginal, and anal openings (Raizada and Mittal, 2008). It is comprised of the levator ani—pubococcygeus, iliococcygeus, and puborectalis—and coccygeus muscles (i.e. *pelvic diaphragm*); compressor urethral muscle; external urethral sphincter; bulbocavernosus; ischiocavernosus; deep and superficial transverse perineal muscles; and perineal membrane (i.e. *urogenital diaphragm*); perineal body; and the endopelvic fascia, arcus tendineus, and ligament supports (Herschorn, 2004). In particular, the female external urethral sphincter is a complex of fibrous tissue and skeletal muscle that wraps around the mid-urethra (i.e., urethral sphincter), passes by the urethra anteriorly, and inserts into the ischial rami (i.e., compressor urethral muscle), and surrounds the urethra and vagina (i.e., urethrovaginal sphincter) (Jung et al., 2012). Together, they form a hammock-like structure underneath the urethra to provide a continence mechanism by pushing the anterior urethral wall towards the posterior urethral wall, compressing the urethra against an underlying layer of endopelvic fascia connected to the anterior vaginal wall, to keep the urethra closed during increases in intraabdominal pressure (i.e., the "hammock hypothesis") (Abrams et al., 2016; Ashton-Miller et al., 2001; DeLancey, 1994b; Patel and Chapple, 2008). The integral theory also hypothesizes this passive urethral closure mechanism pertaining to women (Petros and Ulmsten, 1990).

Two recent studies by Routzong et al. proposed an alternative mechanism of female urinary continence, termed the swing theory, which demonstrated urethral motion and shape changes during passive urethral closure that contradicted those described by the hammock and integral theory (Routzong et al., 2021a,b). Using a computational, biomechanical model of urethral passive closure during an increase in intraabdominal pressure, it was demonstrated that the proximal and distal urethra swing in a coordinated fashion, which longitudinally compresses the bladder neck by pushing it against the perineal membrane. This results in shortening and kinking of the urethra and urethral walls, respectively, and collectively contribute to passive closure of the urethra, where the extent of swinging and kinking of the urethra may be dependent on both urethral muscle tone and connective tissue integrity (Routzong et al., 2021b). In addition, it was found that the *in vivo* motions and deformations of the urethra could be modeled without the "hammock," comprised of the vagina and ligamentous urethral supports, as described by the hammock and integral theories (DeLancey, 1994b; Petros and Ulmsten, 1990; Routzong et al., 2021b).

2.4.1.2 Lower urinary tract function

The lower urinary tract function is two-fold: storing (filling) and expelling urine via coordinated contraction and relaxation of the bladder and urethral sphincter muscles (Zachoval et al., 2000). Urine storage requires the bladder to hold increasing volumes at low pressure (compliance) with appropriate sensation as it fills, and closure of the bladder neck and urethra at rest and during increases in intraabdominal pressure (i.e., high outlet pressure), and no involuntary bladder contractions. Urine expulsion requires sustained synchronized bladder contractions of adequate strength and reduction in sphincter resistance (Yang, 2003). When the bladder reaches storage capacity and it is a socially appropriate time to urinate, sensations of fullness are sent to initiate bladder contraction and relaxation of the internal urethral sphincter (Jung et al., 2012). This contraction is coordinated with relaxation of the external urethral sphincter to permit the passage of urine through the urethral opening. After urination is completed, the urethral sphincters contract to return to their resting, closed state (Yang, 2003).

2.4.1.2.1 The female urethral sphincter mechanism

The urethral sphincters act to maintain continence. The internal urethral sphincter, located at the proximal urethra, is formed by a submucosa layer, a longitudinal smooth muscle layer coincident with the inner detrusor muscle, and a circumferential smooth muscle layer continuous with the external detrusor muscle of the bladder neck (Mangera et al., 2013). The external urethral sphincter is composed of striated muscle, primarily slow twitch fibers, that sustains constant tone for extended periods of time; with sudden increases in intraabdominal pressure, fast twitch fibers are engaged to provide additional urethral resistance to prevent urine from leaking (Ashton-Miller et al., 2001; Yang, 2003). The compressor urethral muscle and urethrovaginal sphincter compress the distal urethra and are integral for the continence mechanism of the urethral sphincters (Ashton-Miller et al., 2001). Damage to, or malfunction of, one or more of the muscular, supportive (connective tissue, vagina, pelvic floor), or neurological components of the urethral sphincters reduces their efficacy of maintaining continence and predisposes to urinary incontinence (UI) (Patel and Chapple, 2008).

2.4.2 Physiological mechanisms of UI

2.4.2.1 Stress UI

Stress UI (SUI), the involuntary urination during physical activity or exertion (e.g., sneezing, coughing, lifting), which increases intraabdominal pressure, is the most common subtype of UI that disproportionately affects women, especially among multiparous women following prolonged or traumatic vaginal deliveries that result in pelvic floor, pelvic/pudendal nerve, or urinary tract injury (Bardsley, 2016; Khandelwal and Kistler, 2013; Patel and Chapple, 2008). To date, there are two prevailing pathological mechanisms of SUI, both of which may coexist: (1) urethral hypermobility and (2) intrinsic sphincter deficiency (Reus et al., 2020).

2.4.2.1.1 Urethral hypermobility

Urethral hypermobility (UH) is characterized by inadequate bladder neck and urethral support. With increased intraabdominal pressure, the weakened supports are unable to keep the urethra closed, resulting in decent of the urethra, which causes an unequal pressure distribution to form along the bladder body and proximal urethra, where the bladder is under more pressure than the urethra (Patel and Chapple, 2008). This reversal of the normal pressure gradient often results in a lower urethral closure pressure and subsequently urine leakage. Improvement or restoration of urethral support with surgical correction is often associated with lower rates of UH.

2.4.2.1.2 Intrinsic sphincter deficiency

Intrinsic sphincter deficiency (ISD) occurs when there is weakness of the urethral sphincter that renders it unable to coapt and generate adequate urethral closing pressure to prevent urine leakage in response to increases in intraabdominal pressure (Lose and Brostrøm, 2014). In severe cases, a slight increase in pressure during small movement can lead to sudden involuntary leakage or even nearly continual leakage if the sphincter is fully open (i.e., relaxed). Damage to the sympathetic and somatic neural pathway from neurodegenerative disorders or from surgical/therapeutic interventions can cause or increase the risk of ISD (Patel and Chapple, 2008).

2.4.2.2 Urge UI

Urge UI (UUI) is the involuntary leakage of urine accompanied with, or preceded by, a strong desire to urinate and is also involved in overactive bladder (OAB) syndrome (i.e., urinary urgency usually accompanied by frequent voiding and nocturia with or without UUI) (Bardsley, 2016; Haylen et al., 2010; Patel and Chapple, 2008). In fact, one in three patients with OAB syndrome have UUI (Patel and Chapple, 2008). Both UUI and OAB syndrome are often characterized by detrusor overactivity.

2.4.2.2.1 Detrusor overactivity

Detrusor overactivity (DO), the involuntary contraction of the detrusor muscles during the storage phase, is frequently associated with symptomatic urinary urgency and urinary incontinence. In particular, when the intravesical pressure exceeds the urethral pressure, involuntary detrusor contractions result in incontinence (Patel and Chapple, 2008). There are two types of DO: (1) neurogenic DO, which occurs in patients with neurological disorders or injury that impact lower urinary tract function and (2) idiopathic (i.e., nonneurogenic) DO, which occurs with patients without neurological abnormalities that may or may not have an evident underlying cause (e.g., bladder outlet obstruction, bladder infection, bladder tumor) (Patel and Chapple, 2008). There are several pathophysiological mechanisms of DO that have been postulated, such as increased afferent nerve activity, which elicits both the urinary reflex and involuntary detrusor contraction, enhanced spontaneous contractility of the detrusor muscle bundles, increased mechanosensitivity of the detrusor receptors, reduced central inhibition, patchy denervation of the detrusor muscle fibers, and poor bladder compliance (Patel and Chapple, 2008).

2.4.2.3 Mixed UI

Mixed UI (MUI), the complaint of involuntary urination with urgency and exertion, is a combination of SUI and UUI symptoms and involves both their physiological mechanisms (Bardsley, 2016; D'Ancona et al., 2019; Patel and Chapple, 2008). Approximately one in three women with urinary incontinence have MUI, and surgical correction of SUI is associated with a UUI cure rate between 50% and 85% (Karram and Bhatia, 1989).

2.4.2.4 Overflow incontinence

Overflow incontinence is defined as the involuntary urine loss caused by overdistension of the bladder, possibly due to impaired detrusor contractility and/or bladder outlet obstruction (DuBeau et al., 2010). It is characterized by a sense of urinary dribbling, slow-flowing urine, and a sense of incomplete emptying. Additionally, it may be associated with patchy denervation of the detrusor muscles (Patel and Chapple, 2008). With overflow incontinence, the intravesical pressure is greater than the resting urethral pressure, causing urine overflows without the contraction of the detrusors. Patients with this form of UI also often experience frequency or urgency symptoms as the detrusors try to expel urine in response to small increases in intraabdominal pressure.

2.4.3 Clinical diagnosis & evaluation

A thorough patient history, including comorbidities, is vital for the evaluation of UI, particularly in identifying the type, severity, and degree of debilitation attributable to the disorder (DeMaagd and Davenport, 2012; Weiss, 1998). Typically, multiple office visits are necessary to conduct the physical examination and required assessments for UI (Frank et al., 2010; Khandelwal and Kistler, 2013).

The clinical evaluation of UI involves several steps. The first is the identification of transient or reversible causes of UI (e.g., medication-induced) (Dowling-Castronovo and Specht, 2009). If determined to be an acute condition, correction of the transient or reversible causes may alleviate symptoms; if determined to be chronic, additional assessments are necessary (Imam, 2004; Khandelwal and Kistler, 2013).

Many patient questionnaires are used to classify the type of chronic UI present (Holroyd-Leduc et al., 2008). Additionally, assessments of other medical conditions in relation to the time of UI symptoms, gynecologic history, and comorbidities are performed to develop appropriate treatment plans to resolve or lessen the severity of symptoms (DeMaagd and Davenport, 2012; McKertich, 2008; Weiss, 1998).

The aggressiveness of treatment is dependent on symptom severity and its impact on quality of life (e.g., activities of daily living, sleep, interpersonal relationships, sexual activity, self-perceived health) (Khandelwal and Kistler, 2013; Weiss, 1998). Thus determining and prioritizing the most burdensome symptoms aids direct management and targeted treatment (Khandelwal and Kistler, 2013). Voiding diaries (e.g., three-day diary, frequency-volume voiding diary) are helpful for providing a more comprehensive voiding history, such as the frequency and situational circumstances of incontinence episodes, as well as classifying the type of incontinence. Furthermore, voiding diaries permit representative, reproducible measures of incontinence, which can be used to establish a baseline for assessing the progression in incontinence severity over the course of treatment to determine management efficacy (Bryan and Chapple, 2004).

Physical examination (e.g., neurologic tests, pelvic floor assessments, urinalysis) and urodynamic analysis also assist identifying the precise transient and persistent causes of incontinence necessary for ensuring an accurate diagnosis (DeMaagd and Davenport, 2012; Khandelwal and Kistler, 2013). Neurologic and external gynecologic examination are useful for detecting disorders, injuries, or abnormalities that cause or worsen UI, such as pudendal neuropathy, pelvic organ prolapse, and pelvic floor injury. Urodynamic tests, which involve quantitative measures of urine flow (e.g., bladder pressure, filling cystometry, urethral pressure, flow rate), are helpful for describing lower urinary tract function, treatment planning, and evaluating treatment outcomes (Abrams et al., 2010; DeMaagd and Davenport, 2012).

2.4.4 Treatment & management

Non-drug methods of treatment for UI include weight loss, exercise, and proper nutrition (where weight loss has been associated with reduced frequency of SUI and UUI). Moderate daily exercise was correlated with lower incidence of UI in comparison to a sedentary lifestyle, and a high-fat diet was associated with a higher risk of SUI onset (Dallosso et al., 2004; DeMaagd and Davenport, 2012; Peterson, 2008; Shamliyan et al., 2008; Subak et al., 2009b).

In addition, habit training and prompted or timed voiding, in which patients are aware of their need to void and intentionally delay voiding when needed, are other nonpharmacological interventions for UI, which, when combined with exercise, are associated with moderate, temporary improvements in daytime UI (DeMaagd and Davenport, 2012). In particular, as first-line treatment, pelvic floor exercises (i.e., Kegels) and bladder training have been found to help alleviate UI symptoms by strengthening and retraining the detrusor muscles to improve modulation of urinary function (Aslan et al., 2008; DeMaagd and Davenport, 2012; Kegel, 1948).

Hormone therapy, specifically topical or localized estrogen, is a common pharmacological intervention that has been shown to improve lower urinary tract symptoms (e.g., frequency, urgency) in postmenopausal women with UUI, where atrophy and dysfunction of the urogenital tract is caused by the loss of estrogen during menopause (Carlile et al., 1988; DeMaagd and Davenport, 2012; Verelst et al., 2002). Second-line therapies for UUI include anticholinergic and beta-3 adrenergic agonist medications that treat UUI and improve bladder control by reducing the mechanosensitivity and (over)activity of the detrusor muscles (Abrams and Andersson, 2007; Lam and Hilas, 2007). Beta-3 agonists are currently favored due to the absence of anti-cholinergic side effects particularly those that impact cognition. For women who fail to respond to medication, third-line therapies are available that include peripheral tibial nerve stimulation, sacral neuromodulation, and intravesical onabotulinum toxin A.

For SUI, there are no approved medications for treatment. Instead, surgery is commonly performed (typically after unsuccessful treatment via conservative methods) to enhance urethral resistance to decrease urine leakage and restore normal bladder function (Rovner and Wein, 2004). Sling procedures, such as the mid-urethral sling, are often used to treat urethral hypermobility and sphincter deficiencies by elevating the urethra from below and increasing urethral compression in response to increases in intraabdominal pressure (Harding and Thorpe, 2010). These surgical interventions have been shown to be effective in treating SUI and associated with improved quality of life (DeMaagd and Davenport, 2012; Rovner and Wein, 2004). Nonsurgical treatments of SUI, aside from exercise and dietary changes, include pessaries, electrical stimulation, and the injection of transurethral bulking agents (Appell, 2008; Culligan and Heit, 2000; McIntosh, 2005).

Treatment for MUI most often involves a combination of both pharmacological therapies and surgical procedures discussed previously, where the predominant condition is generally prioritized (DeMaagd and Davenport, 2012; Thüroff et al., 2011).

References

Abrams, P., Andersson, K.E., Birder, L., Brubaker, L., Cardozo, L., Chapple, C., Cottenden, A., Davila, W., De Ridder, D., Dmochowski, R., Drake, M., DuBeau, C., Fry, C., Hanno, P., Hay Smith, J., Herschorn, S., Hosker, G., Kelleher, C., Koelbl, H., et al., 2010. Fourth International Consultation on Incontinence Recommendations of the International Scientific Committee: evaluation and treatment of urinary incontinence, pelvic organ prolapse, and fecal incontinence. Neurourology and Urodynamics 29 (1), 213–240. https://doi.org/10.1002/NAU.20870.

Abrams, Paul, Andersson, K.E., 2007. Muscarinic receptor antagonists for overactive bladder. BJU International 100 (5), 987–1006. https://doi.org/10.1111/J.1464-410X.2007.07205.X.

Abrams, Paul, Cardozo, L., Wagg, A., Wein, A., 2016. Epidemiology of urinary incontinence (UI) and other lower urinary tract symptoms (LUTS), pelvic organ prolapse (POP) and anal incontinence (AI). In: Incontinence. 6th International Consultation on Incontinence. Tokyo, September 2016, Issue September 2016. https://www.ics.org/publications/ici_6/Incontinence_6th_Edition_2017_eBook_v2.pdf.

Alperin, M., Debes, K., Abramowitch, S., Meyn, L., Moalli, P.A., 2008. LOXL1 deficiency negatively impacts the biomechanical properties of the mouse vagina and supportive tissues. International Urogynecology Journal 19 (7), 977–986.

Altman, D., Ekström, Å., Gustafsson, C., López, A., Falconer, C., Zetterström, J., 2006. Risk of urinary incontinence after childbirth: a 10-year prospective cohort study. Obstetrics and Gynecology 108 (4), 873–878.

Altman, D., Falconer, C., Cnattingius, S., Granath, F., 2008. Pelvic organ prolapse surgery following hysterectomy on benign indications. American Journal of Obstetrics and Gynecology 198 (5), 572.e1–572.e6. https://doi.org/10.1016/J.AJOG.2008.01.012.

Appell, R.A., 2008. Transurethral collagen denaturation for women with stress urinary incontinence. Current Urology Reports 9 (5), 373–379. https://doi.org/10.1007/S11934-008-0065-5.

Aronson, M.P., Bates, S.M., Jacoby, A.F., Chelmow, D., Sant, G.R., 1995. Periurethral and paravaginal anatomy: an endovaginal magnetic resonance imaging study. American Journal of Obstetrics and Gynecology 173 (6), 1702–1710.

Ashton-Miller, J.A., Howard, D., DeLancey, J.O.L., 2001. The functional anatomy of the female pelvic floor and stress continence control system. Scandinavian Journal of Urology and Nephrology, Supplement 35 (207), 1–7. https://doi.org/10.1080/003655901750174773.

Aslan, E., Komurcu, N., Beji, N.K., Yalcin, O., 2008. Bladder training and Kegel exercises for women with urinary complaints living in a rest home. Gerontology 54 (4), 224–231. https://doi.org/10.1159/000133565.

Auwad, W., Steggles, P., Bombieri, L., Waterfield, M., Wilkin, T., Freeman, R., 2008. Moderate weight loss in obese women with urinary incontinence: a prospective longitudinal study. International Urogynecology Journal 19 (9), 1251–1259. https://doi.org/10.1007/S00192-008-0616-9/TABLES/4.

Barber, M.D., Maher, C., 2013. Epidemiology and outcome assessment of pelvic organ prolapse. International Urogynecology Journal 24 (11), 1783–1790.

Bardsley, A., 2016. An overview of urinary incontinence. British Journal of Nursing 25 (18), S14–S21. https://doi.org/10.12968/bjon.2016.25.18.S14.

Bolla, S.R., Jetti, R., 2019. Histology, bladder. In: StatPearls. StatPearls Publishing. http://www.ncbi.nlm.nih.gov/pubmed/31082007.

Bradley, C.S., Zimmerman, M.B., Qi, Y., Nygaard, I.E., 2007. Natural history of pelvic organ prolapse in postmenopausal women. Obstetrics and Gynecology 109 (4), 848–854. https://doi.org/10.1097/01.AOG.0000255977.91296.5d.

Braga, A., Caccia, G., 2018. Pelvic organ prolapse: pathophysiology and epidemiology. In: Management of Pelvic Organ Prolapse. Springer, pp. 19–30.

Brown, J.S., Grady, D., Ouslander, J.G., Herzog, A.R., Varner, R.E., Posner, S.F., 1999. Prevalence of urinary incontinence and associated risk factors in postmenopausal women. Obstetrics and Gynecology 94 (1), 66–70. https://doi.org/10.1016/S0029-7844(99)00263-X.

Brown, J.S., Sawaya, G., Thom, D.H., Grady, D., 2000. Hysterectomy and urinary incontinence: a systematic review. The Lancet 356 (9229), 535–539. https://doi.org/10.1016/S0140-6736(00)02577-0.

Bryan, N.P., Chapple, C.R., 2004. Frequency volume charts in the assessment and evaluation of treatment: how should we use them? European Urology 46 (5), 636–640. https://doi.org/10.1016/J.EURURO.2004.07.023.

Budatha, M., Roshanravan, S., Zheng, Q., Weislander, C., Chapman, S.L., Davis, E.C., Starcher, B., Word, R.A., Yanagisawa, H., 2011. Extracellular matrix proteases contribute to progression of pelvic organ prolapse in mice and humans. The Journal of Clinical Investigation 121 (5), 2048–2059.

Bump, R.C., 1993. Racial comparisons and contrasts in urinary incontinence and pelvic organ prolapse. Obstetrics and Gynecology 81 (3), 421–425. https://doi.org/10.1016/0020-7292(93)90361-y.

Bump, R.C., Mattiasson, A., Bo, K., Brubaker, L.P., DeLancey, J.O.L., Klarskov, P., Shull, B.L., Smith, A.R.B., 1996. The standardization of terminology of female pelvic organ prolapse and pelvic floor dysfunction. American Journal of Obstetrics and Gynecology 175 (1), 10–17. https://doi.org/10.1016/S0002-9378(96)70243-0.

Bump, Richard C., Sugerman, H.J., Fantl, J.A., McClish, D.K., 1992. Obesity and lower urinary tract function in women: effect of surgically induced weight loss. American Journal of Obstetrics and Gynecology 167 (2), 392–399. https://doi.org/10.1016/S0002-9378(11)91418-5.

Carlile, A., Davies, I., Rigby, A., Brocklehurst, J.C., 1988. Age changes in the human female urethra: a morphometric study. The Journal of Urology 139 (3), 532–535. https://doi.org/10.1016/S0022-5347(17)42512-2.

Chaliha, C., 2009. Postpartum pelvic floor trauma. Current Opinion in Obstetrics and Gynecology 21 (6), 474–479.

Chen, C.C.G., Walters, M.D., 2008. Pelvic organ prolapse: clinical diagnosis and presentation. Female Urology, 556–563. https://doi.org/10.1016/B978-1-4160-2339-5.50103-8.

Cheung, R.Y.K., Lee, J.H.S., Lee, L.L., Chung, T.K.H., Chan, S.S.C., 2016. Vaginal pessary in women with symptomatic pelvic organ prolapse: a randomized controlled trial. Obstetrics and Gynecology 128 (1), 73–80. https://doi.org/10.1097/AOG.0000000000001489.

Chiaffarino, F., Chatenoud, L., Dindelli, M., Meschia, M., Buonaguidi, A., Amicarelli, F., Surace, M., Bertola, E., Di Cintio, E., Parazzini, F., 1999. Reproductive factors, family history, occupation and risk of urogenital prolapse. European Journal of Obstetrics & Gynecology and Reproductive Biology 82 (1), 63–67.

Connolly, T.J., Litman, H.J., Tennstedt, S.L., Link, C.L., McKinlay, J.B., 2007. The effect of mode of delivery, parity, and birth weight on risk of urinary incontinence. International Urogynecology Journal 18 (9), 1033–1042.

Corton, M.M., 2012. Aspects of gynecologic surgery. Chapter 38. Anatomy. In: Williams Gynecology, 2nd ed. McGraw-Hill Medical, New York, pp. 917–947.

Couri, B.M., Lenis, A.T., Borazjani, A., Paraiso, M.F.R., Damaser, M.S., 2012. Animal models of female pelvic organ prolapse: lessons learned. Expert Review of Obstetrics & Gynecology 7 (3), 249–260.

Culligan, P.J., 2012. Nonsurgical management of pelvic organ prolapse. Obstetrics and Gynecology 119 (4), 852–860. https://doi.org/10.1097/AOG.0B013E31824C0806.

Culligan, P.J., Heit, M., 2000. Urinary incontinence in women: evaluation and management. American Family Physician 62 (11), 2433–2444.

da Cunha, M.G.M.C.M., Mackova, K., Hympanova, L.H., Bortolini, M.A.T., Deprest, J., 2021. Animal models for pelvic organ prolapse: systematic review. International Urogynecology Journal, 1–14.

Dallosso, H., Matthews, R., McGrother, C., Donaldson, M., 2004. Diet as a risk factor for the development of stress urinary incontinence: a longitudinal study in women. European Journal of Clinical Nutrition 58 (6), 920–926. https://doi.org/10.1038/sj.ejcn.1601913.

Dallosso, H.M., McGrother, C.W., Matthews, R.J., Donaldson, M.M.K., 2003. The association of diet and other lifestyle factors with overactive bladder and stress incontinence: a longitudinal study in women. BJU International 92 (1), 69–77. https://doi.org/10.1046/J.1464-410X.2003.04271.X.

Damron, D.P., Capeless, E.L., 2004. Operative vaginal delivery: a comparison of forceps and vacuum for success rate and risk of rectal sphincter injury. American Journal of Obstetrics and Gynecology 191 (3), 907–910.

D'Ancona, C., Haylen, B., Oelke, M., Abranches-Monteiro, L., Arnold, E., Goldman, H., Hamid, R., Homma, Y., Marcelissen, T., Rademakers, K., Schizas, A., Singla, A., Soto, I., Tse, V., de Wachter, S., Herschorn, S., 2019. The International Continence Society (ICS) report on the terminology for adult male lower urinary tract and pelvic floor symptoms and dysfunction. Neurourology and Urodynamics 38 (2), 433–477. https://doi.org/10.1002/nau.23897.

Danforth, K.N., Townsend, M.K., Lifford, K., Curhan, G.C., Resnick, N.M., Grodstein, F., 2006. Risk factors for urinary incontinence among middle-aged women. American Journal of Obstetrics and Gynecology 194 (2), 339–345. https://doi.org/10.1016/J.AJOG.2005.07.051.

de Sam Lazaro, S., Nardos, R., Caughey, A.B., 2016. Obesity and pelvic floor dysfunction: battling the bulge. Obstetrical & Gynecological Survey 71 (2), 114–125.

DeLancey, J.O.L., 1994a. The anatomy of the pelvic floor. Current Opinion in Obstetrics and Gynecology 6 (4), 313–316. https://doi.org/10.1097/00001703-199408000-00003.

DeLancey, John O.L., 1992. Anatomie aspects of vaginal eversion after hysterectomy. American Journal of Obstetrics and Gynecology 166 (6), 1717–1728. https://doi.org/10.1016/0002-9378(92)91562-O.

DeLancey, John O.L., 1994b. Structural support of the urethra as it relates to stress urinary incontinence: the hammock hypothesis. American Journal of Obstetrics and Gynecology 170 (5), 1713–1723.

DeLancey, John O.L., Kearney, R., Chou, Q., Speights, S., Binno, S., 2003. The appearance of levator ani muscle abnormalities in magnetic resonance images after vaginal delivery. Obstetrics and Gynecology 101 (1), 46–53. https://doi.org/10.1016/S0029-7844(02)02465-1.

DeLancey, John O.L., Morgan, D.M., Fenner, D.E., Kearney, R., Guire, K., Miller, J.M., Hussain, H., Umek, W., Hsu, Y., Ashton-Miller, J.A., 2007. Comparison of levator ani muscle defects and function in women with and without pelvic organ prolapse. Obstetrics and Gynecology 109 (2 PART 1), 295–302. https://doi.org/10.1097/01.AOG.0000250901.57095.ba.

DeMaagd, G.A., Davenport, T.C., 2012. Management of urinary incontinence. Pharmacy and Therapeutics 37 (6), 345. /pmc/articles/PMC3411204/.

Dietz, H.P., 2003. Do Asian women have less pelvic organ mobility than Caucasians? International Urogynecology Journal 14 (4), 250–253.

Dietz, H.P., Simpson, J.M., 2008. Levator trauma is associated with pelvic organ prolapse. BJOG. An International Journal of Obstetrics and Gynaecology 115 (8), 979–984.

Dietz, Hans Peter, 2015. Pelvic organ prolapse-a review. Australian Family Physician 44 (7), 446–452.

Dietz, Hans Peter, Franco, A.V.M., Shek, K.L., Kirby, A., 2012. Avulsion injury and levator hiatal ballooning: two independent risk factors for prolapse? An observational study. Acta Obstetricia et Gynecologica Scandinavica 91 (2), 211–214.

Dietz, Hans Peter, Lanzarone, V., 2005. Levator trauma after vaginal delivery. Obstetrics and Gynecology 106 (4), 707–712.

Doshani, A., Teo, R.E.C., Mayne, C.J., Tincello, D.G., 2007. Uterine prolapse. BMJ. British Medical Journal 335 (7624), 819. https://doi.org/10.1136/BMJ.39356.604074.BE.

Dowling-Castronovo, A., Specht, J.K., 2009. Assessment of transient urinary incontinence in older adults. American Journal of Nursing 109 (2), 62–71. https://doi.org/10.1097/01.NAJ.0000345392.52704.6D.

DuBeau, C.E., Kuchel, G.A., Johnson, T., Palmer, M.H., Wagg, A., 2010. Incontinence in the frail elderly: report from the 4th International Consultation on Incontinence. Neurourology and Urodynamics 29 (1), 165–178. https://doi.org/10.1002/NAU.20842.

Dyer, O., 2019. Transvaginal mesh: FDA orders remaining products off US market. BMJ. British Medical Journal 365, l1839. https://doi.org/10.1136/BMJ.L1839.

Feola, A., Abramowitch, S., Jones, K., Stein, S., Moalli, P., 2010. Parity negatively impacts vaginal mechanical properties and collagen structure in rhesus macaques. American Journal of Obstetrics and Gynecology 203 (6), 595.e1. https://doi.org/10.1016/J.AJOG.2010.06.035.

Foldspang, A., Hvidman, L., Mommsen, S., Nielsen, J.B., 2004. Risk of postpartum urinary incontinence associated with pregnancy and mode of delivery. Acta Obstetricia et Gynecologica Scandinavica 83 (10), 923–927.

Frank, C., Fcfp, M.D., Szlanta, A., Ccfp, M.D., 2010. Office management of urinary incontinence among older patients. Canadian Family Physician 56 (11), 1115. /pmc/articles/PMC2980426/.

Fritsch, H., Zwierzina, M., Riss, P., 2012. Accuracy of concepts in female pelvic floor anatomy: facts and myths! World Journal of Urology 30 (4), 429–435. https://doi.org/10.1007/S00345-011-0777-X/FIGURES/2.

Fry, C.H., Vahabi, B., 2016. The role of the mucosa in normal and abnormal bladder function. In: Basic and Clinical Pharmacology and Toxicology, vol. 119. Blackwell Publishing Ltd., pp. 57–62.

Garriga, J.C., Isern, A.P., Pons, M.E., Retamal, M.D., Fabregas, A.F., Rodriguez-Carballeira, M., 2011. Tridimensional sonographic anatomical changes on pelvic floor muscle according to the type of delivery. International Urogynecology Journal 22 (8), 1011–1018.

Ghetti, C., Gregory, W.T., Clark, A.L., 2007. Risk factors for surgically managed pelvic organ prolapse and urinary incontinence. International Journal of Gynaecology and Obstetrics 98 (1), 63–64.

Giugale, L.E., Moalli, P.A., Canavan, T.P., Meyn, L.A., Oliphant, S.S., 2021. Prevalence and predictors of urinary incontinence at 1 year postpartum. Female Pelvic Medicine & Reconstructive Surgery 27 (2), e436–e441.

Glazener, C., Elders, A., MacArthur, C., Lancashire, R.J., Herbison, P., Hagen, S., Dean, N., Bain, C., Toozs-Hobson, P., Richardson, K., 2013. Childbirth and prolapse: long-term associations with the symptoms and objective measurement of pelvic organ prolapse. BJOG. An International Journal of Obstetrics and Gynaecology 120 (2), 161–168.

Grodstein, F., Fretts, R., Lifford, K., Resnick, N., Curhan, G., 2003. Association of age, race, and obstetric history with urinary symptoms among women in the Nurses' Health Study. American Journal of Obstetrics and Gynecology 189 (2), 428–434. https://doi.org/10.1067/S0002-9378(03)00361-2.

Gyhagen, M., Bullarbo, M., Nielsen, T.F., Milsom, I., 2013. Prevalence and risk factors for pelvic organ prolapse 20 years after childbirth: a national cohort study in singleton primiparae after vaginal or caesarean delivery. BJOG. An International Journal of Obstetrics and Gynaecology 120 (2), 152–160.

Hall, A.F., Theofrastous, J.P., Cundiff, G.W., Harris, R.L., Hamilton, L.F., Swift, S.E., Bump, R.C., Porges, R.F., 1996. Interobserver and intraobserver reliability of the proposed International Continence Society, Society of Gynecologic Surgeons, and American Urogynecologic Society pelvic organ prolapse classification system. American Journal of Obstetrics and Gynecology 175 (6), 1467–1471. https://doi.org/10.1016/S0002-9378(96)70091-1.

Handa, V.L., Blomquist, J.L., McDermott, K.C., Friedman, S., Muñoz, A., 2013. Pelvic floor disorders after vaginal birth: effect of episiotomy, perineal laceration, and operative birth. Obstetric Anesthesia Digest 33 (1), 40–41.

Handa, Victoria L., Blomquist, J.L., Knoepp, L.R., Hoskey, K.A., McDermott, K.C., Muñoz, A., 2011. Pelvic floor disorders 5-10 years after vaginal or cesarean childbirth. Obstetrics and Gynecology 118 (4), 777.

Hannah, M.E., Whyte, H., Hannah, W.J., Hewson, S., Amankwah, K., Cheng, M., Gafni, A., Guselle, P., Helewa, M., Hodnett, E.D., 2004. Maternal outcomes at 2 years after planned cesarean section versus planned vaginal birth for breech presentation at term: the international randomized Term Breech Trial. American Journal of Obstetrics and Gynecology 191 (3), 917–927.

Hannestad, Y., 2003. Are smoking and other lifestyle factors associated with female urinary incontinence? The Norwegian EPINCONT study. BJOG. An International Journal of Obstetrics and Gynaecology 110 (3), 247–254. https://doi.org/10.1016/S1470-0328(02)02927-0.

Hannestad, Y.S., Rortveit, G., Sandvik, H., Hunskaar, S., 2000. A community-based epidemiological survey of female urinary incontinence: the Norwegian EPINCONT study. Epidemiology of Incontinence in the County of Nord-Trøndelag. Journal of Clinical Epidemiology 53 (11), 1150–1157. https://doi.org/10.1016/S0895-4356(00)00232-8.

Harding, C.K., Thorpe, A.C., 2010. The surgical treatment of female stress urinary incontinence. IJU. Journal of the Urological Society of India 26 (2), 257. https://doi.org/10.4103/0970-1591.65401.

Haylen, B.T., de Ridder, D., Freeman, R.M., Swift, S.E., Berghmans, B., Lee, J., Monga, A., Petri, E., Rizk, D.E., Sand, P.K., Schaer, G.N., 2010. An international urogynecological association (IUGA)/international continence society (ICS) joint report on the terminology for female pelvic floor dysfunction. Neurourology and Urodynamics 29 (1), 4–20. https://doi.org/10.1002/nau.20798.

Heilbrun, M.E., Nygaard, I.E., Lockhart, M.E., Richter, H.E., Brown, M.B., Kenton, K.S., Rahn, D.D., Thomas, J.V., Weidner, A.C., Nager, C.W., 2010. Correlation between levator ani muscle injuries on magnetic resonance imaging and fecal incontinence, pelvic organ prolapse, and urinary incontinence in primiparous women. American Journal of Obstetrics and Gynecology 202 (5), 488-e1.

Heliövaara-Peippo, S., Halmesmäki, K., Hurskainen, R., Teperi, J., Grenman, S., Kivelä, A., Tomas, E., Tuppurainen, M., Paavonen, J., 2010. The effect of hysterectomy or levonorgestrel-releasing intrauterine system on lower urinary tract symptoms: a 10-year follow-up study of a randomised trial. BJOG. An International Journal of Obstetrics and Gynaecology 117 (5), 602–609. https://doi.org/10.1111/J.1471-0528.2010.02505.X.

Hendrix, S.L., Clark, A., Nygaard, I., Aragaki, A., Barnabei, V., McTiernan, A., 2002. Pelvic organ prolapse in the Women's Health Initiative: gravity and gravidity. American Journal of Obstetrics and Gynecology 186 (6), 1160–1166.

Herschorn, S., 2004. Female pelvic floor anatomy: the pelvic floor, supporting structures, and pelvic organs. Reviews in Urology 6 Suppl 5 (5), S2–S10. http://www.ncbi.nlm.nih.gov/pubmed/16985905.

Hirsch, A.G., Minassian, V.A., Dilley, A., Sartorius, J., Stewart, W.F., 2010. Parity is not associated with urgency with or without urinary incontinence. International Urogynecology Journal 21 (9), 1095–1102.

Holroyd-Leduc, J.M., Tannenbaum, C., Thorpe, K.E., Straus, S.E., 2008. What type of urinary incontinence does this woman have? JAMA 299 (12), 1446–1456. https://doi.org/10.1001/JAMA.299.12.1446.

Howard, D., Davies, P.S., Delancey, J.O.L., Small, Y., 2000. Differences in perineal lacerations in black and white primiparas. Obstetrics and Gynecology 96 (4), 622–624. https://doi.org/10.1016/S0029-7844(00)00956-X.

Hunskaar, S., Arnold, E.P., Burgio, K., Diokno, A.C., Herzog, A.R., Mallett, V.T., 2000. Epidemiology and natural history of urinary incontinence. International Urogynecology Journal and Pelvic Floor Dysfunction 11 (5), 301–319. https://doi.org/10.1007/S001920070021.

Hunskaar, Steinar, 2008. A systematic review of overweight and obesity as risk factors and targets for clinical intervention for urinary incontinence in women. Neurourology and Urodynamics 27 (8), 749–757. https://doi.org/10.1002/NAU.20635.

Hvidman, L., Foldspang, A., Mommsen, Sø, Nielsen, J.B., 2003. Postpartum urinary incontinence. Acta Obstetricia et Gynecologica Scandinavica 82 (6), 556–563.

Imam, K.A., 2004. The role of the primary care physician in the management of bladder dysfunction. Reviews in Urology 6 (Suppl 1), S38. /pmc/articles/PMC1472846/.

Jackson, R.A., Vittinghoff, E., Kanaya, A.M., Miles, T.P., Resnick, H.E., Kritchevsky, S.B., Simonsick, E.M., Brown, J.S., 2004. Urinary incontinence in elderly women: findings from the health, aging, and body composition study. Obstetrics and Gynecology 104 (2), 301–307. https://doi.org/10.1097/01.AOG.0000133482.20685.d1.

Jelovsek, J.E., Barber, M.D., Norton, P., Brubaker, L., Gantz, M., Richter, H.E., Weidner, A., Menefee, S., Schaffer, J., Pugh, N., Meikle, S., 2018. Effect of uterosacral ligament suspension vs sacrospinous ligament fixation with or without perioperative behavioral therapy for pelvic organ vaginal prolapse on surgical outcomes and prolapse symptoms at 5 years in the OPTIMAL randomized clinical trial. JAMA. Journal of the American Medical Association 319 (15), 1554–1565. https://doi.org/10.1001/jama.2018.2827.

Jones, K.A., Moalli, P.A., 2010. Pathophysiology of pelvic organ prolapse. Female Pelvic Medicine & Reconstructive Surgery 16 (2), 79–89.

Jung, H.J., Jeon, M.J., Yim, G.W., Kim, S.K., Choi, J.R., Bai, S.W., 2009. Changes in expression of fibulin-5 and lysyl oxidase-like 1 associated with pelvic organ prolapse. European Journal of Obstetrics & Gynecology and Reproductive Biology 145 (1), 117–122.

Jung, J., Ahn, H.K., Huh, Y., 2012. Clinical and functional anatomy of the urethral sphincter. International Neurourology Journal 16 (3), 102–106. https://doi.org/10.5213/inj.2012.16.3.102.

Karram, M.M., Bhatia, N.N., 1989. Management of coexistent stress and urge urinary incontinence. Obstetrics and Gynecology 73 (1), 4–7. https://europepmc.org/article/med/2909042.

Kearney, R., Miller, J.M., Ashton-Miller, J.A., DeLancey, J.O.L., 2006. Obstetrical factors associated with levator ani muscle injury after vaginal birth. Obstetrics and Gynecology 107 (1), 144.

Kearney, R., Sawhney, R., DeLancey, J.O.L., 2004. Levator ani muscle anatomy evaluated by origin-insertion pairs. Obstetrics and Gynecology 104 (1), 168.

Kegel, A.H., 1948. Progressive resistance exercise in the functional restoration of the perineal muscles. American Journal of Obstetrics and Gynecology 56 (2), 238–248. https://doi.org/10.1016/0002-9378(48)90266-X.

Kerkhof, M.H., Hendriks, L., Brölmann, H.A.M., 2009. Changes in connective tissue in patients with pelvic organ prolapse - a review of the current literature. International Urogynecology Journal 20 (4), 461–474. https://doi.org/10.1007/S00192-008-0737-1/TABLES/3.

Khandelwal, C., Kistler, C., 2013. Diagnosis of urinary incontinence. American Family Physician 87 (8). http://www.aafp.org/afpAmericanFamilyPhysician543.

Kim, E.J., Chung, N., Park, S.H., Lee, K.-H., Kim, S.W., Kim, J.Y., Bai, S.W., Jeon, M.J., 2013. Involvement of oxidative stress and mitochondrial apoptosis in the pathogenesis of pelvic organ prolapse. The Journal of Urology 189 (2), 588–594.

Kim, S., Harvey, M.-A., Johnston, S., 2005. A review of the epidemiology and pathophysiology of pelvic floor dysfunction: do racial differences matter? Journal of Obstetrics and Gynaecology Canada 27 (3), 251–259.

Klutke, J., Ji, Q., Campeau, J., Starcher, B., Carlos Felix, J., Stanczyk, F.Z., Klutke, C., 2008. Decreased endopelvic fascia elastin content in uterine prolapse. Acta Obstetricia et Gynecologica Scandinavica 87 (1), 111–115.

Knight, K.M., Moalli, P.A., Nolfi, A., Palcsey, S., Barone, W.R., Abramowitch, S.D., 2016. Impact of parity on ewe vaginal mechanical properties relative to the nonhuman primate and rodent. International Urogynecology Journal 27 (8), 1255–1263.

Koelbl, H., Nitti, V., Baessler, K., Salvatore, S., Sultan, A., Yamaguchi, O., 2009. Pathophysiology of urinary incontinence, faecal incontinence and pelvic organ prolapse. In: Abrams, P., Cardozo, L., Khoury, S., Wein, A. (Eds.), Incontinence, pp. 255–330.

Kudish, B.I., Iglesia, C.B., Sokol, R.J., Cochrane, B., Richter, H.E., Larson, J., Hendrix, S.L., Howard, B.V., 2009. Effect of weight change on natural history of pelvic organ prolapse. Obstetrics and Gynecology 113 (1), 81.

Kupelian, V., McVary, K.T., Barry, M.J., Link, C.L., Rosen, R.C., Aiyer, L.P., Mollon, P., McKinlay, J.B., 2009. Association of C-reactive protein and lower urinary tract symptoms in men and women: results from Boston area community health survey. Urology 73 (5), 950–957. https://doi.org/10.1016/J.UROLOGY.2008.12.012.

Lam, S., Hilas, O., 2007. Pharmacologic management of overactive bladder. Clinical Interventions in Aging 2 (3), 337. https://doi.org/10.2165/00019053-200523100-00003.

Landefeld, C.S., Bowers, B.J., Feld, A.D., Hartmann, K.E., Hoffman, E., Ingber, M.J., King, J.T., McDougal, W.S., Nelson, H., Orav, E.J., Pignone, M., Richardson, L.H., Rohrbaugh, R.M., Siebens, H.C., Trock, B.J., 2008. National Institutes of Health State-of-the-Science conference statement: prevention of fecal and urinary incontinence in adults. Annals of Internal Medicine 148 (6), 449–458. https://doi.org/10.7326/0003-4819-148-6-200803180-00210.

Lazarou, G., Grigorescu, B., 2021. Pelvic organ prolapse treatment & management. https://emedicine.medscape.com/article/276259-treatment#showall.

Leach, G.E., Dmochowski, R.R., Appell, R., Blaivas, J.G., Hadley, H.R., Luber, K.M., Mostwin, J.L., O'Donnell, P.D., Roehrborn, C.G., 1997. Female stress urinary incontinence clinical guidelines panel summary report on surgical management of female stress urinary incontinence. The Journal of Urology 158 (3), 875–880.

Liguori, I., Russo, G., Curcio, F., Bulli, G., Aran, L., Della-Morte, D., Gargiulo, G., Testa, G., Cacciatore, F., Bonaduce, D., 2018. Oxidative stress, aging, and diseases. Clinical Interventions in Aging 13, 757.

Liu, X., Zhao, Y., Gao, J., Pawlyk, B., Starcher, B., Spencer, J.A., Yanagisawa, H., Zuo, J., Li, T., 2004. Elastic fiber homeostasis requires lysyl oxidase–like 1 protein. Nature Genetics 36 (2), 178–182.

Liu, X., Zhao, Y., Pawlyk, B., Damaser, M., Li, T., 2006. Failure of elastic fiber homeostasis leads to pelvic floor disorders. The American Journal of Pathology 168 (2), 519–528. https://doi.org/10.2353/AJPATH.2006.050399.

Lose, G., Brostrøm, S., 2014. Low-Pressure Urethra in Women: what does it mean and what can it be used for? International Urogynecology Journal 13 (4), 215–217. https://doi.org/10.1007/S001920200047.

Luber, K.M., Boero, S., Choe, J.Y., 2001. The demographics of pelvic floor disorders: current observations and future projections. American Journal of Obstetrics and Gynecology 184 (7), 1496–1503.

Lukacz, E.S., Lawrence, J.M., Contreras, R., Nager, C.W., Luber, K.M., 2006. Parity, mode of delivery, and pelvic floor disorders. Obstetrics and Gynecology 107 (6), 1253–1260.

Madhu, C., Swift, S., Moloney-Geany, S., Drake, M.J., 2018. How to use the Pelvic Organ Prolapse Quantification (POP-Q) system? Neurourology and Urodynamics 37 (S6), S39–S43. https://doi.org/10.1002/NAU.23740.

Mahadevan, V., 2016. Anatomy of the lower urinary tract. Surgery (United Kingdom) 34 (7), 318–325. https://doi.org/10.1016/j.mpsur.2016.04.001.

Maher, C.M., Feiner, B., Baessler, K., Glazener, C.M.A., 2011. Surgical management of pelvic organ prolapse in women: the updated summary version Cochrane review. International Urogynecology Journal 22 (11), 1445–1457. https://doi.org/10.1007/s00192-011-1542-9.

Mangera, A., Osman, N.I., Chapple, C.R., 2013. Anatomy of the lower urinary tract. Surgery (United Kingdom) 31 (7), 319–325. https://doi.org/10.1016/j.mpsur.2013.04.013.

Mant, J., Painter, R., Vessey, M., 1997. Epidemiology of genital prolapse: observations from the Oxford Family Planning Association study. BJOG. An International Journal of Obstetrics and Gynaecology 104 (5), 579–585. https://doi.org/10.1111/J.1471-0528.1997.TB11536.X.

Marshall, K., Thompson, K.A., Walsh, D.M., Baxter, G.D., 1998. Incidence of urinary incontinence and constipation during pregnancy and postpartum: survey of current findings at the Rotunda Lying-in Hospital. BJOG. An International Journal of Obstetrics and Gynaecology 105 (4), 400–402.

Matsuoka, H., Wexner, S.D., Desai, M.B., Nakamura, T., Nogueras, J.J., Weiss, E.G., Adami, C., Billotti, V.L., 2001. A comparison between dynamic pelvic magnetic resonance imaging and videoproctography in patients with constipation. Diseases of the Colon & Rectum 44 (4), 571–576.

McGrother, C.W., Donaldson, M.M.K., Hayward, T., Matthews, R., Dallosso, H.M., Hyde, C., the Leicestershire MRC Incontinence Study Team, 2006. Urinary storage symptoms and comorbidities: a prospective population cohort study in middle-aged and older women. Age and Ageing 35 (1), 16–24. https://doi.org/10.1093/ageing/afi205.

McIntosh, L., 2005. The role of the nurse in the use of vaginal pessaries to treat pelvic organ prolapse and/or urinary incontinence: a literature review. Urologic Nursing 25 (1), 41–48.

McKertich, K., 2008. Urinary incontinence: assessment in women: stress, urge or both? Australian Family Physician 37 (3).

Minassian, V.A., Hagan, K.A., Erekson, E., Austin, A.M., Carmichael, D., Bynum, J.P.W., Grodstein, F., 2020. The natural history of urinary incontinence subtypes in the Nurses' Health Studies. American Journal of Obstetrics and Gynecology 222 (2), 163.e1–163.e8. https://doi.org/10.1016/j.ajog.2019.08.023.

Mishra, G.D., Cardozo, L., Kuh, D., 2010. Menopausal transition and the risk of urinary incontinence: results from a British prospective cohort. BJU International 106 (8), 1170. https://doi.org/10.1111/J.1464-410X.2010.09321.X.

Moalli, P.A., Klingensmith, W.L., Meyn, L.A., Zyczynski, H.M., 2002. Regulation of matrix metalloproteinase expression by estrogen in fibroblasts that are derived from the pelvic floor. American Journal of Obstetrics and Gynecology 187 (1), 72–79. https://doi.org/10.1067/MOB.2002.124845.

Morgan, D.M., Larson, K., Lewicky-Gaupp, C., Fenner, D.E., Delancey, J.O.L., 2011. Vaginal support as determined by levator ani defect status 6 weeks after primary surgery for pelvic organ prolapse. International Journal of Gynecology and Obstetrics 114 (2), 141–144. https://doi.org/10.1016/j.ijgo.2011.02.020.

Mørkved, S., Bø, K., 1999. Prevalence of urinary incontinence during pregnancy and postpartum. International Urogynecology Journal 10 (6), 394–398.

Occelli, B., Narducci, F., Hautefeuille, J., Francke, J.P., Querleu, D., Crépin, G., Cosson, M., 2001. Anatomic study of arcus tendineus fasciae pelvis. European Journal of Obstetrics & Gynecology and Reproductive Biology 97 (2), 213–219. https://doi.org/10.1016/S0301-2115(00)00527-3.

Olsen, A.L., Smith, V.J., Bergstrom, J.O., Colling, J.C., Clark, A.L., 1997. Epidemiology of surgically managed pelvic organ prolapse and urinary incontinence. Obstetrics and Gynecology 89 (4), 501–506. https://doi.org/10.1016/S0029-7844(97)00058-6.

Otsubo, A., Miyata, Y., Matsuo, T., Mukae, Y., Mitsunari, K., Ohba, K., Sakai, H., 2021. Excessive accumulation of visceral fat is associated with lower urinary symptoms including overactive bladder in female patients. International Journal of Urology 28 (4), 397. https://doi.org/10.1111/IJU.14476.

Pandey, S., Bhattacharya, S., 2010. Impact of obesity on gynecology. Women's Health 6 (1), 107–117.

Patel, A.K., Chapple, C.R., 2008. Pathophysiology of urinary incontinence. Surgery 26 (5), 188–192. https://doi.org/10.1016/j.mpsur.2008.03.009.

Patel, D.A., Xu, X., Thomason, A.D., Ransom, S.B., Ivy, J.S., DeLancey, J.O.L., 2006. Childbirth and pelvic floor dysfunction: an epidemiologic approach to the assessment of prevention opportunities at delivery. American Journal of Obstetrics and Gynecology 195 (1), 23–28.

Persu, C., Chapple, C.R., Cauni, V., Gutue, S., Geavlete, P., 2011. Pelvic Organ Prolapse Quantification System (POP-Q) - a new era in pelvic prolapse staging. Journal of Medicine and Life 4 (1), 75–81. /pmc/articles/PMC3056425/?report=abstract.

Peterson, J.A., 2008. Minimize urinary incontinence: maximize physical activity in women. Urologic Nursing 28 (5), 351–356.

Petros, P., 2011. The integral system. Central European Journal of Urology 64 (3), 110. https://doi.org/10.5173/CEJU.2011.03.ART1.

Petros, Peter E.P., Woodman, P.J., 2008. The integral theory of continence. International Urogynecology Journal 19 (1), 35–40. https://doi.org/10.1007/S00192-007-0475-9/FIGURES/2.

Petros, Peter E. Papa, Ulmsten, U.I., 1990. An integral theory of female urinary incontinence. Acta Obstetricia et Gynecologica Scandinavica 69 (S153), 7–31. https://doi.org/10.1111/J.1600-0412.1990.TB08027.X.

Pizzol, D., Demurtas, J., Celotto, S., Maggi, S., Smith, L., Angiolelli, G., Trott, M., Yang, L., Veronese, N., 2021. Urinary incontinence and quality of life: a systematic review and meta-analysis. Aging Clinical and Experimental Research 33 (1), 25–35. https://doi.org/10.1007/s40520-020-01712-y.

Press, J.Z., Klein, M.C., Kaczorowski, J., Liston, R.M., Von Dadelszen, P., 2007. Does cesarean section reduce postpartum urinary incontinence? A systematic review. Birth 34 (3), 228–237.

Raizada, V., Mittal, R.K., 2008. Pelvic floor anatomy and applied physiology. Gastroenterology Clinics of North America 37 (3), 493–509. https://doi.org/10.1016/j.gtc.2008.06.003. NIH Public Access.

Ramalingam, K., Monga, A., 2015. Obesity and pelvic floor dysfunction. Best Practice & Research Clinical Obstetrics & Gynaecology 29 (4), 541–547.

Razzak, L., 2018. Pathophysiology of pelvic organ prolapse. In: Pelvic Floor Disorders. InTech.

Reus, C.R., Phé, V., Dechartres, A., Grilo, N.R., Chartier-Kastler, E.J., Mozer, P.C., 2020. Performance and safety of the artificial urinary sphincter (AMS 800) for non-neurogenic women with urinary incontinence secondary to intrinsic sphincter deficiency: a systematic review. European Urology Focus 6 (2), 327–338. https://doi.org/10.1016/J.EUF.2018.10.009.

Rortveit, G., Hannestad, Y.S., Daltveit, A.K., Hunskaar, S., 2001. Age-and type-dependent effects of parity on urinary incontinence: the Norwegian EPINCONT study. Obstetrics and Gynecology 98 (6), 1004–1010.

Routzong, M.R., Chang, C., Goldberg, R.P., Abramowitch, S.D., Rostaminia, G., 2021a. Urethral support in female urinary continence part 1: dynamic measurements of urethral shape and motion. International Urogynecology Journal. https://doi.org/10.1007/s00192-021-04765-3.

Routzong, M.R., Martin, L.C., Rostaminia, G., Abramowitch, S., 2021b. Urethral support in female urinary continence part 2: a computational, biomechanical analysis of Valsalva. International Urogynecology Journal, 1–11.

Rovner, E.S., Wein, A.J., 2004. Treatment options for stress urinary incontinence. Reviews in Urology 6 (Suppl 3), S29.

Rynkevic, R., Martins, P., Hympanova, L., Almeida, H., Fernandes, A.A., Deprest, J., 2017. Biomechanical and morphological properties of the multiparous ovine vagina and effect of subsequent pregnancy. Journal of Biomechanics 57, 94–102.

Salvatore, S., De Bastiani, S., Del Deo, F., 2021. Epidemiology and etiology of pelvic organ prolapse. In: Pelvic Floor Disorders. Springer, pp. 547–554.

Sekhar, A., Eberhardt, L., Lee, K.S., 2019. Imaging of the female urethra. Abdominal Radiology 44 (12), 3950–3961. https://doi.org/10.1007/s00261-019-02212-y.

Shahryarinejad, A., Vardy, M.D., 2008. Comparison of human to macaque uterosacral–cardinal ligament complex and its relationship to pelvic organ prolapse. Toxicologic Pathology 36 (7_suppl), 101S–107S. https://doi.org/10.1177/0192623308327115.

Shamliyan, T.A., Kane, R.L., Wyman, J., Wilt, T.J., 2008. Systematic review: randomized, controlled trials of nonsurgical treatments for urinary incontinence in women. Annals of Internal Medicine 148 (6), 459–473. https://doi.org/10.7326/0003-4819-148-6-200803180-00211.

Siccardi, M.A., Bordoni, B., 2021. Anatomy, abdomen and pelvis, perineal body. StatPearls. https://www.ncbi.nlm.nih.gov/books/NBK537345/.

Singh, K., Jakab, M., Reid, W.M.N., Berger, L.A., Hoyte, L., 2003. Three-dimensional magnetic resonance imaging assessment of levator ani morphologic features in different grades of prolapse. American Journal of Obstetrics and Gynecology 188 (4), 910–915. https://doi.org/10.1067/mob.2003.254.

Śliwa, J., Rosner-Tenerowicz, A., Kryza-Ottou, A., Ottou, S., Wiatrowski, A., Pomorski, M., Sozański, L., Zimmer, M., 2018. Analysis of prevalence of selected anamnestic factors among women with pelvic organ prolapse. Advances in Clinical and Experimental Medicine 27 (2), 179–184.

Smith, F.J., Holman, C.D.A.J., Moorin, R.E., Tsokos, N., 2010. Lifetime risk of undergoing surgery for pelvic organ prolapse. Obstetrics and Gynecology 116 (5), 1096–1100. https://doi.org/10.1097/AOG.0b013e3181f73729.

Srivastava, S., 2017. The mitochondrial basis of aging and age-related disorders. Genes 8 (12), 398.

Strohbehn, K., Jakary, J.A., Delancey, J.O.L., 1997. Pelvic organ prolapse in young women. Obstetrics and Gynecology 90 (1), 33–36. https://doi.org/10.1016/S0029-7844(97)00218-4.

Subak, L.L., Richter, H.E., Hunskaar, S., 2009a. Obesity and urinary incontinence: epidemiology and clinical research update. The Journal of Urology 182 (6 SUPPL), 2–7. https://doi.org/10.1016/J.JURO.2009.08.071.

Subak, L.L., Wing, R., West, D.S., Franklin, F., Vittinghoff, E., Creasman, J.M., Richter, H.E., Myers, D., Burgio, K.L., Gorin, A.A., Macer, J., Kusek, J.W., Grady, D., 2009b. Weight loss to treat urinary incontinence in overweight and obese women. The New England Journal of Medicine 360 (5), 481–490. https://doi.org/10.1056/NEJMOA0806375.

Swift, S., Woodman, P., O'Boyle, A., Kahn, M., Valley, M., Bland, D., Wang, W., Schaffer, J., 2005. Pelvic Organ Support Study (POSST): the distribution, clinical definition, and epidemiologic condition of pelvic organ support defects. American Journal of Obstetrics and Gynecology 192 (3), 795–806.

Tähtinen, R.M., Cartwright, R., Tsui, J.F., Aaltonen, R.L., Aoki, Y., Cardenas, J.L., El Dib, R., Joronen, K.M., Al Juaid, S., Kalantan, S., 2016. Long-term impact of mode of delivery on stress urinary incontinence and urgency urinary incontinence: a systematic review and meta-analysis. European Urology 70 (1), 148–158.

Tai, H.C., Chung, S.D., Ho, C.H., Tai, T.Y., Yang, W.S., Tseng, C.H., Wu, H.P., Yu, H.J., 2010. Metabolic syndrome components worsen lower urinary tract symptoms in women with type

2 diabetes. The Journal of Clinical Endocrinology & Metabolism 95 (3), 1143–1150. https://doi.org/10.1210/JC.2009-1492.

Takacs, P., Nassiri, M., Viciana, A., Candiotti, K., Fornoni, A., Medina, C.A., 2009. Fibulin-5 expression is decreased in women with anterior vaginal wall prolapse. International Urogynecology Journal 20 (2), 207–211.

Tegerstedt, G., Miedel, A., Mæhle-Schmidt, M., Nyrén, O., Hammarström, M., 2006. Obstetric risk factors for symptomatic prolapse: a population-based approach. American Journal of Obstetrics and Gynecology 194 (1), 75–81.

Thom, D., 1998. Variation in estimates of urinary incontinence prevalence in the community: effects of differences in definition, population characteristics, and study type. Journal of the American Geriatrics Society 46 (4), 473–480.

Thüroff, J.W., Abrams, P., Andersson, K.E., Artibani, W., Chapple, C.R., Drake, M.J., Hampel, C., Neisius, A., Schröder, A., Tubaro, A., 2011. EAU guidelines on urinary incontinence. European Urology 59 (3), 387–400. https://doi.org/10.1016/J.EURURO.2010.11.021.

Tso, C., Lee, W., Austin-Ketch, T., Winkler, H., Zitkus, B., 2018. Nonsurgical treatment options for women with pelvic organ prolapse. Nursing for Women's Health 22 (3), 228–239. https://doi.org/10.1016/J.NWH.2018.03.007.

Ulrich, D., Edwards, S.L., Su, K., White, J.F., Ramshaw, J.A.M., Jenkin, G., Deprest, J., Rosamilia, A., Werkmeister, J.A., Gargett, C.E., 2014. Influence of reproductive status on tissue composition and biomechanical properties of ovine vagina. PLoS ONE 9 (4), e93172.

Urbankova, I., Callewaert, G., Blacher, S., Deprest, D., Hympanova, L., Feola, A., De Landsheere, L., Deprest, J., 2019. First delivery and ovariectomy affect biomechanical and structural properties of the vagina in the ovine model. International Urogynecology Journal 30 (3), 455–464.

Uzun, H., Zorba, O.Ü., 2012. Metabolic syndrome in female patients with overactive bladder. Urology 79 (1), 72–75. https://doi.org/10.1016/J.UROLOGY.2011.08.050.

Verelst, M., Maltau, J.M., Ørbo, A., 2002. Computerised morphometric study of the paraurethral tissue in young and elderly women. Neurourology and Urodynamics 21 (6), 529–533. https://doi.org/10.1002/NAU.10089.

Viktrup, L., Rortveit, G., Lose, G., 2006. Risk of stress urinary incontinence twelve years after the first pregnancy and delivery. Obstetrics and Gynecology 108 (2), 248–254.

Volløyhaug, I., Mørkved, S., Salvesen, Ø., Salvesen, K.Å., 2015a. Forceps delivery is associated with increased risk of pelvic organ prolapse and muscle trauma: a cross-sectional study 16–24 years after first delivery. Ultrasound in Obstetrics & Gynecology 46 (4), 487–495.

Volløyhaug, I., Mørkved, S., Salvesen, Ø., Salvesen, K.Å., 2015b. Pelvic organ prolapse and incontinence 15–23 years after first delivery: a cross-sectional study. BJOG. An International Journal of Obstetrics and Gynaecology 122 (7), 964–971.

Volløyhaug, Ingrid, Mørkved, S., Salvesen, K.Å., 2016. Association between pelvic floor muscle trauma and pelvic organ prolapse 20 years after delivery. International Urogynecology Journal 27 (1), 39–45.

Vu, D., Haylen, B.T., Tse, K., Farnsworth, A., 2010. Surgical anatomy of the uterosacral ligament. International Urogynecology Journal 21 (9), 1123–1128.

Waetjen, L.E., Liao, S., Johnson, W.O., Sampselle, C.M., Sternfield, B., Harlow, S.D., Gold, E.B., 2007. Factors associated with prevalent and incident urinary incontinence in a cohort of midlife women: a longitudinal analysis of DataStudy of women's health across the nation. American Journal of Epidemiology 165 (3), 309–318. https://doi.org/10.1093/AJE/KWK018.

Walker, G.J.A., Gunasekera, P., 2011. Pelvic organ prolapse and incontinence in developing countries: review of prevalence and risk factors. International Urogynecology Journal 22 (2), 127–135.

Wang, X., Wang, X., Zhou, Y., Peng, C., Chen, H., Lu, Y., 2019. Mitofusin2 regulates the proliferation and function of fibroblasts: the possible mechanisms underlying pelvic organ prolapse development. Molecular Medicine Reports 20 (3), 2859–2866.

Washington, B.B., Erekson, E.A., Kassis, N.C., Myers, D.L., 2010. The association between obesity and stage II or greater prolapse. American Journal of Obstetrics and Gynecology 202 (5), 503-e1.

Weber, A.M., Richter, H.E., 2005. Pelvic organ prolapse. Obstetrics and Gynecology 106 (3), 615–634. https://doi.org/10.1097/01.AOG.0000175832.13266.BB.

Weintraub, A.Y., Glinter, H., Marcus-Braun, N., 2020. Narrative review of the epidemiology, diagnosis and pathophysiology of pelvic organ prolapse. International Brazilian Journal of Urology 46 (1), 5–14. https://doi.org/10.1590/S1677-5538.IBJU.2018.0581.

Weiss, B.D., 1998. Diagnostic evaluation of urinary incontinence in geriatric patients. American Family Physician 57 (11), 2675.

Wesnes, S.L., Hunskår, S., Bo, K., Rortveit, G., 2009. The effect of urinary incontinence status during pregnancy and delivery mode on incontinence postpartum. A cohort study. BJOG. An International Journal of Obstetrics and Gynaecology 116 (5), 700–707.

Whitcomb, E.L., Rortveit, G., Brown, J.S., Creasman, J.M., Thom, D.H., Van Den Eeden, S.K., Subak, L.L., 2009. Racial differences in pelvic organ prolapse. Obstetrics and Gynecology 114 (6), 1271–1277. https://doi.org/10.1097/AOG.0B013E3181BF9CC8.

Wing, R.R., West, D.S., Grady, D., Creasman, J.M., Richter, H.E., Myers, D., Burgio, K.L., Franklin, F., Gorin, A.A., Vittinghoff, E., MacEr, J., Kusek, J.W., Subak, L.L., 2010. Effect of weight loss on urinary incontinence in overweight and obese women: results at 12 and 18 months. Journal of Urology 184 (3), 1005–1010. https://doi.org/10.1016/J.JURO.2010.05.031.

Word, R.A., Pathi, S., Schaffer, J.I., 2009. Pathophysiology of pelvic organ prolapse. Obstetrics and Gynecology Clinics 36 (3), 521–539.

Wu, J.M., Hundley, A.F., Fulton, R.G., Myers, E.R., 2009. Forecasting the prevalence of pelvic floor disorders in US Women: 2010 to 2050. Obstetrics and Gynecology 114 (6), 1278–1283.

Wu, J.M., Matthews, C.A., Conover, M.M., Pate, V., Jonsson Funk, M., 2014a. Lifetime risk of stress urinary incontinence or pelvic organ prolapse surgery. Obstetrics and Gynecology 123 (6), 1201–1206. https://doi.org/10.1097/AOG.0000000000000286.

Wu, J.M., Vaughan, C.P., Goode, P.S., Redden, D.T., Burgio, K.L., Richter, H.E., Markland, A.D., 2014b. Prevalence and trends of symptomatic pelvic floor disorders in US women. Obstetrics and Gynecology 123 (1), 141.

Yang, C.C., 2003. Micturition. In: Encyclopedia of the Neurological Sciences. Elsevier, pp. 150–152.

Zacharin, R.F., 1977. Abdominoperineal urethral suspension: a ten-year experience in the management of recurrent stress incontinence of urine. Obstetrics and Gynecology 50 (1), 1–8.

Zachoval, R., Záleský, M., Lukes, M., Mares, J., Urban, M., Palascak, P., 2000. Lower urinary tract function and its disorders. Ceskoslovenská fysiologie / Ústrední ústav biologický 49 (3), 134–144. https://europepmc.org/article/med/11039243.

Zhao, B., Zhou, J., 2012. Decreased expression of elastin, fibulin-5 and lysyl oxidase-like 1 in the uterosacral ligaments of postmenopausal women with pelvic organ prolapse. Journal of Obstetrics and Gynaecology Research 38 (6), 925–931.

CHAPTER

3 Current surgical treatments for women's genital prolapse

Celine Giordano, Geraldine Giraudet, and Michel Cosson
Centre Hospitalier Universitaire Lille, Université Lille, Lille, France

3.1 Introduction

The aim of this chapter is to describe the main techniques for pelvic organ prolapse (POP) surgeries and expose their mechanical and anatomical impact on surrounding perineal structures. We do not intend to be exhaustive, as there is a significant number of surgical techniques for the treatment of POP, most of which are obsolete or reserved for very specific indications. In addition, operative techniques differ between groups and countries with variable descriptions in the literature and there is a lack of proper scientific evaluation for most of them. In POP surgery, the patient's sense of success may not entirely reflect anatomical results, rendering objective evaluation and comparison between techniques difficult. For example, a post-operative vaginal bulge may still be visible on examination, while the patient reports complete resolution of her symptoms.

A biomechanical view of the consequences of these different surgeries is offered in the last part of the chapter.

3.2 Signs and symptoms of genital prolapse

Physiopathology and clinical symptoms and evaluation of pelvic organ prolapse have been presented in the chapter authored by J.O. Delancey (Chapter 1).

3.3 Surgical indications and alternative treatments

Therapeutic options for pelvic organ prolapse include expectant management, pelvic floor physiotherapy, pessaries, and surgery. Management of POP is generally neither urgent nor essential and is guided by the patient's discomfort. Potential exceptions include severe prolapse, causing urinary retention and kidney damage, recurrent urinary

Biomechanics of the Female Reproductive System: Breast and Pelvic Organs. https://doi.org/10.1016/B978-0-12-823403-7.00014-2

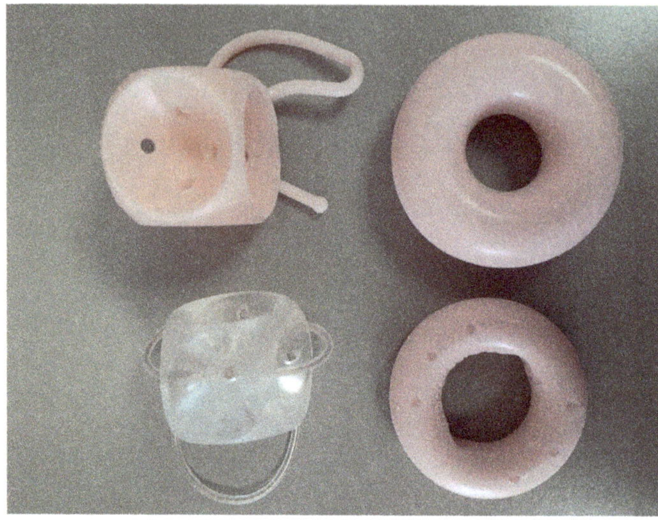

FIGURE 3.1

Different shapes of pessaries.

tract infections, and refractory vaginal ulcers. Non-invasive treatments are favored as first line, but all options should be presented to the patient with the risks and benefits of each.

3.3.1 Pelvic floor physiotherapy

Pelvic floor physiotherapy aims at reinforcing the pelvic floor, as POP occurs secondary to weakness of the pelvic floor muscles and ligaments. Repeated contractions of selected muscles can improve their strength and increase the support of pelvic organs in their normal anatomical position. Pelvic floor physiotherapy is recommended for most patients, because it is minimally invasive and was shown to improve symptoms and slow the progression of POP. Physiotherapy can be used alone or as an adjunct to pessaries or surgery.

3.3.2 Pessaries

Pessaries (see Fig. 3.1) are silicone devices of various shapes and sizes that are placed into the vagina to provide support. Depending on the type, they act by either occupying the vaginal space (e.g., cube and doughnut pessaries) or providing support to the prolapsing component (e.g., ring and ring with support pessaries). Space occupying pessaries create a suction with the vaginal wall, while support pessaries are held in place by levator ani muscle tone. In general, ring pessaries are used for smaller prolapses (stage I–II), while space occupying pessaries are reserved for more significant prolapses (stage III–IV). Pessaries have the advantage of providing immediate relief of symptoms with minimal side effects. They can be used as a temporary solution for patient awaiting surgery or as a long-term option, providing a healthy vaginal mucosa. Risk for failure of pessaries is due to bad tolerance of the device, with local infections or lack of comfort.

3.3.3 Surgery

Surgery is a definitive management option but remains an invasive solution with variable long-term success and potential complications. Relapse rates at 2 years can reach 40% (1) and may require repeat interventions. Complications can be related to anesthesia or to the surgery itself (bleeding, infection, injury to bowel, bladder, ureters, vessels, nerves, *etc.*). Surgery should be considered in patients after failure of conservative management or as first line treatment in patients with bothersome bulge symptoms, any POP-related urinary or rectal symptom, or a significantly impaired quality of life. Childbearing should be completed before most surgical procedures with some exceptions.

A complete understanding of the patient's concern is paramount before opting for surgery, because the optimal procedure may vary according to symptoms but also age, level of activity, comorbidities, previous surgeries, etc. In addition to evaluation of POP bulge symptoms, information about urinary and defecatory dysfunction should be sought. Urinary retention is likely to improve with POP surgery, while the effect on urinary incontinence is variable. In patients with pre-existing stress urinary incontinence, a concomitant continence surgery can be considered. Occult urinary incontinence, that is stress incontinence unmasked by surgery, is harder to manage, because there is no validated way to predict its development post-operatively. Defecatory dysfunction in the form of constipation, incomplete rectal emptying, fecal incontinence, or obstructive symptoms requiring maneuvers for defecation should also be carefully evaluated. Any symptom suggestive of rectal prolapse should be further investigated as concomitant management may be achieved. Imaging studies, such as magnetic resonance imaging (MRI) or defecography, may be required in complex cases, and opinions from gastroenterologist or general surgeons can be valuable. In patients with chronic pelvic pain or dyspareunia, pre-operative referral to pain specialists or sexologist should be considered as surgery is unlikely to improve pain symptoms.

3.4 Surgical techniques

3.4.1 Pelvic floor support

Support to the pelvic floor is provided by interaction between pelvic floor muscles and endopelvic fascia attachments. Condensations of the endopelvic fascia lead to stronger structures that play a key role in support and physiology of pelvic organs. These structures include the uterosacral, cardinal, round and broad ligaments, the meso-ovarium and the mesosalpinx. Among these, only the uterosacral/cardinal ligament complex are supporting structures to the pelvic floor, despite being histologically very different from true skeletal ligaments. These mesentery-like visceral ligaments are composed of blood and lymphatic vessels, nerves, loose connective tissue and adipose tissue (2). They lack bands of strong connective tissue made of collagen type I fiber, as seen in skeletal ligaments (3).

As discussed in DeLancey's chapter, support to the pelvic floor can be divided in three levels. Level 1 corresponds to the support provided by the uterosacral/cardinal ligament complex at the level of the vaginal apex or cervix if present. Level 2 provides support to the middle part of the vagina by lateral attachment of the vagina to the arcus tendinous fascia pelvis and the levator muscles. Level 3 support is achieved by the urogenital diaphragm and perineal body, providing support to the distal part of the vagina. Failure of one or multiple of the supporting structures can lead to pelvic organ prolapse.

Simplistically, three compartments can prolapse through the vaginal wall: the anterior compartment containing the bladder; the apical compartment corresponding to the uterine cervix (or the vaginal vault in cases of previous total hysterectomy), and the posterior compartment, involving the rectum. In hysterectomized patients, the peritoneum and small bowel can also protrude through the vaginal wall in all three compartments (enterocele), although in this case apical and posterior prolapses are more common. In cases of complete procidentia (stage 4 prolapse), all three compartments are exteriorized.

Apical/cervical prolapses are often accompanied by either anterior (more commonly) or posterior descent, while isolated anterior and posterior defects exist but are less common. Biomechanical studies have demonstrated the importance of apical support on anterior vaginal wall. MRI measures have shown that over 50% of anterior wall descent can be explained by apical or cervical descent, while 77% can be explained by the combination of apical decent and vaginal wall length (4). Other factors that contribute to cystocele occurrence and size include paravaginal support and the diameter of the urogenital hiatus in the levator ani muscle (5). Simulated restoration of apical support corrected 55% of anterior prolapse and 30% of posterior ones (6). It remains controversial whether correction of the apical prolapse alone is sufficient to completely restore the anterior and posterior anatomy. On the other hand, repair of anterior and posterior prolapses alone appear to have lower success rates than procedures done with concomitant apical support.

3.4.2 General surgical principles

The aim of POP surgeries is to restore anatomy by suspending prolapsing structures and/or reinforcing supporting tissues. Restauration of level 1 support is achieved by suspension of the apex to the uterosacral/cardinal ligament complex or strong alterative ligaments, such as the sacrospinous or anterior longitudinal ligament. Restoration of level 2 support is achieved by suspension of the vagina to the arcus tendinous fascia pelvis, while level 3 support is restored mainly by plication or superposition of level 3 tissues.

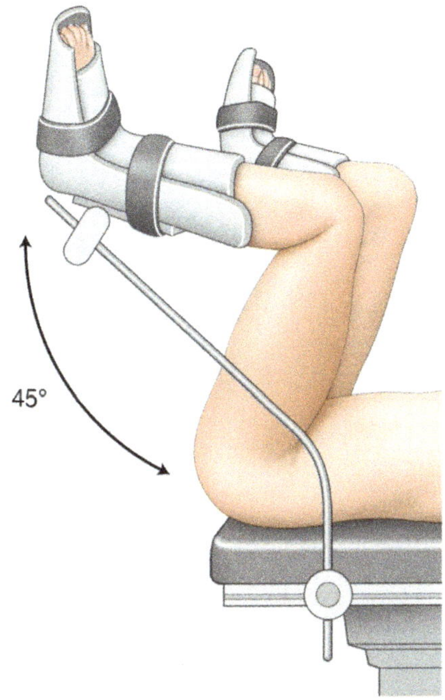

FIGURE 3.2

Position of the patient for vaginal surgery.

Pelvic organ prolapse surgeries can be performed vaginally or abdominally, using native tissue or synthetic meshes. The vaginal route involves a vaginal incision followed by dissection of different tissue planes, with or without entry in the abdominal cavity. The abdominal route offers access from above and can be performed through laparotomy, laparoscopy, or robotics. Laparotomies and vaginal procedures can be performed under either general or local anesthesia (spinal). General anesthesia is mandatory for laparoscopic and robotic procedures, because carbon dioxide is used to inflate the abdominal cavity. For both vaginal and abdominal surgeries, the patient is placed in dorsal lithotomy position, providing necessary access to the vaginal and perineal area (Figs. 3.2 and 3.3). See Figs. 3.4, 3.5, and 3.6.

In the following sections, we will attempt to describe the most common procedures used to correct POP based on DeLancey's levels of support. We will remain general with regards to surgical techniques and specifics on the type of suture used, the location and number of knots or the tension applied to the sutures are voluntarily omitted.

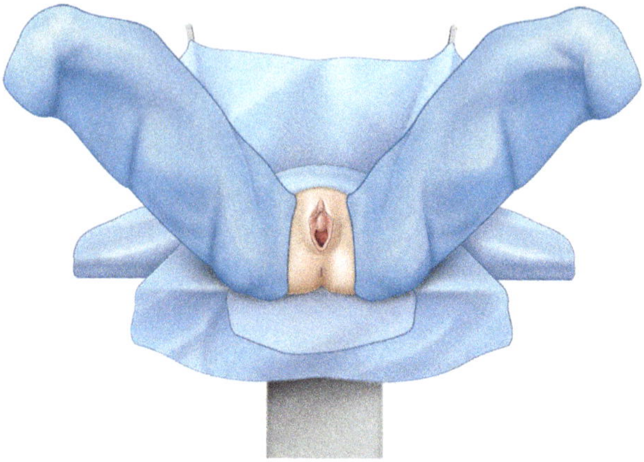

FIGURE 3.3

Vaginal surgery: frontal view after installation.

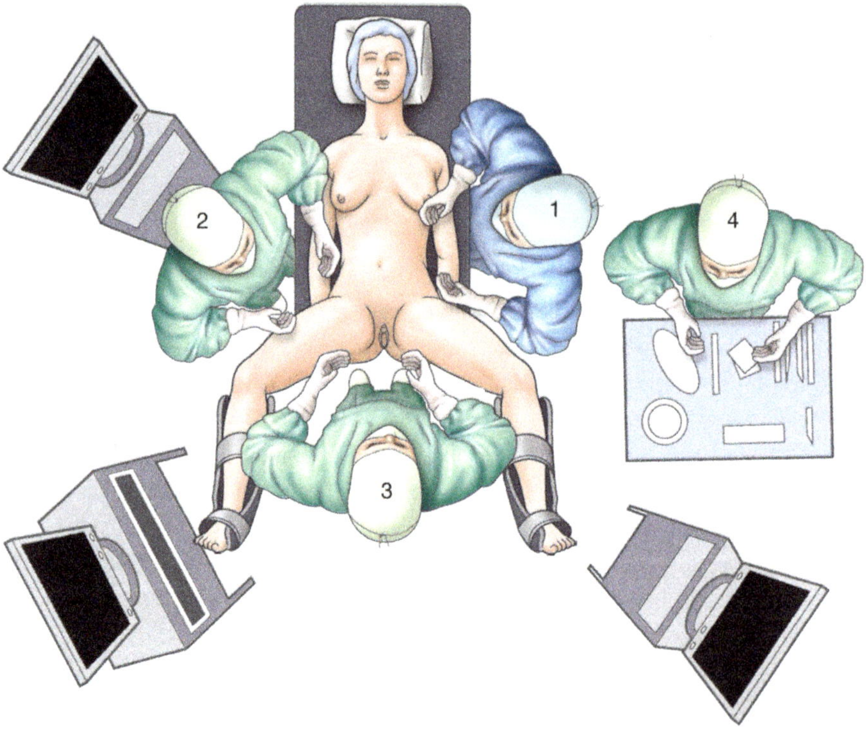

FIGURE 3.4

Installation for laparoscopic surgery. 1: surgeon, 2 and 3: assistants, 4: nurse.

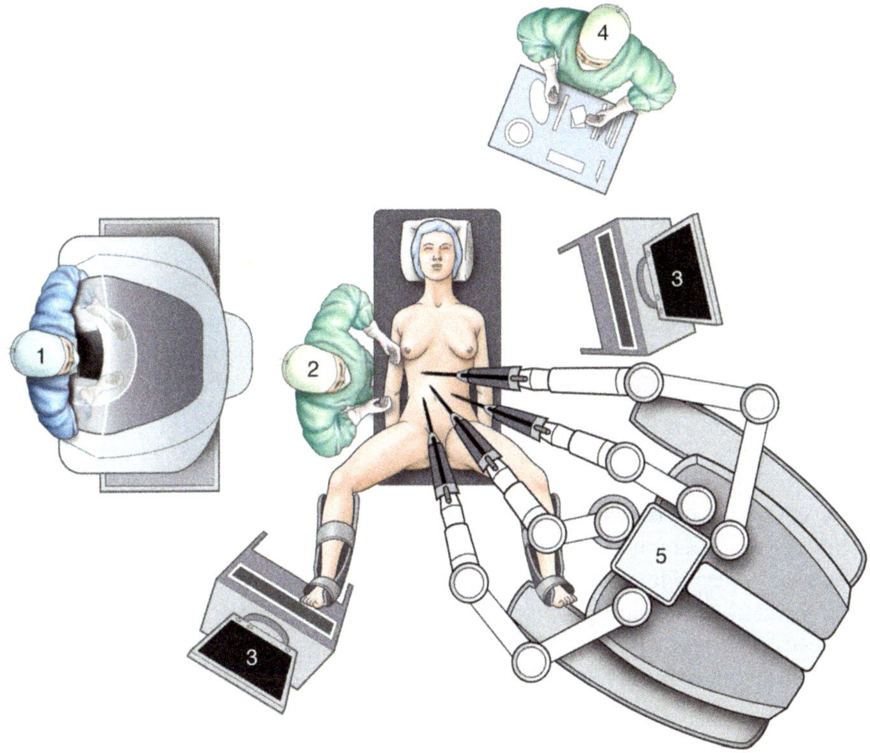

FIGURE 3.5

Installation for robotic surgery. 1: surgeon, 2: assistant, 3: screen for assistant, 4: nurse, 5: robotic arms.

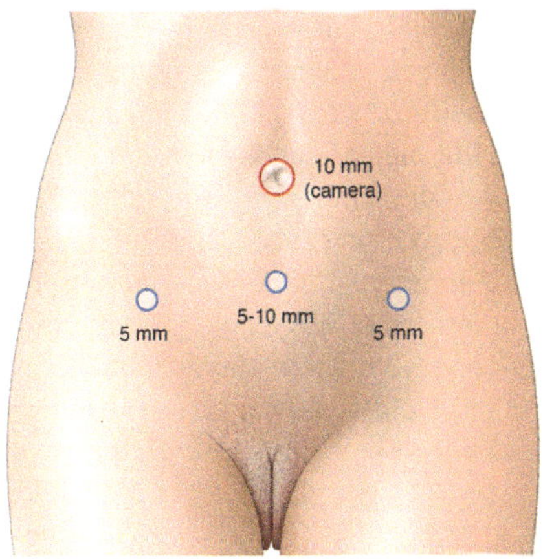

FIGURE 3.6

Main skin incisions for laparoscopic surgery.

3.4.3 Apical suspensions

Treatment of apical descent involves the suspension of the apex to one of three ligaments: 1) the anterior longitudinal ligament at the level of the sacrum; 2) the sacrospinous ligaments, extending from the sacrum to the ischial spine bilaterally, or 3) the utero-sacral ligaments, extending from the cervix to the sacrum bilaterally. The anterior longitudinal ligament and the uterosacral ligaments are accessed from the inside the abdominal cavity, while the sacrospinous ligaments lie within the pelvic floor muscles.

AUGS IUGA describe the different apical suspensions with hysterectomy and grafts: sacrocolpopexy, sacrocervicopexy, and sacrospinous ligament fixation.

Without graft: uterosacral ligament suspension, sacrospinous ligament fixation and iliococcygeus fixation. In case of uterine conservation they add the manchester procedure.

3.4.3.1 Sacrocolpopexy

Sacrocolpopexy is considered the gold standard for apical prolapse treatment with excellent long term anatomical results. It involves the suspension of the vaginal apex or cervix to the anterior longitudinal ligament using a synthetic mesh. The anterior longitudinal ligament is a strong skeletal ligament running from the sacrum to the atlas (first cervical vertebrae) on the anterior surface of the vertebral bodies. The current recommendation for mesh fixation is on the anterior surface of the sacral promontory, that is, at the level of the body of the first sacral vertebrae (S1) (10). Careful dissection of the ligament is mandatory, because major vessels (aortic bifurcation, left common iliac vein), nerves (hypoplastic plexus), the right ureter, and smaller vessels (medial sacral artery) lie very close to the promontory and may be injured during mesh fixation. On the vaginal end, a mesh has traditionally been sutured to the vagina anteriorly and posteriorly after dissection of the vesico-vaginal and recto-vaginal spaces, respectively. Fixation to the levator ani muscles instead of the posterior vagina has also been described (11,12). Isolated anterior or posterior fixation are more rarely performed as outcomes appear to be worse. Following attachment of the mesh to the vagina, a vaginal aid pushes the vagina caudally to restore its anatomical axis, and the mesh is attached at the level of the promontory. Care should be taken not to overcorrect the defect by putting too much tension on the mesh. Peritonization then follows to prevent the formation of abdominal adhesions. Initially described for treatment of vaginal vault prolapse in hysterectomized patient, variations of the technique with concomitant total or supra-cervical hysterectomy or with uterine conservation (sacrohysteropexy) have been described. In cases of concomitant hysterectomy, a subtotal hysterectomy reduces risks of mesh erosion through the healing vaginal vault. See Fig. 3.7.

3.4.3.2 Uterosacral ligaments suspension

Uterosacral ligament suspension (USLS) can be achieved through both an abdominal and a vaginal approach. Less commonly performed, the abdominal approach allows

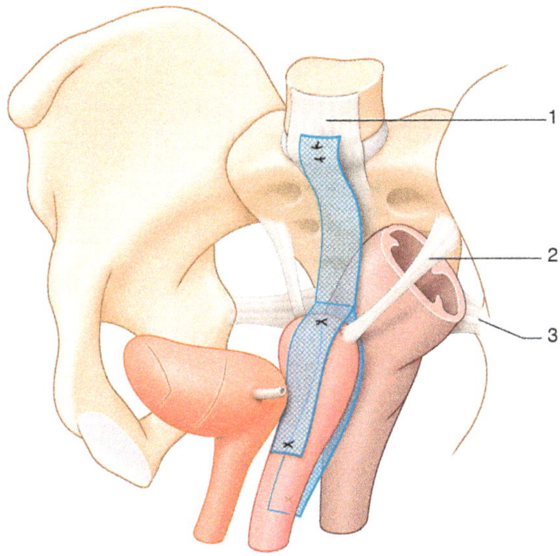

FIGURE 3.7

Complete sacropexy with anterior and posterior mesh, suspended to the prevertebral ligament.

for uterine preservation and simply consists of plication of the uterosacral ligaments bilaterally. Starting distally, a suture is ran through the length of the USL and reattached to the vaginal vault or to itself if the uterus is present, considerably reducing the length of the ligament. When done vaginally, the first step is a hysterectomy providing access to the inside of the abdominal cavity. The uterosacral ligaments are then identified bilaterally on the pelvic side walls, and either the posterior vaginal wall alone or both the anterior and posterior vaginal walls are suspended to them using sutures. Care needs to be taken to avoid the ureters that run very close to the USL. In the high USLS described by Shull (13), the suspension is posterior and medial to the ischial spine on the sacral side of the uterosacral ligaments. The McCall culdoplasty and its modifications also involve suspension of the vaginal vault to the uterosacral ligaments but with the additional objective of treating enteroceles or deep cul-de-sacs. A series of sutures are placed as high as possible from one uterosacral ligament to the other, incorporating the pre-rectal peritoneum, the round ligaments, and anterior peritoneum. This is closed in the purse-string fashion, obliterating the cul-de-sac. The vaginal vault is also attached to the uterosacral ligaments, suspending it when the sutures are tied. McCall culdoplasties are often performed prophylactically following a vaginal hysterectomy. See Fig. 3.8.

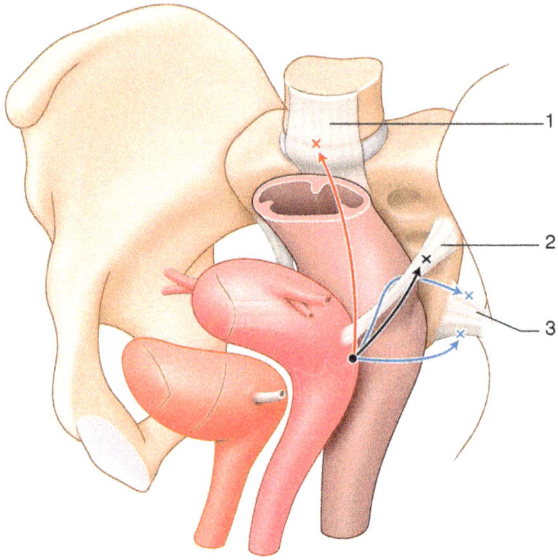

FIGURE 3.8

Apical suspension with uterine conservation to the Uterosacral or Sacrospinous Ligament.

3.4.3.3 Sacrospinous ligament suspension

Sacrospinous ligament suspensions are done exclusively vaginally. Initially described for vault prolapse post-hysterectomy, multiple modifications of this technique have now been described, incorporating hysterectomized and non-hysterectomized patients and using an anterior or posterior approach. To understand the anterior and posterior approach, one has to picture the vagina with the cervix or vaginal apex in the middle, with the anterior compartment lying above the cervix/apex and the posterior compartment lying below the cervix/apex. Access to the ischial spines and therefore the sacrospinous ligaments can be achieved through both an anterior and posterior approach. In the traditional posterior approach described by Richter, a longitudinal incision is made from the apex to the introitus, and the rectum is dissected off the vaginal mucosa. The pararectal fossa are opened by a combination of blunt and sharp dissection, and the sacrospinous ligaments are attained. Using a suture-passing device, such as the Capio or multiple retractors and a conventional needle driver, a stitch is passed through the sacrospinous ligament and anchored to the posterior aspect of the vaginal apex. Tying of this suture brings the vault to the sacrospinous ligament, thereby suspending the apex. This suspension is most commonly performed on the right side because of the absence of the sigmoid colon, but this is based on surgeon's preference. Bilateral suspension is also performed by some teams. The sacrospinous suture should be place 2–3 cm medial from the ischial spine and on middle of the ligament to prevent blood vessel and nerve injury.

When performed from an anterior approach, the vaginal incision extends from the 2–3 cm below the urethrae to 1 cm anterior from the vaginal apex. The bladder is dissected from the vaginal mucosa and the paravesical fossa are opened bluntly bilaterally to access the sacrospinous ligament. Because the paravesical fossa are narrower than the pararectal fossa, a Capio device has to be used. The suspension is usually bilateral, and the sutures are anchored on the anterior aspect of the vault, rather than the posterior aspect. It is important to mention that an anterior SSLF is only possible if a cystocele is present. In the case of an isolated vault prolapse, a posterior approach should be used. Arguments in favor of an anterior approach is a more natural suspension angle; a slightly longer vagina and less anterior compartment decompensation in the form of cystoceles they tend to happen with posterior SSLS (8). See Fig. 3.9.

3.4.4 Level 2 suspensions

Level 2 suspensions are mainly used for treatment of bladder prolapses and aim at restoring the lateral attachments of the vagina to the arcus tendinous fascia pelvis. In the Plastron procedure which was developed for large cystoceles, excess vaginal mucosa is used to create a little hammock under the bladder thereby increasing support. As shown in Fig. 3.10 (lateral view of the vaginal plastron), and depending on the size of the cystocele, a rectangle of varying size is incised on the vaginal mucosa of the anterior defect, the paravesical fossa are opened bilaterally, and access to the ATFP is secured. Using a suspension device such as the Capio (Boston Scientific) the Plastron is suspended to the ATFP in three points bilaterally. The anterior corner is suspended to the ATFP close to the pubic bone, the edge in the mid portion of the Plastron is suspended to the middle of the ATFP and the posterior corner is suspended to the ATFP close to the ischial spine. In an attempt to provide a higher level of suspension and correct the previously described apical defect that often accompanies cystoceles, a modification of the original Plastron procedure was developed. Dissection of the paravesical fossa are carried out more posteriorly and the sacrospinous ligament is accessed. The two posterior corners of the Plastron are suspended to the sacrospinous ligaments instead of the lower portion of the ATFP, similarly to an anterior sacrospinous ligament suspension. The final suspension level remains lower than a true sacrospinous ligament suspension as the Plastron is distal to the apex and part of the suspension remains lateral (Fig. 3.10). In the less common cases of anterior defects stemming from weakened lateral but not apical attachments, a paravaginal suspension can be performed. Access to the ATFP is obtained through dissection of the paravesical fossa following a midline colpotomy. Six sutures are placed in the ATFPs similarly to the original Plastron but they are then attached to the vaginal mucosa laterally rather then the Plastron, thereby suspending the vaginal cul-de-sacs.

3.4.5 Restauration of organ support: level 3

Level 3 support can be restored by techniques that lead to densification and reinforcement of the tissue lying between the organ and the vaginal mucosa or by simply

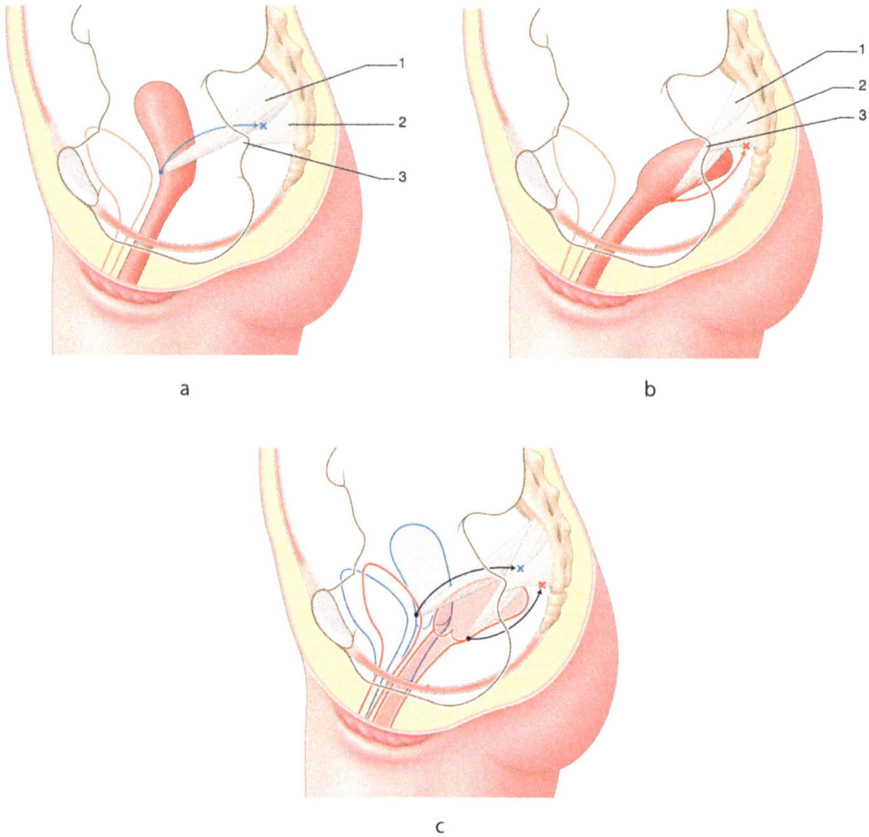

FIGURE 3.9

SSL suspension. a: anterior SSL fixation with uterine conservation; b: posterior SSL fixation with uterine conservation; c: anterior and posterior SSL fixation with different vaginal and bladder axis after correction.

trimming excess vaginal mucosa as in a colpectomy. We can see on this simplified view the anatomical structures involved in the level 3 defect (Fig. 3.11).

3.4.5.1 Anterior prolapses (cystoceles)

The pubocervical fascia, also known as the Halban fascia lies between the bladder and the vaginal mucosa (Fig. 3.12a). Rather than a true fascia, it is a densification of the deepest part of the vagina. In the anterior colporrhaphy, a midline vaginal incision is performed and the pubocervical fascia separated from the vaginal mucosa in the midline and laterally. The fascia, which remains adherent to the bladder, is then plicated from one side to the other. Multiple rows of sutures are placed reducing the cystocele and creating a denser layer of connective tissue to support the bladder. Colpectomy

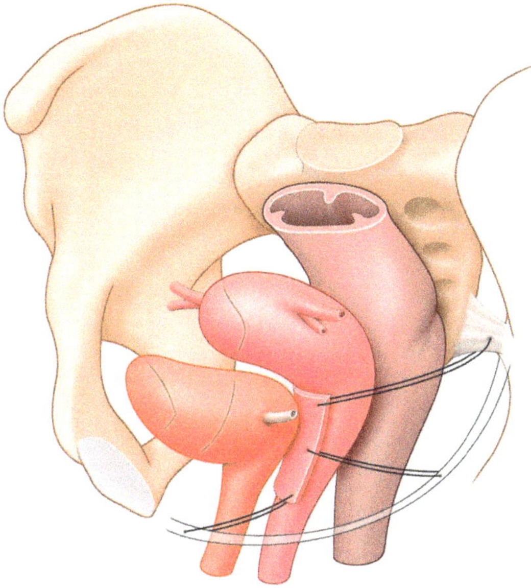

FIGURE 3.10

Lateral view of the vaginal plastron.

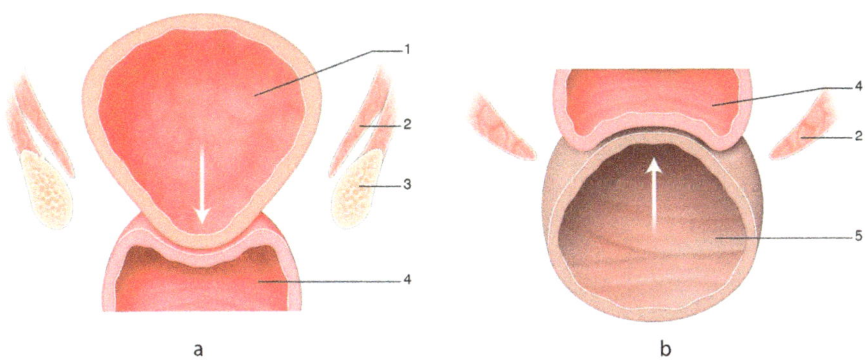

FIGURE 3.11

Anatomical structures involved in the level 3 defect. a: view of the cystocele: bladder 1, lateral muscle 2, pubic bone 3 and vagina 4. b: view of the rectocele: rectum 5.

is optional depending on the extent of the anterior defect. In the Paletot procedure or Halban's Fascia Plication, the pubocervical fascia is isolated (that is separated from the vaginal mucosa and the bladder) on either side of the midline vaginal incision, overlapped and sutured to the opposite lateral vaginal cul de sacs (Fig. 3.12a). In a

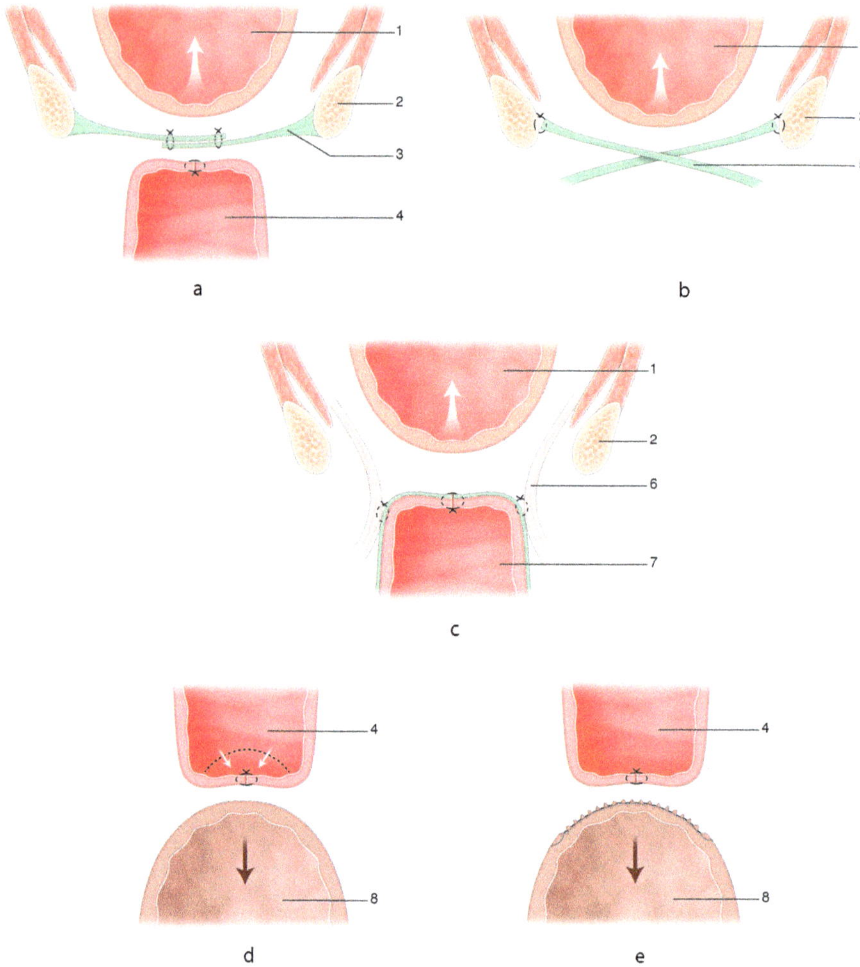

FIGURE 3.12

Treatment of level 3 defects. 1: bladder, 2: pubic bone, 3: halban fascia (deep part of the vagina), 4: vagina, 5: uterosacral ligament, 6: arcus tendineous of the pelvic fascia, 7: vagina, 8: rectum. a: paletot with overlap of the halban fascia under the bladder; b: campbell support of the bladder using both uterosacral ligaments; c: paravaginal suspension to the arcus tendineous of the pelvic fascia and colpectomy; d: posterior vaginal colpectomy to reduce rectocele; e: posterior vaginal colpectomy and plication of the rectal wall to reduce rectocele.

pre-vesical plication, the pubocervical fascia is left adherent to the vaginal mucosa and the bladder is plicated following the same technique as an anterior repair. In the paravaginal suspension (Fig. 3.12c), the lateral suspension to the arcus tendineous is associated with a colpectomy. In the Campbell's procedure (Fig. 3.12b) which has to be associated to a hysterectomy, the uterosacral ligaments (USL) are attached under the bladder to the contralateral subpubic ligament. A colpectomy can be performed if excess vaginal tissue is present.

3.4.5.2 Posterior prolapses (rectoceles)

In the posterior compartment, debate exists on the existence of a true fascia between the rectum and the vagina, limiting the number of surgical procedures to support the rectum. A posterior repair corresponds to the plication of the rectal wall (Fig. 3.12e) after a posterior midline or triangular vaginal incision starting at the hymen. It decreases the width of the posterior vaginal wall and increases the density of tissue between the rectum and the vagina. If excess vaginal tissue is present, a colpectomy can also be performed (Fig. 3.12d). Posterior repairs are often associated to perineorrhaphies that aim at rebuilding the perineal body to provide additional pelvic support. A horizontal incision is made along the hymen between the new designated midline points and extended to the apex on the perineum starting from both extremities. This triangular piece of skin is removed and plication of the bulbocavernosus and transverse perineal muscles is carried out, strengthening the perineal body.

3.4.6 Obliterative procedures

Colpocleisis refers to the complete closure of the vagina. This procedure is typically reserved for frail patients, in which traditional repair techniques are felt to be too aggressive, or in cases of multiple previous treatment failures. Following this operation, vaginal intercourse is not possible, and this needs to be thoroughly discussed with the patient prior to surgery. This surgery is often associated to a perineorrhaphy to narrow the genital hiatus and strengthen the perineal body. Total colpocleisis is used in patients with a previous hysterectomy, whereas Lefort partial colpocleisis is reserved for patients with a uterus and cervix. In both cases, the principle is the same, namely removal of a large portion of vaginal mucosa (Fig. 3.13), progressive reintegration of the prolapse secured by sutures (Fig. 3.14), and vaginal closure (Fig. 3.15). The particularity of the LeFort partial colpocleisis are the bilateral gutters extending from the cervix to the outside and allowing evacuation of cervical secretions and detection of vaginal bleeding. Colpocleisis are highly successful, but failure would present with recurrence of a vaginal bulge.

3.4.7 Autologous vs synthetic mesh

The principle of mesh use for POP treatment is to replace the patient's weak tissues by a strong material to restore support to the organ or to use it for suspension. Sacrocolpopexy and sacrohysteropexy, discussed earlier, are examples of synthetic meshes used abdominally for suspension. When used vaginally, synthetic meshes require tissue dissection and fixation of the material under the organ and to lateral attachment points. They usually replace any plications, overlap, or any native tissue lateral suspension. First commercialized in 2004, these vaginal mesh kits were re-

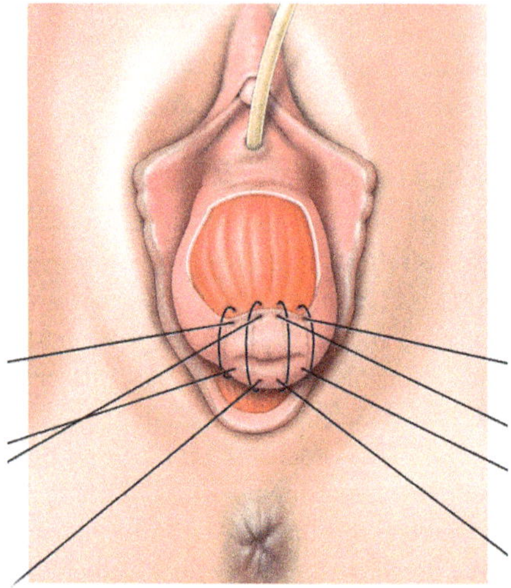

FIGURE 3.13

Closure of the vagina: large resection of anterior and posterior vagina and beginning of closure in front of the cervix.

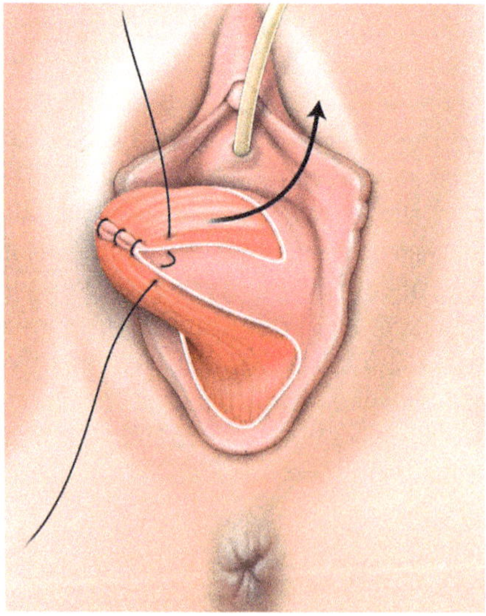

FIGURE 3.14

Closure of the vagina: suture of the anterior and posterior compartment.

cently removed from the market due to high rates of complications in the form of vaginal mesh exposition.

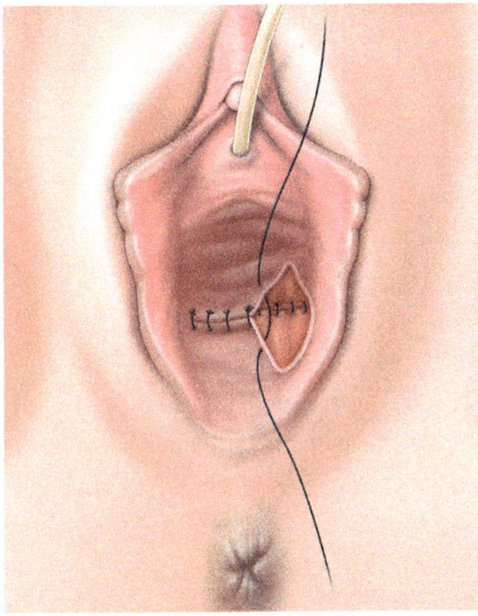

FIGURE 3.15

End of the vaginal closure with reduction of the genital prolapse.

3.5 Surgical strategies

Pelvic organ prolapse often involves the prolapse of multiple compartment within the vagina. As discussed earlier, anterior and posterior prolapses are often the consequence of some level of apical descent. For this reason, it appears logical that restoration of the apical support will significantly improve anterior or posterior prolapses. Sometimes, an apical suspension alone is sufficient, and other times a dedicated anterior or posterior compartment repair needs to be added. In this section, we will attempt to describe common surgical operations, bearing in mind that significant variability exists between groups and that these operations have often not been scientifically evaluated. There is also a lack of clear recommendations on the surgery to perform for specific patient population or based on the type of prolapse. We have tried to summarize different associations in Table 3.1.

Historically, hysterectomies have been associated with POP surgeries, but recent trials have shown non-inferiority of techniques preserving the uterus. Advantages include shorter operative times and less blood loss, quicker recovery and return to work, and fewer complications (12). Nonetheless, one of the common surgical operation used for apical descent include a vaginal hysterectomy associated with either a McCall or a high uterosacral ligament suspension for apical fixation. If anterior or posterior prolapses are present, anterior and posterior repairs plus or minus a perineorrhaphy are performed. Depending on the surgeon's preference, a posterior SSLF can alternatively be used for apical suspension with the same options for anterior and posterior defects.

Table 3.1 Surgical strategies according to prolapse evaluation.

Defect type	Surgery
Isolated anterior	- Anterior colporrhaphy (Fig. 3.16-2) - Pre-vesical plication (Fig. 3.16-1) - Paletot - Paravaginal suspension - Plastron (Fig. 3.18) - Colpocleisis
Isolated apical (Fig. 3.19e)	- Vaginal hysterectomy + McCall - Vaginal hysterectomy + High USLS (Fig. 3.19-a) - Manchester (Fig. 3.19-b) - Sacrocolpopexy with anterior + posterior mesh ± hysterectomy - Posterior SSLF with vaginal strap (Fig. 3.19-c) or vagina in contact with the ligament (Fig. 3.19-d) - Colpocleisis
Anterior + Apical	- Vaginal hysterectomy + McCall + Anterior repair - Vaginal hysterectomy + High USLS + Anterior repair (Fig. 3.17-b) - Sacrocolpopexy with anterior mesh at least ± hysterectomy - Vaginal mesh with SSF (Fig. 3.19-f) - Anterior SSLF ± Paletot or pre-vesical plication (Fig. 3.17-a) - Plastron (Fig. 3.18) - Posterior SSLF + anterior repair - Colpocleisis (Figs. 3.13, 3.14, and 3.15)
Isolated posterior	- Posterior repair ± perineorrhaphy
Posterior + Apical	- Vaginal hysterectomy + McCall + Posterior repair ± perineorrhaphy - Vaginal hysterectomy + High USLS + Posterior repair ± perineorrhaphy - Sacrocolpopexy with posterior mesh at least ± hysterectomy - Posterior SSLF ± posterior repair ± perineorrhaphy - Colpocleisis
Anterior + apical + posterior	- Vaginal hysterectomy + McCall + Anterior repair + Posterior repair ± perineorrhaphy - Vaginal hysterectomy + High USLS + Anterior repair + Posterior repair ± perineorrhaphy (Fig. 3.17-b) - Sacrocolpopexy with anterior and posterior mesh ± hysterectomy - Anterior SSLF + pre-vesical plication or paletot + Posterior SSLF + posterior repair ± perineorrhaphy (Fig. 3.17-a) - Plastron + Posterior SSLF + posterior repair ± perineorrhaphy - Colpocleisis (Figs. 3.13, 3.14, and 3.15)

At our institution, we favor uterine conservation and thus perform sacrospinous ligament suspensions instead of uterosacral ligament suspensions. In the case of apical descent associated with a cystocele, an anterior SSLS can be combined to a Paletot or a pre-vesical plication. Of note, it is complicated to associate an anterior SSLF to an anterior colporrhaphy as the paravesical fossa are accessed from the plane between the bladder and the pubocervical fascia. The fascia is therefore left adherent to the vaginal mucosa and not to the bladder, as required for an anterior colporrhaphy. For large cystoceles, a modified Plastron can be done restoring both apical and lateral suspension. If a concomitant large posterior defect is present, a posterior SSLS can be added to the previous surgeries and, in cases of smaller defects, a simple posterior repair can be undertaken.

Sacrocolpopexies can be used to treat apical descent associated with cystoceles, isolated rectocele or both. Very successful, this technique has traditionally been reserved for younger patients and patients with an active life style, but some groups offer it as first line treatment for most patients. In older and frail patients with any type of prolapse, colpocleisis are very successful and often require the shortest operative time.

3.6 Mechanical consequences of the different management options

3.6.1 Expectant management

With progression of the prolapse, the ligaments and vaginal wall will lengthen, whereas surrounding tissues will lose their elasticity. The evolution POP happens over multiple years and is unpredictable in time. Periods of worsening and improvement follow each other depending on the patient's level of activity. Overall, with time, there is a progressive increase in the severity of the prolapse and its related symptoms, with significant variation from one patient to another.

3.6.2 Autologous surgeries

Determining the immediate post-operative result of an operation is easy to predict if the technique is well described and well performed. Predicting the long-term results is harder, as multiple factors, such as the behavior of the autologous structures used for repair, the extent of surgical tissue damage, and the healing pattern of tissues, are unknown. In addition, if modelization of a single surgical techniques is possible, the combination of multiple techniques as it is often done in POP surgeries significantly complicates the task.

3.6.2.1 Suspensions

In theory, modelization of an apical suspension is relatively simple: it results in the suspension under tension, using either a synthetic mesh or non-absorbable sutures, of the vaginal apex/cervix to a specific anatomical structure. However, multiple *in*

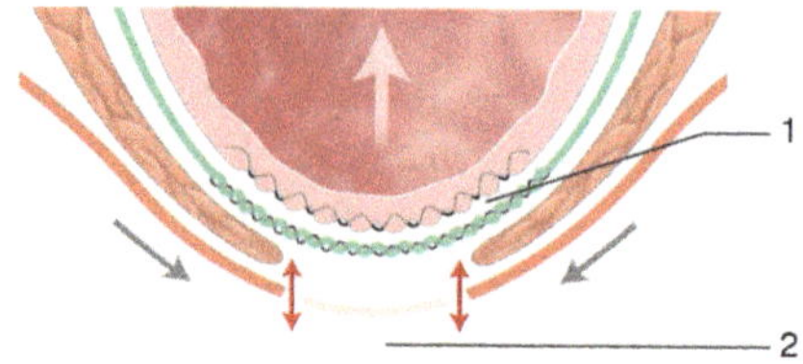

FIGURE 3.16

Treatment of the cystocele with plication of the bladder and the Halban fascia (1) and colpectomy (2).

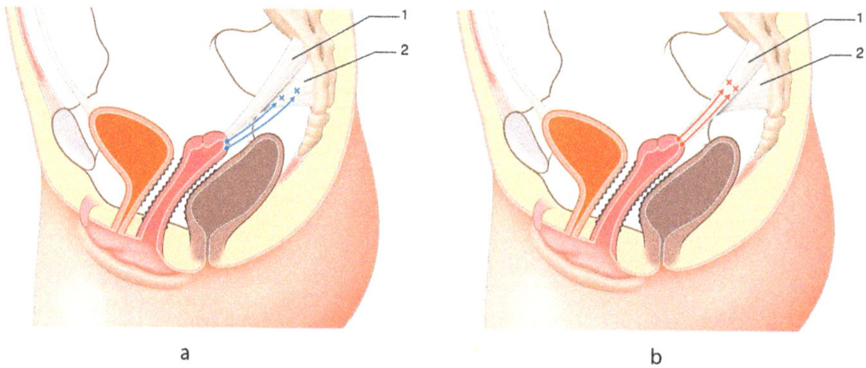

FIGURE 3.17

Vaginal repair with anterior and posterior plications and colpectomies associated with apical suspension to the sacrospinous ligament (a) or the uterosacral ligament (b).

vivo parameters that are hard to evaluate will modify the initial modelization. For example, the solidity of the structure used for suspension, the exact location of the suspension, and the tension applied to the suspension sutures are important. A high tension will lead to direct contact between the apex/cervix and the suspension area, whereas lower tension will bring the apex/cervix to an intermediate location. The recommendation is not to overcorrect the defect, but rather to reposition the apex in its anatomical position without excessive tension; this is, however, hard to evaluate and is very dependent on the surgeon's judgement.

A frequent operation will be (Fig. 3.17a and b) a combination of anterior and posterior plications and colpectomies with apical suspension to the sacrospinous ligament (Fig. 3.17a) or to the uterosacral ligament (Fig. 3.17b).

In case of a large cystocele with apical defect, we propose an anterior sacrospinous suspension associated to a paravaginal suspension of a vaginal patch to the arcus tendineous of the pelviec fascia (Fig. 3.18).

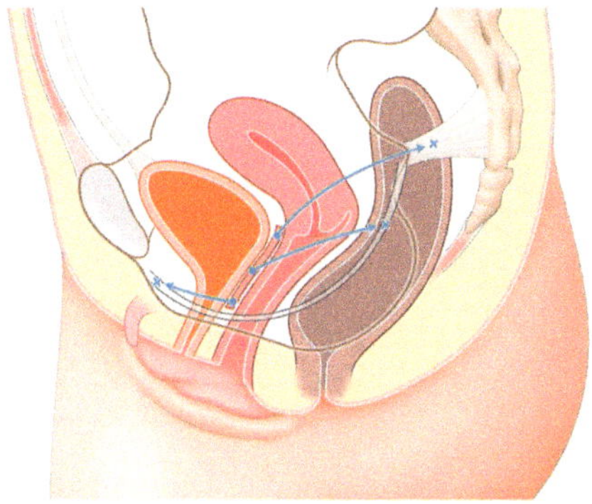

FIGURE 3.18

Plastron under the bladder with anterior sacrospinous suspension and paravaginal suspension to the arcus tendineous of the pelvic fascia.

Many options can be proposed to combine the different procedures at the different levels, according to the preoperative evaluation (Fig. 3.19 and Table 3.1). ICI 2017 describes the preferred apical suspensions as laparoscopic sacrocolpopexy + repair and in case of conservation of the uterus vaginal sacrospinous hysteropexy.

3.6.2.2 Plications

Plications such as pre-vesical (Fig. 3.16) and pre-rectal plications and overlap of tissues, such as Paletots (Fig. 3.12a), will usually lead to a combination of increased resistance of the plicated or overlapped tissue and a decreased surface area of the structure that is plicated. For adequate modelization, the plicated area would need to be estimated and data on the mechanical changes caused by a plication would need to be available. Unfortunately, this data is not available in the literature. See Fig. 3.10d and 3.10e.

3.6.2.3 Colpectomy

The removal of vaginal mucosa is one of the most popular surgical technique due to its simplicity, but its common association to different plications complicate modelization. In itself, the removal of the tissue under the bladder or in front of the rectum will increase the tension on the remaining vagina and relocate the organs to a more anatomical position. See Fig. 3.16.

3.6.3 Mesh surgery

Synthetic mesh have been used for multiple years in abdominal hernia surgeries before their use was expanded to vaginal surgeries. Their mechanical behavior is well understood but can unfortunately not be extrapolated to vaginal meshes as in vivo

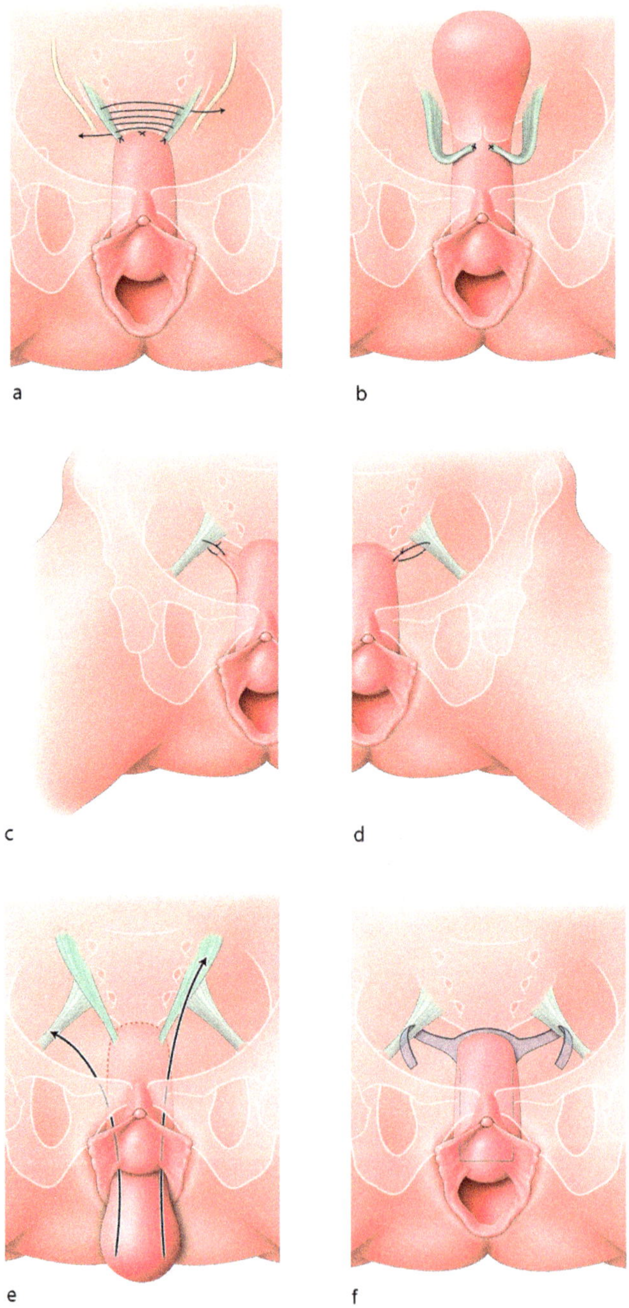

FIGURE 3.19

a: plication USL; b: manchester amputation cervix and plication of USL; c: right SSF with vaginal strap; d: left SSF with vagina in contact with the ligament; e: apical suspension to the USL left side of the patient or SSL right side of the patient; f: synthetic vaginal mesh suspended to the SSL.

tensile forces are very different from one location in the body to another. Mesh mechanical properties are well characterized before implantation, but their in vivo behavior remain largely unknown. While the impact of uniaxial loading may be suitable to understand mesh mechanics in sacrocolpopexy (Fig. 3.7), mesh response to vaginal multiaxial forces need to be better characterized.

CHAPTER

Physiology and physiopathology of pregnancy and delivery

4

Bertrand Gachon[a,b,c], Xavier Fritel[a,b], Zdenek Rusavy[d,e], Fabrice Pierre[a], Khaled M. Ismail[d,e], and Vladimir Kalis[d,e]

[a]*Poitiers University Hospital, Department of Obstetric and Gynecology, Poitiers, France*
[b]*Poitiers University, INSERM CIC 1402, Poitiers University Hospital, Poitiers, France*
[c]*Nantes University, Movement Interaction Performance Lab, MIP EA4334, Nantes, France*
[d]*Department of Obstetrics and Gynecology, Faculty of Medicine in Pilsen, Pilsen, Czech Republic*
[e]*Biomedical center, Faculty of Medicine in Pilsen, Pilsen, Czech Republic*

4.1 Introduction

Childbirth is a unique condition during a woman's lifetime. Indeed, during the delivery the human body is submitted to huge constraints that it would not be able to assume in another context. As an example, during a vaginal delivery pelvic floor muscles are stretched up to 3 times their initial length (Ashton-Miller and DeLancey, 2007; Ashton-Miller and Delancey, 2009). The same expansion-related principle applies to joint mobility, especially for pelvic joints, to allow the progression of the fetus into the mother's pelvis (Ashton-Miller and DeLancey, 2007; Ashton-Miller and Delancey, 2009). This phenomenon, considered a physiologic adaptation during pregnancy, is essential for permitting a normal vaginal delivery, regarded as a spontaneous vaginal delivery without perineal trauma involving levator ani muscles and/or anal sphincters (Gachon et al., 2016, 2019). It is likely that this adaptation during pregnancy is dependent on intrinsic characteristics women's characteristics (tissue's biomechanical characteristics, genetics) and on the course of pregnancy (normal or complicated). However, it appears that there exists an individual variability from one woman to another to have a normal delivery, which explains why predictive approaches (for the mode of delivery and for pelvic floor trauma at childbirth) are disappointing (Jelovsek et al., 2018; Meister et al., 2016; Schuit et al., 2012; Jeong et al., 2015; Kim et al., 2010). Indeed, these approaches are focused on data about anthropometrics data from the mother and the baby, and not on individual characteristics of woman. In this chapter, we will first describe changes in the intrinsic women's characteristics during pregnancy, report data from animal experimentation, the physiological process associated with these changes, and present data about the association between these intrinsic characteristics and the risk of perineal trauma at

Biomechanics of the Female Reproductive System: Breast and Pelvic Organs. https://doi.org/10.1016/B978-0-12-823403-7.00015-4

childbirth. Then, we will describe the process of vaginal delivery using a biomechanical approach with a specific focus in perineal trauma prevention and spontaneous vaginal delivery optimization.

4.2 Changes in biomechanical intrinsic characteristics during pregnancy

4.2.1 Changes in joint mobility in women

There is some inconsistency about changes in joint mobility in the literature, according to the mode of evaluation considered. There are two ways of investigating joint mobility in humans (often excessively considered as ligamentous laxity): a general assessment of joint mobility using the Beighton's score or a focused assessment of a considered joint (Gachon et al., 2016; Beighton et al., 2012).

The approach with a general assessment using Beighton's score is simple and reliable, but weak to report changes across the time. Studies that used this tool to investigate changes in joint mobility during pregnancy failed to report significant ones (Calguneri et al., 1982; Vollestad et al., 2012).

Changes in joint mobility during pregnancy have been under investigation since 1934 when Abramson et al. reported an increase in joint mobility in the pubic symphysis (Abramson and Wilson, 1934). More recently, studies looked for changes in several joint mobility (metacarpo-phalangeal joint, abduction of the fourth finger, anterior drawer test for the knee and other ones) (Calguneri et al., 1982; Gachon et al., 2017; Dumas and Reid, 1997; Schauberger et al., 1996; Lindgren and Kristiansson, 2014).

It appears that several joints, upper and lower limbs joints, of the woman body increase their mobility during pregnancy. This is often considered as an increase in ligamentous laxity, though none of the evocated studies directly investigated ligaments. One limitation is that there is no recent study that investigated specifically joints directly involved with childbirth (pelvic joints). Furthermore, these observations are contradictory to a recent observation, which failed to report significant changes in elastic properties of the patellar tendon during pregnancy using ultrasound (Bey et al., 2019).

Despite these research-related limitations, the hypothesis that joint mobility increases during pregnancy, especially for pelvic joints as a preparative mechanism for allowing a vaginal delivery, is widely accepted. Another interpretation is that this increase in joint mobility is the reflect of a global change in biomechanical tissues of pregnant women that could involve other tissues, such as pelvic floor muscles, with a potential impact in the mode of delivery and the risk of pelvic floor disorders (Gachon et al., 2019).

4.2.2 Changes in spinal curve in women

During pregnancy, different changes are reported about pregnant women's spinal curve: changes in lumbar lordosis, changes in lateral inclination, etc. (Betsch et al., 2015; Okanishi et al., 2012; Hay et al., 2015). The role of these changes is to maintain the women's center of gravity at the center of her support polygon (Okanishi et al., 2012). It is likely that these changes in spinal curve are allowed by the increase in joint mobility and/or in ligamentous laxity that were reported in the previous paragraph. There is no data in the literature in this regard, but it is possible that these changes in spinal curve may have an impact on the mode of delivery when considering that it can induce changes in the inclination in the pelvic inlet and so a hypothetic impact on the mode of the delivery (Desseauve et al., 2019b). It is also reported that the spinal curve is associated with pelvic floor disorders occurrence in non-pregnant women, and it is possible that this outcome could influence the risk of postnatal pelvic floor disorders and/or pelvic floor trauma at childbirth (Mattox et al., 2000).

4.2.3 Change in pelvic organ mobility

We reported in the previous paragraphs changes that occurred in biomechanical characteristics of women during pregnancy, but for anatomical sites (peripheral joints, spinal curve) that could be considered as not directly involved with childbirth. We do not have any recent data about pelvic bones or joint assessments, but there is important literature about changes in pelvic organ mobility. This can be easily assessed clinically with a standardized approach, thanks to the pelvic organ prolapse quantification (POP-Q) (Bump et al., 1996). It brings information about the mobility of several fixed points at the anterior/posterior vaginal wall, the cervix during a Valsalva maneuver with reference to the hymen. It also assesses the length of several segments at the following: total vaginal length at rest (tvl), genital hiatus (gh, distance between the urinary meatus and the posterior midline hymen) and perineal body (pb, distance between the posterior margin of the genital hiatus and the midanal opening); the last two measures being performed during Valsalva maneuver (Bump et al., 1996). Another possibility to investigate pelvic organ mobility is to use ultrasounds and especially transperineal ultrasound, which allows a 3D assessment of women's pelvic floor (Dietz, 2010).

4.2.3.1 Clinical considerations

There is consistent literature reporting an increase in pelvic organ mobility during pregnancy (Gachon et al., 2016, 2017; O'Boyle et al., 2003, 2002; Reimers et al., 2016). These studies report that through pregnancy, the position of each POP-Q points become progressively lower. A similar observation is reported about total vaginal length, which progressively increases, and about perineal distension with an increase in genital hiatus and perineal body respective lengths (Gachon et al., 2016, 2017; O'Boyle et al., 2003, 2002; Reimers et al., 2016). These changes in pelvic organ mobility explain the significant incidence of pelvic organ prolapse (stage 2 or more with a pelvic descent to the hymen level), reported in up to 30% of pregnant women (Chen et al., 2013). This increase in mobility and distension appears continuous through pregnancy with a progressive recovery in the postpartum period. This recovery appears faster in the case of cesarean section compared to vaginal delivery, but without any significant difference between these two groups 12 months after the delivery (Reimers et al., 2016).

4.2.3.2 Ultrasound considerations

There is abundant literature reporting an increase in pelvic organ mobility during pregnancy. This is reported for the bladder neck descent, which becomes more and more important (up to more than 20 mm) through pregnancy; this is in accordance with the observation of a progressively lower position of anterior POP-Q points and with the increasing prevalence of stress urinary incontinence through pregnancy. Indeed, an important bladder neck descent (more than 15 mm) is associated with stress urinary incontinence (Gachon et al., 2017; Dietz and Bennett, 2003; Dietz et al., 2004, 2003; Shek et al., 2010). This bladder neck descent can be easily measured in 2D transperineal ultrasound by measuring the distance between the bladder neck and the pubic bone at rest, then on the basis of the Valsalva maneuver (Fig. 4.1). The bladder neck descent is represented by the difference between the measure at rest and the measure during the Valsalva maneuver. Pelvic floor ultrasound investigations in pregnant women mainly focus on levator hiatus area measurements. The levator hiatus consists of a space delimited laterally by the right and left levator ani muscles (LAM), forward by the pubic symphysis, and backward by the external anal sphincter (Fig. 4.2) (Ashton-Miller and DeLancey, 2007). It is through this space that the fetuses must progress during a vaginal delivery with a massive stretch of the levator ani muscles (up to 300%), explaining the risk of levator ani muscle trauma (Ashton-Miller and DeLancey, 2007). It is reported that women having the lowest area in late pregnancy are more incline to deliver with an operative delivery (cesarean section or instrumental delivery) (Dietz et al., 2003; Toozs-Hobson et al., 2008; van Veelen et al., 2015). In non-pregnant women this levator hiatus area is associated with pelvic organ prolapse occurrence, especially when larger than 25 cm^2 (Dietz et al., 2008). During pregnancy, several authors reported an increase in levator hiatus area, whether the measure is done at rest, during the Valsalva maneuver, or perineal contraction (Gachon et al., 2017; Shek et al., 2010; van Veelen et al., 2015; Bo et al., 2014; Siafarikas et al., 2014; Staer-Jensen et al., 2015). Here again, this phenomenon could be considered as a preparative process in view of accepting the constraints applied during vaginal delivery. The main limitation for the interpretation of these data is that it remains an indirect assessment of the biomechanical characteristics of pelvic floor muscles (such as joint mobility not being a direct measure of ligamentous laxity). In the near future, innovative non-invasive ultrasound techniques will probably be able to report changes in elastic properties of pelvic floor muscles during pregnancy. Indeed, ultrasound shear wave elastography is reported as non-invasive exam, allowing a direct and quantitative measurement of LAM stiffness (Gachon et al., 2018b). There is currently a prospective study investigating changes in elastic

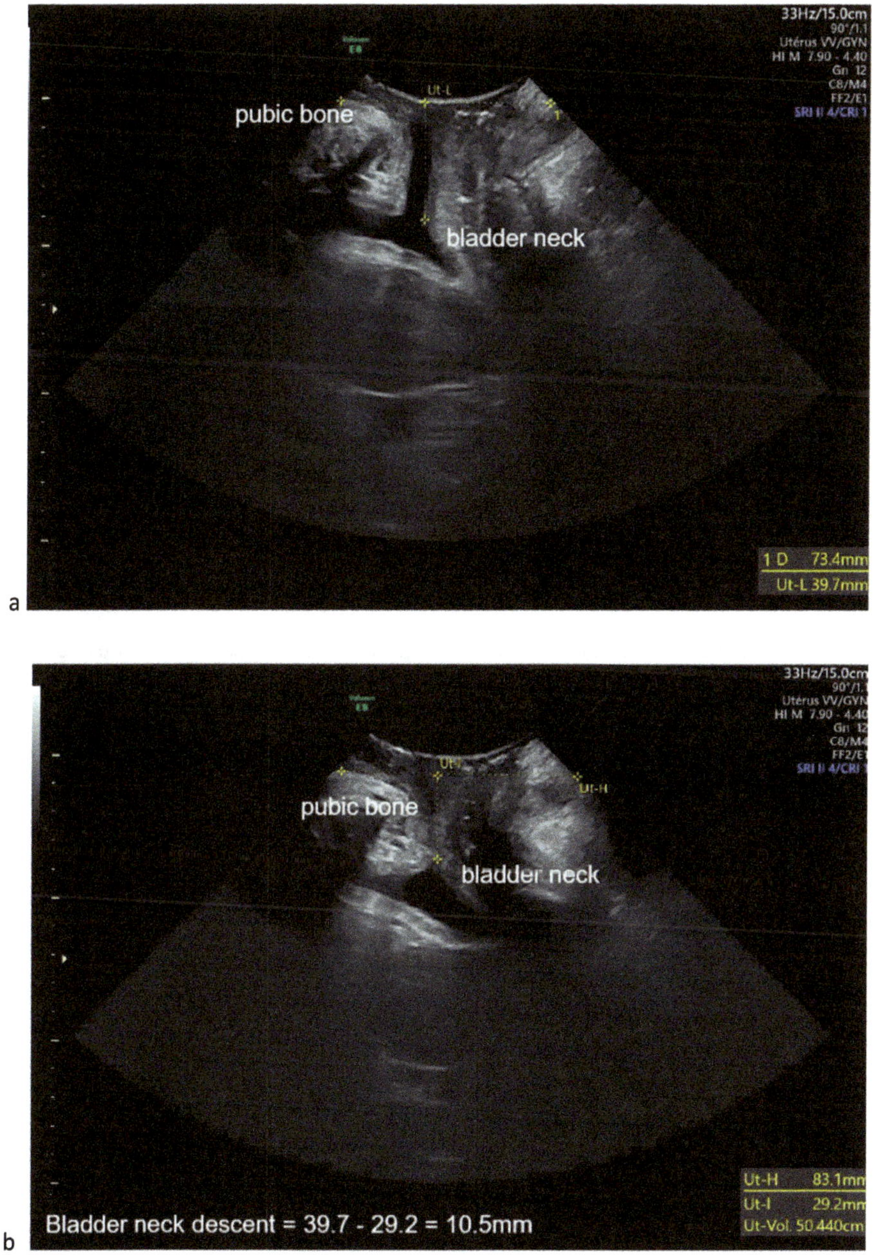

FIGURE 4.1

Bladder neck descent assessment using 2D transperineal ultrasound: (a) at rest; (b) during Valsalva maneuver).

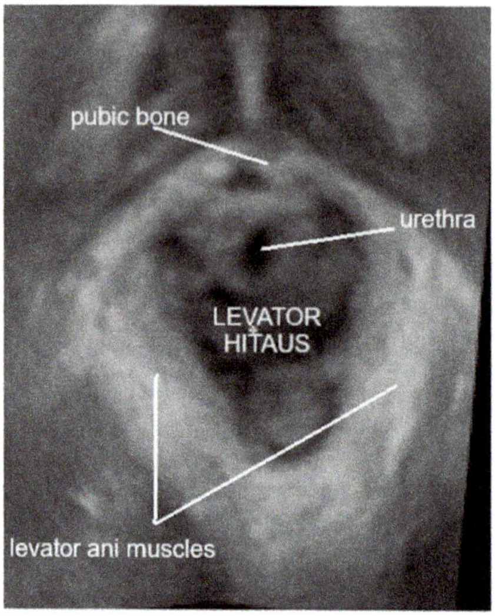

FIGURE 4.2

Levator hiatus assessment using transperineal 3D ultrasound at rest.

properties of the LAM and the external anal sphincter in pregnant women using this technique (Fig. 4.3) (Gachon et al., 2020).

4.3 Pathophysiological process

Though there is abundant literature reporting changes in the intrinsic biomechanical characteristics of women during pregnancy, the pathophysiological process associated with these changes remains partially known. It is likely that hormonal changes that occur during pregnancy can impact tissues' characteristics, though data are conflicting. At a deeper level of assessment, changes in biomechanical characteristics are probably due to a remodeling of the tissue's architecture, and especially a collagen remodeling considering that collagen is highly involved in tissue's stiffness.

4.3.1 Hormonal considerations

There are several changes in hormonal profile during pregnancy (relaxin, estrogen, progesterone). One of the most often reported hypothesis is the role of relaxin, which is a hormone produced by the ovaries, the mammary tissue, and the placenta. This hormone is reported as being involved in conjunctive tissue remodeling (Schauberger et al., 1996). Data are conflicting about the association between the level of relaxin during pregnancy and changes in joint mobility, some studies reporting that a high level of relaxin is associated with high joint mobility, whereas others failed to report

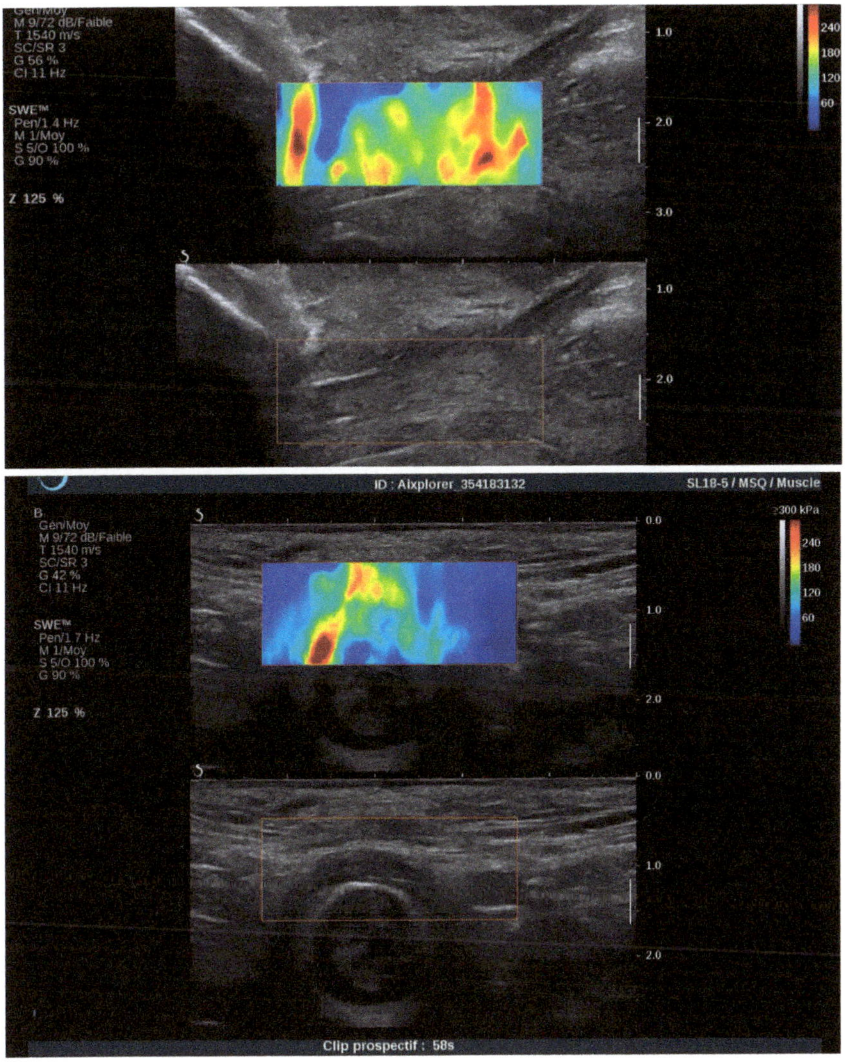

a

b

FIGURE 4.3

Transperineal ultrasound elastography assessment of the levator ani muscle (a) and external anal sphincter (b) during Valsalva maneuver.

any association (Vollestad et al., 2012; Schauberger et al., 1996; Marnach et al., 2003; MacLennan et al., 1986). Pregnancy is also associated with an increase in the level of expression of sexual hormones, especially estradiol. Regarding data available in the literature, it is not possible to formulate any conclusions, considering that some studies report an increase in stiffness in muscles or tendons, whereas others report a decrease in stiffness, and others no association at all (Marnach et al., 2003; Hansen and Kjaer, 2014; Burgess et al., 2009). Finally, even if a hormonal hypothesis appears safe regarding the importance of changes in hormones expression, and the fact that biomechanical changes involved upper and lower limbs as well as pelvic floor, there is no evidence available to safely support this hypothesis. Further works are required in this thematic to improve our knowledge about the association of hormonal impregnation during pregnancy and changes in women's biomechanical characteristics.

4.3.2 Conjunctive tissue's remodeling

As previously stated, the role of hormones during pregnancy remains hardly known, but it is expected to influence the conjunctive tissue's metabolism, especially that of collagen. Collagen is one of the most important components of the extracellular matrix and is highly involved in tissue's mechanical characteristics (Gachon et al., 2019; Alperin et al., 2015). There are several types of collagen with different biological impact. As an example, the type 1 collagen is mainly involved in stiffness, whereas the type 3 is mainly involved in elasticity. One hypothesis for explaining the increase in joint mobility, pelvic organ mobility, and perineal distension during pregnancy is a collagen remodeling with a decrease in the type 1/type 3 ratio (Gachon et al., 2019; Alperin et al., 2015). Nevertheless, this remains as a hypothesis; data available are issued from animal experimentation and so hardly applicable to women. A condensation of data from animal experimentation is provided just below.

4.3.3 Data from animal experimentation

The rat is the animal model widely used for assessing pelvic floor regarding anatomical similitudes with human (Alperin et al., 2015).

Different research teams investigated the biomechanical properties of the vaginal wall and its changes during pregnancy. Rahn et al. report that in late pregnant rats, the vaginal circumference was twice that of non-pregnant rats (Rahn et al., 2008). These authors hypothesized the role of the elastin, considering that in their experience non-pregnant rats knock-out for *fbln5* implied that elastin's metabolism had comparable biomechanical characteristics than in late pregnant ones (Rahn et al., 2008). Lowder reported that compared with virgin rats, the vagina's stiffness and ultimate load at failure were lower than in pregnant rats. The maximal distension was observed at the time of delivery (Lowder et al., 2007). Feola et al. reported an increase in vaginal compliance through pregnancy using clinical and ultrasound measurements (Feola et al., 2014). This report appears in accordance with clinical observations during pregnancy, mentioned above, with an increase in total vaginal length and a lower position of POP-Q points (Gachon et al., 2017; O'Boyle et al., 2003, 2002; Reimers et al., 2016).

Other research teams provided data about pelvic floor muscles, especially the levator ani muscle. Alperin et al. reported an increase in muscular fiber length during pregnancy, which is allowed by an increase in the number of sarcomeres in series. Interestingly, these authors reported an opposite observation for the vagina. Indeed, they reported an increase in pelvic floor muscles stiffness, which seems to be associated with an increase collagen 5 expression (Alperin et al., 2015, 2016). The other interesting point is that such biomechanical change only involved pelvic floor muscles and not peripheral muscles, suggesting the importance of the local environment and the importance of the constraints applied by the gravid uterus (Alperin et al., 2015, 2016). The authors hypothesized that this increase in stiffness might be a physiologic adaptation of the pelvic floor to protect it from perineal trauma at childbirth (Alperin et al., 2015, 2016). The same research team compared the impact of simulated vaginal birth (by inducting a vaginal distension that replicates fetal crowning) on pelvic floor muscles in pregnant and non-pregnant rats. They reported a much more significant increase in muscular fiber length, and so a much higher distensibility, in the group of pregnant rats (Catanzarite et al., 2018).

In addition, animal experimentations reported significant changes in the elastic properties of the pelvic floor with opposite behaviors between the vagina, which decrease its stiffness, whereas the muscle seems to increase its stiffness with an increase in the number and the length of muscular fibers, which also allows a more significant distensibility. The pathophysiological process for these changes remains hardly known, but the metabolism of collagen and elastin appears to be involved. It is likely that these changes during pregnancy affect the ability to deliver vaginally, without important perineal trauma.

4.4 Association with the mode of delivery and the risk of perineal trauma at childbirth

4.4.1 Mode of delivery

It is likely that some of the biomechanical changes reported above have an impact on the mode of delivery. As an example, the increase in ligamentous laxity probably affects changes in spinal curvature. Some women may have a major increase in lumbar lordosis, and so a verticalization of the pelvic inlet that could lead to difficulties for the fetal descent due to a discordance between its axis of progression and the axis of the pelvic inlet (Desseauve et al., 2019b). These associations between women's biomechanical characteristics, pelvic inlet, and the mode of delivery will be developed later in this chapter, but at this point, we can safely hypothesize that they are associated. Another related example—bearing on changes that occurs in pelvic floor—is the increase in levator hiatus area. Indeed, we reported above that the fetus will have to progress within this hiatus and that women with the smallest hiatus will be more incline to deliver in an operative way (Toozs-Hobson et al., 2008; van Veelen et al., 2015).

4.4.2 Perineal trauma

The hypothesis of an association between the elastic properties of pelvic floor muscles and their changes during pregnancy appears to make safe the major stretch that is applied to these muscles (300%) during vaginal delivery. It is likely that their intrinsic elastic properties will affect their ability to sustain this strain. Though there is no clinical evidence supporting this hypothesis, a reported association between peripheral joint mobility and the risk of obstetric anal sphincter injuries at childbirth, support the same (Gachon et al., 2018a).

4.5 Biomechanics of spontaneous normal delivery

Upright posture and bipedal locomotion with their numerous advantages enabled the evolution of the mankind. However, the necessary adaptations have posed some challenges to reproduction in general, and childbirth in particular. The curvature of the spine, needed for absorbing shocks while walking upright, has significant implications on the course of a vaginal delivery. The promontory changes the shape of the pelvic inlet to such extent that the head of a mature fetus can engage in the pelvis only when rotated with its sagittal suture in the transverse or oblique diameters. Furthermore, the stable upright posture demands shortening of the distance in between the legs/hip joints. This adaptation led to a narrowing of the pelvis in the mid-pelvic plane. Consequently, the fetal head cannot pass through the midpelvis with its sagittal suture in the transverse diameter, which mandates rotation inside the true pelvis. Furthermore, the pelvic axis curves to enable the pelvic floor to hold the organs within the abdominal cavity against the gravity. This curvature impedes a straight passage of the fetus through the pelvis. Finally, given the size of the brain, the fetal head compared to the body has considerably larger dimensions than in animals. All evolutionary adaptations and implications, taken together, make childbirth in humans, unlike animals, a very complicated biomechanical process. This chapter will briefly review the mechanism of a normal cephalic vertex delivery necessary and other considerations for modeling of a vaginal delivery.

4.5.1 Delivery of the head

Since the passage through the pelvis is narrow, the fetus must always assume positions that are associated with smallest passing diameters. The positional changes required to navigate the pelvic canal constitute the seven cardinal movements that constitute the mechanism of labor, i.e., engagement, descent, flexion, internal rotation, extension, external rotation, and expulsion. All these movements occur sequentially, but with a high degree of overlap. The smallest and the most convenient diameter for the delivery of the fetal head is the sub-occipito-bregmatic diameter. This is the passing diameter in a normal delivery of a fetus in vertex (occiput) presentation. To engage into the small pelvis with this diameter, the fetal head must undergo flexion. The posterior fontanel becomes the leading part, i.e., the lowermost point of the fetus, entering the pelvis first. Once the fetal head engages into the pelvis, the uterine contractions and pressure from the birth canal induce changes in the fetal attitude. These changes consist mainly of fetal straightening, i.e., loss of dorsal convexity and closer application of the extremities to the body. The fetal body shape is transformed so that it passes with the smallest possible cross section. After the head has descended through the pelvic inlet with its sagittal suture being in the oblique or transverse diameter, it must undergo an internal rotation to be able to enter the mid-pelvis. The head rotates with the leading part under the symphysis pubis. The funnel-shaped LAM usually facilitates this rotation. Apart from change of shape of the pelvis between individual planes, which the fetus negotiates by a series of rotations, the fetus has also to overcome the curved pelvic axis. Hence, once passing the mid-pelvis, the flexed fetal head must undergo an extension with the base of the occiput towards the pubic symphysis. The pelvic floor, again, facilitates this movement. Once the head passes through the mid-pelvis, it does not recede back in between the contractions, and the vaginal opening with the perineum are stretched tightly around it. This state is termed crowning. With progressive distension of the perineum and vaginal opening, the head is gradually born as the occiput, bregma, forehead, nose, mouth, and finally the chin. The delivery through slow and gradual distention of the perineum may be accelerated by its traumatization.

4.5.2 Delivery of the shoulder

The shoulders naturally follow the head during the delivery, although the largest passing diameter (biacromial diameter) is perpendicular to that of fetal head (sagittal suture). By the time the fetal head passes through the mid-pelvis the shoulders pass though the pelvic inlet and engage in the pelvis. To pass through the mid-pelvis, the shoulders must rotate with the anterior shoulder under the symphysis pubis. This rotation is observed by the accoucheur as an external rotation of the fetal head. This movement is brought about by the same pelvic factors that produced internal rotation of the head, but now acting on the shoulders. The shoulders further progress along the curvature of the axis of the birth canal leading to lateral flexion of the fetal trunk. Once the anterior shoulder is delivered, the deltoid muscle insertion of the anterior fetal arm is brought into a tight contact with the symphysis pubis, and the perineum is distended by the posterior shoulder. After the delivery of the shoulders, the rest of the fetus quickly passes leading to its expulsion.

4.6 Peripartum biomechanical considerations

4.6.1 Physiological changes

4.6.1.1 Hormonal changes

As aforementioned, several changes in hormonal metabolism are reported during pregnancy with a potential impact on women's tissues intrinsic characteristics, whereas intrapartum hormonal changes are mainly responsible for the onset and regulation of uterine contractions. Therefore there impact is more pronounced on the "powers," rather than the "passages". The ratio between estrogen and progesterone increases, thus diminishing the progesterone inhibiting effect on the myometrium

(Vannuccini et al., 2016). The estrogens seem to be involved in a process of increasing the number of receptors for prostaglandins and oxytocin and in the activity of enzymes involved in myometrial contraction (Challis et al., 2000; Kamel, 2010). Oxytocin directly causes myometrial contractions and facilitates prostaglandin production (Vannuccini et al., 2016; Lopez Bernal, 2003). Prostaglandins are the essential substance to ripen the cervix and increase myometrial contractions (Challis et al., 1997).

4.6.1.2 Changes to muscles and connective tissue

Fetal growth and increase in intrauterine volume increase myometrial tension (Shynlova et al., 2020). This mechanical distention along with physiologic uterine inflammation induces cytokine expression, ultimately triggering myometrial contractions (Osman et al., 2003; Shynlova et al., 2013). From a gross biomechanical perspective, the thickness of myometrium changes from 8.1–8.8 mm (non-labor) to 5.8–6.8 mm (active labor) and to 10.6–12.1 mm (second stage of labor) (Buhimschi et al., 2003). The muscles and connective tissue surrounding the birth canal undergo a substantial degree of stretch. Indeed, it is reported that the levator hiatus circumference increases by three-folds (Ashton-Miller and DeLancey, 2007; Ashton-Miller and Delancey, 2009). As already stated, there are changes that occur in the levator hiatus area through pregnancy and probably into the intrinsic tissular characteristics of the LAM as a preparation for the delivery. Despite this, the levator ani muscle is torn from its insertion to the pubic bone in 15% of women after a spontaneous vaginal delivery (van Delft et al., 2015). The trauma to the tissues surrounding the birth canal may significantly alter the course and the speed of progress of the vaginal delivery. Connective tissue changes during labor, i.e., changes in collagen cross-linking and composition of extracellular matrix, are also essential for the cervical ripening and intrapartum dilation (Akins et al., 2011). These changes lead to a reduction in tissue stiffness and resistance.

4.6.2 Powers

4.6.2.1 Contractions

Forceful, frequent, and synchronized myometrial contractions are needed for a successful spontaneous vaginal delivery. The contraction of the myocytes is coordinated with a descending pressure gradient (from the fundus to the isthmus) (Hellman et al., 1950), resulting in an indirect pressure on the cervix by the fetus and forcing the cervix to dilate and stretch upwards (Caldeyro-Barcia and Poseiro, 1955). From the biomechanical point of view, it is important that the forces of labor are intermittent, leaving time for restitution and reoxygenation of the placenta. Given the viscoelastic properties of the birth passages, the fetus recedes back in between the contractions. Hence, the longer the inter-contraction interval, the less efficient is the dilatation. This interval gradually decreases during the labor. In a normal second stage of labor, the contractions repeat every two to four minutes and last approximately one minute (Ashton-Miller and Delancey, 2009; Bok et al., 2020). The intrauterine pressure determined by the contractions increases in the course of labor and is further potentiated by maternal pushing in the second stage of labor (Grimm, 2021).

Apart from intrauterine pressure, the myometrial tension cannot be precisely measured. Therefore fetal head-to-cervix (vagina, perineum) contact pressure, or expulsive force, is more relevant for biomechanical depiction of the labor process. Indeed, clinically this force is a better predictor of a successful vaginal birth (Allman et al., 1996). However, the differences in measured values in these forces between the first and second stage of labor are immense.

4.6.2.2 Maternal pushing

Apart from uterine contractions, the main natural power in the final stages of vaginal delivery is the involuntary urge to push, triggered by fetal pressure on the pelvic floor (Lemos et al., 2017). In contrast, voluntary maternal pushing is a modifiable obstetrical intervention. The commonest type of instructed (coached) pushing is a Valsalva maneuver. The other possible variants, open-glottis (exhaled) pushing, is not as established (Lemos et al., 2017; de Tayrac and Letouzey, 2016).

In "immediate" pushing, the woman is asked to commence pushing once the second stage is diagnosed, whereas, in "delayed" pushing (or passive descent method), the woman is advised to await the spontaneous urge to push (Lemos et al., 2017). Pushing also varies depending on the relationship between the timing of pushing and the contraction, where the parturient either starts pushing from the beginning of a contraction (Parnell et al., 1993) or waits until the contraction reaches its peak. In both cases, this voluntary push stops at the end of contraction. The number of breaths during one contraction differ between three to five, each of these breaths lasts approximately 10 seconds (Lemos et al., 2017; Yeates and Roberts, 1984).

Finally, just in the final moments before and during fetal head expulsion, the woman might either continue to push during the final contraction, stop pushing during, and push between contractions or stop pushing all together and rely on the reflexive pushing (Laine et al., 2008).

4.6.2.3 Fundal pressure

Fundal pressure (e.g., the Kristeller's maneuver) is an exertion of pressure on the uterine fundus towards the birth canal to accelerate the delivery of the neonate by shortening the final moments of the second stage of labor (Kline-Kaye and Miller-Slade, 1990). This is considered a dangerous intervention by many practitioners and professional bodies, even banned in some countries, and considered a physical assault; therefore the maneuver did not get rigorous evaluation. An inflatable fundal belt was used instead of manual pressure and reported non-inferiority of the technique with regards to pelvic floor trauma (Hofmeyr et al., 2017). However, more recent non-randomized ultrasonographic studies found a higher prevalence of LAM avulsion after Kristeller (Youssef et al., 2019; Takmaz et al., 2020).

Table 4.1 Types of the female pelvis.

Type	Shape	Comment
Gynecoid (mesatipellic)	round, shallow, and open	Formerly considered the most common female pelvis shape. However, this type was found only in 14.3% in a recent study.
Android (brachypellic)	narrower, heart-shaped	This type is similar to the male pelvis. It is currently considered the most common type amongst females (59.3%).
Anthropoid (dolichopellic)	narrow and deep pelvis with a shape of an "upright" oval	
Platypelloid (platypellic)	wide and shallow "lying" oval	It is the least common type.

4.6.3 Passages

4.6.3.1 Type of pelvis

Obstetrically, the bony pelvis is horizontally divided into planes: inlet (brim), mid-pelvis, and outlet. At the time of the vaginal delivery, there is increased mobility of the pelvic joints because of relaxin. This increased mobility allows better adaptation to the passing fetus. Maternal position also seems to play a role. Computational simulation demonstrated that the outlet transverse and antero-posterior diameters increase by 1.1 cm and 0.6 cm in a squatting position, respectively (Hemmerich et al., 2019). However, other measures are less dynamic.

The mid-pelvis is the plane with the narrowest transverse interspinal diameter, which has been shown to vary in women delivering vaginally and by cesarean (Perlman et al., 2019). Additionally, the pubic arch angle is an important measure of the pelvic outlet and ranges from 65.6° to 131.6° (Albrich et al., 2015). There are four main types of female pelvis (see Table 4.1).

4.6.3.2 Birth canal

The birth canal consists of vaginal walls, surrounding endopelvic fascia, pelvic floor musculature (involving mainly PVM and perineal muscles), and perineal body. Anatomically, vagina is a thin highly elastic tube that can be divided into the upper (supra levator) part and lower (infra levator) part that is surrounded proximally by PVM (creating a U-shaped loop called levator hiatus) and distally by the perineum. LAM trauma, induced by its stretch, may alter the course and the outcome of childbirth. Microtrauma (irreversible overdistension) leads to the levator hiatus enlargement. This overdistension could happen due to intrapartum stretching or partial denervation. Macro trauma (levator avulsion) is a mechanical intrapartum partial or complete detachment of the antero-inferior margin of the muscle from the pubic bone. This avulsion may happen unilaterally or bilaterally, most frequently unilaterally and on the right side (van Delft et al., 2015, 2014).

The perineum is a complex fibromuscular structure with extreme variability in the overall shape, composition, and size (Kreklau et al., 2018; Cechova et al., 2021). The high interindividual variability in the tissue characteristics may be determined by maternal age, ethnicity, BMI, parity and/or genetically. The perineal body is an integral part of the perineum since this is a structure where all perineal components confluence. Recent modeling studies have stressed the importance of this structure; as its absence in a finite element model of the vaginal delivery has led to a dramatic change in all measured tissue strain parameters, including LAM (Cechova et al., 2021). It is a pyramidal structure. The perineal body length is defined as the distance from the posterior margin of the hymen to the midanal opening (Haylen et al., 2016).Nevertheless, the perineal dimensions significantly change at crowning and immediately following birth (Kalis et al., 2010; Zemcik et al., 2012). Due to the anal dilation (2.8 ± 0.9 cm) at the end of vaginal delivery, it is suggested that the perineal length should be measured from the fourchette to the border between peri-anal epidermis and anal epithelium at 12 o'clock (Kalis et al., 2010).

Several factors affect severity of perineal trauma, including the biomechanical properties and dimensions of the pelvic floor structures, e.g., the length of the perineum (Lane et al., 2017) and the extent of perineal edema (Samuelsson et al., 2000). Currently, there is no reliable clinical tool for the intrapartum assessment of perineal elasticity. Moreover, intrapartum measurement of the biomechanical characteristics of the perineum and LAM is extremely challenging. Within research settings, attempts have been made to evaluate the elasticity of pelvic floor structures (LAM, perineum) prior childbirth using static elastography (Chen et al., 2015) shear wave elastography (Gachon et al., 2018b, 2020) and tactile imaging (Rusavy et al., 2020; Brandt et al., 2020).

4.6.4 Passenger

4.6.4.1 Fetal size

Fetal size is an important determinant of childbirth pelvic floor trauma. Birthweight > 4000 g and shoulder dystocia are recognized risk factors for OASIS (Gynaecologists RCoo, 2015). The head is the largest structure of the fetal body and, normally, the leading part exerting the pressure on surrounding structures. Thus from a biomechanical point of view, the fetal head size and shape is more important than the birthweight itself, unless in some circumstances that result in head/shoulder disproportion (e.g., in fetal growth restriction or macrosomia), where the biomechanical impact of the shoulder is more pronounced. Larger head circumference is associated with prolonged second stage and levator trauma (Valsky et al., 2009; Aydin and Aydin, 2015). Similarly, larger fetal head size is a risk factor for vaginal wall trauma (Jansson et al., 2020) and increases the maximum strain in the perineum (Jansova et al., 2017).

4.6.4.2 Malpositions and malpresentations

The fetal body orientation is defined by its lie, position, presentation, and attitude of the fetus (Table 4.2). The most common physiologic orientation is longitudinal lie, left position, with left occiput anterior presentation (OA) and flexion in all fetal

Table 4.2 Description of the fetal body orientation.

Parameter	Explanation	Norm	Variation of norm	Abnormality	Comment
Lie	orientation of the fetal axis in uterus	longitudinal	-	transverse	not vaginally deliverable
				oblique	
Position	position of the fetal back	left	left-anterior	(transverse) anterior	not vaginally deliverable
			left-posterior	(transverse) posterior	
		right	right-anterior	(transverse) superior	
			right-posterior	(transverse) inferior	
Presentation	part of the fetus is expected to be delivered first	cephalic	vertex occiput anterior	vertex (persistent) occiput posterior	
				sinciput (military)	usually occiput posterior
				brow	not vaginally deliverable
				face	vaginal delivery possible but currently not recommended
		breech	complete	footling	footling not vaginally deliverable
			frank		
Attitude	relationship of fetal parts to one another	flexed	-	extended	

joints. Any deviation from this fetal orientation is associated with either alteration in the aforementioned delivery mechanism or increase in the passing diameters of the fetus. Occiput posterior (OP) presentation is consistently considered a risk factor for LAM avulsion and perineal trauma, given that the head passes with larger dimensions as the biparietal diameter is located more posteriorly (Gynaecologists RCoo, 2015). Indeed, in OP, PVM stretch is 1.1-fold and stress is 3.6-fold higher than in OA (Parente et al., 2009; Havelkova et al., 2020). Another malpresentation, called asynclitism, means that one parietal bone is tilted towards shoulder. Therefore the other parietal bone occurs lower in the birth canal, hence increasing the passing circumference of the fetal head. Usually, a spontaneous rectification happens during labor. However, in some cases asynclitism persists (Bamberg et al., 2012) causing asymmetry in (maximum) stress in pelvic floor structures (Cechova et al., 2021).

4.6.4.3 Fetal head shape

The fetal skull consists of bones, sutures, and fontanelles. The sutures and fontanelles are connective tissue structures, hence easily deformable when fetal head passes through the pelvis. It has been documented that the margins of the fetal skull bones approximate during vaginal birth (molding or configuration) and that overlapping of these margins is not uncommon (Ami et al., 2019; Bamberg et al., 2017). The molding happens mainly in antero-posterior diameter between occipital and parietal bones (in lambdoid suture) (Ami et al., 2019; Iversen et al., 2020). Initially, the occipital bone is depressed in relation to parietal bones (Iversen et al., 2020), and parietal bones override each other. At the time of crowning and expulsion, the head reconfigures its shape; the fronto-occipital diameter changes from the initial 10.3 cm to 10.8 cm during extension and 11.2 cm at expulsion (Bamberg et al., 2017). Concomitantly, the distance from vertex to the base of skull in the neonate with head circumference of 33.5 cm shortens from 6.4 cm to 5.6 cm (Bamberg et al., 2017).

4.6.5 Birth position

There are several maternal positions during the final moments of labor, most frequently divided into horizontal and vertical positions (Desseauve et al., 2017) (Table 4.3). Flexion of the thighs has been demonstrated to be more important than leg abduction (Desseauve et al., 2019b, 2020). Although it can be achieved by squatting, this position is not considered biomechanically optimal (Desseauve et al., 2020, 2019a). Upright positions, in general, might increase the risk of second-degree perineal tears, nonetheless, there is no clear evidence regarding OASIS yet (Gupta et al., 2017). Of note, the risk of OASIS might be confounded by the ability to visualize the perineum, and hence effectiveness of manual perineal protection (Laine et al., 2008; Samuelsson et al., 2000). To our knowledge there is no study evaluating maternal positions association to LAM trauma.

Table 4.3 Maternal positions in labor.

Vertical/horizontal position	Category	Description	Comment
Horizontal	Supine	woman lying on her back, position of legs not described	often considered as lithotomy perineal support depending on legs position
	Lithotomy	woman lying on her back, legs elevated and supported, hips and knees flexed, thighs apart	perineal support optimally feasible
	Lateral (Sim's position)	woman lying on her side, upper leg usually elevated hips and knees flexed, thighs apart	usually left lateral
	Semi-recumbent (French birthing position)	woman in supine (or lithotomy) position with her head and neck elevated 30° to 60° to the horizontal line [Desseauve2017]	legs usually elevated and supported, hips and knees flexed, thighs apart
Semi-recumbent	recumbent semi-sitting	leaning on her back (or side) at an angle 45° from horizontal line	legs usually elevated and supported, hips and knees flexed, thighs apart perineal support feasible depending on position of legs and elevation of the bed
Upright (vertical)	sitting	sitting on a bed, on a chair, or on a ball	perineal support difficult to perform
	squatting	woman with bended knees and legs apart	use of a birthing bar is common perineal support complicated but feasible
	kneeling all fours	kneeling and bent forward	perineal support feasible
	standing	the woman is in an upright position standing by herself or against to a support (bed, chair, or partner)	perineal support difficult to perform
Other	**waterbirth**	any maternal position	perineal support feasibility depending on maternal position and shape of the bathtub (cask)

4.6.6 Practitioner

4.6.6.1 Manual perineal protection

Manual perineal protection (MPP) is a method known for nearly 2000 years. The goal is to protect the perineum from redundant trauma by reducing the maximum strain (Zemcik et al., 2012). However, its execution is not widely reported. Furthermore, it is a very complex intervention, occurring within a very short time (mean 13.6 s) (Kalis et al., 2020) and the proper knowledge and execution regarding its application is suboptimal (Kleprlikova et al., 2020).

Currently the Finnish and Viennese techniques are considered as the most effective (Kleprlikova et al., 2020; Kalis and Rusavy, 2017). This was based on biomechanical principles that demonstrated the following achievements of biomechanical modeling in relation to MPP:

- applying the fingers to the perineal surface (skin) together with the contraction of palmar muscles to disperse the tension has the effect of reducing this strain/tension in deeper layers of the perineum (Jansova et al., 2014b,a)
- the extent of this reduction depends on the modification used, i.e., extent of the movement and on the final position of the fingers (Jansova et al., 2014b,a)
- difference of one cm in fingers application might result in 30% difference in maximum perineal strain (Jansova et al., 2014b,a)
- in the most efficient modification, the thumb and index-finger are applied anterolaterally to the fourchette (2 cm anteriorly, 6 cm from the midline) and approximated by one cm on each side (Jansova et al., 2014b,a)
- the best MPP modification in the case of a large head may reduce the perineal strain below the strain achieved by hands-off in the case of a small fetal head, hence MPP should be executed for the delivery of all babies (Jansova et al., 2017)
- the mean forces executed by thumb, index-finger, and middle finger were 26.7 N, 25.5 N, and 20.2 N, whereas maximum forces achieved 34.3 N, 32.6 N, and 27.6 N, respectively (Kalis et al., 2020).

4.6.6.2 Operative vaginal delivery

Currently, vacuum-extraction is preferred over forceps for the sake of maternal pelvic floor (Sultan et al., 2019; de Leeuw et al., 2019). From a biomechanical perspective, the use of forceps significantly increases the passing diameter of the fetal head and eliminates time for restitution of tissues between the contraction and slow gradual dilatation during contractions. Use of vacuum-extraction is associated with similar effect, without the increase in size of the passing object. Forceps is consistently recognized as a risk factor for both LAM trauma and OASIS. The rates of LAM avulsion between vacuum and spontaneous vaginal delivery are consistently comparable. In the latest review, the incidence of LAM avulsion was 1%, 15%, 21%, and 52% following cesarean, spontaneous, vacuum, and forceps deliveries, respectively (van Delft et al., 2015, 2014). The situation is similar regarding OASIS. A review of RCTs revealed an OASIS rate of 14% after forceps compared to 7.5% after vacuum (O'Mahony et al., 2010). Currently, a proper mediolateral or lateral episiotomy

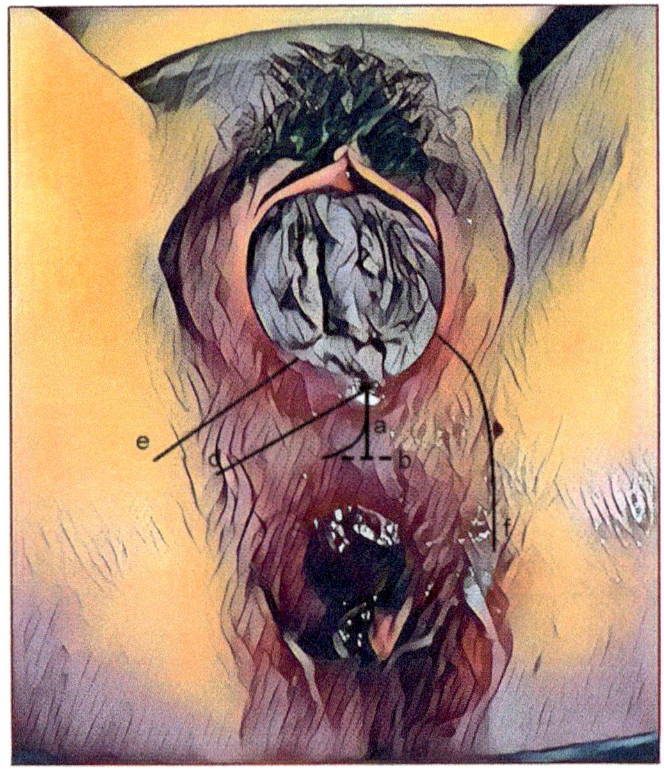

FIGURE 4.4

Types of episiotomy.

during operative vaginal birth to decrease the risk of OASIS is supported by the literature, however, this recommendation is not routinely followed (Sultan et al., 2019; de Leeuw et al., 2019). Large observational studies found that execution of a proper episiotomy may reduce risk of both instruments. However, it remains consistently higher after forceps (van Bavel et al., 2018; Gurol-Urganci et al., 2013). There is currently a large national prospective study, which is to involve 15,000 primiparous women who delivered by operative vaginal delivery; its aim is to investigate the role of episiotomy on the risk of OASI in this context (INSTRUMODA) (Gachon et al., 2021).

4.6.6.3 Other interventions

There are several interventions that should be considered in biomechanical simulations because of their known impact on the risk of perineal trauma. These interventions may decrease the risk of perineal trauma by either increasing tissue elasticity (perineal massage, warmth), reducing friction (obstetric gel), or enlarging the introitus (episiotomy). Several types of episiotomy have been described (Fig. 4.4), and these will have different biomechanical ramifications that should be considered (Kalis et al., 2012). Because of the range of perineal deformation and edema during crowning of the fetal head (Kalis et al., 2010), there is evidence that location (type), length, and timing of episiotomy are important factors in anal sphincter trauma/integrity (Kalis and Prka, 2017). Given this, the terms "incision angle" and "suture angle" of episiotomy were introduced. 50% relative reduction in the risk of OASIS for every 6° away from the perineal midline was calculated (Eogan et al., 2006). Another study found that a combination of episiotomy length > 17 mm, point of beginning > 9 mm from the midline and suture angle of 30–60° significantly reduces OASIS (Stedenfeldt et al., 2012).

4.6.6.4 Shoulder delivery

After the fetal head, the shoulders are the second largest passing structure. Since perineal trauma may occur during shoulder delivery after a head expulsion with intact perineum (Hehir et al., 2018; Nesbitt et al., 1998), an optimal model of vaginal delivery should include this element (Hulot et al., 2020; Mottet et al., 2017).

4.7 Model development, testing, and validation

It must be acknowledged that a biomechanical simulation always leads to a degree of simplification. Therefore it is of utmost importance that the team involved in such studies is multidisciplinary, involving biomechanicians, obstetricians, midwives, and anatomists. To develop a realistic biomechanical (computational) model, several entry data must be used: appropriate anatomical dimensions, angles, and landmarks of female pelvis; tissue characteristics; realistic fetal head shape, size, and position; depiction of molding; real life birth trajectories and forces. Once all these aspects are fulfilled, it is then feasible to assess the impact of deviations from normality and obstetric interventions.

4.8 Conclusion

In this chapter, we reported that pregnancy is associated with significant changes in the intrinsic biomechanical characteristics of a women. Even if the mechanisms remain hardly understood, it is likely that these changes occur as a preparation to vaginal delivery and the unique biomechanical challenge that it represents. Considering these intrinsic women's characteristics in the management of childbirth may improve the mechanisms of delivery with a better protection of women's pelvic floor. Nevertheless, this step is not yet reached in our current practices. Biomechanical, particularly computational studies, hardly fit into the hierarchy of evidence. Nevertheless, in obstetrics, there are a substantial proportion of variables and interventions, where the level of evidence is difficult to be obtained by means of RCTs, due to ethical reasons, clinical measurement imprecision, difficulty to set standardized conditions, or the number of confounders (Parnell et al., 2001; Raisanen et al., 2014). In such cases, biomechanical simulation studies have the potential to overcome such shortfalls. Moreover, biomechanical studies can achieve data that are clinically unobtainable, e.g., LAM stretch or the effect of MPP on deeper perineal structures (Jansova et al., 2014b,a; Lien et al., 2004). Furthermore, the biomechanical explanation of the process of vaginal delivery helps clinicians understand why some procedures should be preferred whereas others avoided if possible.

4.9 List of Abreviations

LAM – Levator ani muscle
OASI – Obstetric anal sphincter injury

References

Abramson, D.R., Wilson, P.D.S.M., 1934. Relaxation of the pelvic joints in pregnancy. Surg. Gynecol. Obstet. 58, 595–613.

Akins, M.L., Luby-Phelps, K., Bank, R.A., Mahendroo, M., 2011. Cervical softening during pregnancy: regulated changes in collagen cross-linking and composition of matricellular proteins in the mouse. Biol. Reprod. 84 (5), 1053–1062.

Albrich, S.B., Shek, K., Krahn, U., Dietz, H.P., 2015. Measurement of subpubic arch angle by three-dimensional transperineal ultrasound and impact on vaginal delivery. Ultrasound Obstet. Gynecol. 46 (4), 496–500.

Allman, A.C., Genevier, E.S., Johnson, M.R., Steer, P.J., 1996. Head-to-cervix force: an important physiological variable in labour. 2. Peak active force, peak active pressure and mode of delivery. Br. J. Obstet. Gynaecol. 103 (8), 769–775.

Alperin, M., Kaddis, T., Pichika, R., Esparza, M.C., Lieber, R.L., 2016. Pregnancy-induced adaptations in intramuscular extracellular matrix of rat pelvic floor muscles. Am. J. Obstet. Gynecol.

Alperin, M., Lawley, D.M., Esparza, M.C., Lieber, R.L., 2015. Pregnancy-induced adaptations in the intrinsic structure of rat pelvic floor muscles. Am. J. Obstet. Gynecol. 213 (2), 191.

Ami, O., Maran, J.C., Gabor, P., Whitacre, E.B., Musset, D., Dubray, C., Mage, G., Boyer, L., 2019. Three-dimensional magnetic resonance imaging of fetal head molding and brain shape changes during the second stage of labor. PLoS ONE 14 (5), e0215721.

Ashton-Miller, J.A., DeLancey, J.O., 2007. Functional anatomy of the female pelvic floor. Ann. N.Y. Acad. Sci. 1101, 266–296.

Ashton-Miller, J.A., Delancey, J.O., 2009. On the biomechanics of vaginal birth and common sequelae. Annu. Rev. Biomed. Eng. 11, 163–176.

Aydin, S., Aydin, C.A., 2015. Evaluation of labor-related pelvic floor changes 3 months after delivery: a 3D transperineal ultrasound study. Int. Urogynecol. J. 26 (12), 1827–1833.

Bamberg, C., Deprest, J., Sindhwani, N., Teichgraberg, U., Guttler, F., Dudenhausen, J.W., Kalache, K.D., Henrich, W., 2017. Evaluating fetal head dimension changes during labor using open magnetic resonance imaging. J. Perinat. Med. 45 (3), 305–308.

Bamberg, C., Rademacher, G., Guttler, F., Teichgraber, U., Cremer, M., Buhrer, C., Spies, C., Hinkson, L., Henrich, W., Kalache, K.D., et al., 2012. Human birth observed in real-time open magnetic resonance imaging. Am. J. Obstet. Gynecol. 206 (6), 505.

Beighton, P., Grahame, R., Bird, H.A., 2012. Hypermobility of Joints, fourth edition. Springer, London, New York.

Betsch, M., Wehrle, R., Dor, L., Rapp, W., Jungbluth, P., Hakimi, M., Wild, M., 2015. Spinal posture and pelvic position during pregnancy: a prospective rasterstereographic pilot study. Eur. Spine J. 24 (6), 1282–1288.

Bey, M.E., Marzilger, R., Hinkson, L., Arampatzis, A., Legerlotz, K., 2019. Patellar tendon stiffness is not reduced during pregnancy. Front. Physiol. 10, 334.

Bo, K., Hilde, G., Tennfjord, M.K., Staer-Jensen, J., Siafarikas, F., Engh, M.E., 2014. Pelvic floor muscle variables and levator hiatus dimensions: a 3/4D transperineal ultrasound cross-sectional study on 300 nulliparous pregnant women. Int. Urogynecol. J. 25 (10), 1357–1361.

Bok, S.M., Carmona, G.E.P., Crawford, J., Eskander, R., Desai, M., Ross, M.G., 2020. Redefining second stage of labor: number of pushing contractions. AJP Rep. 10 (2), e183–e186.

Brandt, J.S., Rosen, T., Van Raalte, H., Kurtenos, V., Egorov, V., 2020. Characterization of perineum elasticity and pubic bone-perineal critical distance with a novel tactile probe: results of an intraobserver reproducibility study. Open J. Obstet. Gynecol. 10 (4), 493–503.

Buhimschi, C.S., Buhimschi, I.A., Malinow, A.M., Weiner, C.P., 2003. Myometrial thickness during human labor and immediately post partum. Am. J. Obstet. Gynecol. 188 (2), 553–559.

Bump, R.C., Mattiasson, A., Bo, K., Brubaker, L.P., DeLancey, J.O., Klarskov, P., Shull, B.L., Smith, A.R., 1996. The standardization of terminology of female pelvic organ prolapse and pelvic floor dysfunction. Am. J. Obstet. Gynecol. 175 (1), 10–17.

Burgess, K.E., Pearson, S.J., Onambele, G.L., 2009. Menstrual cycle variations in oestradiol and progesterone have no impact on in vivo medial gastrocnemius tendon mechanical properties. Clin. Biomech. (Bristol, Avon) 24 (6), 504–509.

Caldeyro-Barcia, R.A.H., Poseiro, J., 1955. Normal and abnormal uterine contractility in labor. Triangle 2, 41.

Calguneri, M., Bird, H.A., Wright, V., 1982. Changes in joint laxity occurring during pregnancy. Ann. Rheum. Dis. 41 (2), 126–128.

Catanzarite, T., Bremner, S., Barlow, C.L., Bou-Malham, L., O'Connor, S., Alperin, M., 2018. Pelvic muscles' mechanical response to strains in the absence and presence of pregnancy-induced adaptations in a rat model. Am. J. Obstet. Gynecol. 218 (5), 512e511–512e519.

Cechova, H., Kalis, V., Havelkova, L., Rusavy, Z., Fiala, P., Rybarova, M., Hyncik, L., Krofta, L., Ismail, K.M., 2021. Finite element modeling of maximum stress in pelvic floor structures during the head expulsion (FINESSE) study. Int. Urogynecol. J.

Challis, J.R., Lye, S.J., Gibb, W., 1997. Prostaglandins and parturition. Ann. N.Y. Acad. Sci. 828, 254–267.

Challis, J.R.G., Matthews, S.G., Gibb, W., Lye, S.J., 2000. Endocrine and paracrine regulation of birth at term and preterm. Endocr. Rev. 21 (5), 514–550.

Chen, L., Low, L.K., DeLancey, J.O., Ashton-Miller, J.A., 2015. In vivo estimation of perineal body properties using ultrasound quasistatic elastography in nulliparous women. J. Biomech. 48 (9), 1575–1579.

Chen, Y., Li, F.Y., Lin, X., Chen, J., Chen, C., Guess, M.K., 2013. The recovery of pelvic organ support during the first year postpartum. BJOG 120 (11), 1430–1437.

de Leeuw, J.W., Kalis, V., Ismail, K.M., Thakar, R., Sultan, A.H., 2019. Not only residents, but gynecologists and obstetricians, too. Acta Obstet. Gynecol. Scand. 98 (7), 945.

de Tayrac, R., Letouzey, V., 2016. Methods of pushing during vaginal delivery and pelvic floor and perineal outcomes: a review. Curr. Opin. Obstet. Gynecol. 28 (6), 470–476.

Desseauve, D., Fradet, L., Gachon, B., Cherni, Y., Lacouture, P., Pierre, F., 2020. Biomechanical comparison of squatting and "optimal" supine birth positions. J. Biomech. 105, 109783.

Desseauve, D., Fradet, L., Lacouture, P., Pierre, F., 2017. Position for labor and birth: state of knowledge and biomechanical perspectives. Eur. J. Obstet. Gynecol. Reprod. Biol. 208, 46–54.

Desseauve, D., Fradet, L., Lacouture, P., Pierre, F., 2019a. Is there an impact of feet position on squatting birth position? An innovative biomechanical pilot study. BMC Pregnancy Childbirth 19 (1), 251.

Desseauve, D., Pierre, F., Fernandez, A., Panjo, H., Decatoire, A., Lacouture, P., Fradet, L., 2019b. Assessment of pelvic-lumbar-thigh biomechanics to optimize the childbirth position: an "in vivo" innovative biomechanical study. Sci. Rep. 9 (1), 15968.

Dietz, H.P., 2010. Pelvic floor ultrasound: a review. Am. J. Obstet. Gynecol. 202 (4), 321–334.

Dietz, H.P., Bennett, M.J., 2003. The effect of childbirth on pelvic organ mobility. Obstet. Gynecol. 102 (2), 223–228.

Dietz, H.P., Eldridge, A., Grace, M., Clarke, B., 2004. Does pregnancy affect pelvic organ mobility? Aust. N. Z. J. Obstet. Gynaecol. 44 (6), 517–520.

Dietz, H.P., Moore, K.H., Steensma, A.B., 2003. Antenatal pelvic organ mobility is associated with delivery mode. Aust. N. Z. J. Obstet. Gynaecol. 43 (1), 70–74.

Dietz, H.P., Shek, C., De Leon, J., Steensma, A.B., 2008. Ballooning of the levator hiatus. Ultrasound Obstet. Gynecol. 31 (6), 676–680.

Dumas, G.A., Reid, J.G., 1997. Laxity of knee cruciate ligaments during pregnancy. J. Orthop. Sports Phys. Ther. 26 (1), 2–6.

Eogan, M., Daly, L., O'Connell, P.R., O'Herlihy, C., 2006. Does the angle of episiotomy affect the incidence of anal sphincter injury? BJOG 113 (2), 190–194.

Feola, A., Endo, M., Deprest, J., 2014. Biomechanics of the rat vagina during pregnancy and postpartum: a 3-dimensional ultrasound approach. Int. Urogynecol. J. 25 (7), 915–920.

Gachon, B., Desgranges, M., Fradet, L., Decatoire, A., Poireault, F., Pierre, F., Fritel, X., Desseauve, D., 2018a. Is increased peripheral ligamentous laxity in term pregnant women associated with obstetric anal sphincter injury? Int. Urogynecol. J. 29 (11), 1589–1595.

Gachon, B., Desseauve, D., Fradet, L., Decatoire, A., Lacouture, P., Pierre, F., Fritel, X., 2016. Changes in pelvic organ mobility and ligamentous laxity during pregnancy and postpartum. Review of literature and prospects. Prog. Urol. 26 (7), 385–394.

Gachon, B., Fritel, X., Fradet, L., Decatoire, A., Lacouture, P., Panjo, H., Pierre, F., Desseauve, D., 2017. Is levator hiatus distension associated with peripheral ligamentous laxity during pregnancy? Int. Urogynecol. J. 28 (8), 1223–1231.

Gachon, B., Fritel, X., Pierre, F., Nordez, A., 2020. In vivo assessment of the elastic properties of women's pelvic floor during pregnancy using shear wave elastography: design and protocol of the ELASTOPELV study. BMC Musculoskelet. Disord. 21 (1), 305.

Gachon, B., Nordez, A., Pierre, F., Fradet, L., Fritel, X., Desseauve, D., 2018b. In vivo assessment of the levator ani muscles using shear wave elastography: a feasibility study in women. Int. Urogynecol. J.

Gachon, B., Nordez, A., Pierre, F., Fritel, X., 2019. Tissue biomechanical behavior should be considered in the risk assessment of perineal trauma at childbirth. Arch. Gynecol. Obstet. 300 (6), 1821–1826.

Gachon, B., Schmitz, T., Artzner, F., Parant, O., De Tayrac, R., Ducarme, G., Le Ray, C., Pizzoferrato, A.C., Garabedian, C., Riethmuller, D., et al., 2021. A core outcome set development for a French national prospective study about the effect of mediolateral episiotomy on obstetric anal sphincter injury during operative vaginal delivery (INSTRUMODA). BMC Pregnancy Childbirth 21 (1), 251.

Grimm, M.J., 2021. Forces involved with labor and delivery-a biomechanical perspective. Ann. Biomed. Eng.

Gupta, J.K., Sood, A., Hofmeyr, G.J., Vogel, J.P., 2017. Position in the second stage of labour for women without epidural anaesthesia. Cochrane Database Syst. Rev. 5, CD002006.

Gurol-Urganci, I., Cromwell, D.A., Edozien, L.C., Mahmood, T.A., Adams, E.J., Richmond, D.H., Templeton, A., van der Meulen, J.H., 2013. Third- and fourth-degree perineal tears among primiparous women in England between 2000 and 2012: time trends and risk factors. BJOG 120 (12), 1516–1525.

Gynaecologists RCoo, 2015. Green-top Guideline No. 29. The management of third- and fourth-degree perineal tears.

Hansen, M., Kjaer, M., 2014. Influence of sex and estrogen on musculotendinous protein turnover at rest and after exercise. Exerc. Sport Sci. Rev. 42 (4), 183–192.

Havelkova, L., Krofta, L., Kochova, P., Liska, V., Kalis, V., Feyereisl, J., 2020. Persistent occiput posterior position and stress distribution in levator ani muscle during vaginal delivery computed by a finite element model. Int. Urogynecol. J. 31 (7), 1315–1324.

Hay, O., Dar, G., Abbas, J., Stein, D., May, H., Masharawi, Y., Peled, N., Hershkovitz, I., 2015. The lumbar lordosis in males and females, revisited. PLoS ONE 10 (8), e0133685.

Haylen, B.T., Maher, C.F., Barber, M.D., Camargo, S., Dandolu, V., Digesu, A., Goldman, H.B., Huser, M., Milani, A.L., 2016. An international urogynecological association (IUGA)/international continence society (ICS) joint report on the terminology for female pelvic organ prolapse (POP). Int. Urogynecol. J. 27 (4), 655–684.

Hehir, M.P., Rubeo, Z., Flood, K., Mardy, A.H., O'Herlihy, C., Boylan, P.C., D'Alton, M.E., 2018. Anal sphincter injury in vaginal deliveries complicated by shoulder dystocia. Int. Urogynecol. J. 29 (3), 377–381.

Hellman, L.M., Harris, J., Reynolds, S.R., 1950. Characteristics of the gradients of uterine contractility during the first stage of true labor. Bull. Johns Hopkins Hosp. 86 (4), 234–248.

Hemmerich, A., Bandrowska, T., Dumas, G.A., 2019. The effects of squatting while pregnant on pelvic dimensions: a computational simulation to understand childbirth. J. Biomech. 87, 64–74.

Hofmeyr, G.J., Vogel, J.P., Cuthbert, A., Singata, M., 2017. Fundal pressure during the second stage of labour. Cochrane Database Syst. Rev. 3, CD006067.

Hulot, G., Ramanah, R., Riethmuller, D., Mottet, N., 2020. The impact of active delivery of the anterior arm during vacuum-assisted vaginal delivery on perineal tears: a clinical practice evaluation. J. Matern. Fetal Neonatal. Med. 33 (19), 3308–3312.

Iversen, J.K., Kahrs, B.H., Torkildsen, E.A., Eggebo, T.M., 2020. Fetal molding examined with transperineal ultrasound and associations with position and delivery mode. Am. J. Obstet. Gynecol. 223 (6), 909e901–909e908.

Jansova, M., Kalis, V., Lobovsky, L., Hyncik, L., Karbanova, J., Rusavy, Z., 2014a. The role of thumb and index finger placement in manual perineal protection. Int. Urogynecol. J. 25 (11), 1533–1540.

Jansova, M., Kalis, V., Rusavy, Z., Raisanen, S., Lobovsky, L., Laine, K., 2017. Fetal head size and effect of manual perineal protection. PLoS ONE 12 (12), e0189842.

Jansova, M., Kalis, V., Rusavy, Z., Zemcik, R., Lobovsky, L., Laine, K., 2014b. Modeling manual perineal protection during vaginal delivery. Int. Urogynecol. J. 25 (1), 65–71.

Jansson, M.H., Franzen, K., Hiyoshi, A., Tegerstedt, G., Dahlgren, H., Nilsson, K., 2020. Risk factors for perineal and vaginal tears in primiparous women - the prospective POPRACT-cohort study. BMC Pregnancy Childbirth 20 (1), 749.

Jelovsek, J.E., Chagin, K., Gyhagen, M., Hagen, S., Wilson, D., Kattan, M.W., Elders, A., Barber, M.D., Areskoug, B., MacArthur, C., et al., 2018. Predicting risk of pelvic floor disorders 12 and 20 years after delivery. Am. J. Obstet. Gynecol. 218 (2), 222e1–222e19.

Jeong, E.H., Park, K.H., Ryu, A., Oh, K.J., Lee, S.Y., Kim, A., 2015. Clinical and sonographic parameters at mid-trimester and the risk of cesarean delivery in low-risk nulliparas. J. Clin. Ultrasound 43 (4), 235–242.

Kalis, V., Karbanova, J., Bukacova, Z., Bednarova, B., Rokyta, Z., Kralickova, M., 2010. Anal dilation during labor. Int. J. Gynaecol. Obstet. 109 (2), 136–139.

Kalis, V., Laine, K., de Leeuw, J.W., Ismail, K.M., Tincello, D.G., 2012. Classification of episiotomy: towards a standardisation of terminology. BJOG 119 (5), 522–526.

Kalis, V., Rusavy, Z., Havelkova, L., Zitka, T., Tolar, D., Ismail, K.M., 2020. Metrics of perineal support (MOPS) study. BMC Pregnancy Childbirth 20 (1), 361.

Kalis, V.J.M., Rusavy, Z., 2017. Chapter 3: Perineal Mapping. Publishing SI.

Kalis, V.R.Z., Prka, M., 2017. Chapter 6: episiotomy. In: Childbirth Trauma. Springer-Verlag L.

Kamel, R.M., 2010. The onset of human parturition. Arch. Gynecol. Obstet. 281 (6), 975–982.

Kim, S.N., Park, K.H., Jung, H.J., Hong, J.S., Shin, D.M., Kang, W.S., 2010. Clinical and sonographic parameters at 37 weeks' gestation for predicting the risk of primary Cesarean delivery in nulliparous women. Ultrasound Obstet. Gynecol. 36 (4), 486–492.

Kleprlikova, H., Kalis, V., Lucovnik, M., Rusavy, Z., Blaganje, M., Thakar, R., Ismail, K.M., 2020. Manual perineal protection: the know-how and the know-why. Acta Obstet. Gynecol. Scand. 99 (4), 445–450.

Kline-Kaye, V., Miller-Slade, D., 1990. The use of fundal pressure during the second stage of labor. J. Obstet. Gynecol. Neonatal. Nurs. 19 (6), 511–517.

Kreklau, A., Vaz, I., Oehme, F., Strub, F., Brechbuhl, R., Christmann, C., Gunthert, A., 2018. Measurements of a 'normal vulva' in women aged 15-84: a cross-sectional prospective single-centre study. BJOG 125 (13), 1656–1661.

Laine, K., Pirhonen, T., Rolland, R., Pirhonen, J., 2008. Decreasing the incidence of anal sphincter tears during delivery. Obstet. Gynecol. 111 (5), 1053–1057.

Lane, T.L., Chung, C.P., Yandell, P.M., Kuehl, T.J., Larsen, W.I., 2017. Perineal body length and perineal lacerations during delivery in primigravid patients. Proc. (Bayl. Univ. Med. Cent.) 30 (2), 151–153.

Lemos, A., Amorim, M.M., Dornelas de Andrade, A., de Souza, A.I., Cabral Filho, J.E., Correia, J.B., 2017. Pushing/bearing down methods for the second stage of labour. Cochrane Database Syst. Rev. 3, CD009124.

Lien, K.C., Mooney, B., DeLancey, J.O., Ashton-Miller, J.A., 2004. Levator ani muscle stretch induced by simulated vaginal birth. Obstet. Gynecol. 103 (1), 31–40.

Lindgren, A., Kristiansson, P., 2014. Finger joint laxity, number of previous pregnancies and pregnancy induced back pain in a cohort study. BMC Pregnancy Childbirth 14, 61.

Lopez Bernal, A., 2003. Mechanisms of labour–biochemical aspects. BJOG 110 (Suppl 20), 39–45.

Lowder, J.L., Debes, K.M., Moon, D.K., Howden, N., Abramowitch, S.D., Moalli, P.A., 2007. Biomechanical adaptations of the rat vagina and supportive tissues in pregnancy to accommodate delivery. Obstet. Gynecol. 109 (1), 136–143.

MacLennan, A.H., Nicolson, R., Green, R.C., Bath, M., 1986. Serum relaxin and pelvic pain of pregnancy. Lancet 2 (8501), 243–245.

Marnach, M.L., Ramin, K.D., Ramsey, P.S., Song, S.W., Stensland, J.J., An, K.N., 2003. Characterization of the relationship between joint laxity and maternal hormones in pregnancy. Obstet. Gynecol. 101 (2), 331–335.

Mattox, T.F., Lucente, V., McIntyre, P., Miklos, J.R., Tomezsko, J., 2000. Abnormal spinal curvature and its relationship to pelvic organ prolapse. Am. J. Obstet. Gynecol. 183 (6), 1381–1384. Discussion 1384.

Meister, M.R., Cahill, A.G., Conner, S.N., Woolfolk, C.L., Lowder, J.L., 2016. Predicting obstetric anal sphincter injuries in a modern obstetric population. Am. J. Obstet. Gynecol. 215 (3), 310.

Mottet, N., Bonneaud, M., Eckman-Lacroix, A., Ramanah, R., Riethmuller, D., 2017. Active delivery of the anterior arm and incidence of second-degree perineal tears: a clinical practice evaluation. BMC Pregnancy Childbirth 17 (1), 141.

Nesbitt, T.S., Gilbert, W.M., Herrchen, B., 1998. Shoulder dystocia and associated risk factors with macrosomic infants born in California. Am. J. Obstet. Gynecol. 179 (2), 476–480.

O'Boyle, A.L., O'Boyle, J.D., Ricks, R.E., Patience, T.H., Calhoun, B., Davis, G., 2003. The natural history of pelvic organ support in pregnancy. Int. Urogynecol. J. Pelvic Floor Dysfunct. 14 (1), 46–49. Discussion 49.

O'Boyle, A.L., Woodman, P.J., O'Boyle, J.D., Davis, G.D., Swift, S.E., 2002. Pelvic organ support in nulliparous pregnant and nonpregnant women: a case control study. Am. J. Obstet. Gynecol. 187 (1), 99–102.

Okanishi, N., Kito, N., Akiyama, M., Yamamoto, M., 2012. Spinal curvature and characteristics of postural change in pregnant women. Acta Obstet. Gynecol. Scand. 91 (7), 856–861.

O'Mahony, F., Hofmeyr, G.J., Menon, V., 2010. Choice of instruments for assisted vaginal delivery. Cochrane Database Syst. Rev. (11), CD005455.

Osman, I., Young, A., Ledingham, M.A., Thomson, A.J., Jordan, F., Greer, I.A., Norman, J.E., 2003. Leukocyte density and pro-inflammatory cytokine expression in human fetal membranes, decidua, cervix and myometrium before and during labour at term. Mol. Hum. Reprod. 9 (1), 41–45.

Parente, M.P., Jorge, R.M., Mascarenhas, T., Fernandes, A.A., Martins, J.A., 2009. The influence of an occipito-posterior malposition on the biomechanical behavior of the pelvic floor. Eur. J. Obstet. Gynecol. Reprod. Biol. 144 (Suppl 1), S166–S169.

Parnell, C., Langhoff-Roos, J., Iversen, R., Damgaard, P., 1993. Pushing method in the expulsive phase of labor. A randomized trial. Acta Obstet. Gynecol. Scand. 72 (1), 31–35.

Parnell, C., Langhoff-Roos, J., Moller, H., 2001. Conduct of labor and rupture of the sphincter ani. Acta Obstet. Gynecol. Scand. 80 (3), 256–261.

Perlman, S., Raviv-Zilka, L., Levinsky, D., Gidron, A., Achiron, R., Gilboa, Y., Kivilevitch, Z., 2019. The birth canal: correlation between the pubic arch angle, the interspinous diameter, and the obstetrical conjugate: a computed tomography biometric study in reproductive age women. J. Matern. Fetal Neonatal. Med. 32 (19), 3255–3265.

Rahn, D.D., Ruff, M.D., Brown, S.A., Tibbals, H.F., Word, R.A., 2008. Biomechanical properties of the vaginal wall: effect of pregnancy, elastic fiber deficiency, and pelvic organ prolapse. Am. J. Obstet. Gynecol. 198 (5), 590.

Raisanen, S., Selander, T., Cartwright, R., Gissler, M., Kramer, M.R., Laine, K., Heinonen, S., 2014. The association of episiotomy with obstetric anal sphincter injury–a population based matched cohort study. PLoS ONE 9 (9), e107053.

Reimers, C., Staer-Jensen, J., Siafarikas, F., Saltyte-Benth, J., Bo, K., Ellstrom Engh, M., 2016. Change in pelvic organ support during pregnancy and the first year postpartum: a longitudinal study. BJOG 123 (5), 821–829.

Rusavy, Z., Kalis, V., Aglyamov, S., Egorov, V., 2020. Feasibility and safety of antepartum tactile imaging. Int. Urogynecol. J.

Samuelsson, E., Ladfors, L., Wennerholm, U.B., Gareberg, B., Nyberg, K., Hagberg, H., 2000. Anal sphincter tears: prospective study of obstetric risk factors. BJOG 107 (7), 926–931.

Schauberger, C.W., Rooney, B.L., Goldsmith, L., Shenton, D., Silva, P.D., Schaper, A., 1996. Peripheral joint laxity increases in pregnancy but does not correlate with serum relaxin levels. Am. J. Obstet. Gynecol. 174 (2), 667–671.

Schuit, E., Kwee, A., Westerhuis, M.E., Van Dessel, H.J., Graziosi, G.C., Van Lith, J.M., Nijhuis, J.G., Oei, S.G., Oosterbaan, H.P., Schuitemaker, N.W., et al., 2012. A clinical prediction model to assess the risk of operative delivery. BJOG 119 (8), 915–923.

Shek, K.L., Dietz, H.P., Kirby, A., 2010. The effect of childbirth on urethral mobility: a prospective observational study. J. Urol. 184 (2), 629–634.

Shynlova, O., Lee, Y.H., Srikhajon, K., Lye, S.J., 2013. Physiologic uterine inflammation and labor onset: integration of endocrine and mechanical signals. Reprod. Sci. 20 (2), 154–167.

Shynlova, O., Nadeem, L., Zhang, J., Dunk, C., Lye, S., 2020. Myometrial activation: novel concepts underlying labor. Placenta 92, 28–36.

Siafarikas, F., Staer-Jensen, J., Hilde, G., Bo, K., Ellstrom Engh, M., 2014. Levator hiatus dimensions in late pregnancy and the process of labor: a 3- and 4-dimensional transperineal ultrasound study. Am. J. Obstet. Gynecol. 210 (5), 484.

Staer-Jensen, J., Siafarikas, F., Hilde, G., Benth, J.S., Bo, K., Engh, M.E., 2015. Postpartum recovery of levator hiatus and bladder neck mobility in relation to pregnancy. Obstet. Gynecol. 125 (3), 531–539.

Stedenfeldt, M., Pirhonen, J., Blix, E., Wilsgaard, T., Vonen, B., Oian, P., 2012. Episiotomy characteristics and risks for obstetric anal sphincter injuries: a case-control study. BJOG 119 (6), 724–730.

Sultan, A.H., Thakar, R., Ismail, K.M., Kalis, V., Laine, K., Raisanen, S.H., de Leeuw, J.W., 2019. The role of mediolateral episiotomy during operative vaginal delivery. Eur. J. Obstet. Gynecol. Reprod. Biol. 240, 192–196.

Takmaz, T., Aydin, S., Gorchiyeva, I., Karasu, A.F.G., 2020. The usual suspect: cross-sectional study of fundal pressure at second stage of delivery and the association with pelvic floor damage. Int. Urogynecol. J.

Toozs-Hobson, P., Balmforth, J., Cardozo, L., Khullar, V., Athanasiou, S., 2008. The effect of mode of delivery on pelvic floor functional anatomy. Int. Urogynecol. J. Pelvic Floor Dysfunct. 19 (3), 407–416.

Valsky, D.V., Lipschuetz, M., Bord, A., Eldar, I., Messing, B., Hochner-Celnikier, D., Lavy, Y., Cohen, S.M., Yagel, S., 2009. Fetal head circumference and length of second stage of labor are risk factors for levator ani muscle injury, diagnosed by 3-dimensional transperineal ultrasound in primiparous women. Am. J. Obstet. Gynecol. 201 (1), 91.

van Bavel, J., Hukkelhoven, C., de Vries, C., Papatsonis, D.N.M., de Vogel, J., Roovers, J.W.R., Mol, B.W., de Leeuw, J.W., 2018. The effectiveness of mediolateral episiotomy in preventing obstetric anal sphincter injuries during operative vaginal delivery: a ten-year analysis of a national registry. Int. Urogynecol. J. 29 (3), 407–413.

van Delft, K., Sultan, A.H., Thakar, R., Schwertner-Tiepelmann, N., Kluivers, K., 2014. The relationship between postpartum levator ani muscle avulsion and signs and symptoms of pelvic floor dysfunction. BJOG 121 (9), 1164–1171. Discussion 1172.

van Delft, K., Thakar, R., Sultan, A.H., Kluivers, K.B., 2015. Does the prevalence of levator ani muscle avulsion differ when assessed using tomographic ultrasound imaging at rest vs on maximum pelvic floor muscle contraction? Ultrasound Obstet. Gynecol. 46 (1), 99–103.

van Veelen, G.A., Schweitzer, K.J., van Hoogenhuijze, N.E., van der Vaart, C.H., 2015. Association between levator hiatal dimensions on ultrasound during first pregnancy and mode of delivery. Ultrasound Obstet. Gynecol. 45 (3), 333–338.

Vannuccini, S., Bocchi, C., Severi, F.M., Challis, J.R., Petraglia, F., 2016. Endocrinology of human parturition. Ann. Endocrinol. (Paris) 77 (2), 105–113.

Vollestad, N.K., Torjesen, P.A., Robinson, H.S., 2012. Association between the serum levels of relaxin and responses to the active straight leg raise test in pregnancy. Man. Ther. 17 (3), 225–230.

Yeates, D.A., Roberts, J.E., 1984. A comparison of two bearing-down techniques during the second stage of labor. J. Nurse Midwifery 29 (1), 3–11.

Youssef, A., Salsi, G., Cataneo, I., Pacella, G., Azzarone, C., Paganotto, M.C., Krsmanovic, J., Montaguti, E., Cariello, L., Bellussi, F., et al., 2019. Fundal pressure in second stage of labor (Kristeller maneuver) is associated with increased risk of levator ani muscle avulsion. Ultrasound Obstet. Gynecol. 53 (1), 95–100.

Zemcik, R., Karbanova, J., Kalis, V., Lobovsky, L., Jansova, M., Rusavy, Z., 2012. Stereophotogrammetry of the perineum during vaginal delivery. Int. J. Gynaecol. Obstet. 119 (1), 76–80.

PART

2

Mechanical properties – constitutive laws – experimental characterizations

CHAPTER

Inverse problems in the characterization of soft connective tissue: perspective for reproduction system

5

Stéphane Avril

Mines Saint-Étienne, Université Jean Monnet, INSERM, U 1059 SAINBIOSE, Saint-Étienne, France

5.1 Introduction

Computational biomechanics aims at predicting the mechanical response of living systems: given for instance a complete description of the aorta and of the surrounding tissues, we can predict the deformations that are going to be induced by the deployment of a stent graft during surgical repair of an aneurysm (Perrin et al., 2016, 2015a,b). More generally, we could predict the deformations of a hollow organ under the action of radial forces that could be applied for a measurement purpose using an appropriate intraluminal device. This problem of predicting the result of measurements is called the simulation problem, or the forward problem.

The inverse problem consists of using the actual result of some measurements to infer the values of the parameters that characterize the system, such as unknown parameters of the material model, unknown elements of the boundary conditions, or even sometimes the unknown initial geometry of the solid before the application of any mechanical action (loadfree configuration in finite deformations).

Whereas the forward problem has a unique solution, the inverse problem does not. As an example, consider measurements of deformations in the wall of a hollow sphere, representing an idealized hollow organ, after the application of internal pressure. Given the precise description of the wall and its mechanical behavior, and given the value of support stiffness surrounding the hollow organ (boundary conditions), we can uniquely predict the values of the radial deformation (forward problem), but there are different combinations of wall mechanical properties and surrounding support stiffness that give exactly the same radial deformation. Therefore the inverse problem—inferring the mechanical properties of the organ and its surrounding tissues from observations of the radial deformation provoked by a change of internal pressure—has multiple solutions (in fact, an infinite number). Because of this, in the

Biomechanics of the Female Reproductive System: Breast and Pelvic Organs. https://doi.org/10.1016/B978-0-12-823403-7.00017-8

inverse problem, one needs to make explicit any available *a priori* information on the model parameters.

Formally, an inverse problem implies the reconstruction of complete unknown fields (for instance the field of unknown material properties) and is usually illposed, which means that existence and uniqueness of the solution are not always guaranteed. The illposedness may be due to a lack of reliable data and/or to an overcomplexity of the model. When access to more reliable data and complexity reduction of the model are not possible in practice, illposedness may be overcome mathematically by resorting to regularization approaches. Solving such inverse problems implies the definition of a cost function, estimating the distance between the model predictions and the measurements. The cost function is minimized in the least-squares sense. In general situations, the model is solved numerically using a finite-element model updating technique (FEMU).

A subcategory of inverse problem is made by identification problems. An identification problem has a finite number of unknowns and may become well-posed. This may happen in very specific situations, for instance, when full-field measurements are available (Avril et al., 2008a). In that specific case, an alternative to FEMU is possible: the virtual fields method (VFM), which has been shown to be more robust and efficient in these situations (Pierron and Grédiac, 2012).

In this chapter, we first review different situations in computational biomechanics, where an inverse problem or an identification problem has to be solved, with a particular focus on the reproduction system. After this presentation, we present in detail the FEMU technique, which is traditionally used to solve inverse problems. We then present sequential method based on the principle of Kalman filters for dynamic problems. Finally, we present in detail the VFM and show different cases of its possible application. Eventually, research directions are proposed for the emerging field of pelvic biomechanics and identification of biomechanical properties using advanced imaging techniques of soft tissue strains.

5.2 Common sources of inverse problem in soft tissue biomechanics

For the sake of illustration and to facilitate focus on problems on which a significant research effort has been focused, the current chapter is restricted to two main situations:

1. Reconstruction of the field of material properties in soft tissues using medical imaging techniques that are applied *in vivo*. This situation is named "elastography of soft tissue biomechanics";
2. Identification of material constants driving the constitutive equations of soft tissues using dedicated testing, where a stress strain curve fitting cannot be simply performed by reading load and displacements measurements. This situation is

named "non-agreed standard testing of soft tissue biomechanics" and mostly concerns soft tissues in *ex situ* conditions.

5.2.1 Elastography of soft tissue biomechanics

This situation occurs when one wants to derive mechanical properties of soft tissues *in vivo* and *in situ*. The basic steps are the following: (1) subject the tissues (organs or region of the body) to a deformation, (2) measure the displacement field in the entire domain, and (3) compute the mechanical properties by solving an inverse problem.

5.2.1.1 Subject the tissues to a deformation

The most common elastography technique, strain elastography (or compression elastography), relies on manually deforming a tissue with an ultrasound transducer, and inferring the stiffness of the tissue from the deformation observed in the ultrasound images. An extension of ultrasound strain elastography is acoustic radiation force impulse (ARFI) elastography. Instead of manually compressing the tissue, ARFI creates a focused ultrasound pulse that travels through the tissue. Using rapid imaging ultrasound pulse echoes, the tissue deformation can be tracked, and analyzed to provide a qualitative stiffness map. Its main advantage over strain elastography is that it is not limited to superficial tissues that can be manually deformed, although it is still limited by ultrasound penetration depths.

The most common alternative to ultrasound elastography is magnetic resonance elastography (MRE), which is a magnetic resonance imaging technique that works by measuring the shear wave propagation through soft tissues (Muthupillai and Ehman, 1996; Bensamoun et al., 2006). The shear wave is generated by an MR-compatible mechanical transducer, which is synchronized to the MR image acquisition.

In vivo, noninvasive stimuli can also be employed; the most used one being the natural blood action on the arterial wall (Avril et al., 2009b).

5.2.1.2 Measure the displacement fields

The displacement field may be determined using either ultrasound (see Ophir et al., 1991, for an example), computed tomography (CT) (Liu and Morgan, 2007) or nuclear magnetic resonance (NMR) (see Muthupillai et al., 1995, for an example). When using NMR, displacements are calculated from the phase of the measured magnetic field and are required to be time harmonic.

In the case of ultrasound, a speckle image of the specimen in the undeformed state is recorded. Thereafter, the specimen is deformed and another speckle image is recorded. These images are registered to yield the displacement field. More recently, efforts to provide quantitative mechanical property data from ultrasound elastography led to the development of shear wave elastography, which relies on similar physical principles to MRE, where propagation of an externally imposed acoustic wave is tracked, and used to estimate the tissue shear modulus. This stimulus can be generated by ARFI, and its propagation imaged with high frame rate ultrasound, as in the supersonic elastography technique (Bercoff et al., 2004).

5.2.1.3 Compute the mechanical properties

Strain or compression static elastography usually assumes that regions of smaller strains indicate higher elastic modulus (Bilston and Tan, 2015). It is rather a technique for detecting variations in relative stiffness. The use of inversion techniques has been shown to be more robust (Van Houten et al., 2000; Avril et al., 2008c), but the uncertainty in the boundary conditions and in the viscoelastic effects led static elastography to be preferred by transient elastography.

Transient US or MR elastography attempts to estimate the elastic properties (Young's modulus or shear modulus) of a tissue based on the relationship between mechanical properties and the propagation characteristics of mechanical vibration waves (Ophir et al., 1991; Van Houten et al., 1999). The inverse problem consists in deriving the local tissue modulus from the measured displacement fields using the wave equation written in the context of linear elasticity, such as

$$\rho \frac{\partial^2 \boldsymbol{U}}{\partial t^2} = \mu \nabla^2 \boldsymbol{U} + (\lambda + \mu) \nabla (\nabla \boldsymbol{U}) + \eta \frac{\partial \nabla^2 \boldsymbol{U}}{\partial t} + (\xi + \eta) \frac{\partial \nabla (\nabla \boldsymbol{U})}{\partial t}, \tag{5.1}$$

where $\boldsymbol{U}$ is the displacement vector, ρ is the density of the tissue, μ is the shear modulus, λ the second Lamé coefficient, η the shear viscosity, and ξ is the viscosity of the compressional wave. The variable t refers to the time. Despite some concerns over the validity of this approach, a much simplified version is often used, which estimates local wave speed from the imaging data, and calculates the shear modulus from

$$\mu = \rho \nu^2 \lambda^2, \tag{5.2}$$

where ν is the frequency of the vibration and λ is the local wavelength.

Whereas most applied studies use some form of direct numerical inversion, such as the one of Eq. (5.2), recent work is exploring the use of inversion techniques that have the potential to be more robust (Van Houten et al., 2000; Kauer et al., 2002; Doyley et al., 2000; Oberai et al., 2003; Goenezen et al., 2011, 2012; Drakonaki et al., 2012). They will be presented in detail in the next sections of this chapter.

5.2.1.4 Applications in the pelvic system

Strain elastography of the pelvic floor carries diagnostically important information about the dynamic response of the pelvic floor muscle, which cannot be readily captured and assimilated by the observer during the scanning process. Peng et al. (2007) an ultrasound imaging approach based on motion tracking quantitatively analyze the dynamic parameters of pelvic floor muscles on the ano-rectal angle. Ami et al. (2009) showed that real-time ultrasonic strain elastography could provide detailed mapping and characterization of fibroids. Thyer et al. (2008) introduced a static-state translabial ultrasound method of measuring pubovisceral muscle strain during Valsalva and contraction.

Transient ultrasound (O'Hara et al., 2019; Swiatkowska-Freund and Preis, 2011) and MR (Stewart et al., 2011) elastography is also emerging for the pelvic system. It has been used to identify biological and technical confounders in the non-pregnant cervix when applying shear wave elastography with an endovaginal transducer (O'Hara et al., 2019; Swiatkowska-Freund and Preis, 2011). The frequency-dependent elastic moduli of human uterine tissue have also been characterized by Kiss et al. (2006), and some promising results have been obtained with *ex vivo* uteri by Hobson et al. (2007).

5.2.2 Nondestructive invasive techniques

Whereas elastography seeks at imaging strains or material properties in soft tissues, less sophisticated techniques can simply apply some load on a tissue and measure the induced deformation at a given point. This can also be a source of inverse problems, which will not be presented in this chapter, as we eventually focus on problems involving imaging techniques and full-field measurements. We simply mention the well-known palpation technique and the technique of tissue aspiration, both involved in an interesting study related to the reproduction system conducted by Egorov et al. (2019). Their approach, called vaginal tactile imaging, allowed biomechanical mapping of the female pelvic floor to quantify tissue elasticity, pelvic support, and pelvic muscle functions. An explicit axisymmetric finite-element simulation of the aspiration experiment was used together with a Levenberg–Marquardt algorithm to estimate the material model parameters in an inverse parameter determination process (as described in Section 5.3.3.1).

5.2.3 Nonagreed standard testing of soft tissue biomechanics

Thousands of scientific publications provide material properties of soft tissues that were obtained by characterizing test samples excised from a dead body (cadaver). This is a source of many inverse problems (Fung, 2013; Humphrey, 2013; Avril and Evans, 2017). Indeed, for many technical materials, it is relatively easy to make test samples according to the agreed standards, but for most biological materials this is much more difficult due to a number of reasons listed below:

1. To create a sample, it has to be taken out of a body. If it concerns human tissue, it is either left over material of surgery or post mortem material. Another option is to use tissue from animals. In all situations, the material is taken out of the body, so it loses part of its integrity, the pretension found in the living system is (partly) gone or difficult to maintain, the tissue is no longer supplied with blood, and will start to deteriorate quickly. However, when isolated and preserved in a proper way, the negative effects of making samples of biological materials can be reduced considerably and the parameters may change, but the physical behavior will be similar to the behavior *in vivo*. Also during testing the physical and chemical environment of the material has to be controlled (humidity, temperature).
2. It is extremely difficult to make samples according to the standards (for example a dogbone shape of a soft biological material). The amount of material available is usually small or very small. Clamping is a big issue.

3. Often the material has inhomogeneous properties, so the assumption in standard tests that the stress strain field is homogeneous is not a valid assumption.
4. Especially for soft biological materials, the samples can be stretched to very high deformations; they have a highly nonlinear and visco-elastic behavior so the strain and strain rate history play an important role, and localization effects appear even if they have to be neglected in standard tests.

Despite the difficulties given above, standard tests are done quite often, because they are well defined, and they constitute the only way to gain a good understanding of the physical behavior of the material and to define constitutive equations (Fung, 2013). Some of the above problems can be circumvented or partly solved by using inverse methods. The biggest advantage of inverse methods is the enormous freedom that is created to design experiments, alleviating some of the difficulties encountered in standard tests. The main inverse methods will be presented in detail in the sections that follow.

Concerning the reproduction system, as for other tissues, knowledge of biomechanical properties is critical to further developing accurate surgical techniques and physiological prosthetic materials. The elastic properties of the pubovisceral muscle are likely to be important for pelvic organ support. However, published data are scarce. We mention here the recent biomechanical uniaxial tension tests performed on pelvic floor tissues (ligaments and organs) by Chantereau et al. (2014), who characterized the mechanical properties of young pelvic soft tissues.

5.3 The finite element model updating method

5.3.1 Useful definitions and concepts

5.3.1.1 Deformation tensors

Deformations are mathematically described as functions, which map the material description $\boldsymbol{X}$ (spatial description in the reference configuration) onto the spatial description $\boldsymbol{x}$ in the current configuration, such as

$$\boldsymbol{x} = \boldsymbol{\phi}(\boldsymbol{X}, t), \tag{5.3}$$

$$\boldsymbol{X} = \boldsymbol{\phi}^{-1}(\boldsymbol{x}, t). \tag{5.4}$$

The displacement vector is defined as $\boldsymbol{U}(\boldsymbol{X}, t) = \boldsymbol{x} - \boldsymbol{X} = \boldsymbol{\phi}(\boldsymbol{X}, t) - \boldsymbol{X}$.

The velocity is defined by taking time derivatives of the mapping as follows:

$$\dot{\boldsymbol{U}}(\boldsymbol{X}, t) = \left[\frac{\partial \boldsymbol{\phi}(\boldsymbol{X}, t)}{\partial t}\right]_X. \tag{5.5}$$

It can also be expressed in terms of the spatial description by inserting the inverse mapping

$$\boldsymbol{v}(\boldsymbol{x}, t) = \dot{\boldsymbol{U}}\left(\boldsymbol{\phi}^{-1}(\boldsymbol{x}, t), t\right). \tag{5.6}$$

The deformation gradient for this motion is

$$F(X,t) = \frac{\partial \phi(X,t)}{\partial x}. \tag{5.7}$$

The right Cauchy–Green and left Cauchy–Green stretch tensor are, respectively,

$$C = F^T F, \tag{5.8}$$

$$B = F F^T. \tag{5.9}$$

The Green–Lagrange strain tensor is

$$E = \frac{1}{2}(C - \mathbf{1}), \tag{5.10}$$

where **1** is the identity tensor.

5.3.1.2 Stress tensors

The first stress tensor that has to be introduced is the Cauchy stress tensor σ. It gives us the stress state in the deformed body. It is defined in the spatial configuration. The traction vector t obtained from the application of the surface normal in the spatial (deformed) configuration n is called the Cauchy traction vector

$$t = \sigma . n. \tag{5.11}$$

Since it gives us the actual stress of the body, the Cauchy stress is called the true stress in engineering.

It is convenient to define the second Piola-Kirchhoff stress tensor S through a push-forward, such as

$$S = J F^{-1} \sigma F, \tag{5.12}$$

where $J = \det(F)$. The S tensor is defined for the reasons that it is a material tensor, which is defined in the reference configuration and that it is the work conjugate of the Green–Lagrange strain tensor E, enabling one to define different constitutive equations. It also has the attractive property that it is symmetric.

5.3.1.3 Constitutive equations

For the sake of illustration, we restrict the presentation of inversion techniques of this chapter to the identification of a single specific hyperelastic model often used for soft biological tissues of the reproduction system. In hyperelasticity, the existence of a strain energy density function is assumed, from which a constitutive relation between stress and strain is derived. The total energy that is needed to deform the body is only dependent on the initial and the end state, that is, the state of the body is independent of the loading path.

Let us illustrate this throughout the whole chapter with the following strain energy function defined for incompressible solids (Gasser et al., 2006):

$$\psi = \frac{\mu_1}{2}(I_1 - 3) + \frac{\mu_2}{4\gamma}\left(e^{\gamma(I_\kappa - 1)^2} - 1\right), \tag{5.13}$$

where μ_1, μ_2, and γ are material parameters, I_1 is an invariant defined as

$$I_1 = tr\,(\boldsymbol{C}), \tag{5.14}$$

and I_κ is a compound invariant consisting of isotropic and anisotropic contributions defined as

$$I_\kappa = \boldsymbol{C} : (\kappa \mathbf{1} + (1 - 2\kappa)\boldsymbol{M} \otimes \boldsymbol{M}), \tag{5.15}$$

where the unit vector $\boldsymbol{M} = \cos\theta \boldsymbol{e_1} + \sin\theta \boldsymbol{e_2}$ defines the orientation along which the tissue is stiffest, whereas κ characterizes the degree of anisotropy, varying between 0 and 1.

When $\kappa = 0$, it would model a composite with all the fibers perfectly aligned in the direction $\boldsymbol{M}$, and at $\kappa = 1$ the fibers would be perfectly aligned in the perpendicular direction, $\boldsymbol{M}^{\perp}$. Finally, $\kappa = 1/2$ models the case where fibers would have no preferential direction (isotropic). The parameters μ_1 and μ_2 are the effective stiffnesses of the matrix and fiber phases, respectively, both having dimensions of force per unit length. The γ parameter is a nondimensional parameter that governs the tissue's strain stiffening response.

The second Piola-Kirchhoff stress tensor, $\boldsymbol{S}$, is written as

$$\boldsymbol{S} = 2\frac{\partial \psi}{\partial I_1}\mathbf{1} + 2\frac{\partial \psi}{\partial I_\kappa}(\kappa \mathbf{1} + (1 - 2\kappa)\boldsymbol{M} \otimes \boldsymbol{M}) + p\boldsymbol{C}^{-1}, \tag{5.16}$$

where p is a Lagrange multiplier accounting for the condition of incompressibility.

It gives after substitution

$$\boldsymbol{S} = \mu_1 \mathbf{1} + \mu_2 e^{\gamma(I_\kappa - 1)^2}(\kappa \mathbf{1} + (1 - 2\kappa)\boldsymbol{M} \otimes \boldsymbol{M}) + p\boldsymbol{C}^{-1}. \tag{5.17}$$

The Cauchy stress is given by

$$\boldsymbol{\sigma} = \mu_1 \boldsymbol{B} + \mu_2 e^{\gamma(I_\kappa - 1)^2}(\kappa \boldsymbol{B} + (1 - 2\kappa)\boldsymbol{m} \otimes \boldsymbol{m}) + p\mathbf{1}, \tag{5.18}$$

where

$$\boldsymbol{m} = \boldsymbol{F}\boldsymbol{M}. \tag{5.19}$$

The material parameters μ_1, μ_2, γ, κ, and θ, including parameters describing their possible regional variations, are represented onwards with a unique vector denoted β.

5.3.2 Forward problem

The mechanical fields across the reference domain, denoted Ω_0, are governed by the equations of dynamics. The solution of these governing equations by a numerical method is the so-called forward problem. In this section, we will simplify the problem by leaving aside the acceleration forces (static problem). Dynamic problems will be discussed in the following sub-sections dedicated to sequential methods.

5.3.2.1 Strong form

The strong form of the quasi-static problem is as follows: find the displacement field $\boldsymbol{U}$ and the pressure p such as

$$\mathrm{Div}\,[\boldsymbol{F}\boldsymbol{S}] = \mathbf{0} \text{ on } \Omega_0, \tag{5.20a}$$

$$\boldsymbol{U} = \boldsymbol{g} \text{ on } \Gamma_g, \tag{5.20b}$$

$$\boldsymbol{F}\boldsymbol{S}.\boldsymbol{N} = \boldsymbol{t} \text{ on } \Gamma_t, \tag{5.20c}$$

where Ω_0 is the domain of interest, Γ_t is the boundary of the domain, where tractions are applied while Γ_g is the boundary of the domain where displacements are assigned. The whole boundary of the domain is defined as

$$\partial\Omega_0 = \Gamma_g \cup \Gamma_t. \tag{5.20d}$$

The equations are completed with the incompressibility constraint

$$\det(\boldsymbol{F}) - 1 = 0 \text{ on } \Omega_0. \tag{5.20e}$$

As mentioned above, this problem statement must be augmented by the stress–strain relation to specify the stress in terms of the deformation (Eq. (5.17) or (5.18)).

5.3.2.2 Weak form

The weak form can be derived easily from the strong form by multiplying Eq. (5.20) with a vector test function, integrating-by-parts over the reference domain, and utilizing the traction boundary condition. The incompressibility constraint is multiplied by another scalar test function and integrated over the reference domain. The weak form is given by the following: find $\underline{U} \equiv [\boldsymbol{U}, p]$ such that

$$\mathcal{A}\left(\underline{W}, \underline{U}; \beta\right) - (\boldsymbol{w}, \boldsymbol{t})_{\Gamma_t} = 0 \ \forall \underline{W} \equiv (\boldsymbol{w}, q) \in \mathrm{V} \times \mathrm{P}, \tag{5.21}$$

where $\mathcal{A}$ is an operator defined as

$$\mathcal{A}\left(\underline{W}, \underline{U}, \beta\right) = \int_{\Omega_0} \nabla\boldsymbol{w} : [\boldsymbol{F}\boldsymbol{S}]\,\mathrm{d}\Omega_0 + \int_{\Omega_0} (J-1)\,q\,\mathrm{d}\Omega_0, \tag{5.22a}$$

and $(.,.)_{\Gamma_t}$ is the $\mathrm{L}^2(\Gamma_t)$ inner product evaluated over the Γ_t reference boundary domain such as

$$(\boldsymbol{w}, \boldsymbol{t})_{\Gamma_t} = \int_{\Gamma_t} \boldsymbol{w}.\boldsymbol{t}\,\mathrm{d}S_0, \tag{5.22b}$$

whereas V, S, and P are vectorial spaces of second-order tensors defined as

$$\mathrm{V}=\left\{\boldsymbol{w} \mid \boldsymbol{w}_i \in \mathrm{H}^1\left(\Omega_0\right), \boldsymbol{w}_i=0 \text{ on } \Gamma_g\right\}, \tag{5.23}$$

$$\mathrm{S}=\left\{\boldsymbol{u} \mid \boldsymbol{u}_i \in \mathrm{H}^1\left(\Omega_0\right), \boldsymbol{u}_i=\boldsymbol{g}_i \text{ on } \Gamma_g\right\}, \tag{5.24}$$

$$P \subseteq \mathrm{L}^2\left(\Omega_0\right). \tag{5.25}$$

The weak form may be discretized by selecting finite-dimensional spaces $\mathrm{S}^h \subset \mathrm{S}$, $\mathrm{V}^h \subset \mathrm{V}$, and $\mathrm{P}^h \subset \mathrm{P}$ and using Galerkin's approach. This leads to a nonlinear algebraic problem for the displacement and pressure degrees of freedom that may be solved using the Newton–Raphson method. It can be solved uniquely for every feasible set of material parameters and arbitrary geometries and leads to a nonlinear system of equations of the type

$$\mathcal{A}(\underline{W}^h, \underline{U}^h; \beta)-\left(\boldsymbol{w}^h, \boldsymbol{t}^h\right)_{\Gamma_t}=0. \tag{5.26}$$

Starting with an initial vector $\underline{U}^h(0)$, the resolution consists in recursively linearizing the nonlinear system of equations around $\underline{U}^h(n)$ assuming

$$\underline{U}^h(n+1)=\underline{U}^h(n)+\epsilon \delta \underline{U}^h(n), \tag{5.27}$$

$$\mathcal{A}\left(\underline{W}^h, \underline{U}^h(n+1); \beta\right)=\mathcal{A}\left(\underline{W}^h, \underline{U}^h(n); \beta\right)+\epsilon \mathcal{B}\left(\underline{W}^h, \delta \underline{U}^h(n); \beta, \underline{U}^h(n)\right), \tag{5.28}$$

where $\mathcal{B}$ is an operator defined as

$$\mathcal{B}(\underline{W}, \delta \underline{U}; \beta, \underline{U})=\lim_{\epsilon \rightarrow 0} \frac{d}{d \epsilon} \mathcal{A}(\underline{W}, \underline{U}+\delta \underline{U}; \beta), \tag{5.29}$$

$$\mathcal{B}(\underline{W}, \delta \underline{U}; \beta, \underline{U})=\left(\underline{W}, \underline{\underline{K}}(\beta, \underline{U}) . \delta \underline{U}\right)_{\Omega_0}=\left(\underline{\underline{K}}^T(\beta, \underline{U}) . \delta \underline{U}, \underline{W}\right)_{\Omega_0}, \tag{5.30}$$

where $(.,.)_{\Omega_0}$ is the $\mathrm{L}^2(\Omega_0)$ inner product evaluated in the Ω_0 reference domain, $\underline{\underline{K}}(\beta, \underline{U})$ is a linear operator defined in the S × P space transforming any field $\underline{W}$ in $\underline{\underline{K}}(\beta, \underline{U}) . \underline{W}$, whereas $\underline{\underline{K}}^T(\beta, \underline{U})$ is the adjoint of $\underline{\underline{K}}(\beta, \underline{U})$.

5.3.3 Inverse problem

5.3.3.1 Definition of the cost function

The inverse problem consists in the reconstruction of the material and geometric parameters, β, given the relevant (measured) displacements fields $\boldsymbol{U}_{\text{meas}}$.

The inverse problem is stated as follows: given n_{meas} measured displacement fields $\boldsymbol{U}^1_{\text{meas}}, \boldsymbol{U}^2_{\text{meas}}, \ldots \boldsymbol{U}^n_{\text{meas}}$, find the N_β material properties $\beta=\left[\beta_1, \beta_2, \ldots, \beta_{N_\beta}\right]$

such that the objective function

$$\pi(\beta) = \frac{1}{2}\sum_{i=1}^{n_{\text{meas}}} w_i \left\| \boldsymbol{T}\boldsymbol{U}^i - \boldsymbol{T}\boldsymbol{U}^i_{\text{meas}} \right\|^2 + \frac{1}{2}\sum_{j=1}^{N_\beta} \alpha_j R(\beta_j) \tag{5.31}$$

is minimized under the constraint that the displacement fields satisfy the equilibrium equations written in their weak form in Eq. (5.21), or after discretization in Eq. (5.26) (for the sake of simplification, no difference is made anymore in the notation between discretized and continuous variables onwards). In Eq. (5.31), each field $\boldsymbol{U}^i$ corresponds to a different set of boundary conditions $\boldsymbol{t}^i$, for instance, corresponding to different time steps t_i of overall loading. Since we expect their magnitudes to be quite different at each step i, each displacement field is multiplied by a weighting factor w_i, which is selected to ensure that the contributions to the objective function from all measurements are of the same order. The tensor $\boldsymbol{T}$ is selected to possibly weight different components of displacement differently.

The second term is the regularization term, where α_j is the regularization parameter. This has to be chosen appropriately, depending on the noise level in the measured displacements and on the illposedness of the problem. The regularization term can be thought of as a penalty term in the objective function that ensures a certain smoothness to the reconstructed material properties. A possible function proposed by the group of Oberai and Barbone (Oberai et al., 2003; Goenezen et al., 2011, 2012) is

$$R(\beta_j) = \int_{\Omega_0} \sqrt{\left|\nabla \beta_j\right|^2 + c^2}\, d\Omega_0, \tag{5.32}$$

where $\nabla \beta_j$ is the gradient of material properties, and c is a coefficient controlling the regularization (Oberai et al., 2003; Goenezen et al., 2011, 2012).

The objective function defined in Eq. (5.31) is usually minimized using a gradient-based optimization method (Oberai et al., 2003; Goenezen et al., 2011, 2012). For this, we need an efficient way to compute the gradient of this function with respect to the material properties. If we represent the regional variations of material properties using N piece-wise linear finite-element shape functions, whose nodal values are the parameters of the inverse problem, and if we consider the use of gradient-based algorithms, we can show that a straightforward calculation of the gradient requires N solves the forward elasticity problem. This cost is computationally prohibitive for typical values of N. To circumvent this difficulty, the adjoint method, which requires only two solves (independent of N) to compute the gradient, is the most appropriate.

5.3.3.2 The adjoint method

To evaluate the gradient vector, the adjoint equations can be derived at the continuous or the discrete level. Changing β by a small (infinitesimal) amount $\delta\beta$ will cause the displacement fields $\boldsymbol{U}^i$ to change a small amount $\boldsymbol{\delta U}^i$, and the pressure p will change accordingly of a small amount δp. The relationship between $\boldsymbol{\delta U}^i$ and $\delta\beta$ can be obtained by differentiating Eq. (5.21) with respect to $\boldsymbol{U}$ and β. One obtains

$$\mathcal{B}\left(\underline{W}, \delta\underline{U}^i; \beta, \underline{U}^i\right) + \mathcal{C}\left(\underline{W}, \delta\beta; \underline{U}^i, \beta\right) = 0, \quad (5.33)$$

where $\mathcal{C}$ is an operator defined as

$$\mathcal{C}\left(\underline{W}, \delta\beta; \underline{U}^i, \beta\right) = \lim_{\epsilon \to 0} \frac{d}{d\epsilon} \mathcal{A}\left(\underline{W}, \underline{U}^i; \beta + \epsilon\delta\beta\right). \quad (5.34)$$

Similarly, differentiating the objective function gives

$$\mathrm{D}_\beta \pi\left(\beta, \delta\beta\right) = \sum_{i=1}^{n_{\text{meas}}} \left(w_i \boldsymbol{T}\boldsymbol{\delta}\boldsymbol{U}^i, \boldsymbol{T}\boldsymbol{U}^i - \boldsymbol{T}\boldsymbol{U}^i_{\text{meas}}\right)_{\Omega_0} + \frac{1}{2}\sum_{j=1}^{N_\beta} \alpha_j D_\beta R\left(\beta_j, \delta\beta_j\right), \quad (5.35)$$

where D_β denotes the differentiation with respect to the components of β.

To compute the gradient of our objective function, one could select a basis on which to represent $\delta\beta$. Then for each basis vector, one solves Eq. (5.33) for $\boldsymbol{\delta U}^i$. Finally, Eq. (5.35) can be used to compute the gradient for that single component of β. One would then select the next basis vector for $\delta\beta$, and repeat the process.

A particularly efficient alternative to the brute force method just described is to use the adjoint equations. The following (linear) boundary value problems is introduced for the n_{meas} functions $\underline{W}^i \in \mathrm{V} \times \mathrm{P}$ such as

$$\mathcal{B}\left(\underline{W}^i, \underline{V}; \beta, \underline{U}^i\right) + \left(w_i \boldsymbol{T}\boldsymbol{V}, \boldsymbol{T}\boldsymbol{U}^i - \boldsymbol{T}\boldsymbol{U}^i_{\text{meas}}\right)_{\Omega_0} = 0 \; \forall \underline{V} \equiv [\boldsymbol{V}, q] \in \mathrm{V} \times \mathrm{P}. \quad (5.36)$$

Introducing the linear operator $\underline{\underline{K}}^T\left(\beta, \underline{U}^i\right)$, one obtains

$$\left(\underline{\underline{K}}^T\left(\beta, \underline{U}^i\right).\underline{W}^i, \underline{V}\right)_{\Omega_0} + \left(w_i \boldsymbol{T}\boldsymbol{V}, \boldsymbol{T}\boldsymbol{U}^i - \boldsymbol{T}\boldsymbol{U}^i_{\text{meas}}\right)_{\Omega_0} = 0 \; \forall \underline{V} \equiv [\boldsymbol{V}, q] \in \mathrm{V} \times \mathrm{P}, \quad (5.37)$$

which finally can be written as

$$\underline{\underline{K}}^T\left(\beta, \underline{U}^i\right).\underline{W}^i = \boldsymbol{T}\boldsymbol{U}^i - \boldsymbol{T}\boldsymbol{U}^i_{\text{meas}}. \quad (5.38)$$

In terms of $\underline{W}^i$, we now may compute the gradient as follows: First, we note that since $\delta\underline{U}^i \in \mathrm{V} \times \mathrm{P}$, we may replace $\underline{V}$ in Eq. (5.37) with $\delta\underline{U}^i$. This gives

$$\mathcal{B}\left(\underline{W}^i, \delta\underline{U}^i; \beta, \underline{U}^i\right) + \left(w_i \boldsymbol{T}\boldsymbol{V}, \boldsymbol{T}\boldsymbol{U}^i - \boldsymbol{T}\boldsymbol{U}^i_{\text{meas}}\right)_{\Omega_0} = 0. \quad (5.39)$$

Similarly, since $\underline{W}^i \in \mathrm{V} \times \mathrm{P}$, we may replace $\underline{W}$ in Eq. (5.33) with $\underline{W}^i$. This gives

$$\mathcal{B}\left(\underline{W}^i, \delta\underline{U}^i; \beta, \underline{U}^i\right) + \mathcal{C}\left(\underline{W}^i, \delta\beta; \underline{U}^i, \beta\right) = 0. \quad (5.40)$$

We now subtract Eq. (5.40) from Eq. (5.39) to find

$$\left(w_i \boldsymbol{T}\boldsymbol{V}, \boldsymbol{T}\boldsymbol{U}^i - \boldsymbol{T}\boldsymbol{U}^i_{\text{meas}}\right)_{\Omega_0} = \mathcal{C}\left(\underline{W}^i, \delta\beta; \underline{U}^i, \beta\right). \tag{5.41}$$

Eq. (5.41) shows that once we have computed $\underline{W}^i$, we can directly evaluate the gradient of our data matching term for any number of $\delta\beta$ directions, without solving another boundary value problem. The final expression for the gradient results by substituting Eq. (5.41) into Eq. (5.35) to obtain

$$\mathrm{D}_\beta \pi\left(\beta, \delta\beta\right) = \sum_{i=1}^{n_{\text{meas}}} \mathcal{C}\left(\underline{W}^i, \delta\beta; \underline{U}^i, \beta\right) + \frac{1}{2}\sum_{j=1}^{N_\beta} \alpha_j D_\beta R\left(\beta_j, \delta\beta_j\right). \tag{5.42}$$

5.3.3.3 Minimization of the cost function and resolution of the inverse problem

In a finite-element implementation all variables, including displacement, pressure, and material properties, can be represented by linear finite element basis functions. The optimization variables are the nodal values of the material parameters, and the gradient with respect to these variables is given by Eq. (5.42) such as

$$\mathrm{g} = \mathrm{D}_\beta \pi\left(\beta, \delta\beta\right). \tag{5.43}$$

More precisely, we can represent with the same shape functions as for the displacement and the pressure fields. Then the discrete gradient vector $\nabla\pi(\beta)$ may be given by each of its j_A components, which are defined as

$$\begin{aligned} \mathrm{g}_{j_A}(\beta) &= \mathrm{D}_\beta \pi\left(\beta, \mathrm{N}_A(\mathrm{X})\,\mathrm{e}_j\right) \\ &= \sum_{i=1}^{n_{\text{meas}}} \mathcal{C}\left(\underline{W}^i, \mathrm{N}_A(\mathrm{X})\,\mathrm{e}_j; \underline{U}^i, \beta\right) + \frac{1}{2}\sum_{j=1}^{N_\beta} \alpha_j D_\beta R\left(\beta_j, \mathrm{N}_A(\mathrm{X})\,\mathrm{e}_j\right), \end{aligned} \tag{5.44}$$

where e_j is an N_β-dimensional vector with 1 in the jth component and zeros everywhere else.

In the case of a linear behavior, one can define the linear operator $\underline{\underline{K_j}}(\beta, \underline{U}^i)$, which operates over the S × P space according to

$$\lim_{\epsilon\to 0} \frac{d}{d\epsilon}\mathcal{A}\left(\underline{W}^i, \underline{U}^i; \beta + \epsilon \mathrm{N}_A(\mathrm{X})\,\mathrm{e}_j\right) = \left(\underline{W}^i, \underline{\underline{K_j}}(\beta, \underline{U}^i).\underline{U}^i\right)_{\Omega_0}. \tag{5.45}$$

Having an efficient method to derive the gradient of the cost function, an iterative algorithm can be implemented to find the optimum value. Starting with an initialization β_0, the algorithm of the steepest descent tries to find a scalar α such that the cost function $\pi(\alpha) = \pi(\beta_n + \alpha \mathrm{g}_{j_A}(\beta))$ reaches its minimum at $\beta_{n+1} = \beta_n + \alpha \mathrm{g}_{j_A}(\beta)$ (line search).

The initial guess β_0 for the material parameters is usually homogeneous, and then the gradient is calculated using the procedure described above, the whole process being repeated until convergence.

A more efficient algorithm than the steepest descent is the BFGS method. In numerical optimization, the Broyden–Fletcher–Goldfarb–Shanno (BFGS) algorithm is an iterative method for solving unconstrained nonlinear optimization problems. The search direction p_n at stage n is given by the solution of the analogue of the Newton equation

$$B_n p_n = -\nabla\pi\left(\beta_n\right), \tag{5.46}$$

where B_n is an approximation to the Hessian matrix, which is updated iteratively at each stage, and $\nabla\pi(\beta_n)$ is the gradient of the cost function evaluated at β_n. A line search in the direction p_n is then used to find the next point β_{n+1}. Instead of requiring the full Hessian matrix at the point β_{n+1} to be computed as B_{n+1}, the approximate Hessian at stage n is updated by the addition of two matrices according to

$$B_{n+1} = B_n + U_n + V_n. \tag{5.47}$$

Both U_n and V_n are symmetric rank-one matrices, but their sum is a rank-two update matrix. From an initial guess β_0 and an approximate Hessian matrix B_0, the following steps are repeated as β_n converges to the solution.

1. Obtain a direction p_n by solving

$$B_n p_n = -\nabla\pi\left(\beta_n\right). \tag{5.48}$$

2. Set $s_n = \alpha_n p_n$.
3. Perform a line search to find an acceptable stepsize α_n in the direction found in the first step, then update $\beta_{n+1} = \beta_n + s_n$.
4.

$$y_k = \nabla\pi\left(\beta_{n+1}\right) - \nabla\pi\left(\beta_n\right). \tag{5.49}$$

5.

$$B_{n+1} = B_n + \frac{y_n y_n^T}{y_n^T s_n} - \frac{B_n s_n s_n^T B_n}{s_n^T B_n s_n}. \tag{5.50}$$

Practically, B_0 can be initialized with $B_0 = I$, so that the first step will be equivalent to a gradient descent, but further steps are more and more refined by B_n, the approximation to the Hessian.

5.3.4 Applications to the pelvic system

FEMU was previously applied to the pelvic system by Silva et al. (2016, 2019). They used an inverse finite element analysis to calculate the Mooney–Rivlin constitutive model parameters for the passive mechanical behavior of the pelvic floor muscles.

The numerical model of the pelvic floor muscles and bones was built from magnetic resonance axial images acquired at rest. Note also that Kauer et al. (2002) applied FEMU with the tissue aspiration technique to estimate in vivo soft tissue material model parameters in the pelvic system.

5.4 Sequential methods

The FEMU approach presented in the previous section, which consists in minimizing a least-squares criterion, which includes a regularization term and the difference between the observations and the model prediction, is often called a variational approach. When one has to model a **dynamic** system, as in the case of fluid-structure interactions (FSI) for hemodynamics, one of the main difficulties of a variational approach, such as the FEMU method, lies in the iterative evaluation of the criterion, involving many solutions of the forward problem, and its gradient, typically adjoint-based, usually requires a laborious implementation.

Alternatively, a sequential approach, based on a generalization of the Kalman filter, can be implemented. With sequential algorithms, the model prediction is improved at every time instant by analyzing the discrepancy between the actual measurements and the model observation outputs (Bertoglio et al., 2012). In the fully linear case, it was proved that the Kalman filter gives the same result as a variational approach based on a least squares criterion.

Let $\boldsymbol{U}^i = \boldsymbol{U}(t_i)$ be the state variable at every time step t_i (typically $\boldsymbol{U}^i$ denotes in this section the finite element approximation of the displacements and the velocities in the soft tissue and possibly in a surrounding fluid). We assume that the state variable satisfies nonlinear dynamics equations, without any model uncertainties, and that this may bring the following equation:

$$\boldsymbol{U}^{i+1} = \boldsymbol{A}\left(\boldsymbol{U}^i, \boldsymbol{t}^{i+1}; \beta\right), \tag{5.51}$$

where $\boldsymbol{t}^{i+1}$ represents the loading and β represents the vector of model parameters to be identified.

We suppose that measurement observations are available at t_i,

$$\boldsymbol{U}_{\text{meas}}(t_i) = \boldsymbol{U}^i_{\text{meas}} = \boldsymbol{H}\boldsymbol{U}^i + \boldsymbol{\zeta}(t_i), \tag{5.52}$$

where $\boldsymbol{H}$ represents an observation operator applied to the real state variable $\boldsymbol{U}^i = \boldsymbol{U}(t_i)$, and $\boldsymbol{\zeta}(t_i)$ includes the measurement noise and the discretization error. Measurements are assumed to be available at every time step t_i, $1 \le i \le n_{\text{meas}}$.

In a sequential approach, the inverse problem can be viewed as minimizing a cost function at every time step t_i,

$$\pi(\boldsymbol{U}^{i+1}, \beta^{i+1}) = \frac{1}{2}\left\| \boldsymbol{U}^{i+1}_{\text{meas}} - \boldsymbol{H}\boldsymbol{U}^{i+1} \right\|^2_{W^{-1}} + \frac{1}{2}\left\| \beta^{i+1} - \beta^i \right\|^2_{\left(P_i^\beta\right)^{-1}}$$

$$
+\frac{1}{2}\left\|U^{i+1}-A\left(U^{i},t^{i+1};\beta^{i}\right)\right\|^{2}_{\left(P_i^U\right)^{-1}}. \tag{5.53}
$$

In this expression, $\| \|^2_{W^{-1}}$, $\| \|^2_{\left(P_i^\beta\right)^{-1}}$, and $\| \|^2_{\left(P_i^U\right)^{-1}}$ denote some norms used to measure the observations, the parameters, and the state, respectively. These norms give a different weight to the different terms and therefore account for the "confidence" in the different quantities. From a statistical viewpoint, the "confidence" can be viewed as the inverse of covariance matrices (W, P_i^β and P_i^U), which explains the notation. It is important to notice that P_i^β and P_i^U are updated at every time step as the confidence in the model increases. The cost function may be rewritten as

$$
\begin{aligned}
\pi\left(U^{i+1},\beta^{i+1}\right)=&\frac{1}{2}\left(U^{i+1}_{\text{meas}}-HU^{i+1}\right)^{T}W^{-1}\left(U^{i+1}_{\text{meas}}-HU^{i+1}\right)\\
&+\frac{1}{2}\left(\beta^{i+1}-\beta^{i}\right)^{T}\left(P_i^\beta\right)^{-1}\left(\beta^{i+1}-\beta^{i}\right)\\
&+\frac{1}{2}\left(U^{i+1}-A\left(U^{i},t^{i+1};\beta^{i}\right)\right)^{T}\left(P_i^U\right)^{-1}\\
&\times\left(U^{i+1}-A\left(U^{i},t^{i+1};\beta^{i}\right)\right).
\end{aligned} \tag{5.54}
$$

The sequential approach, also known as filtering, addresses the minimization problem in the following way: it modifies the predictions ($\beta^{i+1}=\beta^i$ and $U^{i+1}=A\left(U^i,t^{i+1};\beta^i\right)$) with a correction term that takes into account the discrepancy between actual measurements and observations generated by the model such as

$$
U^{i+1}=A\left(U^{i},t^{i+1};\beta^{i}\right)+K^{U}\left(U^{i+1}_{\text{meas}}-HA\left(U^{i},t^{i+1};\beta^{i}\right)\right), \tag{5.55}
$$

$$
\beta^{i+1}=\beta^{i}+K^{\beta}\left(U^{i+1}_{\text{meas}}-HA\left(U^{i},t^{i+1};\beta^{i}\right)\right). \tag{5.56}
$$

The quantity $U^{i+1}_{\text{meas}}-HA\left(U^i,t^{i+1};\beta^i\right)$ is known as the innovation, and the operators K^U and K^β depend on the method. For linear problems, the most famous sequential approach is the Kalman filter.

Concerning the computational complexity: whereas the variational method has to solve several forward and adjoint problems on the whole interval $[t_0, t_{n_{\text{meas}}}]$, the estimation in the sequential algorithm is computed by solving only once the filtered dynamics. However, the optimal the operators K^U and K^β are determined by operations (multiplications, inversions, etc.) involving full matrices of the size of the state and the observations, which may make Kalman-based filters prohibitive for discrete problems derived from partial differential equations (PDEs).

The Kalman filter presented above is only valid when the dynamics and the observation operator are linear. In case of nonlinear operator, as it is the case for the hyperelastic identification problem considered in this chapter, the most straightforward extension consists in deriving tangent operators at every time step. This is the extended Kalman filter. However, it has two drawbacks: the computation of tangent operators and the precision of the estimated values. The latter is critical as it can be shown that when $\boldsymbol{\zeta}(t_i)$ is a white Gaussian noise, the nonlinear operators propagate a bias in the results. This is commonly circumvented by using the unscented Kalman Filter (Bertoglio et al., 2012), which instead of deriving tangent operators, approximates the operators $\boldsymbol{K}^U$ and $\boldsymbol{K}^\beta$ from the values taken by the nonlinear operators at a set of vectors chosen initially and called particles.

5.5 The virtual fields method

5.5.1 General introduction

The virtual fields method (VFM) was developed to identify the parameters governing constitutive equations, the experimental data processed for this purpose being displacement or strain fields. It will be shown in this chapter that one of its main advantages is the fact that, in several cases, the sought parameters can be directly found from the measurements, without resorting to finite-element analysis.

The VFM relies on the principle of virtual power (PVP), which is written with particular virtual fields. It represents in fact the weak form the local equations of equilibrium, which are classically introduced in mechanics of deformable media. Assuming a quasi-static transformation (absence of acceleration forces) and assuming the absence of body forces, the PVP can be written as follows for any domain defined by its volume Ω_{t_i} in the current configuration and by its external boundary $\partial\Omega_{t_i}$:

$$\underbrace{-\int_{\Omega_{t_i}} \boldsymbol{\sigma} : (\nabla \otimes \boldsymbol{v}^*) d\omega}_{W^*_{int}} + \underbrace{\int_{\partial\Omega_{t_i}} \boldsymbol{t}^i . \boldsymbol{v}^* ds}_{W^*_{ext}} = 0, \tag{5.57}$$

where $\boldsymbol{\sigma}$ is the Cauchy stress tensor, $\boldsymbol{v}^*$ is a virtual velocity field defined across the volume of the solid, $\nabla \otimes \boldsymbol{v}^*$ is the gradient of $\boldsymbol{v}^*$, $\boldsymbol{t}^i$ are the tractions across the boundary (surface denoted $\partial\Omega_{t_i}$), W^*_{int} is the virtual power of internal forces, and W^*_{ext} is the virtual power of external forces.

A very important property is in fact that the equation above is satisfied for any kinematically admissible (KA) virtual field $\boldsymbol{v}^*$. By definition, a KA virtual field must satisfy the boundary conditions of the actual velocity field to cancel the contribution of the resulting forces on the portion of the boundary along which actual displacement are prescribed. It must be pointed out that this requirement is not really necessary in all cases, but this point is not discussed here for the sake of simplicity. KA virtual fields are also assumed to be $\mathfrak{C}^0$ functions.

The principle of virtual power (PVP) has been used for the identification of material properties since 1990 (Grédiac, 1989) through the virtual fields method (VFM), which is an inverse method based on the use of full-field deformation data. The first step of the VFM consists of introducing the constitutive equations. In the case of hyperelasticity, and still neglecting acceleration forces, Eq. (5.57) becomes

$$-\int_{\Omega_{t_i}} \left(J^i\right)^{-1} \boldsymbol{F}^i \frac{\partial \psi}{\partial \boldsymbol{E}^i} \left(\boldsymbol{F}^i\right)^T : \left(\nabla \otimes \boldsymbol{v}^*\right) d\Omega_{t_i} + \int_{\partial \Omega_{t_i}} \boldsymbol{t}^i . \boldsymbol{v}^* ds = 0, \qquad (5.58)$$

where $\boldsymbol{F}^i$ and $\boldsymbol{E}^i$ are, respectively, the Cauchy–Green stretch and the Green–Lagrange strain tensors derived from the measured displacement field $\boldsymbol{U}^i$ at time step t_i, and $J^i = \det(\boldsymbol{F}^i)$. This supposes implicitly that $\boldsymbol{F}^i$ can be reconstructed across the whole domain using the measured displacement fields (Avril et al., 2008b, 2009a).

Eq. (5.58) being satisfied for any KA virtual field, any new KA virtual field provides a new equation. The VFM relies on this property by writing this equation above with a set of KA virtual fields chosen *a priori* (Pierron and Grédiac, 2012). The number of virtual fields and their type depend on the nature of the strain energy function. Two different cases can be distinguished:

1. the strain energy density function depends linearly on the sought parameters. Writing Eq. (5.58) with as many virtual fields as unknowns leads to a system of linear equations, which provides the sought parameters after inversion.
2. the strain energy density function involve nonlinear relations with respect to the constitutive parameters. In this case, identification must be performed by minimizing a cost-function derived from Eq. (5.58).

With our constitutive equation defined in Eq. (5.13), we obtain

$$\begin{aligned} &-\int_{\Omega_{t_i}} \mu_1 \boldsymbol{B}^i : \left(\nabla \otimes \boldsymbol{v}^*\right) d\Omega_{t_i} \\ &\quad -\int_{\Omega_{t_i}} \mu_2 e^{\gamma\left(I_\kappa^i - 1\right)^2} \left(\kappa \boldsymbol{B}^i + (1-2\kappa)\, \boldsymbol{m}^i \otimes \boldsymbol{m}^i\right) : \left(\nabla \otimes \boldsymbol{v}^*\right) d\Omega_{t_i} \\ &\qquad -\int_{\Omega_{t_i}} p^i \nabla . \boldsymbol{v}^* d\Omega_{t_i} = \int_{\partial \Omega_{t_i}} \boldsymbol{t}^i . \boldsymbol{v}^* ds. \end{aligned} \qquad (5.59)$$

Measurements usually provide the deformed configuration, so it is convenient that this equation is written in the deformed configuration.

The equation is valid for any virtual fields. As we are not interested in p^i, it has been shown that $\boldsymbol{v}^*$ should be chosen such that $\nabla . \boldsymbol{v}^* = 0$.

Other rules for choosing appropriate virtual fields are very specific to each problem. A systematic choice was proposed for linear elastic constitutive equations (Avril and Pierron, 2007; Avril et al., 2004) and further extended to elastoplasticity (Pierron et al., 2010) and hyperelasticity (Mei et al., 2021b). These virtual fields were optimized for minimizing noise effects in the case of linear elasticity. In the case of hyperelastic constitutive equations, noise in the data is less an issue as tissues undergo large strains. In that case, we have recently proposed virtual fields that would be appropriate for different types of loading (tension inflation test (Avril et al., 2010), bulge inflation test (Kim et al., 2012)). They permit obtaining equations such as

$$\mu_1 A_{ij} + \mu_2 \kappa B_{ij}(\gamma) + \mu_2 (1 - 2\kappa) C_{ij}(\gamma, \theta) = L_{ij}, \tag{5.60}$$

where A_{ij}, B_{ij}, C_{ij}, and L_{ij} can be evaluated directly from the experimental measurements. Index i is for the different time steps, for which deformations and loads are measured and index j is for different possible choices of virtual fields.

Eq. (5.60) is an equation of the unknown material parameters for each choice of virtual field i and at every stage j of the test. The equation is linear in μ_1, $\kappa\mu_2$, and $\mu_2(1 - 2\kappa)$, but it is nonlinear in γ and θ. The solution is found by minimizing a cost function defined as

$$\pi(\beta) = \sum_j \sum_i \left(\mu_1 A_{ij} + \mu_2 \kappa B_{ij}(\gamma) + \mu_2 (1 - 2\kappa) C_{ij}(\gamma, \theta) - L_{ij}\right)^2. \tag{5.61}$$

This cost function can be minimized by the simplex method or using a genetic algorithm in case of multiple minima.

In summary, the VFM is an efficient method for the identification of material properties, but it requires specific conditions for its application, especially the availability of deformation gradients across the whole domain of interest. Such requirement is not needed by variational or sequential methods presented before.

5.5.2 Applications to soft tissues

Regarding applications to soft tissues, the VFM was first applied to the identification of uniform material properties in arterial walls (Avril et al., 2010). Recent extension in the same field were proposed for the inverse characterization of regional, nonlinear, anisotropic properties of murine aortas with different types of lesions (Bersi et al., 2016, 2019, 2020).

The VFM was recently applied successfully in biaxial tension by Kazerooni et al. (2020) to identify the parameters of a Holzapfel model (Holzapfel et al., 2000) for the skin. In ophtalmology, it was used by Girard et al. (2015), Zhang et al. (2020), and further by Mei et al. (2021b) to identify material properties of the lamina cribosa, which is a connective tissue structure in the optical nerve head, which is of great interest to researchers studying development and progression of glaucoma.

There are no applications so far to pelvic biomechanics. However, recent developments in the VFM for elastography applications (Mei et al., 2021a; Mei and Avril, 2019) are promising for future applications in that field.

5.6 Conclusions and future directions emerging field of pelvic biomechanics

Inverse problems consist of using the actual result of some measurements to infer the values of the parameters that characterize the system, such as unknown parameters of the material model, unknown elements of the boundary conditions, or even sometimes the unknown initial geometry of the solid before the application of any mechanical action (loadfree configuration in finite deformations). There are many other situations in soft tissue biomechanics where inverse problems have to be resolved. For the sake of illustration, the current chapter was intentionally restricted to two main situations: elastography of soft tissue biomechanics and nonstandard testing of soft tissues. First we presented in details the FEMU technique, which is traditionally used to solve inverse problems. We then presented sequential method based on the principle of Kalman filters for dynamic problems. Finally, we presented in detail the VFM and showed different cases of its possible application.

For each approach, the main objective was to introduce the principles required to understand the theory of the methods. This is the first time that a book chapter presents these different methods under a unified framework. Specific applications of each method may be found in the referred journal publications.

There are many directions related to soft tissue biomechanics and where inverse problems have to be resolved; such directions are the topic of intense research at the moment. Without wanting to be exhaustive, we can point the 3 following directions that may require research efforts in the inverse problem community in the near future, and which may have a significant impact in reproduction biomechanics:

1. Development of robust inversion techniques to estimate the viscoelastic properties from elastography imaging data. This is an ongoing area of research, as this is an inherently ill-conditioned numerical problem. Poor quality datasets with low signal-to-noise ratio tend to provide unreliable tissue property estimates, with different biases depending on the inversion algorithms applied. The viscoelastic properties of the pubovisceral muscle or other soft tissues of the pelvic system are likely to be important for organ support and probably affect progress in labor (Jing et al., 2012).
2. Development of robust inversion techniques to estimate the parameters of growth and remodeling models. The migration of cells through the fibrous network of the extracellular matrix is an integral part of many biological processes, including tissue morphogenesis, wound healing, and cancer metastasis. For instance, the dramatic changes in the material behavior of the cervix over the normal period of gestation by considering the turnover of collagen from mature crosslinked fibers to immature loosely connected fibrils (Myers and Ateshian, 2014). Accurate measurements of constitutive parameters for models aimed at predicting these effects will deserve significant research efforts in the future.
3. Representation of the data uncertainties. The most general (and simple) theory is obtained when using a probabilistic point of view, where the *a priori* information on the model parameters is represented by a probability distribution over the "model space." A great challenge of computational soft tissue biomechanics is to transform this a priori probability distribution into a range of uncertainty in the predictions. This may be important for treatment planning based on biomechanical deformable image registration for instance (Velec et al., 2017).

References

Ami, O., Lamazou, F., Mabille, M., Levaillant, J., Deffieux, X., Frydman, R., Musset, D., 2009. Real-time transvaginal elastosonography of uterine fibroids. Ultrasound in Obstetrics & Gynecology 34, 486–488.

Avril, S., Badel, P., Duprey, A., 2010. Anisotropic and hyperelastic identification of in vitro human arteries from full-field optical measurements. Journal of Biomechanics 43, 2978–2985.

Avril, S., Bonnet, M., Bretelle, A.-S., Grédiac, M., Hild, F., Ienny, P., Latourte, F., Lemosse, D., Pagano, S., Pagnacco, E., et al., 2008a. Overview of identification methods of mechanical parameters based on full-field measurements. Experimental Mechanics 48, 381.

Avril, S., Evans, S., 2017. Material Parameter Identification and Inverse Problems in Soft Tissue Biomechanics. Springer.

Avril, S., Feissel, P., Pierron, F., Villon, P., 2008b. Estimation of the strain field from full-field displacement noisy data: comparing finite elements global least squares and polynomial diffuse approximation. European Journal of Computational Mechanics/Revue Européenne de Mécanique Numérique 17, 857–868.

Avril, S., Feissel, P., Pierron, F., Villon, P., 2009a. Comparison of two approaches for differentiating full-field data in solid mechanics. Measurement Science & Technology 21, 015703.

Avril, S., Grédiac, M., Pierron, F., 2004. Sensitivity of the virtual fields method to noisy data. Computational Mechanics 34, 439–452.

Avril, S., Huntley, J., Pierron, F., Steele, D., 2008c. 3D heterogeneous stiffness reconstruction using MRI and the virtual fields method. Experimental Mechanics 48, 479–494.

Avril, S., Huntley, J.M., Cusack, R., 2009b. In vivo measurements of blood viscosity and wall stiffness in the carotid using PC-MRI. European Journal of Computational Mechanics/Revue Européenne de Mécanique Numérique 18, 9–20.

Avril, S., Pierron, F., 2007. General framework for the identification of constitutive parameters from full-field measurements in linear elasticity. International Journal of Solids and Structures 44, 4978–5002.

Bensamoun, S.F., Ringleb, S.I., Littrell, L., Chen, Q., Brennan, M., Ehman, R.L., An, K.-N., 2006. Determination of thigh muscle stiffness using magnetic resonance elastography. Journal of Magnetic Resonance Imaging 23, 242–247.

Bercoff, J., Tanter, M., Fink, M., 2004. Supersonic shear imaging: a new technique for soft tissue elasticity mapping. IEEE Transactions on Ultrasonics, Ferroelectrics, and Frequency Control 51, 396–409.

Bersi, M.R., Bellini, C., Di Achille, P., Humphrey, J.D., Genovese, K., Avril, S., 2016. Novel methodology for characterizing regional variations in the material properties of murine aortas. Journal of Biomechanical Engineering 138, 071005.

Bersi, M.R., Bellini, C., Humphrey, J.D., Avril, S., 2019. Local variations in material and structural properties characterize murine thoracic aortic aneurysm mechanics. Biomechanics and Modeling in Mechanobiology 18, 203–218.

Bersi, M.R., Santamaría, V.A.A., Marback, K., Di Achille, P., Phillips, E.H., Goergen, C.J., Humphrey, J.D., Avril, S., 2020. Multimodality imaging-based characterization of regional material properties in a murine model of aortic dissection. Scientific Reports 10, 1–23.

Bertoglio, C., Moireau, P., Gerbeau, J.-F., 2012. Sequential parameter estimation for fluid–structure problems: application to hemodynamics. International Journal for Numerical Methods in Biomedical Engineering 28, 434–455.

Bilston, L.E., Tan, K., 2015. Measurement of passive skeletal muscle mechanical properties in vivo: recent progress, clinical applications, and remaining challenges. Annals of Biomedical Engineering 43, 261–273.

Chantereau, P., Brieu, M., Kammal, M., Farthmann, J., Gabriel, B., Cosson, M., 2014. Mechanical properties of pelvic soft tissue of young women and impact of aging. International Urogynecology Journal 25, 1547–1553.
Doyley, M., Meaney, P., Bamber, J., 2000. Evaluation of an iterative reconstruction method for quantitative elastography. Physics in Medicine and Biology 45, 1521.
Drakonaki, E., Allen, G., Wilson, D., 2012. Ultrasound elastography for musculoskeletal applications. The British Journal of Radiology 85, 1435–1445.
Egorov, V., Lucente, V., Van Raalte, H., Murphy, M., Ephrain, S., Bhatia, N., Sarvazyan, N., 2019. Biomechanical mapping of the female pelvic floor: changes with age, parity and weight. Pelviperineology 38, 3.
Fung, Y., 2013. Biomechanics: Mechanical Properties of Living Tissues. Springer Science & Business Media.
Gasser, T.C., Ogden, R.W., Holzapfel, G.A., 2006. Hyperelastic modelling of arterial layers with distributed collagen fibre orientations. Journal of the Royal Society Interface 3, 15–35.
Girard, M.J., Tan, D., Ang, M., Mehta, J.S., Zhang, L., Chung, C.W., Mani, B., Tun, T.A., Aung, T., 2015. An engineering-based methodology to characterize the in vivo nonlinear biomechanical properties of the cornea with application to glaucoma subjects. Investigative Ophthalmology & Visual Science 56, 1099.
Goenezen, S., Barbone, P., Oberai, A.A., 2011. Solution of the nonlinear elasticity imaging inverse problem: the incompressible case. Computer Methods in Applied Mechanics and Engineering 200, 1406–1420.
Goenezen, S., Dord, J.-F., Sink, Z., Barbone, P.E., Jiang, J., Hall, T.J., Oberai, A.A., 2012. Linear and nonlinear elastic modulus imaging: an application to breast cancer diagnosis. IEEE Transactions on Medical Imaging 31, 1628–1637.
Grédiac, M., 1989. Principe des travaux virtuels et identification. Comptes Rendus de L'Académie Des Sciences. Série 2, Mécanique, Physique, Chimie, Sciences de l'univers, Sciences de la Terre 309, 1–5.
Hobson, M.A., Kiss, M.Z., Varghese, T., Sommer, A.M., Kliewer, M.A., Zagzebski, J.A., Hall, T.J., Harter, J., Hartenbach, E.M., Madsen, E.L., 2007. In vitro uterine strain imaging: preliminary results. Journal of Ultrasound in Medicine 26, 899–908.
Holzapfel, G.A., Gasser, T.C., Ogden, R.W., 2000. A new constitutive framework for arterial wall mechanics and a comparative study of material models. Journal of Elasticity and the Physical Science of Solids 61, 1–48.
Humphrey, J.D., 2013. Cardiovascular Solid Mechanics: Cells, Tissues, and Organs. Springer Science & Business Media.
Jing, D., Ashton-Miller, J.A., DeLancey, J.O., 2012. A subject-specific anisotropic visco-hyperelastic finite element model of female pelvic floor stress and strain during the second stage of labor. Journal of Biomechanics 45, 455–460.
Kauer, M., Vuskovic, V., Dual, J., Székely, G., Bajka, M., 2002. Inverse finite element characterization of soft tissues. Medical Image Analysis 6, 275–287.
Kazerooni, N.A., Wang, Z., Srinivasa, A., Criscione, J., 2020. Inferring material parameters from imprecise experiments on soft materials by virtual fields method. Annals of Solid and Structural Mechanics, 1–14.
Kim, J.-H., Avril, S., Duprey, A., Favre, J.-P., 2012. Experimental characterization of rupture in human aortic aneurysms using a full-field measurement technique. Biomechanics and Modeling in Mechanobiology 11, 841–853.
Kiss, M.Z., Hobson, M.A., Varghese, T., Harter, J., Kliewer, M.A., Hartenbach, E.M., Zagzebski, J.A., 2006. Frequency-dependent complex modulus of the uterus: preliminary results. Physics in Medicine and Biology 51, 3683.

Liu, L., Morgan, E.F., 2007. Accuracy and precision of digital volume correlation in quantifying displacements and strains in trabecular bone. Journal of Biomechanics 40, 3516–3520.

Mei, Y., Avril, S., 2019. On improving the accuracy of nonhomogeneous shear modulus identification in incompressible elasticity using the virtual fields method. International Journal of Solids and Structures 178, 136–144.

Mei, Y., Deng, J., Guo, X., Goenezen, S., Avril, S., 2021a. Introducing regularization into the virtual fields method (VFM) to identify nonhomogeneous elastic property distributions. Computational Mechanics, 1–19.

Mei, Y., Liu, J., Guo, X., Zimmerman, B., Nguyen, T.D., Avril, S., 2021b. General finite-element framework of the virtual fields method in nonlinear elasticity. BioRxiv.

Muthupillai, R., Ehman, R.L., 1996. Magnetic resonance elastography. Nature Medicine 2, 601–603.

Muthupillai, R., Lomas, D., Rossman, P., Greenleaf, J.F., Manduca, A., Ehman, R.L., 1995. Magnetic resonance elastography by direct visualization of propagating acoustic strain waves. Science 269, 1854–1857.

Myers, K., Ateshian, G.A., 2014. Interstitial growth and remodeling of biological tissues: tissue composition as state variables. Journal of the Mechanical Behavior of Biomedical Materials 29, 544–556.

Oberai, A.A., Gokhale, N.H., Feijóo, G.R., 2003. Solution of inverse problems in elasticity imaging using the adjoint method. Inverse Problems 19, 297.

O'Hara, S., Zelesco, M., Sun, Z., 2019. Shear wave elastography on the uterine cervix: technical development for the transvaginal approach. Journal of Ultrasound in Medicine 38, 1049–1060.

Ophir, J., Cespedes, I., Ponnekanti, H., Yazdi, Y., Li, X., 1991. Elastography: a quantitative method for imaging the elasticity of biological tissues. Ultrasonic Imaging 13, 111–134.

Peng, Q., Jones, R., Shishido, K., Constantinou, C.E., 2007. Ultrasound evaluation of dynamic responses of female pelvic floor muscles. Ultrasound in Medicine & Biology 33, 342–352.

Perrin, D., Badel, P., Orgeas, L., Geindreau, C., du Roscoat, S., Albertini, J.-N., Avril, S., 2016. Patient-specific simulation of endovascular repair surgery with tortuous aneurysms requiring flexible stent-grafts. Journal of the Mechanical Behavior of Biomedical Materials 63, 86–99.

Perrin, D., Badel, P., Orgéas, L., Geindreau, C., Dumenil, A., Albertini, J.-N., Avril, S., 2015a. Patient-specific numerical simulation of stent-graft deployment: validation on three clinical cases. Journal of Biomechanics 48, 1868–1875.

Perrin, D., Demanget, N., Badel, P., Avril, S., Orgéas, L., Geindreau, C., Albertini, J.-N., 2015b. Deployment of stent grafts in curved aneurysmal arteries: toward a predictive numerical tool. International Journal for Numerical Methods in Biomedical Engineering 31, e02698.

Pierron, F., Avril, S., Tran, V.T., 2010. Extension of the virtual fields method to elasto-plastic material identification with cyclic loads and kinematic hardening. International Journal of Solids and Structures 47, 2993–3010.

Pierron, F., Grédiac, M., 2012. The Virtual Fields Method: Extracting Constitutive Mechanical Parameters from Full-Field Deformation Measurements. Springer Science & Business Media.

Silva, E., Parente, M., Brandão, S., Mascarenhas, T., Natal Jorge, R., 2019. Characterizing the biomechanical properties of the pubovisceralis muscle using a genetic algorithm and the finite element method. Journal of Biomechanical Engineering 141.

Silva, M., Brandao, S., Parente, M., Mascarenhas, T., Natal Jorge, R., 2016. Establishing the biomechanical properties of the pelvic soft tissues through an inverse finite element analysis using magnetic resonance imaging. Proceedings of the Institution of Mechanical Engineers. Part H, Journal of Engineering in Medicine 230, 298–309.

Stewart, E.A., Taran, F.A., Chen, J., Gostout, B.S., Woodrum, D.A., Felmlee, J.P., Ehman, R.L., 2011. Magnetic resonance elastography of uterine leiomyomas: a feasibility study. Fertility and Sterility 95, 281–284.

Swiatkowska-Freund, M., Preis, K., 2011. Elastography of the uterine cervix: implications for success of induction of labor. Ultrasound in Obstetrics & Gynecology 38, 52–56.

Thyer, I., Shek, C., Dietz, H., 2008. New imaging method for assessing pelvic floor biomechanics. Ultrasound in Obstetrics & Gynecology 31, 201–205.

Van Houten, E.E., Paulsen, K.D., Miga, M.I., Kennedy, F.E., Weaver, J.B., 1999. An overlapping subzone technique for MR-based elastic property reconstruction. Magnetic Resonance in Medicine 42, 779–786.

Van Houten, E.E., Weaver, J., Miga, M., Kennedy, F., Paulsen, K., 2000. Elasticity reconstruction from experimental MR displacement data: initial experience with an overlapping subzone finite element inversion process. Medical Physics 27, 101–107.

Velec, M., Moseley, J.L., Svensson, S., Hårdemark, B., Jaffray, D.A., Brock, K.K., 2017. Validation of biomechanical deformable image registration in the abdomen, thorax, and pelvis in a commercial radiotherapy treatment planning system. Medical Physics 44, 3407–3417.

Zhang, L., Beotra, M.R., Baskaran, M., Tun, T.A., Wang, X., Perera, S.A., Strouthidis, N.G., Aung, T., Boote, C., Girard, M.J., 2020. In vivo measurements of prelamina and lamina cribrosa biomechanical properties in humans. Investigative Ophthalmology & Visual Science 61, 27.

CHAPTER

Mechanical properties of women pelvic soft tissues

6

Shelby White[a], Mathias Brieu[b], and Kristin Miller[c]

[a]*Tulane University, New Orleans, LA, United States*

[b]*College of Engineering, Computer Science and Technology, Dept. of Mechanical Engineering, California State University, Los Angeles, CA, United States*

[c]*University of Texas at Dallas, Dallas, TX, United States*

6.1 Introduction

Biomechanics is defined as the application of mechanics to biology and medicine to characterize the body's response to physical stressors (Humphrey, 2002; Damaser and Hoyte, 2016). These physical stressors may be external, such as gravity, or internal processes, such as aging or movement from everyday activities (Damaser and Hoyte, 2016). Biomechanics is a crucial, albeit often understudied, field that aids in understanding how the body responds to both our internal and external environment. This understanding is crucial for medical advancement as the body is intelligently designed and is continually evolving to the dynamic functional demands of the changing mechanical environment. For example, those who strength train constantly apply stimuli to their muscle due to repetitive and consistent engagement (Bassel-Duby and Olson, 2006). Alternatively, if a person refrains from physical activity, for example, due to bed rest, the muscle is not engaged for an extended period of time, inducing atrophy as a result (Berg et al., 2007). Though less intuitive, the female pelvic floor undergoes significant fluctuations during a patient's lifetime and as such has similar cellular and microstructural remodeling processes that result in changes in the functional or biomechanical aspect of the tissues. Regardless of the organ system, this unique quality of cells seeking to establish, maintain, or restore a preferred biomechanical state in response to evolving environmental demands is unique to biological materials and perfectly exemplifies why understanding the mechanics of tissues is vital to the progression of medicine.

To quantify these mechanical demands, experimental designs are carefully crafted and carried out in either the benchtop or clinical setting. Furthermore, mechanical testing provides two classes of experimental data: data related to the structural properties of the tissue and data associated with the material properties. Structural properties describe how a material, in this case, a biological soft tissue, responds to an applied force or deformation. These properties are dependent on the type of material, the amount of the material, and the geometry. Examples of structural properties of tissues include the maximum force or load, ultimate elongation, stiffness, and energy

Biomechanics of the Female Reproductive System: Breast and Pelvic Organs. https://doi.org/10.1016/B978-0-12-823403-7.00018-X

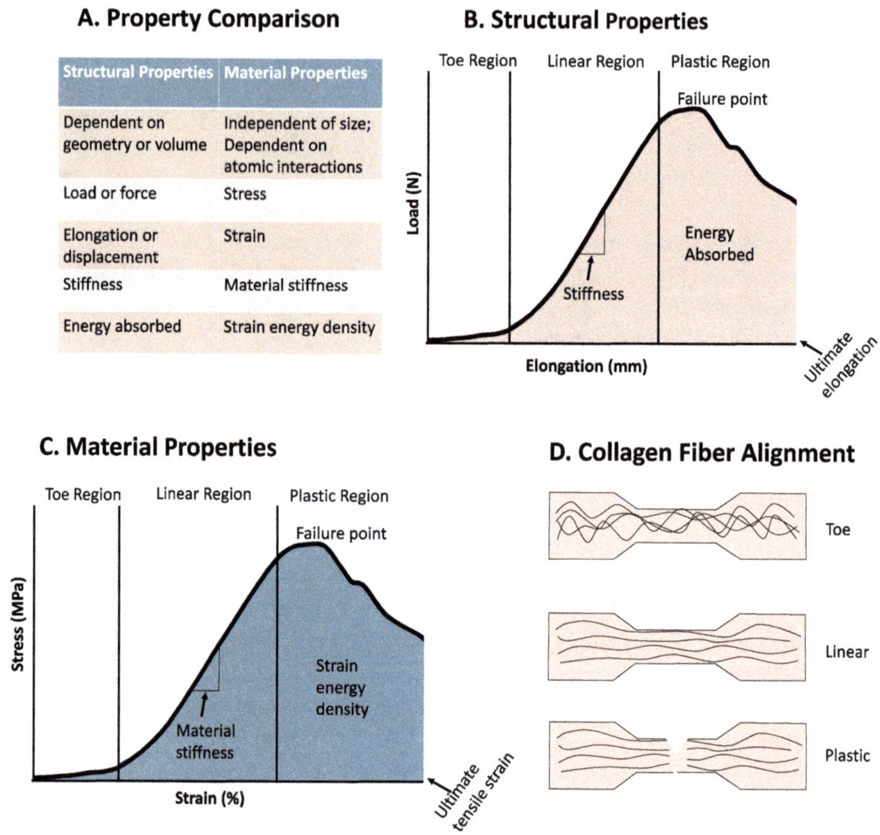

FIGURE 6.1

No permission required. Comparison of the structural and material properties commonly measured during mechanical testing. Structural properties are dependent on geometry and volume, whereas material properties are independent of size and dependent on atomic interactions (A). Both structural (B) and material (C) properties demonstrate the nonlinear behavior of biological soft tissues when displayed as a function of load vs. elongation or stress vs strain. Each has a toe, linear, and plastic region or nonreversible on the curve with a corresponding alignment of collagen fibers (D).

absorbed. On the other hand, material properties are based purely on the material's internal structure and are independent of volume and geometry. Examples of material properties include stress, strain, material stiffness, and strain energy density. Often the structural properties are related to the material properties; for example, *stress* or the applied force or load divided by the cross-sectional area. See Fig. 6.1A for a summary and comparison of these two property types. However, the interpretation of the two property types are quite different; for example, the ultimate force observed in a mechanical test would vary based on whether a surgeon excised a whole uterosacral ligament or only half; however, the stress observed would be the same once the cross-sectional area normalizes the sample.

Additionally, load-elongation and stress-strain curves represent the characterization of the structural and material properties, Fig. 6.1B, 6.1C, respectively. Though these curves look similar at first glance, the critical differences of geometry dependence lie in the axes and the quantities measured. However, there are similar attributes to both: primarily the toe, reversible linear or nonlinear, and plastic regions found on the curve. With soft tissues, the toe region is a nonlinear response associated with a significant elongation and small changes in force. For most soft tissues, the toe region is also associated with resistance to loading provided by the elastic fibers, whereas the undulated collagen fibers realign toward the direction of loading and uncrimp. The first region is characterized by a proportional relationship between load and deformation if the response observed is linear, or a nonproportional relationship for nonlinear response. However, in both cases the response is reversible. Such behavior is generally associated with collagen fibers becoming fully recruited and resisting deformation. The third portion, the plastic or nonreversible region, is where the specimen experiences permanent structural damage, likely due to collagen fibers breaking, which will finally lead to rupture (Fig. 6.1D).

Carefully designed and controlled mechanical testing experiments biomechanically characterize the structural and material properties. During the testing, either a force (load) or a deformation (change in geometry) is applied to the tissue. Examples of these experiments include planar tensile testing, extension-inflation, shear, creep testing, and many more. Whereas each testing method measures different aspects of either structural or material properties, each has unique advantages and limitations. For a summary, see Fig. 6.2.

Planar tensile testing is categorized as either uniaxial or biaxial, depending on whether the tissue sample is being pulled in one or two directions (Fig. 6.2A, 6.2B). The purpose of planar tensile testing is to apply tension to the tissue sample in either one primary axis or two axes and measure the corresponding response of the tissue. For planar testing, the tissue is securely clamped and loaded at a constant rate until failure, wherein mechanical testing *failure* occurs when tissue tears or breaks completely. Force data may be obtained by utilizing force transducers and specialized software, and strain data is often measured from video measurements that calculate displacement between two marked points or through speckle-coating. Though planar testing is the most commonly used form of mechanical testing, cutting the tissue limits the clinical relevance. To properly clamp the tissue in the planar setting, the tissue needs to have a section excised in the case of human samples or cut from the original tubular geometry in animals, such as the rat. This cutting process results in altered cell-matrix interactions that may not recapitulate the in vivo arrangement (Timmons et al., 2009; Yellon et al., 2009).

Whereas planar testing is limited by disrupting the geometry, biaxial extension inflation testing allows for the in vivo geometry and the native cell and extracellular matrix interactions to be maintained (Caulk et al., 2015; Amin et al., 2012; Sokolis et al., 2011) (Fig. 6.2C). During biaxial extension-inflation testing, the tissue is cannu-

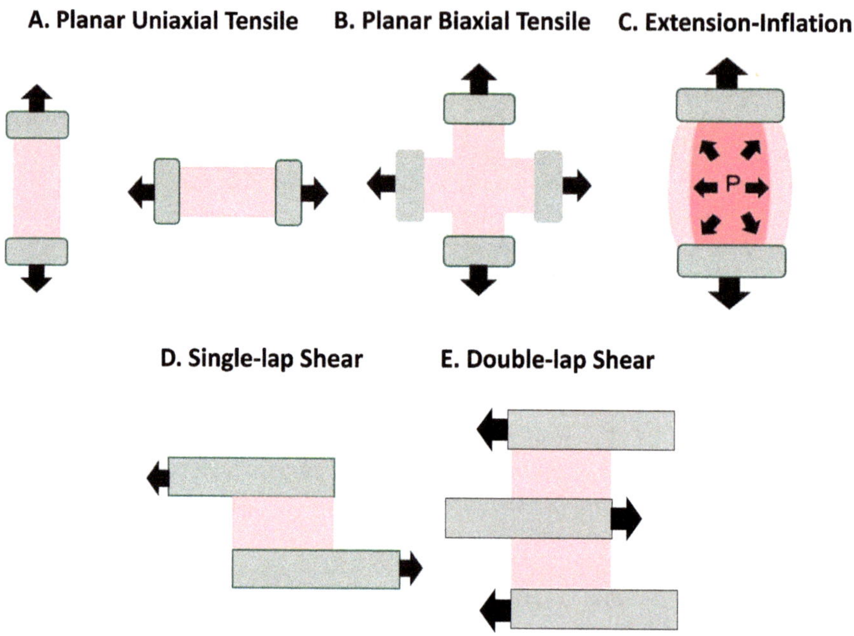

FIGURE 6.2

No permission required. The various methods of common mechanical testing techniques. Techniques include planar testing both in a uniaxial (A) and biaxial fashion (B), biaxial extension-inflation (C), and single-lap shear (D), and double-lap shear (E). Each testing technique uses different methods to recapitulate different loading environments and is equipped with unique advantages and limitations.

lated and pressurized over a range of physiologic pressures (load). The resulting outer diameter is then optically tracked with a camera (deformation). Though this testing method allows for recapitulating the in vivo geometry and testing over physiologic loads, no current testing system allows for whole human tissue to be cannulated and tested. In terms of the female pelvic floor, studies employed extension-inflation testing to investigate the rodent vagina and cervix (Conway et al., 2019; Robison et al., 2017; Clark et al., 2019).

In shear testing, tangential forces are applied to the tissue in either a single or double lap schematic (Fig. 6.2D, 6.2E). During single lap testing, the tissue sample is placed between two plates, and a load is applied to failure. To account for non-symmetric, or eccentric loading, double-lap shear testing uses a similar testing setup but has two tissue specimens between three plates as the load is applied until failure. A common use of shear-lap testing is measuring the *in-situ* strength of the tissue, particularly of joints (Da Silva et al., 2009; Kendall, 1975; Camanho et al., 2005). Applications of shear-lap testing for the pelvic floor includes the human premenopausal vagina (Feola et al., 2013).

The *elastic response* corresponds to the fibrous structural components of the tissue (Lees and Davidson, 1977), and may be assessed using the previous testing methods. However, a subset of biomechanics studies the viscous nature of materials or the time-dependent response proportional to how quickly a material is deformed. Tissues are classified as "viscoelastic," owing to a portion of the mechanical properties to the elastic or solid structures and a portion to the fluid or viscous components, such as glycosaminoglycans or water. This unique nature of biological soft tissues was first characterized by Y.C. Fung, the father of biomechanics, wherein "bio-viscoelastic solids" exhibit viscoelastic properties given the suitable testing methods and conditions (Fung, 1993). Previously mentioned testing setups are commonly used to measure the elastic nature of tissues and may be adapted to measure the viscoelastic response, namely stress-relaxation and creep testing.

Stress-relaxation, or the ability of the tissue to relax, when constantly deformed, is correlated with the number of proteoglycans and smooth muscle cells that comprise the tissues (Humphrey et al., 2014; Connizzo, 2013; Eshel and Lanir, 2001). For example, the vagina has a greater stress-relaxation response than that of the uterosacral ligament. This is particularly advantageous considering the functions of the two organs; the vagina will undergo a large deformation during events such as menstruation or childbirth, and it needs to be able to sustain these changes without injury. To measure this relaxation ability, the tissues are loaded instantaneously to a set strain level and held at this length over time with an observed subsequent decrease in force. However, stress relaxation testing is highly sensitive to the speed and precision of the testing equipment, so comparisons across tests and organs are difficult.

Creep testing is another commonly used mechanical testing technique to quantify the viscoelastic properties of tissues. Herein creep is defined as the time-dependent deformation when tissue is subjected to a constant level of stress. Standard testing methods to measure the creep response of tissue includes suspending the tissue and allowing a known weight to hang from the sample and measuring the change in length over time with strain gauges or optical tracking. Another creep testing technique utilizes a sample similar to that of the extension-inflation to apply a set pressure and measure the change in the outer diameter.

One common characteristic of each of these mechanical testing set up is that they have the ability to quantify the contributions from various microstructural components, such as collagen and elastic fibers, smooth muscle cells, and the remaining ground matrix, such as the proteoglycans, glycosaminoglycans, and water within the tissue. For this chapter, the contractile contribution from smooth muscle cells is also known as the active component, as it is actively contracting and generating a force within the tissue. Similarly, the collagen and elastic fibers largely make up the passive contribution. The remaining ground matrix contributes mostly to the viscoelasticity of the tissue, though it should be noted that viscoelasticity may be active or passive, depending on whether the smooth muscle cells are contractile. The ground matrix contribution for each organ and tissue type is considered the passive viscoelastic mechanics within this chapter.

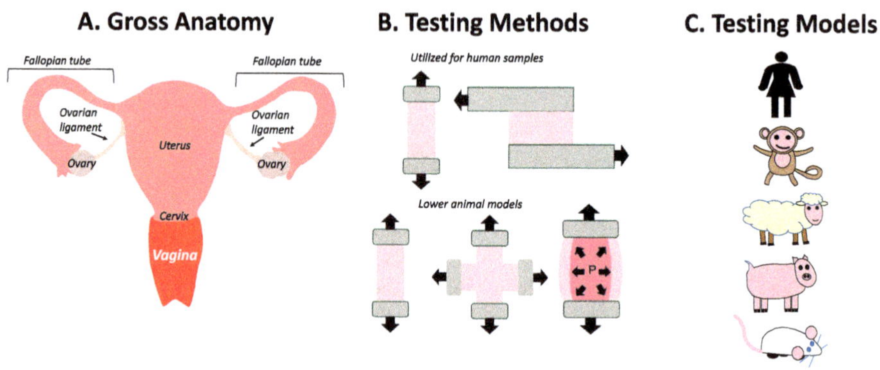

FIGURE 6.3

No permission required. The mechanical overview of the vagina. The vagina is located below the uterus and cervix (A). Uniaxial and biaxial tensile testing, single-lap shear, and extension-inflation testing methods are employed to characterize the mechanics of the vagina (B). In addition to uniaxial and single-lap shear testing having been used to quantify the human vaginal mechanics, non-human primates, ewes, swine, and rodents have also been used to quantify the mechanics of the vagina.

In summary, the female pelvic floor undergoes large changes in mechanical loading throughout the lifetime of a patient. The chapter presented below focuses on the mechanics of the vagina, uterus (including cervix), levator ani, and uterosacral ligament. For each organ mentioned, the sections are divided into contributions from the contracting smooth muscle cells, contributions from collagen and elastic fibers, and contributions from the nonfibrous ground matrix. To optimize clinical relevance, studies that utilize human-derived samples are the primary focus of this chapter and supplemented with notable studies that utilize animal models to give a basis for the field of biomechanics and the female pelvic floor.

6.2 Vagina

6.2.1 Overview

The vagina is a soft, elastic, muscular organ that acts as a canal to the female reproductive organs (Fig. 6.3A). The vagina functions to provide lubrication and sensation during sexual activity and serves as a passageway for sperm, menstrual blood and tissue, and childbirth. Furthermore, the vagina provides mechanical support for the entire pelvic floor. The human vagina is approximately tubular in shape and averages to be 7.5 cm along the anterior wall and 9 cm along the posterior wall of the vagina (McLoughlin and Wylie, 2006). In addition, the vagina is approximately 4.2 cm in diameter (Barnhart et al., 2006). However, these values are averages, and the vaginal geometry changes not only from patient to patient but as a function of menstruation and menopausal status, age, parity, and preexisting conditions.

The vagina comprises four distinct layers: the epithelium, subepithelium, muscularis, and the adventitia. The epithelium is composed of epithelial cells and serves as a protective barrier against potential infections and pathogens. The subepithelium is responsible for the passive mechanical properties of the vagina and contains dense layers of collagen and elastic fibers that are aligned randomly. Additionally, circumferentially and axially oriented smooth muscle cells located in the muscularis contribute to the active or contractile mechanics of the vagina. The subepithelium and muscularis are considered the primary contributors to the structural and mechanical properties of the vagina. The adventitia connects the reproductive system to the rectum and bladder and is composed of loose connective tissue of randomly oriented collagen and elastic fibers, nerves, and capillaries. Additionally, glycosaminoglycans are located in the subepithelium, muscularis, and adventitia and contribute to the characteristic viscoelastic nature of the vagina.

Previous investigations employed both uniaxial and biaxial testing techniques to mechanically characterize the vagina (Rubod et al., 2007; Pena et al., 2009; Lopez et al., 2015; Skoczylas et al., 2013). In addition to the planar testing methods that extend the tissue in tension, biaxial shear lap testing and extension-inflation protocols characterized the mechanical properties in animal models, such as the mouse (Robison et al., 2017; Clark et al., 2019; Akintunde et al., 2018; Capone et al., 2019), rat (Moalli et al., 2005; Basha et al., 2009; Feloa et al., 2010), sheep (Rubod et al., 2007; Knight et al., 2016), and nonhuman primate (Fig. 6.3B, 6.3C) (Feola et al., 2010).

Pioneers in the field of biomechanics provided the groundwork to describe general characteristics of soft biological tissues: elasticity, nonlinearity, anisotropy, and incompressibility (Fung, 1993; Green and Naghdi, 1995; Rivlin, 1948; Skalak et al., 1973). Rubod et al. led one of the first investigations to quantify the vagina as part of these characteristics. Rubod was the first to demonstrate that human vaginal tissue is nonlinear, elastic, and can undergo large deformations (Jean-Charles et al., 2010). This was accomplished by collecting vaginal samples from women undergoing prolapse repair surgeries and cadavers who exhibited no pelvic floor dysfunctions.

6.2.2 Contributions from collagen and elastic fibers

Due to the significant mechanical contributions of the elastic fibers, collagen fibers, and nonfibrous components of the extracellular matrix, the passive mechanical properties of the vagina were of high interest and among the first to be characterized with regards to the biomechanics of the female pelvic floor. The most prevalent testing method for characterization was uniaxial tensile testing. Tissue types utilized in uniaxial testing include human vaginal samples (Peña et al., 2011), nonhuman primates (Feola et al., 2010), ewes (Knight et al., 2014), and rats (Feloa et al., 2010). Additionally, the biaxial mechanical properties of the vagina are studied using animal models, such as swine (Women's Health Weekly, 2019) and mice (Clark et al., 2019). Because the vagina is loaded multiaxially in the body, biaxial characterization offers the benefit of more accurately recapitulating the native loading environment by measuring the response of the tissue in two directions. Due to logistics in obtaining human samples and the current limitations in testing devices, the passive properties of the human vagina have yet to be characterized biaxially. Acknowledging that this is of interest to current and future studies, human vaginal samples tested with uniaxial methods will be the priority of this chapter given the clinical relevance.

Examples that utilized uniaxial tensile testing include Pena et al., wherein vaginal samples from postmenopausal women with prolapse were excised and frozen once in a saline solution before mechanical testing (Pena et al., 2009). Undergoing one freeze-thaw cycle is a commonly used practice for investigating the passive mechanical properties, as Rubod et al. demonstrated that the passive mechanical behavior of ewe vaginal tissue was unaffected by one freeze-thaw cycle; however, the smooth muscle cells are not able to contract, thus disabling contributions from the active biomechanics (Rubod et al., 2007). Additional findings from uniaxial tensile testing are the statistical link between a patient's obesity level and the material stiffness of the vaginal wall in the passive state. This development in reproductive biomechanics launched further studies that would aid in determining obesity as a risk factor for the development of pelvic floor disorders (Lopez et al., 2015).

Though the Lopez study mentioned earlier (Lopez et al., 2015) quantified the vaginal wall material stiffness along the anterior portion of the vagina, regions of the human vaginal tissue may have different compositions and, thus, varying mechanical properties. Whereas the ewe (Rubod et al., 2007), rat (Women's Health Weekly, 2019), and mouse model (Clark et al., 2019) demonstrated regional variations, apparent locational differences in the mechanical postmenopausal human vagina were not observed (Feola et al., 2013). However, the authors suggest this may be due to sample size and noted that vaginal samples from premenopausal women were excluded from the study, both of which warrant further investigation and should be a key hallmark moving forward (Ulrich et al., 2012).

6.2.3 Contributions from smooth muscle cells

The active properties of the vagina are dictated primarily by the circumferentially and axially aligned smooth muscle cells located in the muscularis. The contractility of the smooth muscle cells, particularly within the female pelvic floor, is influenced by several biologic factors, such as steroid hormone profile, autonomic nervous system contributions, and physical factors, such as the stretching the vagina experiences throughout a woman's everyday activities. For example, the hormone oxytocin induces vaginal contraction during sexual stimulation (Gravina et al., 2014). Consequently, during menopause, there is a decline in steroid hormones, namely estrogen, which results in vaginal thinning (Castelo-Branco et al., 2005). Additionally, a decrease in vaginal smooth muscle cells abruptly occurs with the onset of menopause (Semmens and Wagner, 1982). Estrogen replacement therapies aid in alleviating some of the dryness and discomfort associated with changes in the epithelial layer, but the effect these treatments have on the vaginal smooth muscle cells in the muscularis is unclear. Animal models, such as the rat (Basha et al., 2013) and mouse (Clark et al., 2019), were used to determine the effect of these biological factors, but the implications on the active properties of the vagina remain unknown.

The scarcity of human samples involved in active mechanical testing is likely due to the logistical hurdles that currently exist for maintaining smooth muscle cell viability. Utilizing animal models in the benchtop setting allows researchers to immediately explant the tissue, minimizing the time between sample acquisition and contractility testing. Animal model findings allow for the groundwork to be made concerning the human vaginal mechanics. For example, previous studies utilizing animal models investigating contractility pointed to the vagina being responsive to adrenergic nerve receptors (Owman and Sjöberg, 1966; Kim et al., 2002). Applying this previous knowledge, the role of cholinergic nerves in the human was first investigated by Basha et al., wherein one human sample was tested and contracted in response to carbochol and potassium (Kim et al., 2002).

Though much remains to be discovered concerning the baseline vaginal contractility, certain pelvic floor disorders, such as pelvic organ prolapse, and the relation to smooth muscle content are of scientific interest. Despite the prevalence of pelvic organ prolapse in women, only one study has directly evaluated the effect of pelvic organ prolapse on human vaginal contractility along the anterior vaginal wall (Northington et al., 2011). Vaginal samples from patients with prolapse had a smaller proportion of muscle within the longitudinal strip, however, the prolapse patients generated higher contractile forces in response to potassium chloride. Though the increased response to potassium chloride was surprising, the prolapse samples did not contract in response to phenylephrine stimulation, unlike the control samples, suggesting a potential difference in the cholinergic and adrenergic vaginal receptors as a function of prolapse development. Immunofluorescence microscopy confirmed fewer adrenergic receptors in vaginal samples with prolapse, indicating that prolapse induces a decrease in vaginal contractility in response to adrenergic stimulation (Northington et al., 2011).

6.2.4 Contributions from the nonfibrous ground matrix

Biological soft tissues are expected to exhibit viscoelastic properties given the suitable testing methods and conditions (Fung, 1993). For the human vagina, the viscoelastic nature was confirmed for the first time by excising sections of the vaginal muscularis from women diagnosed with pelvic organ prolapse (Ettema et al., 1998). Previous reports that investigated the viscoelastic nature of the vagina utilized human obtained samples (Pena et al., 2009; Feola et al., 2010; Chuong et al., 2014) as well as swine (Pack et al., 2019) and rat (Weli et al., 2017) animal models.

To investigate both elastic and viscoelastic properties of the human vagina, Pena et al. (2009) further investigated the viscoelastic properties of the vagina by using previously obtained uniaxial data and developing a mathematical model to describe the viscoelastic response of vaginal tissue. The model formulated was an essential first step towards elucidating the viscoelastic nature of the human vagina and paved the way for future endeavors. Additionally, Feola et al. utilized single-lap shearing testing techniques (Feola et al., 2013, 2010). Samples included in the study are premenopausal women with and without prolapse and postmenopausal women on and off hormone therapy replacement. Interestingly, this study measured the *complex modulus*, or measurements of both the elastic and loss modulus contributions, where postmenopausal women with prolapse on hormone therapy demonstrated the highest complex modulus. Future investigations are required to adequately appreciate the role steroid hormones and hormone replacement therapy have on the vaginal viscoelastic properties.

Studies that have characterized the elastic passive and active properties of the vagina utilized *ex vivo* techniques, where the organ samples were taken from the body and tested in human-controlled conditions that simulated the body. *Ex vivo* testing methods are limited by the necessity of explanting the tissue from the body. Even under ideal conditions, connective tissues are disrupted, and the geometry of the organ is altered. Testing conditions that allow for *in vivo* methods, or methods that allow for mechanical testing directly from the body, negate this limitation, albeit are less common. *In vivo* techniques to characterize the viscoelastic properties of the vagina include applying a suction pressure to the anterior vaginal wall, such as that conducted in the Chuong et al. study, which that determined the vaginal wall was more compliant than the suprapubic region. In addition, the vaginal wall did not store as much recoverable energy as the suprapubic region (Chuong et al., 2014).

6.2.5 Summary

To summarize, previous uniaxial tensile testing methods quantified the mechanical properties of the human vagina with regards to contributions from collagen and elastic fibers and the nonfibrous ground matrix. Due to the sensitivity of smooth muscle cells, however, the previous tests did not account for the active contractile function of the human vagina. To remedy this, uniaxial tension, biaxial tension, and extension-inflation testing are employed to characterize vaginas from animal models, such as nonhuman primates, ewe, swine, and rodents. For a summary of the mechanical testing techniques used to study the vagina, see Fig. 6.3. Future investigations would greatly benefit from adopting these animal model techniques to better understand the human vaginal mechanical properties regarding smooth muscle cell contractility. These investigations would be instrumental in improving the current therapeutics for pelvic floor disorders, such as pelvic organ prolapse, which is associated with a decrease in the smooth muscle cell and glycosaminoglycan content. Furthermore, most previously mentioned studies categorized patients as either pre-or post-menopausal, often excluding one group to mechanically characterize the other. This binary paradigm has essentially grouped an 18-year-old patient like a 40-year-old patient, even though recent evidence suggests that the hormonal profile, glycosaminoglycan, and smooth muscle content change as a patient ages closer to menopause but has not reached complete menopausal status (Semmens and Wagner, 1982; Pandit and Ouslander, 1997; Berman et al., 1998; Anand et al., 2012; Ripperda et al., 2017). Expanding the ages of patients included in the mechanical characteriza-

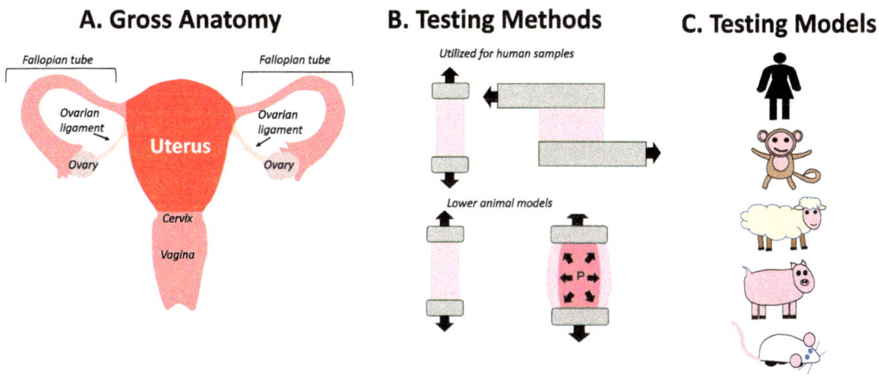

FIGURE 6.4

No permission required. The mechanical overview of the uterus. The uterus is located above the cervix and medial to the fallopian tubes (A). Investigations employed uniaxial tensile testing, shear-lap, and extension inflating testing to investigate the uterine properties of the uterus (B). In addition to the human-derived samples, animal models are used, such as the nonhuman primate, ewe, swine, and rodent.

tion is essential for fully understanding the mechanical properties of the vagina and bettering the treatment options for both young and older patients.

6.3 Uterus

6.3.1 Overview

The uterus is a hollow, muscular, organ located in the female pelvic floor between the bladder and the rectum. The human uterus is a fibrous pear-shaped organ that is approximately 7.6 cm long, 4.5 cm wide, and 3.0 cm thick (DeFranco and Evans, 2014) (Fig. 6.4A). Given that these values are averages, it is imperative to note that they change with age, parity, and preexisting conditions. The uterus primarily functions to house the developing fetus during gestation, but the organ also underlies the hormonal fluctuations that accompany the typical 28-day menstrual cycle for the nonpregnant woman. Furthermore, the uterus has an immense mechanical function to contract forcefully during menstruation as well as for safe parturition and postpartum healing. From an engineering standpoint, the hollow organ undergoes an immense amount of growth and remodeling from birth to puberty, during the 28-day menstrual cycle, during pregnancy and postpartum healing, and during the onset of menopause.

To accomplish these diverse and critical functions, the uterus is comprised of three layers: the endometrium, myometrium, and peritoneum. The endometrium is composed of glandular cells that produce secretions; the endometrium is shed during menstruation. The myometrium is the thickest layer of the uterine wall and is primarily smooth muscle cells; it expands during gestation and causes contractions during labor. The outermost layer, the serosa, is primarily composed of epithelial cells that create a smooth surface and allows natural movement in the pelvic floor (Metaxa-Mariatou et al., 2002; Leppert, 1991).

Despite this clear mechanical function and constant remodeling, the uterus is vastly understudied from the biomechanical perspective. However, recent investigations employed uniaxial tensile and compression techniques and indentation testing methods to characterize the biomechanics of the uterus (Fang, 2021; Conrad et al., 1966; Pearsall and Roberts, 1978; Myers et al., 2008) (Fig. 6.4B). More physiological testing methods, such as biaxial tensile testing and biaxial extension inflation employed animal models, such as the mouse, rat, and, swine to mechanically test the uterus (Conway et al., 2019; Manoogian et al., 2008; Harkness and Harkness, 1959) (Fig. 6.4C), but also in vivo nondestructive aspiration or indentation experiments (Bauer et al., 2007; Shi et al., 2019). Just as previously mentioned regarding the vagina, the mechanical testing and data analysis techniques are based on the previous work of early biomechanical engineers and largely under the assumption that the organ is elastic, nonlinear, anisotropic, and incompressible. Some aspects were studied in human obtained samples and some are addressed in animal models as detailed below with an emphasis on the human obtained mechanical properties.

6.3.2 Contributions from collagen and elastic fibers

Much like the other organs in the pelvic floor, the uterus has significant mechanical contributions from collagen and elastic fibers; thus these properties are the most in-depth facet to human uterine mechanics. In one of the first studies to mechanically test human uterus tissues, Conrad et al. obtained 92 pregnant and nonpregnant samples from patients through uniaxial tensile testing techniques (Conrad et al., 1966). Nonpregnant patients experienced greater stress at lower strains, indicating a higher elastic modulus or increased material stiffening; age, parity, and menstrual phase did not affect the material properties.

In addition to the tensile testing methods above, Pearsall and Roberts tested samples from the human uterus in both a tensile and compression fashion to demonstrate that the uterus is anisotropic and nonlinear (Pearsall and Roberts, 1978). Specifically, the group investigated the myometrium, or the muscle layer, of the uterus in both tension and compression and found that the mechanical properties of the human uterus are region-dependent, specifically the outer myometrium was more compliant as compared to the inner myometrium during compression testing, but the exact opposite pattern occurred during tensile testing (Pearsall and Roberts, 1978). This study was instrumental in shedding light on the anisotropic and nonlinear behavior of the tissue due to the collagen and muscle fiber orientation. Such information is vital for formulating mathematical models that predict mechanical behavior in the clinical setting.

The Fang et al. study further observed this nonlinear behavior by mechanically testing nonpregnant human uterine samples (Fang, 2021). Within the Fang study, optical tracking and spherical indentation techniques mechanically characterized the uterine samples. Essentially, the tissue samples are speckled with ink to help optically track the location of the dots throughout the tests (as a measure of deformation), and then a spherical indenter creates a load on the tissue, which results in deformation. Interestingly, the uterine samples in the Fang study had Young's modulus values two orders of magnitude lower than that of the Pearsall study, though the authors of the Fang investigation suspected that this may be due to analyzing the samples at equilibrium, rather than instantaneously throughout the testing period (Fang, 2021; Pearsall and Roberts, 1978). Myers et al. (2008) proposed a study comparing the mechanical response of cervical tissue in three groups: nonpregnant hysterectomy without previous vaginal delivery, nonpregnant hysterectomy with previous vaginal delivery, and pregnant hysterectomy at time of cesarean section. Samples were tested in confined and nonconfined compression and tension. It was shown that nonpregnant hysterectomy tissue was significantly stiffer than pregnant hysterectomy and that nonpregnant hysterectomy without previous vaginal delivery was stiffer than nonpregnant hysterectomy with previous vaginal delivery. These results lead to suspect that the cervical tissues have a significant change of mechanical properties during the pregnancy, and that delivery impacts the mechanism of remodeling of tissue and the induced mechanical properties.

Though the aforementioned tensile and compression tests were completed in a uniaxial fashion, the human uterus is loaded multiaxially in the body. As such, utilizing advanced testing techniques, such as biaxial tensile testing would be more physiologically relevant. Biaxial extension-inflation and biaxial tensile testing were previously employed using the mouse (Conway et al., 2019) and swine (Myers et al., 2008) model, however, how the human uterus is biaxially characterized is not fully understood. This limitation is likely due to the difficulty in obtaining the samples, lack of non-pathologic controls, and the size of testing equipment needed to test the whole human uterus. Moving forward, investing in advanced mechanical testing tools would remedy this limitation and improve the physiologic relevance of the current benchtop testing techniques.

6.3.3 Contributions from smooth muscle cells

The primary aspect of studying the active or contractile properties of the uterus has primarily been concerned with changes during pregnancy and labor, which will be discussed in a following chapter. With the clinical emphasis being on the pregnant state, little is understood about the baseline contractile state of the human uterus. Though pregnancy studies may have implications for the nonpregnant state, which was the case in Embrey and Mollison, and the insertion of a foley balloon that measured uterine contractions during labor (Embrey and Mollison, 1967), the Embrey study determined that with mechanical distention of the uterine wall, contractions increased and became more coordinated. Furthermore, the study found that pressure increased up to 80 mmHg with the distension, which, according to Laplace's Law, is correlated with an increase in uterine wall tension or stress, though only noted and not directly measured within the study (Embrey and Mollison, 1967). Despite this study having been primarily concerned with pregnant patients, the relations with pressure have additional implications for the nonpregnant state. For example, the pressure within the pelvic floor acutely increases with activities such as walking, exercise, coughing, and chronically with increasing age. How the contractile properties change with these factors is unknown but has a scientific basis for future studies, justified by clinical need.

Furthermore, mechanical characterization of the nonpregnant uterus is primarily completed in animal models due to the logistics in obtaining human samples from nonpregnant patients. One example of studies completed in the nonpregnant uterus is Sigger et al., which compared the electrical activity and contractility of uterine samples from pregnant and nonpregnant ewe samples (Sigger et al., 1984). Herein, the observation that electrical continuity was required for contraction and shed doubt that mechanotransduction was the primary source for uterine contractions; this conclusion was later opposed four years later by Parkington et al. by the use of only pregnant ewe samples (Ferlin et al., 1988).

6.3.4 Contributions from the nonfibrous ground matrix

Being a biological tissue, the uterus is a "bio-viscoelastic solid" due to the glycosaminoglycans, water content, and the secretions from the glandular cells of the endometrium. *In vivo* testing techniques, such as those performed in Mizrahi et al. (1980) are often utilized to characterize the viscoelastic nature of tissues. Mizrahi analyzed the uterine contractions of the human uterus during labor and used a strain rosette analysis to experimentally collect and computationally calculate the strain of the uterine wall during labor. One primary result of this study was that the uterus becomes more anisotropic as a function of labor progression, however, whereas this is relevant information to obstetrics and gynecologic advancement, the study did not quantify the anisotropy of the nonpregnant uterus. One potential remedy of this limitation is utilizing the same setup used by Mizrahi in annual gynecologic visits. Additionally, as this study utilized an *in vivo* experimental protocol, the difficulty in obtaining nonpregnant uterine samples may be bypassed, as it does not require the patient to donate or lose any uterine tissue.

Additional *in vivo* techniques completed by Kauer et al. served as a comparison for preciously assessed tissue in the lab benchtop setting (Kauer et al., 2002). Both investigations led by Kauer utilized aspiration tests on the same anatomical locations: the ventral, dorsal, and fundus portions of the uterus. By evaluating these locations both *in vivo* and *ex vivo*, the field was able to assess how accurate the benchtop experimental setups are. The *in vivo* testing techniques increased in stiffness and decreased stretch values compared to the benchtop *ex vivo* setups. Whereas there are mechanical differences noted between the two set up conditions, there are several reasons for this and still offer validation for the benchtop setting, such as disrupted geometry and cell-extracellular matrix interactions. One primary way to combat this difference in testing methods is advancements in protocol techniques and further investments for *in vivo* techniques, which quantify the active viscoelastic properties as neither of the mentioned studies calculated contractile and viscoelastic responses.

Harkness and Harkness (1959) present some results of loading-unloading tension and confined and nonconfined compression coupled to cervical tissue hydration and sulfated glycosaminoglycan (GAG) quantification. The hysteresis between loading and unloading and the relaxation during the compression tests showed a very significant difference between the nonpregnant hysterectomy groups and the pregnant hysterectomy group. Concurrently, it has been shown that sulfated GAG content and hydration were significantly higher in the pregnant tissue. These results allow exhibiting a huge change in terms of viscoelastic properties, with a similar viscoelastic response for both nonpregnant hysterectomy groups and almost no viscoelastic effects for the pregnant hysterectomy group.

6.3.5 Summary

Previous investigations employed testing techniques, such as uniaxial and shear testing, to mechanically characterize the human uterus. Additional testing methods to characterize the uterus include spherical indentation and a Foley catheter to measure the uterine wall distension. In addition to testing uterine samples from humans, nonhuman primates, ewes, swine, and rodents are used to study the uterine mechanics under uniaxial tensile and extension-inflation techniques, Fig. 6.4. Much of the field's current state is focused on characterizing the uterus during pregnancy and postpartum healing. Though this is relevant, studies that focus on baseline characterization are required to improve the livelihood of menstruating nonpregnant patients or postmenopausal patients with a concern of pelvic floor disorders. Additionally, the human-derived studies are only able to test sections of the uterus due to size limitations. Therefore there is a great need for further investments in advancing the available mechanical testing tools. The field would greatly benefit from whole-organ testing, as regional differences may exist. Also, whole organ testing under pressurization would greatly aid the field as the uterus undergoes immense changes in pressure in a patient's lifetime.

6.4 Uterosacral ligaments

6.4.1 Overview

In addition to the pelvic organs, the pelvic floor consists of supportive ligaments and muscles. These supportive tissues work in tandem to distribute loads on the organs and the pelvic floor in place as the body moves (Nortion, 1993). These pelvic floor ligaments and supportive muscles may become damaged with age due to chronic loading or in response to highly mechanical events, such as pregnancy or childbirth. When the supportive structures are damaged, the pelvic floor organs may move, or descend, which may result in pelvic floor disorders, such as pelvic organ prolapse (Wu et al., 2014; Diwadkar et al., 2009; Barber and Maher, 2013). To better understand these clinical situations, which may weaken or damage the pelvic floor support, we must first understand the baseline characteristics of the ligaments and muscles. Much

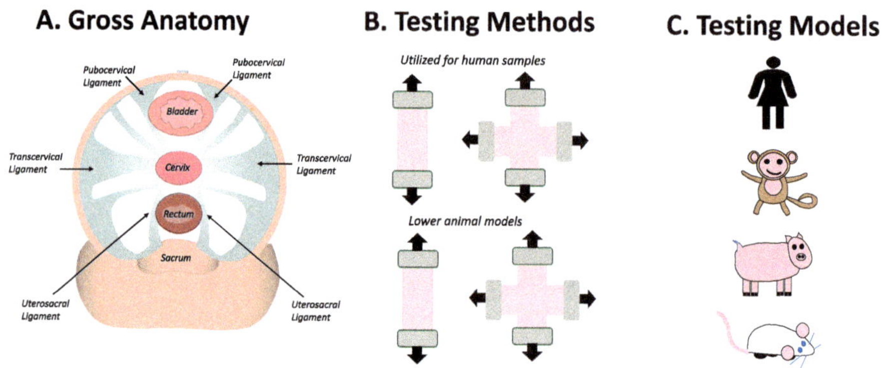

FIGURE 6.5

No permission required. Mechanical overview of the Uterosacral ligament (USL). The USL is associated with the rectum and inferior to the transcervical ligament (A). Uniaxial and biaxial testing methods characterized the mechanics of the USL (B). Each testing method was utilized in human and animal models, such as nonhuman primates, swine, and rodents (C).

like the pelvic organs, this characterization is completed by carefully planned out mechanical testing and analysis. Herein, we discuss the mechanical characterization of two of the most supportive structures: the uterosacral ligaments (USLs) and levator ani muscle.

Within the pelvic floor support system, the USLs are a pair of fan-shaped structures that are approximately 13 cm long and attach to the sacral spine and extend around the rectum to connect to the posterior cervix and apical vagina (Siddique et al., 2006; Campbell, 1950; Vu et al., 2010) (Fig. 6.5A). These tissues are highly collagenous structures composed of parallel bundles of collagen arranged in multidirectional layers, smooth muscle cells and elastic fibers between the layers of collagen, and various nerves, blood vessels, and lymphatics (Campbell, 1950; Word et al., 2009). Interestingly, the USL is noted to have regional changes in the microstructure. Specifically, the ligament is composed of dense smooth muscle cells at the cervical attachment site, which decreases in density away from the cervix (Campbell, 1950). Furthermore, the median portion of the USL is predominately connective tissue and fewer smooth muscle cells. The sacral third is primarily loose connective tissue with intermingled fat cells (Campbell, 1950).

From a functional perspective, the USL is one of the most crucial supportive structures in the pelvic floor. In fact, in humans, the USL can support more than 17 kg of weight before failure (Buller et al., 2001). To provide optimal support of the pelvic floor, the USL is vital in allowing flexibility and movement for daily activities and adjusting to the significant changes in mechanical demands that occur during pregnancy and menstruation. The USL must also counterbalance the range of extension allowed in the pelvic floor to maintain the proper length and arrangement of organs when faced with increasing pressures through coughing, movement, age, exercise, or gravity (Chapron et al., 2003; Ludwig et al., 2020).

6.4.2 Mechanical contributions from collagen and elastic fibers

The passive mechanics of the USL is largely contributed to by extracellular matrix components of the ligament, namely the collagen and elastic fibers. Similar to most soft tissues in the body, uniaxial tensile conditions first mechanically characterized the USL. Jones et al. first investigated the resilience of the USL as it relates to symptomatic prolapse, vaginal delivery, menopausal status, and age (Reay Jones et al., 2003). *Resilience* in this context refers to the ability to absorb and release energy while deforming elastically. Within the study, resilience decreased as prolapse symptoms developed, leading to one of the first notions that the supportive ligaments, not just the pelvic organs, may play a role in the development of POP. Interestingly, the Jones study was one of the first studies to investigate the mechanical properties of the human USL and was motivated by a pathology, rather than elucidating some of the baseline characteristics of the ligament.

To best answer this need of baseline characterization, the Rivaux group dissected the left and right broad ligament, round ligament, and USL from 13 fresh cadavers and underwent uniaxial tensile testing, similar to that of the Jones study (Rivaux et al., 2013). For the first time, researchers reported a nonlinear stress-strain relation and elastic mechanical response in the USL. Furthermore, compared to the other ligaments, the USL was the most rigid under high and low deformation. Martins et al. furthered this baseline characterization by grouping patients by *gravidity* or previous pregnancy status. This was accomplished by categorizing patients as either parous or nonparous (Martins et al., 2013). Essentially, researchers excised ligaments from 15 parous and nonporous cadavers, which did not have any physical signs of development or progression of pelvic organ prolapse. Similar testing methods of uniaxial tensile testing were completed and confirmed that barring signs of pelvic floor disorder progression, nulliparous patients have stiffer ligaments and lower maximum stress than parous women.

It is noteworthy that all of the above-mentioned studies characterized the USL in a uniaxial fashion. Though not fully understood, previous findings suggest that the USL may be loaded multiaxially in the body (Tan et al., 2016; Jing et al., 2012; Becker and Vita, 2015). Danso et al. sought to further investigate multiaxial loading by testing 24 postmenopausal human USL samples in a planar biaxial fashion (Tan et al., 2016; Danso et al., 2020). Prior to mechanical testing, samples were taken from the insertion point into the uterine body and categorized as either nonprolapse controls, prolapse stage I/II, and stage III/IV. *Stage* in this context refers to the progression of prolapse and is measured according to the pelvic organ prolapse quantification (POP-Q) system, a standard measuring tool used to enhance communication between clinicians, benchtop researchers, and the general public (Bump et al., 1996; Bump, 2014). Along the main *in vivo* loading direction, the grade III/IV prolapse groups were more compliant and had a higher peak stretch as compared to the lower grade and control samples. Interestingly, the researchers reported no significant differences in mechanical properties in the perpendicular loading direction. The authors suggested this may be due to sample size and the high variability in testing human samples. Though this is a crucial step to elucidating the physiologic mechanical properties of the USL, it did not account for the smooth muscle cell composition of the ligament.

6.4.3 Contributions from contractile smooth muscle cells

The active or contractile properties of the USL are not fully understood or investigated in current experimental setups. This is more than likely due to the obstacles in obtaining fresh uterosacral ligaments, as they would need to be procured in immediate cadaveric surgeries or during corrective surgeries where the ligament is already damaged or considered pathologic. Adapting methods from animal model investigations, such as those completed by the De Vita group, which has extensively studied the mechanics of the USL, may offer a remedy to this limitation. In addition, investigations in other organ models, such as the vagina, may offer insight into characterizing the active mechanical properties of the USL (Tan et al., 2016; Feola et al., 2012; Huntington et al., 2018; Tan et al., 2015; Baah-Dwomoh et al., 2018).

6.4.4 Contributions from the nonfibrous ground matrix

Similar to the Danso study, Baah-Dwomoh et al. investigated the passive viscoelastic properties of the USL via planar biaxial tensile methods (Baah-Dwomoh et al., 2018). Interestingly, the study in question set out to examine the validity of animal models, such as swine, when investigating the USL. No noticeable differences in peak strain, secant modulus of the pre-creep response, or strain at the end of the creep tests were observed between the swine and cadaveric human samples, thus giving merit to the use of swine for USL mechanical testing. Comparative mechanical studies across species are crucial to better select and justify animal models and help bridge the gap between laboratory testing and clinical application (Declan, 2008).

This validation was furthered by Tan et al. to investigate the viscoelastic properties of the USL and cardinal ligament (CL) during planar biaxial testing (Tan et al., 2016). Herein, the USL was characterized to have a significant creep response over a two-hour time period. Similar to the results noted by Danso et al., the tissue deformed to a greater degree in the main loading direction as compared to the perpendicular loading direction. Whereas this may be due to sample size, it merits further investigation on how the USL is loaded in the body and if this loading environment changes with heavy impact situations, such as aging or childbirth. Understanding this natural loading environment will further advance our clinical therapeutics and improve vaginal health practices.

Whereas the passive elastic testing methods were primarily studied in a uniaxial fashion, the passive viscoelastic properties of the USL are primarily characterized using biaxial testing methods. This shift in testing techniques is particularly advantageous in improving the physiologic relevance to the experimental design and better understanding the mechanics of the USL. However, it should be noted that to date no investigations have characterized the active elastic or active viscoelastic studies of the USL. Because it is possible that smooth muscle cells may influence the ground matrix, future studies would benefit from including this critical microstructural component that is widely understudied.

6.4.5 Summary

Previous investigations employed uniaxial and biaxial tensile techniques to mechanically characterize the human USL, particularly with regards to the elastic and collagen fiber contributions and the mechanics of the nonfibrous ground matrix. Additional studies used animal models, such as the nonhuman primate, swine, and rodent and characterized animal USL samples in similar uniaxial and biaxial tensile setups, Fig. 6.5. Interestingly, even though the USL is composed of smooth muscle cells, the active contractile properties of the USL are not fully understood. Future investigations of the USL would benefit from adapting testing techniques from other organs, such as the vagina, to characterize the contractile contribution. As this would require immediate dissection and testing, this is likely best carried out in animal models initially to gain a baseline understanding. An additional limitation of the current state of the field is that the USL is often too large to test as a whole tissue sample. As a result, only regions of the USL are tested within each investigation, making comparisons and conclusions difficult. To remedy this, there is a call for future investigations to develop mechanical testing tools that allow for the whole tissue to be tested at one time. Testing the whole tissue would allow for the known regional differences to be accounted for and increase the clinical relevance of the benchtop experiments.

6.5 Levator ani

6.5.1 Introduction

The levator ani is the striated supportive muscle component of the pelvic floor, including the pubococcygeus, iliococcygeus, and coccygeus muscles. The pubococcygeus muscle runs from the pubic bone to the center of the perineum. This muscle also aids in controlling the flow of urine and weakens for some patients as a function of gravidity, which may result in urinary incontinence (Wu et al., 2010). The iliococcygeus muscle extends more laterally from the obturator internus muscle to the tail bone. The coccygeus muscle begins on either lateral side of the pubic symphysis and encircles the rectum and is associated with the external anal sphincter. Though each muscle is unique in origin and insertion points, they are collectively regarded as one broadsheet of striated muscle in the posterior pelvic floor (Fig. 6.6A).

The levator ani, acting almost like a shelf, supports the reproductive system by the tonic contractions generated by the muscle wall (Baah-Dwomoh et al., 2018). The levator ani also helps manage mechanical pressures during daily movement and distributes the forces generated by the limbs on the pelvic cavity during walking and

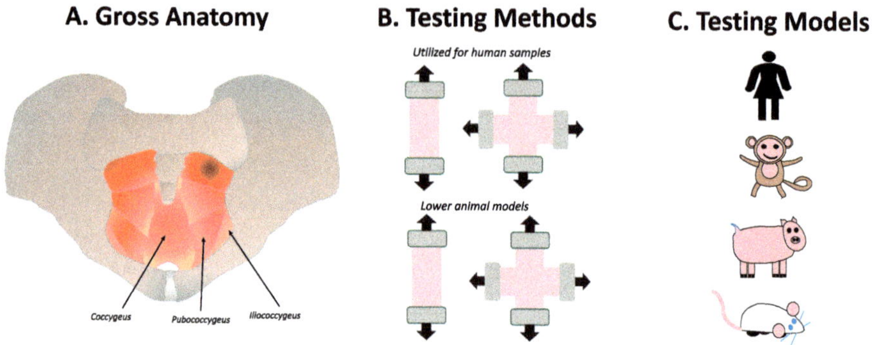

FIGURE 6.6

No permission required. The mechanical overview of the levator ani. The levator ani is comprised of three separate muscles: the coccygeus, pubococcygeus, and iliococcygeus, widely regarded as one broad sheet of supportive muscle (A). Planar uniaxial and biaxial testing techniques were employed to mechanically test the human levator ani (B). In addition to the human samples, animal models tested include nonhuman primates, swine, and rodents.

other movements. More specific to dynamic changes in the female pelvic floor, the levator ani may lose the supportive function due to damage or weakening caused by aging or childbearing, which often leads to pelvic floor disorders, such as pelvic organ prolapse (Ashton-Miller and DeLancey, 2009; Abramowitch et al., 2009; Thyer et al., 2008).

Previous investigations employed planar uniaxial tensile and biaxial tensile testing to mechanically characterize the human levator ani (Fig. 6.6B) (Becker and Vita, 2015; Smith et al., 2013; Luo et al., 2014). As human samples are logistically difficult to obtain, animal models, such as the rodent, swine, and nonhuman primate, are often used to study the mechanical properties of the levator ani (Fig. 6.6C) (Moalli et al., 2005; Tan et al., 2015; Vardy et al., 2005). Interestingly, in the aforementioned organs and tissues, the organ function is generally the same across species. The levator ani, however, has different functions in human and animal models. In animal models, such as rodents, the levator ani is key for holding up and maneuvering the tail. As humans have adapted to walking upright, we lost the need for a tail. As such, the function has evolved from tail support to opposing downward forces, such as abdominal pressure and gravity. With that, it is even more crucial to think critically about the need for refined translational medicine as we utilize animal models to understand the mechanical response and functions of the human levator ani. Because of this limitation, the chapter will focus on the underpinnings for the human levator ani (Abramowitch et al., 2009; Moalli et al., 2005).

6.5.2 Contributions from collagen and elastic fibers

The passive mechanical properties of the levator ani were primarily measured to develop better constitutive models that would aid in using computer models to predict possible damage during childbirth. In an example, Nagle et al. utilized nine levator ani samples from female cadavers in uniaxial tensile testing (Nagle et al., 2014). Herein, the obtained stress-strain data was fit to four different constitutive models. From the models chosen, the donor's age was not a significant factor in either the constitutive parameters or the mechanical testing data, though this is not too surprising, considering all samples derived from postmenopausal patients (64–96 years). Though a great first step in elucidating the mechanics of the levator ani, sample sizes of only postmenopausal patients introduces a major limitation when considering the basic science and clinical applications of the levator ani mechanics. As one of the primary ways the levator ani is damaged is through childbirth, the menopausal status may be of significance when characterizing the mechanical properties of the levator ani and should be a future study.

Further studies, such as Burnett et al., partially mediated this limitation (Burnett et al., 2020). Herein there was a younger (<52 years with median 43.5 years) and an older (>52 years with median 74.9 years) group. Utilizing a passive loading protocol, the older samples were stiffer compared to the younger groupings.

6.5.3 Contributions from skeletal muscle cells

Contrary to the previous organs, the levator ani is composed of skeletal muscle, rather than smooth muscle. Even though a different muscle type than those previously mentioned, the skeletal muscle also aids the tissue in providing support to the pelvic floor. Because of the role that the levator ani plays in vaginal support and pregnancy, it is assumed that the biomechanical response of the levator ani would change as a function of pregnancy. However, the degree to which the levator ani changed was not well understood until Davidson et al. utilized samples from 59 women to quantify the changes in levator ani structural stiffness and active force changes during pregnancy and postpartum (Davidson et al., 2020). Interestingly, there was no change in structural stiffness, but there was among the post-partum time points, depending on whether the patient had a vaginal or cesarean delivery. Furthermore, there were significant changes in the active forces generated by the levator ani in both the pregnancy and post-partum timepoints.

Though the levator ani plays a major role in pregnancy and child birthing, it fundamentally functions as a level for pelvic floor support. To date, however, the active baseline characteristics of the human levator ani have not been quantified. This is likely due to the nature of obtaining samples and maybe remedied as *in vivo* testing methods progress and become more integrated into the routine clinical setting. Possible ways to include this in routine settings are utilizing *in vivo* testing setups in annual gynecological appointments across pre- and post-menopausal women.

Examples of this *in vivo* assessment, though not with healthy levator ani tissue, were demonstrated by Derpapas et al. (2012). The study utilized 3D/4D translabial ultrasound to measure the outer diameter of the pelvic floor, which quantified deformation as a function of the women exhaling to a set pressure (the load). The plotted deformation curve was then used to quantify *compliance*, the inverse of structural stiffness, and found that the levator ani from patients with more pronounced prolapse development was less compliant.

6.5.4 Contributions from the nonfibrous ground matrix

To date, there are no studies that characterize the sole viscoelastic nature of the human levator ani, likely due to the manner of obtaining samples. Though *in vivo* studies characterized by Derpapas and Davidson do have viscoelastic contributions through the smooth muscle cells and proteoglycans, it is difficult to delineate the sole contributions. The viscoelastic properties of the levator ani may be best studied using animal models, such as rodents, but are limited by the key functional differences of the levator ani in animal models and human samples.

To minimize this limitation Jing et al. utilized 6 squirrel monkeys, 3 pregnant and 3 control, to undergo stress-relaxation testing (Jing, 2010). The squirrel monkey is beneficial, because it is an animal model, so samples are logistically easier to obtain and at the same time anatomically similar to humans. Though the sample size was smaller than those reported above, it was noted that pregnancy significantly affected the levator ani's viscosity. However, the material stiffness did not change between the pregnant and control groups, which the authors suggested may be due to the smaller sample size. Moving forward, animal models must be carefully chosen to understand the viscoelastic contributions of the levator ani, with a preference towards bipedal models.

6.5.5 Summary

Previous investigations employed uniaxial and biaxial tensile testing to characterize the elastic and collagen fibers and the nonfibrous ground matrix mechanical contributions to the human levator ani. Additional studies utilized animal models, such as the nonhuman primate, swine, and rodent and characterized animal levator ani samples in similar uniaxial and biaxial tensile setups. For a summary of the mechanical testing techniques used to study the levator ani, see Fig. 6.6. Post-menopausal patients or patients with preexisting pelvic floor disorders are two central populations in the current mechanical investigations of the levator ani. As such, little is known about premenopausal or healthy baseline properties of the levator ani. This knowledge gap is crucial to fill, because the levator ani functions as a supportive shelf for both pre- and post-menopausal women and can be damaged during the process of childbirth. If the treatment options for pelvic floor disorders or postpartum healing are ever to advance, filling gaps of knowledge, such as the baseline mechanical characterization, is crucial and should be a focus moving forward.

References

Abramowitch, S., Feola, A., Jallah, Z., Moalli, P., 2009. Tissue mechanics, animal models, and pelvic organ prolapse: a review. European Journal of Obstetrics & Gynecology and Reproductive Biology 144, 146–158.

Akintunde, A., Robison, K., Capone, D., Desrosiers, L., Knoepp, L., Miller, K., 2018. Effects of elastase digestion on the murine vaginal wall biaxial mechanical response. American Society of Mechanical Engineers 141 (2).

Amin, M., Le, V., Wagenseil, J., 2012. Mechanical testing of mouse carotid arteries: from newborn to adult. Journal of Visual Experiments 60.

Anand, L.M., Wang, U.C., French, U.J., Isaacson-Schmid, U.M., Wall, U.L., Mysorekar, U.I., 2012. Estrogen affects the glycosaminoglycan layer of the murine bladder. Female Pelvic Medicine & Reconstructive Surgery 18 (3), 148–152.

Ashton-Miller, J.A., DeLancey, J.O.L., 2009. On the biomechanics of vaginal birth and common sequelae. Annual Review of Biomedical Engineering 11 (1), 163–176.

Baah-Dwomoh, A., Alperin, M., Cook, M., De Vita, R., 2018. Mechanical analysis of the uterosacral ligament: swine vs human. The Journal of the Biomedical Engineering Society 46 (12), 2036–2047.

Barber, M., Maher, C., 2013. Epidemiology and outcome assessment of pelvic organ prolapse. Including Pelvic Floor Dysfunction 24 (11), 1783–1790.

Barnhart, K.T., Izquierdo, A., Pretorius, E.S., Shera, D.M., Shabbout, M., Shaunik, A., 2006. Baseline dimensions of the human vagina. Human Reproduction 21 (6), 1618–1622.

Basha, M., Labelle, E., Northington, G., Wang, T., Wein, A., Chacko, S., 2009. Functional significance of muscarinic receptor expression within the proximal and distal rat vagina. American Journal of Physiology. Regulatory, Integrative and Comparative Physiology 297 (5).

Basha, M.E., Chang, S., Burrows, L.J., Lassmann, J., Wein, A.J., Moreland, R.S., Chacko, S., 2013. Effect of estrogen on molecular and functional characteristics of the rodent vaginal muscularis. Journal of Sexual Medicine 10 (5), 1219–1230.

Bassel-Duby, R., Olson, E.N., 2006. Signaling pathways in skeletal muscle remodeling. Annual Review of Biochemistry 75, 19.

Bauer, M., Mazza, E., Nava, A., Zeck, W., Eder, M., Bajka, M., Cacho, F., Lang, U., Holzapfel, G.A., 2007. In vivo characterization of the mechanics of human uterine cervices. Annals of the New York Academy of Sciences 1101 (1), 186–202.

Becker, W., Vita, R., 2015. Biaxial mechanical properties of swine uterosacral and cardinal ligaments. Biomechanics and Modeling in Mechanobiology 14 (3), 549–560.

Berg, H., Eiken, O., Miklavcic, L., Mekjavic, I., 2007. Hip, thigh and calf muscle atrophy and bone loss after 5-week bedrest inactivity. European Journal of Applied Physiology 99 (3), 283–289.

Berman, J.R., McCarthy, M.M., Kyprianou, N., 1998. Effect of estrogen withdrawal on nitric oxide synthase expression and apoptosis in the rat vagina. Urology 51 (4), 650–656.

Buller, J.L., Thompson, J.R., Cundiff, G.W., Krueger Sullivan, L., Schön Ybarra, M.A., Bent, A.E., 2001. Uterosacral ligament: description of anatomic relationships to optimize surgical safety. Obstetrics and Gynecology 97 (6), 873–879.

Bump, R., 2014. The POP-Q system: two decades of progress and debate. Including Pelvic Floor Dysfunction 25 (4), 441–443.

Bump, R.C., Mattiasson, A., Bø, K., Brubaker, L.P., Delancey, J.O.L., Klarskov, P., Shull, B.L., Smith, A.R.B., 1996. The standardization of terminology of female pelvic organ prolapse and pelvic floor dysfunction. American Journal of Obstetrics and Gynecology 175 (1), 10–17.

Burnett, L.A., Cook, M., Shah, S., Michelle Wong, M., Kado, D.M., Alperin, M., 2020. Age-associated changes in the mechanical properties of human cadaveric pelvic floor muscles. Journal of Biomechanics 98.

Camanho, P.P., Tavares, C.M.L., de Oliveira, R., Marques, A.T., Ferreira, A.J.M., 2005. Increasing the efficiency of composite single-shear lap joints using bonded inserts. Composites. Part B, Engineering 36 (5), 372–383.

Campbell, R., 1950. The anatomy and histology of the sacrouterine ligaments. American Journal of Obstetrics and Gynecology 59 (1), 1–12.

Capone, D., Clark, G., Bivona, D., Ogola, B., Desroseiers, L., Knoepp, L., Lindsey, S., Miller, K., 2019. Evaluating residual strain throughout the murine female reproductive system. Journal of Biomechanics 82, 299–306.

Castelo-Branco, C., Cancelo, M.J., Villero, J., Nohales, F., Juliá, M.D., 2005. Management of post-menopausal vaginal atrophy and atrophic vaginitis. Maturitas 52 (1), 46–52.

Caulk, A., Nepiyushchikh, Z., Shaw, R., Dixon, B., Gleason, R., 2015. Quantification of the passive and active biaxial mechanical behavior and microstructural organization of rat thoracic ducts. Journal of the Royal Society Interface 12.

Chapron, C., Fauconnier, A., Vieira, M., Barakat, H., Dousset, B., Pansini, V., VacherLavenu, M.C., Dubuisson, J.B., 2003. Anatomical distribution of deeply infiltrating endometriosis: surgical implications and proposition for a classification. Human Reproduction 18 (1), 157–161.

Chuong, C.-J., Ma, M., Eberhart, R.C., Zimmern, P., 2014. Viscoelastic properties measurement of the prolapsed anterior vaginal wall: a patient-directed methodology. European Journal of Obstetrics and Gynecology 173 (1), 106–112.

Clark, G.L., Pokutta-Paskaleva, A.P., Lawrence, D.J., Lindsey, S.H., Desrosiers, L., Knoepp, L.R., Bayer, C.L., Gleason, R.L., Miller, K.S., 2019. Smooth muscle regional contribution to vaginal wall function. Interface Focus 9 (4), 20190025.

Connizzo, B.K., 2013. Effect of age and proteoglycan deficiency on collagen fiber re-alignment and mechanical properties in mouse supraspinatus tendon. Journal of Biomechanical Engineering 135 (2), 021019.

Conrad, J.T., Johnson, W.L., Kuhn, W.K., Hunter, C.A., 1966. Passive stretch relationships in human uterine muscle. American Journal of Obstetrics and Gynecology 96 (8), 1055–1059.

Conway, C.K., Qureshi, H.J., Morris, V.L., Danso, E.K., Desrosiers, L., Knoepp, L.R., Goergen, C.J., Miller, K.S., 2019. Biaxial biomechanical properties of the nonpregnant murine cervix and uterus. Journal of Biomechanics 94.

Da Silva, L.F.M., Carbas, R.J.C., Critchlow, G.W., Figueiredo, M.A.V., Brown, K., 2009. Effect of material, geometry, surface treatment and environment on the shear strength of single lap joints. International Journal of Adhesion and Adhesives 29 (6), 621–632.

Damaser, M.e., Hoyte, L.P.J.e., 2016. Biomechanics of the Female Pelvic Floor. Academic Press Is an Imprint of Elsevier, London, UK, London, UK.

Danso, E.K., Schuster, J.D., Johnson, I., Harville, E.W., Buckner, L.R., Desrosiers, L., Knoepp, L.R., Miller, K.S., 2020. Comparison of biaxial biomechanical properties of post-menopausal human prolapsed and non-prolapsed uterosacral ligament. Scientific Reports 10 (1), 7386.

Davidson, M.J., Nielsen, P.M.F., Taberner, A.J., Kruger, J.A., 2020. Change in levator ani muscle stiffness and active force during pregnancy and post-partum. International Urogynecology Journal 31 (11), 2345.

Declan, B., 2008. Translational research: crossing the valley of death. Nature 453 (7197), 840.

DeFranco, E.e., Evans, A.T.e., 2014. Manual of Obstetrics, eighth edition. Wolters Kluwer Health, Philadelphia.

Derpapas, A., Digesu, A.G., Vijaya, G., Fernando, R., Khullar, V., 2012. Real-time in vivo assessment of levator ani muscle deformation in women. European Journal of Obstetrics & Gynecology and Reproductive Biology 165 (2), 352–356.

Diwadkar, B.G., Barber, D.M., Feiner, E.B., Maher, E.C., Jelovsek, E.J., 2009. Complication and reoperation rates after apical vaginal prolapse surgical repair: a systematic review. Obstetrics and Gynecology 113 (2, Part 1), 367–373.

Embrey, M.P., Mollison, B.G., 1967. The unfavourable cervix and induction of labour using a cervical balloon. BJOG: An International Journal of Obstetrics & Gynaecology 74 (1), 44–48.

Eshel, H., Lanir, Y., 2001. Effects of strain level and proteoglycan depletion on preconditioning and viscoelastic responses of rat dorsal skin. Annals of Biomedical Engineering 29 (2), 164–172.

Ettema, G.J.C., Goh, J.T.W., Forwood, M.R., 1998. A new method to measure elastic properties of plastic-viscoelastic connective tissue. Medical Engineering & Physics 20 (4), 308–314.

Fang, S., 2021. Anisotropic mechanical properties of the human uterus measured by spherical indentation. Annals of Biomedical Engineering.

Feloa, A., Moalli, P., Alperin, M., Duerr, R., Gandley, R., Abramowitch, S., 2010. Impact of pregnancy and vaginal delivery on the passive and active mechanics of the rat vagina. Annals of Biomedical Engineering 39 (1), 549–558.

Feola, A., Abramoitch, S., Stein, Z., Barone, W., Palcsey, S., Moalli, P., 2012. Deterioration in biomechanical properties of the vagina following implantation of a high-stiffness prolapse mesh. International Journal of Obstetrics and Gynaecology 120 (2).

Feola, A., Abramowitch, S., Jones, K., Stein, S., Moalli, P., 2010. Parity negatively impacts vaginal mechanical properties and collagen structure in rhesus macaques. American Journal of Obstetrics and Gynecology 203 (6), 595.e1–595.e8.

Feola, A., Duerr, R., Moalli, P., Abramowitch, S., 2013. Changes in the rheological behavior of the vagina in women with pelvic organ prolapse. Including Pelvic Floor Dysfunction 24 (7), 1221–1227.

Ferlin, A., Pengo, M., Pizzol, D., Carraro, U., Frigo, A.C., Foresta, C., 1988. Co-ordination of electrical activity in the myometrium of pregnant ewes. Journal of Reproduction and Fertility 82 (2), 697–705.

Fung, Y., 1993. Mechanical properties and active remodeling of blood vessels. Biomechanics, 321–391.

Gravina, F.S., van Helden, D.F., Kerr, K.P., de Oliveira, R.B., Jobling, P., 2014. Phasic contractions of the mouse vagina and cervix at different phases of the estrus cycle and during late pregnancy (Research Article). PLoS ONE 9 (10).

Green, A.E., Naghdi, P.M., 1995. A unified procedure for construction of theories of deformable media. I. Classical continuum physics. Proceedings - Royal Society. Mathematical and Physical Sciences 448 (1934), 335–356.

Harkness, M.L., Harkness, R.D., 1959. Changes in the physical properties of the uterine cervix of the rat during pregnancy. The Journal of Physiology 148, 524.

Humphrey, J., 2002. Cardiovascular Solid Mechanics: Cells, Tissues, and Organs. Springer, New York.

Humphrey, J.D., Dufresne, E., Schwartz, M.A., 2014. Mechanotransduction and extracellular matrix homeostasis. Nature Reviews. Molecular Cell Biology, 802–812.

Huntington, A., Rizzuto, E., Abramowitch, S., Prete, Z., De Vita, R., 2018. Anisotropy of the passive and active rat vagina under biaxial loading. Annals of Biomedical Engineering 47, 272–281.

Jean-Charles, C., Rubod, C., Brieu, M., Boukerrou, M., Fasel, J., Cosson, M., 2010. Biomechanical properties of prolapsed or non-prolapsed vaginal tissue: impact on genital prolapse surgery. Including Pelvic Floor Dysfunction 21 (12), 1535–1538.

Jing, D., 2010. Experimental and Theoretical Biomechanical Analyses of the Second Stage of Labor.

Jing, D., Ashton-Miller, J.A., Delancey, J.O.L., 2012. A subject-specific anisotropic visco-hyperelastic finite element model of female pelvic floor stress and strain during the second stage of labor. Journal of Biomechanics 45 (3), 455–460.

Kauer, M., Vuskovic, V., Dual, J., Szekely, G., Bajka, M., 2002. Inverse finite element characterization of soft tissues. Medical Image Analysis 6 (3), 275–287.

Kendall, K., 1975. Crack propagation in lap shear joints. Journal of Physics. D, Applied Physics 8 (5), 512–522.

Kim, N.N., Min, K., Huang, Y.-H., Goldstein, I., Traish, A.M., 2002. Biochemical and functional characterization of alpha-adrenergic receptors in the rabbit vagina. Life Sciences 71 (24), 2909–2920.

Knight, K., Moalli, P., Nolfi, A., Palcsey, S., Barone, W., Abramowitch, S., 2016. Impact of parity on ewe vaginal mechanical properties relative to the nonhuman primate and rodent. Including Pelvic Floor Dysfunction 27 (8), 1255–1263.

Knight, Katrina, Moalli, Pamela, Palcsey, Stacy, Barone, William, Abramowitch, S.D., 2014. Parity Negatively Impacts the Uniaxial Mechanical Properties of the Vagina in the Ewe.

Lees, S., Davidson, C.L., 1977. The role of collagen in the elastic properties of calcified tissues. Journal of Biomechanics 10 (8), 473,477–475,486.

Leppert, P.C., 1991. Three-dimensional structures of uterine elastic fibers: scanning electron microscopic studies. Connective Tissue Research 27 (1), 15.

Lopez, S., Eberhart, R., Zimmern, P., Chuong, C.-J., 2015. Influence of body mass index on the biomechanical properties of the human prolapsed anterior vaginal wall. Including Pelvic Floor Dysfunction 26 (4), 519–525.

Ludwig, S., Göktepe, S., Mallmann, P., Jäger, W., 2020. Evaluation of different 'tensioning' of apical suspension in women undergoing surgery for prolapse and urinary incontinence. In Vivo 34 (3), 1371.

Luo, J., Smith, T.M., Ashton-Miller, J.A., Delancey, J.O.L., 2014. In vivo properties of uterine suspensory tissue in pelvic organ prolapse. Journal of Biomechanical Engineering 136 (2), 021016.

Manoogian, S.J., McNally, C., Stitzel, J.D., Duma, S.M., 2008. Dynamic biaxial tissue properties of pregnant porcine uterine tissue. Stapp Car Crash Journal 52, 167.

Martins, P., Silva-Filho, A., Fonseca, A., Santos, A., Santos, L., Mascarenhas, T., Jorge, R., Ferreira, A., 2013. Strength of round and uterosacral ligaments: a biomechanical study. Archives of Gynecology and Obstetrics 287 (2), 313–318.

McLoughlin, A., Wylie, Linda, 2006. Essential Anatomy and Physiology in Maternity Care, 1st edn. Churchill Livingstone, p. 241. Price £18.95.

Metaxa-Mariatou, V., McGavigan, C.J., Robertson, K., Stewart, C., Cameron, I., Campbell, S., 2002. Elastin distribution in the myometrial and vascular smooth muscle of the human uterus. Molecular Human Reproduction 8 (6), 559–565.

Mizrahi, J., Karni, Z., Polishuk, W.Z., 1980. Isotropy and anisotropy of uterine muscle during labor contraction. Journal of Biomechanics 13 (3), 211–218.

Moalli, P.A., Howden, N.S., Lowder, J.L., Navarro, J., Debes, K.M., Abramowitch, S.D., Woo, S.L.Y., 2005. A rat model to study the structural properties of the vagina and its supportive tissues. American Journal of Obstetrics and Gynecology 192 (1), 80–88.

Myers, K.M., Paskaleva, A.P., House, M., Socrate, S., 2008. Mechanical and biomechanical properties of human cervical tissue. Acta Biomaterial 4 (1), 104–116.

Nagle, A.S., Barker, M.A., Kleeman, S.D., Haridas, B., Douglas Mast, T., 2014. Passive biomechanical properties of human cadaveric levator ani muscle at low strains. Journal of Biomechanics 47 (2), 583–586.

Northington, G., Basha, M., Arya, L., Wein, A., Chacko, S., 2011. Contractile response of human anterior vaginal muscularis in women with and without pelvic organ prolapse. Journal of Reproductive Science 18 (3), 296–303.

Nortion, P., 1993. Pelvic floor disorders: the role of fascia and ligaments. Clinical Obstetrics & Gynaecology 36, 926–938.

Owman, C., Sjöberg, N.-O., 1966. Adrenergic nerves in the female genital tract of the rabbit. With remarks on cholinesterase-containing structures Zeitschrift für Zellforschung und Mikroskopische Anatomie 74 (2), 182–197.

Pack, E., Dubik, J., Snyder, W., Simon, A., Clark, S., De Vita, R., 2019. Biaxial stress relaxation of vaginal tissue in pubertal gilts. Journal of Biomechanical Engineering 142 (3).

Pandit, G.L., Ouslander, G.J., 1997. Postmenopausal vaginal atrophy and atrophic vaginitis. The American Journal of the Medical Sciences 314 (4), 228–231.

Pearsall, G.W., Roberts, V.L., 1978. Passive mechanical properties of uterine muscle (myometrium) tested in vitro. Journal of Biomechanics 11 (4), 167,173–171,176.

Pena, E., Calvo, B., Martinez, M., Martins, P., Mascarenhas, T., Jorge, R., Ferreira, A., 2009. Experimental study and constitutive modeling of the viscoelastic mechanical properties of human prolasped vaginal tissue. Biomechanics and Modeling in Mechanobiology 9 (1), 35–44.

Peña, E., Martins, P., Mascarenhas, T., Natal Jorge, R.M., Ferreira, A., Doblaré, M., Calvo, B., 2011. Mechanical characterization of the softening behavior of human vaginal tissue. Journal of the Mechanical Behavior of Biomedical Materials 4 (3), 275–283.

Reay Jones, N.H.J., Healy, J.C., King, L.J., Saini, S., Shousha, S., Allen-Mersh, T.G., 2003. Pelvic connective tissue resilience decreases with vaginal delivery, menopause and uterine prolapse. British Journal of Surgery 90 (4), 466–472.

Ripperda, C., Maldonado, P., Acevedo, J., Keller, P., Akgul, Y., Shelton, J., Word, R.A., 2017. Vaginal estrogen: a dual-edged sword in postoperative healing of the vaginal wall. North American Menopause Society 24 (7), 838–849.

Rivaux, G., Rubod, C., Dedet, B., Brieu, M., Gabriel, B., Cosson, M., 2013. Comparative analysis of pelvic ligaments: a biomechanics study. Including Pelvic Floor Dysfunction 24 (1), 135–139.

Rivlin, R.S., 1948. Large elastic deformations of isotropic materials. IV. Further developments of the general theory. Philosophical Transactions of the Royal Society of London, Series A: Mathematical and Physical Sciences 241 (835), 379–397.

Robison, K., Conway, C., Desrosiers, L., Knoepp, L., Miller, K., 2017. Biaxial mechanical assessment of the murine vaginal wall using extension-inflation testing. Journal of Biomechanical Engineering 139 (10).

Rubod, C., Boukerrou, M., Brieu, M., Dubois, P., Cosson, M., 2007. Biomechanical properties of vaginal tissue. Part 1: new experimental protocol. The Journal of Urology 178 (1), 320–325.

Semmens, J.P., Wagner, G., 1982. Estrogen deprivation and vaginal function in postmenopausal women. JAMA 248 (4), 445.

Shi, L., Yao, W., Gan, Y., Zhao, L.Y., McKee, W.E., Vink, J., Wapner, R.J., Hendon, C.P., Myers, K., 2019. Anisotropic material characterization of human cervix tissue based on indentation and inverse finite element analysis. Journal of Biomechanical Engineering 141 (9), 091017.

Siddique, S., Gutman, R., Schön Ybarra, M., Rojas, F., Handa, V., 2006. Relationship of the uterosacral ligament to the sacral plexus and to the pudendal nerve. Including Pelvic Floor Dysfunction 17 (6), 642–645.

Sigger, J.N., Harding, R., Jenkin, G., 1984. Relationship between electrical activity of the uterus and surgically isolated myometrium in the pregnant and non-pregnant ewe. Journal of Reproduction and Fertility 70 (1), 103–114.

Skalak, R., Tozeren, A., Zarda, R.P., Chien, S., 1973. Strain energy function of red blood cell membranes. Biophysical Journal 13 (3), 245–264.

Skoczylas, L., Jallah, Z., Sugino, Y., Stein, S., Feel, A., Yoshimura, N., Moalli, P., 2013. Regional differences in rat vaginal smooth muscle contractility and morphology. Reproductive Sciences 20 (4), 382–390.

Smith, T.M., Luo, J., Hsu, Y., Ashton-Miller, J., Delancey, J.O., 2013. A novel technique to measure in vivo uterine suspensory ligament stiffness. American Journal of Obstetrics and Gynecology 209 (5), 484.e1–484.e7.

Sokolis, D., Orfanidis, I., Peroulis, M., 2011. Biomechanical testing and material characterization for the rat large intestine: regional dependence of material parameters. Institute of Physics and Engineering in Medicine 32 (12).

Tan, T., Cholewa, N., Case, S., De Vita, R., 2016. Micro- structural and biaxial creep properties of the swine uterosacral– cardinal ligament complex. The Journal of the Biomedical Engineering Society 44 (11), 3225–3237.

Tan, T., Davis, F.M., Gruber, D.D., Massengill, J.C., Robertson, J.L., De Vita, R., 2015. Histo-mechanical properties of the swine cardinal and uterosacral ligaments. Journal of the Mechanical Behavior of Biomedical Materials 42, 129.

Thyer, I., Shek, C., Dietz, H.P., 2008. New imaging method for assessing pelvic floor biomechanics. Ultrasound in Obstetrics & Gynecology 31 (2), 201–205.

Timmons, B.C., Fairhurst, A.-M., Mahendroo, M.S., 2009. Temporal changes in myeloid cells in the cervix during pregnancy and parturition. The Journal of Immunology 182 (5), 2700.

Ulrich, D., Edwards, S.L., White, J.F., Supit, T., Ramshaw, J.A.M., Lo, C., Rosamilia, A., Werkmeister, J.A., Gargett, C.E., 2012. A preclinical evaluation of alternative synthetic biomaterials for fascial defect repair using a rat abdominal hernia model (alternative biomaterials for fascial defect repair). PLoS ONE 7 (11), e50044.

Vardy, M.D., Gardner, T.R., Cosman, F., Scotti, R.J., Mikhail, M.S., Preiss-Bloom, A.O., Williams, J.K., Cline, J.M., Lindsay, R., 2005. The effects of hormone replacement on the biomechanical properties of the uterosacral and round ligaments in the monkey model. American Journal of Obstetrics and Gynecology 192 (5), 1741–1751.

Vu, D., Haylen, B., Tse, K., Farnsworth, A., 2010. Surgical anatomy of the uterosacral ligament. Including Pelvic Floor Dysfunction 21 (9), 1123–1128.

Weli, H.K., Akhtar, R., Chang, Z., Li, W.-W., Cooper, J., Yang, Y., 2017. Advanced glycation products' levels and mechanical properties of vaginal tissue in pregnancy. European Journal of Obstetrics and Gynecology 214, 78–85.

Findings from Virginia Polytechnic Institute and State University in the Area of Biomechanical Engineering Reported: Women's Health Weekly, 2019. Swine Vagina Under Planar Biaxial Loads: an Investigation of Large Deformations and Tears. Report. p. 142.

Word, R.A., Pathi, S., Schaffer, J.I., 2009. Pathophysiology of pelvic organ prolapse. Obstetrics and Gynecology Clinics of North America 36 (3), 521–539.

Wu, M.J., Matthews, A.C., Conover, M.M., Pate, M.V., Jonsson Funk, M.M., 2014. Lifetime risk of stress urinary incontinence or pelvic organ prolapse surgery. Obstetrics and Gynecology 123 (6), 1201–1206.

Wu, Q., Mao, X., Luo, L., Ying, T., Li, Q., Teng, Y., 2010. Characteristics of pelvic diaphragm hiatus in pregnant women with stress urinary incontinence detected by transperineal three-dimensional ultrasound. Zhonghua Fu Chan Ke Za Zhi 45 (5), 326–330.

Yellon, S.M., Ebner, C.A., Elovitz, M.A., 2009. Medroxyprogesterone acetate modulates remodeling, immune cell census, and nerve fibers in the cervix of a mouse model for inflammation-induced preterm birth. Reproductive Sciences 16 (3), 257–264.

CHAPTER

7 Mechanical properties of breast tissue

Sergio Dempsey[a] and Abbas Samani[a,b,c,d]

[a]*School of Biomedical Engineering, Western University, London, ON, Canada*
[b]*Department of Medical Biophysics, Western University, London, ON, Canada*
[c]*Department of Electrical and Computer Engineering, Western University, London, ON, Canada*
[d]*Imaging Research, Robarts Research Institute, Western University, London, ON, Canada*

7.1 Introduction

In a healthy female reproductive system, breasts play the vital role of nourishing offspring and increasing physical attraction, where the breast tissue mechanical properties (e.g. stiffness) play important roles in these functions. A prevalent disease of the breast in the female population is breast cancer. In 2020, breast cancer was the most diagnosed cancer, globally impacting over 2.2 million lives (Sung et al., 2021). The mechanical properties of breast tissue have become increasingly important in breast tissue disease diagnosis and treatment, including plastic surgeries post breast lumpectomy or mastectomy. This has motivated the scientific community to improve all steps in the breast cancer diagnosis and treatment pipelines. Applications of breast tissue mechanical properties have found an important role in several areas of this pipeline, such as tumor identification (Sigrist et al., 2017), pre-operative surgery planning, including colocation and image registration (Gamage et al., 2017b; García et al., 2018; Hipwell et al., 2016), image-guided procedures, such as biopsy and intraoperative procedures (Carter et al., 2009; Griesenauer et al., 2018), and post-operative surgery outcome prediction of breast augmentation (Lapuebla-Ferri et al., 2011; Myung et al., 2019; Vavourakis et al., 2016a; Haddad et al., 2016). Breast tissue mechanical properties also offer the ability to make biomechanically informed bra design, thereby optimizing comfort and quality of daily life (Cai et al., 2016; Y. Sun et al., 2021; Zhang et al., 2020). In all these applications, accurate mechanical properties will lead to accurate simulation, hence improved accuracy/success of procedures and predictions.

There are multiple, mechanically unique tissue types required to describe the overall mechanics of the breast. Major contributors to breast mechanics are the encapsulating skin, adipose tissue, fibroglandular tissue, Cooper's ligaments, tumor tissue, how these tissues connect to each other, and lastly, how they all connect to the chest wall (Gamage et al., 2019). These contributors are what this chapter will focus on as they are the most influential in deformation and currently definable in a mechanics context. Other considerations which impact the overall mechanics are physiological

Biomechanics of the Female Reproductive System: Breast and Pelvic Organs. https://doi.org/10.1016/B978-0-12-823403-7.00019-1

temperature, vasculature, lymphatics, age, and reproduction factors (Gamage et al., 2017b; Kruse et al., 2000). The latter include pregnancy, menopause, and time in menstrual cycle (Lorenzen et al., 2003).

There are two main approaches to apply knowledge of the breast tissue mechanics in the context of the above-mentioned applications. First, the knowledge of tissue intrinsic mechanics allows for the stratification of breast tissue lesions, fabrication of tissue-like materials for training with phantoms (Solanki and Raja, 2010), fabricating implants in breast augmentation (Haddad et al., 2016; Myung et al., 2019), or full breast prosthetics (Cruz et al., 2018). Second, this knowledge can be used to make reliable breast biomechanical models (Gamage et al., 2019; García et al., 2018; Hipwell et al., 2016). These biomechanical models are physics-based in-silico representations of the breast, which enable the prediction of mechanical properties, deformation under loading, and tissue stresses. The latter may impact comfort and quality of daily life, as will be discussed in later chapters, or can be potentially used as a measure to predict disease development, such as breast cancer.

This chapter serves as a review of knowledge garnered regarding the breast tissue mechanical properties through the past several years with a focus on what is required to capture the true mechanics of the breast for current and future applications. We will begin with fundamentals of how we define the mechanics of the breast. We will then explore several breast tissue mechanics measurement systems, how they compare and to what extent they are translatable. Application examples of breast tissue mechanical properties by the research and industrial community is also presented to show how this knowledge is used for specific applications. We finish the chapter with some final remarks on the needs of the clinic, and the needs of science in the application of mechanics, and future directions for the study of breast tissue mechanics.

7.2 Mechanics of breast tissue

The mechanical properties of the breast define how the breast responds to load. This load could be gravity-induced, e.g., load resulting from reorienting the chest from prone to supine positions, or it could be the result of external perturbation, e.g., breast compression during mammography procedure. This section will serve to describe a brief foundation of the fundamentals of describing deformation, then present the constitutive models used to define breast mechanics that relate deformation to force, before moving on to how to measure the parameters that are unique to breast tissue that define that relationship. We finish by presenting a global perspective of measured mechanical properties and discuss general trends.

7.2.1 A brief introduction to continuum mechanics

To describe breast deformation mathematically, an appropriate choice is to use the theory of continuum mechanics. The theory of continuum mechanics assumes that matter is infinitely divisible with no change to composition, and that the properties of the material, i.e., stiffness, density, temperature, vary by definable *continuous* functions.

With this concept, we describe the entire motion of an object as the new positions of all points of the object *in space* $\overline{x}$ for each *material* coordinate $\overline{X}$ of the original undeformed object. This is concisely put as $\overline{x} = \varphi\left(\overline{X}, t\right)$, which implies that a material point located at $\overline{X}$ will transform based on some continuous scalar function φ, which depends on time t and move to resulting new location $\overline{x}$. Naturally, it is required that at time zero, we have $\varphi\left(\overline{X}, 0\right) = \overline{X}$. This equation of motion can also be used to describe the relative change in position or *deformation* between two material points in time as well. This relative change is described mathematically by the deformation gradient tensor $\overline{\overline{\boldsymbol{F}}}_{ij} = \frac{\partial \overline{x}_i}{\partial \overline{X}_j}$, where we have explicitly shown indices to express that $\overline{\overline{\boldsymbol{F}}}_{ij}$ is a 2nd-order tensor. The deformation gradient tensor contains all the information on the relative change of motion in time of a chosen particle that can be applied in many ways. For instance, one can track differential vectors between material points using $\boldsymbol{F}$ (accents dropped) in $d\overline{x} = \boldsymbol{F} \cdot d\overline{X}$ in time. We can also track the stretch (λ), which is the ratio of the length of a directed segment from a material point to its new length in time t for any unit direction $\overline{T}$ with Eq. (7.1):

$$\lambda = \sqrt{1 - 0.5\overline{T}\left(\boldsymbol{F}^T \cdot \boldsymbol{F} - \boldsymbol{I}\right)\overline{T}}, \tag{7.1}$$

where $\boldsymbol{I}$ is the identity tensor, ^{T} represents the matrix transpose, and $0 < \lambda < \infty$. $\boldsymbol{F}$ can also be used to measure the unit elongation, which is called strain (ε) with $\varepsilon = \lambda - 1$, where if ε is negative, the material is considered compressed, and if ε is positive, it is stretched. Along with stretch, angles between two vectors at local material points can be computed with Eq. (7.2):

$$\cos(\theta) = \frac{\overline{T}_1 \cdot \boldsymbol{F}^T \cdot \boldsymbol{F} \cdot \overline{T}_2}{\lambda_1 \lambda_2}, \tag{7.2}$$

where $\overline{T}_{1,2}$ are unit direction vectors and $\lambda_{1,2}$ are the respective stretches. Observe that $\boldsymbol{F}$ is only used as the dot with itself in the equations that contain relative deformation information ($\boldsymbol{F}^T \cdot \boldsymbol{F}$). We can simplify notation by renaming this product $\boldsymbol{C}$, called the *right Cauchy–Green deformation tensor*, which is a symmetric 2nd-order tensor. Another tensor we will discuss later is the Green–Lagrange strain tensor $\boldsymbol{E} = \frac{1}{2}(\boldsymbol{C} - \boldsymbol{I})$. At small deformations, $\boldsymbol{E}$ is called the infinitesimal strain tensor $\boldsymbol{\varepsilon}$, where the diagonal components of $\boldsymbol{E}$, ε_{ii}, are the strains introduced previously, where ii do not follow Einstein notation but refer to the position in the tensor $\boldsymbol{E}$. The off-diagonal components at small deformations are related to the changes in angles.

Capturing deformation with $\boldsymbol{F}$, $\boldsymbol{C}$, $\boldsymbol{E}$, or $\boldsymbol{\varepsilon}$ depends on convenience and applicability in most cases, but what is important is that each tensor can describe stretching in any direction $\overline{T}$ and the variation of angles between directed segments in material points. This means we can describe all manners of material deformation (with exception of rigid body motion) with either of these tensors alone. However, the question of how tissue intrinsic properties play a role in the deformation is yet to be answered.

Consider the example of breast palpation to potentially identify a lesion. Via manual palpation, the breast will deform and compress, and based on the internal distribution of stiffness, will exert a resultant force on the finger. To palpate the breast, a clinician may choose to apply a constant local deformation to the breast's lesion or a healthy tissue site, which results in sensing two different reaction forces at the tip of the finger. Alternatively, the clinician may apply a constant force to the two breast sites and sense different deformations. In both cases, the different stiffness characteristics of the two sites dictate what is sensed by the clinician.

There are many types of forces that can act on a body. There are classes of forces called *body forces*, which are not associated with contact, such as gravity and magnetic forces. Forces that act on the surface of the body, which are associated with contact, are called *surface forces*. Examples include friction and forces generated by mechanical stimulation in the clinic, such as manual palpation or compression through ultrasound probe. *Internal forces* are another class of forces, which include forces generated through muscle contraction. In the context of the breast, only gravity and surface forces are applicable.

The equation for total sum of body forces is $F_b = \int \rho \overline{\boldsymbol{b}}\, dV$, where material mass density ρ and gravity force per unit mass $\overline{\boldsymbol{b}}$ can vary as functions of position and time, and the integration is over the entire body. The sum of all surface forces acting on a body is $F_s = \int \nabla \cdot \boldsymbol{\sigma} dS$, where we have introduced *Cauchy's stress tensor* $\boldsymbol{\sigma}$, which is a function of position and time. It is a symmetric second-order tensor, which tracks the spatial frame $\overline{\boldsymbol{x}}$ in time t. Cauchy's stress tensor is somewhat analogous to $\boldsymbol{F}$ (and other storing deformation tensors) in that it contains all the information on the state of stress a point is undergoing in any direction in time as opposed to deformation. Together, the body forces and surface forces sum to $\frac{d}{dt} \int \rho \overline{\boldsymbol{v}} dV$, which is a statement of Newton's 2nd law for the forces under consideration, where $\overline{\boldsymbol{v}}$ is the velocity. If density is constant in time or mass is always conserved, which is true with the breast, the derivative can be brought into the integral using Reynold's lemma, giving $\int \rho \overline{\boldsymbol{a}} dV$. Since all integrals in the global form of the equations are over the same body volume, we present the local form for the sum of forces in Eq. (7.3), called Cauchy's equation of motion:

$$\rho \overline{\boldsymbol{b}} + \nabla \cdot \boldsymbol{\sigma} = \rho \overline{\boldsymbol{a}}. \tag{7.3}$$

This is the equation that governs all forces currently considered in breast tissue. Based on the quantitative definition of deformation and stresses, it is now possible to delve into the mechanism of a material constitutive relationship, which relates deformation and stress. This relationship is unique to each material and, in this chapter's context, relates stress to strain for each breast tissue type; namely, skin, adipose, fibroglandular, and tumor.

7.2.2 Constitutive models

7.2.2.1 Constitutive model considerations

This sub-section is meant to discuss requirements of a constitutive model to qualify as a good representative of intrinsic tissue mechanics, in addition to being convenient to use mathematically. The first consideration is how to relate the deformation to stress. A colloquial example is to consider a spring. A force is exerted on a spring to elongate it and in doing so has done work on the spring imparting energy. That energy is said to be stored in the spring elongation (deformation) via the scalar energy $U = \frac{1}{2}kx^2$. This expression is the foundation of 1D Hooke's law. Its derivative with respect to displacement returns the common $F = kx$, where k is the spring-elastic constant. This single-dimensional representation of material mechanics is generally not suitable for formulating breast mechanics, as it is a 3D continuum, which necessitates considering the information carried in the deformation at each point from $\boldsymbol{F}$, $\boldsymbol{C}$, $\boldsymbol{E}$, or $\boldsymbol{\varepsilon}$.

The constitutive relationship for stress and strain postulates the existence of a strain energy density function (SEDF) that depends on the deformation of a material. This SEDF is then tied to stress via equations that involve the partial derivatives of U with respect to deformation parameters. The specific equation will depend on the form of U, convenience, and the sought form of the stress tensor. Cauchy's stress tensor introduced in Section 7.2.1 is the stress tensor in the spatial frame $\overline{x}$. Stress tensors exist for other frames of reference; for example, the first and second Piola-Kirchoff stress tensors, which are defined in $\overline{\boldsymbol{X}}$. Full forms of each relationship will not be presented here, and the reader is referred to (Holzapfel, 2001). What is important for unique tissue mechanics is the form of U. Immediately, it can be recognized that there may be potential dependence of U's on the tensors storing deformation; e.g., $U(\boldsymbol{C})$, $U(\boldsymbol{F})$, $U(\boldsymbol{E})$, and $U(\boldsymbol{\varepsilon})$. Another condition for U is the existence of a rest state for the material at zero deformation (all strains = 0), where the SEDF is zero.

Given the structure of the breast, as described in Chapter 1, it is known that there is a supporting structure of Cooper's ligaments throughout breast tissue. These ligaments are stiffer than surrounding adipose and fibroglandular tissue, because their constituents include a significant amount of stiff collagen. One option to treat these ligaments is to have a constitutive relationship for them and other tissues separately. Alternatively, it is often more convenient to have a constitutive relationship that represents the combined observed effects of ligaments dispersed throughout the background tissues of adipose or fibroglandular. The latter description is called fiber reinforced material, which brings up the concept of material anisotropy.

Isotropic tissues respond to load in an identical way when the load is applied in any arbitrary direction. In this special case, the isotropic material opens more options for the form of U related to chosen deformation tensor parameters. As $\boldsymbol{C}$ is a symmetric second-order tensor, it has special properties that are invariant to rotation and are aptly called invariants. The three principal invariants are shown in Eq. (7.4):

$$\begin{aligned} I_1 &= \lambda_1^2 + \lambda_2^2 + \lambda_3^2 \\ I_2 &= \lambda_1^2\lambda_2^2 + \lambda_2^2\lambda_3^2 + \lambda_3^2\lambda_1^2 \\ I_3 &= \lambda_1^2\lambda_2^2\lambda_3^2, \end{aligned} \tag{7.4}$$

where λ_i^2 $(i = 1, 2, 3)$ are the eigenvalues of $\boldsymbol{C}$. I_i $(i = 1, 2, 3)$ are constant under rotation, and as such it is common to see constitutive relationships based on these invariants. Since the I_i parameters depend on the principal stretches themselves, occasionally a constitutive relationship will depend on them explicitly. Either way, the I_i and λ_i parameters are based on $\boldsymbol{C}$, which contains all the information of deformation. This expands the options for U to include $U(I_1, I_2, I_3)$ or $U(\lambda_1, \lambda_2, \lambda_3)$.

Anisotropy exists when the material exhibits different mechanical properties in different directions. An example mentioned previously is the stiffness direction dependence of fibroglandular tissue, resulting from the embedded Cooper's ligaments. Special cases of anisotropy are transverse isotropic, where in only 1 direction the tissue stiffness is different, and orthotropic anisotropy, where in 3 orthogonal planes the stiffness is different. To describe anisotropy in a continuum mechanics sense, more invariants are needed since I_i $(i = 1, 2, 3)$ are not direction dependent. In the case of transverse isotropy, which is at times assumed for some breast tissue, 2 new pseudo-invariants are introduced, $I_4 = \overline{A} \cdot \boldsymbol{C}\overline{A}$, and $I_5 = \overline{A} \cdot \boldsymbol{C}^2\overline{A}$, where $\overline{A}$ is the direction of the fibers in the undeformed state. I_4 and I_5 allow stiffness to depend on 1 direction preferentially. If multiple families of fibers exist in different directions (for instance, structural anisotropy of collagen in the skin) new pseudo invariants can be made by using the fiber directions $\overline{A}$ and $\overline{B}$, making $I_{4(i,j)} = \overline{A} \cdot \boldsymbol{C}\overline{B}$, $I_{5(i,j)} = \overline{A} \cdot \boldsymbol{C}^2\overline{B}$ $(i, j = 1, 2, \ldots, \text{N})$ and $I_{9(i,j)} = (\overline{A} \cdot \overline{B})^2$, where the I parameters numbering follows (Holzapfel, 2001). Thus anisotropy opens other forms of U, including the transverse isotropic $U(I_1, I_2, I_3, I_4, I_5)$, and $U(I_1, I_2, I_3, I_{4(i,j)}, I_{5(i,j)}, I_{9(i,j)})$.

The U function will also depend on constants representing the material's unique intrinsic properties. These constants can be calculated by fitting experimental data obtained from mechanical testing. Such data is *observed* at purely macroscopic level. This follows the popular phenomenological approach of material characterization, where the premise involves considering an equivalent homogeneous model of a material made of many different constituents structured in a certain way. Another approach is called the structural approach, which seeks to develop U, which captures all intricacies of different constituents explicitly, for instance different intensities of collagen fibers in different directions and their uncoiling characteristics.

7.2.2.2 Popular constitutive models for breast tissues

The simplest constitutive model is called the generalized Hooke's law, which is a linear elastic material in the form of $U(\boldsymbol{\varepsilon}) = \frac{1}{2}\boldsymbol{\varepsilon} : \mathbb{C} : \boldsymbol{\varepsilon}$, where $\mathbb{C}$ is a 4th-order tensor containing 81 coefficients. Due to symmetries, this tensor has at most 21 parameters and in the isotropic case, reduces to 2. The relationship for U to $\boldsymbol{\sigma}$ in this case is $\boldsymbol{\sigma} = \frac{\partial U}{\partial \boldsymbol{\varepsilon}}$ which simplifies to Eq. (7.5):

$$\boldsymbol{\sigma} = \lambda Tr(\boldsymbol{\varepsilon})\boldsymbol{I} + 2\mu\boldsymbol{\varepsilon} \tag{7.5}$$

in the isotropic case. The two parameters λ and μ are called Lamé's constants and are unique to each material. It has become more convenient, however, to use two different expressions derived by solving for strain in terms of stress, which yields the more popular Young's modulus, E, and Poisson's ratio, ν, which can be obtained via

Eqs. (7.6) and (7.7) as follows:

$$E = \frac{\mu(3\lambda + 2\mu)}{\lambda + \mu}, \tag{7.6}$$

$$\nu = \frac{\lambda}{2(\lambda + \mu)} \tag{7.7}$$

In this form, E and ν have much more palpable meaning to engineers and scientists, where the Young's modulus is the slope of the stress-strain curve, and ν is a measure of compressibility (for instance highly compressible porous lung vs. incompressible high water content brain). Conversely, the inverse relationships are shown in Eqs. (7.8) and (7.9):

$$\lambda = \frac{\nu E}{(1+\nu)(1-2\nu)}, \tag{7.8}$$

$$\mu = \frac{E}{2(1+\nu)} \tag{7.9}$$

where μ represents the shear modulus and is at times symbolized as G. More advanced forms of U are typically nonlinear in their relationship to strain. This nonlinearity is attributed to the gradual unwinding and, hence, stiffening of collagen fibers during elongation, which makes the material stiffer under larger strains.

In the following table, we present the constitutive equations that have been used to describe the breast adipose and fibroglandular tissues. We have also included constitutive relationships that assume local linear elastic theory $E = \frac{\partial \sigma}{\partial \varepsilon}$. It is noteworthy that as the emphasis is on breast tissues, we have omitted models related to adipose tissues that were not harvested from the breast. There is evidence that adipose tissue samples harvested from porcine are not an accurate representation of human adipose tissue (Z. Sun et al., 2021). Moreover, human adipose tissue samples harvested from different anatomical sites, such as the lower abdomen or buttock, do not contain the suspensory Cooper's ligaments of the breast, hence are not expected to be good representative of the breast adipose. We have omitted the field of skin mechanics and, in lieu, refer the reader to the comprehensive review done by Limbert (2019). Skin was omitted, because while it has been tested extensively and is known to be nonlinear anisotropic (Jor et al., 2013; Limbert, 2019; Oomens et al., 2017), the stiffness is impacted by epidermal layer thickness, which is location-dependent, whether the patient is sweating/obese (Sandby-Møller et al., 2003), and most importantly, the anisotropic orientations cannot be measured for breast tissue. These issues limit the validity and popularity of skin mechanics data for modeling the breast, as will be shown in Section 7.3.

7.2.3 Constitutive model coefficient determination

As shown in Table 7.1, each model has a variable number of parameters that can add to the complexity of the SEDF. These parameters are generally called hyperelastic

Table 7.1 Constitutive models of breast adipose and fibroglandular tissue discussed in this chapter.

Model name	Reference	Constitutive equation
Generalized Hooke's law (LE)	(Oliver and Agelet de Saracibar, 2017)	$U = \frac{1}{2}\boldsymbol{\varepsilon} : \mathbb{C} : \boldsymbol{\varepsilon}$
neo-Hookean (NH)	(Rivlin and Saunders, 1951) (NH eqn. 19.1 M–R eqn. 1.3 5P eqn 1.2)	$U = \frac{\mu}{2}(I_1 - 3)$
Mooney–Rivlin (M–R)		$U = C_{10}(I_1 - 3) + C_{01}(I_2 - 3)$
5-Term polynomial (5P)		$U = \sum_{i,j=0}^{2} C_{ij}(I_1 - 3)^i (I_2 - 3)^j$, $C_{ij} = 0, i + j \leq 2$
Ogden (OG)	(Ogden, 1972)	$U = \sum_{i=1}^{N} \frac{2\mu}{\alpha_i^2}(\lambda_1^{\alpha_i} + \lambda_2^{\alpha_i} + \lambda_3^{\alpha_i} - 3)$
Veronda–Westmann (VW)	(Veronda and Westmann, 1970)	$U = \mu_0(\frac{1}{\beta}(e^{\beta(I_1-3)} - 1) - \frac{1}{2}(I_2 - 3))$
Exponential (EX)	(Wellman et al., 1999)	$E = 3\mu_0 e^{m\varepsilon}$

coefficients, which can be measured for clinical use. There are many methods to measure the mechanical properties of the breast tissues, including in-vivo measurement using elastography techniques and mechanical testing of ex-vivo tissue specimens and other inversion-based techniques. A brief overview of popular methods will be presented here.

7.2.3.1 Elastography

Elastography is a technique developed to image tissue stiffness in-vivo. Elastography imaging systems consist of tissue mechanical stimulation, displacement data acquisition, and image reconstruction components. The displacement data acquisition component requires a conventional imaging modality, such as ultrasound (US) or magnetic resonance imaging (MRI). Though the primary application of elastography is imaging of heterogeneous distribution of in-vivo tissue stiffness, it is possible to use data pertaining to relatively homogeneous tissue volumes for mechanical characterization of various types of breast tissue.

US elastography (USE) is most common as it requires relatively inexpensive US imaging scanners. It offers four different types of imaging, including strain imaging, relative Young's modulus (linear) imaging, nonlinear hyperelastic coefficient imaging (both using finite element (FE) method), and shear modulus using shear wave elastography. The first three types of imaging require quasi-static loading pertaining to ultrasound probe compression onto the breast, which causes deformation. This deformation field is captured using analysis of radio-frequency signal frames pertaining to two frames of pre- and post-compression. This analysis follows tissue motion tracking techniques further described in Section 7.2.3.1.2.

7.2.3.1.1 US strain imaging

Strain imaging uses the deformation field to compute axial and lateral strains in the image plane. These strains identify lesions that typically deform less than surrounding background tissue. Though this technique is the most common in the clinic

(Sigrist et al., 2017), it will not be discussed further since it does not directly represent constitutive models and violates the theory of continuum mechanics by assuming uniform stress distribution throughout the breast tissue field of view (FOV).

7.2.3.1.2 Linear and nonlinear finite element ultrasound elastography (FE-USE)

Linear and nonlinear FE-USE offers the ability to reconstruct the relative heterogeneous mechanical parameters of the tissue slice being imaged. This is a more advanced and accurate technique than the strain imaging technique, which comes at additional computational cost. This, along with the lack of a standardized protocol and large clinical studies, has limited its use in the clinic to date. However, this method can be very beneficial in the clinic, as it can better stratify different types of tumors or benign lesions based on the ratio of stiffness parameter of the tissue of interest to that of normal tissue. Such ratios are extensively reported in the literature, while their estimation is based on the theory of continuum mechanics, which increases the accuracy of estimated parameters (Mousavi et al., 2014, 2017).

To generate images of linear elasticity or hyperelasticity parameters, an FE model is used to solve Cauchy's equation of motion mentioned at the end of Section 7.2.1. These FE models are physics-based that follow the theory of continuum mechanics. Reconstructing the heterogeneous mechanical properties using the FE model follows an iterative computational technique. First, the model is initialized assuming homogeneous mechanical parameters. The model is then run using measured tissue displacements. The output internal displacement/strain field of the model is then compared with the measured from tissue motion tracking. Based on quantitative agreement, the spatial mechanical properties are updated, and the FE model is run iteratively until the agreement is maximized based on a predefined threshold. It is notable that for nonlinear reconstruction, larger displacements are required, which can be obtained from multiple compressive frames, while the iterative reconstruction procedure remains the same in theory (Oberai et al., 2009).

Linear reconstructions return the relative heterogeneous stiffness in the image plane. The reason for it being relative is because the applied force from a transducer is unknown. For linear elastic Young's modulus reconstruction, this does not limit the ability to stratify lesions based on stiffness ratios (Mousavi et al., 2017). For nonlinear reconstruction, the amount of precompression matters, but certain coefficients of the nonlinear model may not be influenced. The nonlinear constitutive models studied are the NH and VW models (Goenezen et al., 2012; Gokhale et al., 2008). For the NH model and the initial shear modulus of the VW model, the coefficients remain relative, but the nonlinear parameter γ is exact, given the form of the VW model.

Limitations: FE-USE has several limitations that have been actively investigated by the research community. Primarily, FE-USE relies on the accuracy of the deformation field between consecutive compressed frames. Computing this displacement field can be done using various techniques, whereas there is no current consensus on the best methodology. Popular techniques are speckle tracking, optical flow, and time-delay estimation methods (Ashikuzzaman et al., 2019; Wang et al., 2020). The

FE model also makes assumptions that lead to some errors. Since conventional US only images a single slice, the model is 2-dimensional in nature, whereas the tissue is 3-dimensional, as there are significant tissue out of plane displacements. Assumptions used in the model are primarily plane strain or plane stress (Oberai et al., 2009), whereas more recently efforts have been made for more accurate estimation of breast tissue out of plane displacements (Kheirkhah et al., 2020). Present techniques also assume the tissue is isotropic, which is not accurate due to the influence of Cooper's ligaments. There is also error in the compression of the ultrasound probe. If there is any sliding, tilting, or twisting during US probe compression, the displacements and subsequent strain images cannot be measured reliably, causing inaccuracies in stiffness reconstruction. This can be minimized with robotic arms (Zhang et al., 2021), but clinical studies thus far have only considered manual probe compression (Goenezen et al., 2012; Mousavi et al., 2017). Lastly, since breast tissues are hyperelastic in nature, the relative ratio of stiffness between tissues is a function of the compression applied by the US probe (Barr and Zhang, 2012; Matsumura et al., 2009; Umemoto et al., 2014). Thus far, no technique has been developed to correct for this variability. As such, for error minimization, compression magnitudes are kept small to best match infinitesimal linear elasticity theory and to only capture the tissue initial Young's modulus and shear modulus, which have also been measured using mechanical testing techniques, such as indentation discussed in Section 7.2.3.2. Finally, for nonlinear FE reconstruction, a present persistent issue is the impact of gravity loading on the breast. Since the breast is not in its rest state and influenced by gravity before ultrasound compression, there will be induced pre-strains, which the FE model cannot capture, impacting the hyperelastic parameter estimation. This is technically not a source of error if the tissue is truly linear elastic, which is not the case.

7.2.3.1.3 Ultrasound shear wave elastography and computer tomographic elastography

Ultrasound is also capable of measuring speed of shear wave propagation in tissue using ultrasound shear wave elastography (US-SWE) imaging (Davis et al., 2019; Sigrist et al., 2017). First an actuator induces shear waves in the tissue, which are perpendicular to sampling beam before measuring the spatial distribution of the speed of the shear wave. Shear wave speed is directly related to the shear modulus μ through Eq. (7.10):

$$\mu = \rho c_s^2, \tag{7.10}$$

where ρ is the mass density of the tissue, and c_s is the shear wave speed. Through the relationship to shear modulus given in Section 7.2.2.2, the Young's modulus E can also be calculated if the Poisson's ratio ν is known. Shear wave stiffness images are used in the clinic, though only in conjunction with B-mode imaging and strain imaging due to their larger validation and proof of clinical use (Sigrist et al., 2017).

Ultrasound computer tomography (UCT) is a more recent development, which can reconstruct the 3D heterogeneous acoustic properties of breast tissue using a specialized imaging system (Gemmeke et al., 2017; Hopp et al., 2015). This has many benefits, including the full volumetric reconstruction of the speed of sound, which is related to mechanical properties shown in Eq. (7.11):

$$\mu = 9c^2\rho\,(1 - 2\nu)\,. \tag{7.11}$$

Recent work using UCT has been introduced in combination with strain imaging and inversion techniques to reconstruct heterogeneous mechanical properties based on FE modeling, similar to the FE-USE method discussed earlier but extended to 3D (Hopp and Ruiter, 2021).

Limitations: The main limitations of US-SWE and UCT stem from Eq. (7.10) with US-SWE (Davis et al., 2019), and Eq. (7.11) with UCT (Hopp et al., 2015). Since the mass density cannot be directly measured or imaged, it inherently limits reconstruction accuracy to the breast tissue density range of 0.938–1.035 g/ml reported in the literature (Johns and Yaffe, 1987). For UCT, the second issue is the unknown Poisson's ratio, both spatially and in general. The equation for μ for UCT is extraordinarily sensitive to Poisson's ratio. It is often assumed that the material is incompressible, where $\nu = 0.5$, but that does not work in this equation, and a more exact measure is needed. Results suggest that the breast tissue is slightly compressible but is patient-specific, and there is no consensus (Han et al., 2012; Martínez-Martínez et al., 2017; van Houten et al., 2003). Lastly, a major limitation with this method is that it presently only measures linear elastic isotropic properties.

7.2.3.1.4 Magnetic resonance shear wave elastography (MR-SWE)

MR-SWE follows similar principles to US-SWE using a device to generate waves in the tissue that can then be tracked to reconstruct the shear stiffness. Depending on the reconstruction, relationships to stiffness can vary from simple analytical equations that depend on the propagation wavelength to complex shear moduli, which can then be related to shear stiffness (Mariappan et al., 2010). The exact reconstruction methods differ from the equations used in US-SWE and are typically lab specific. A comprehensive review of stiffness reconstruction techniques can be found in (Fovargue et al., 2018). An important consideration is that common MR-SWE theory treats the tissue as viscoelastic, which is advantageous as it can measure tissue stiffness and viscosity in contrast to quasi-static elastography, which only measures stiffness parameters. MR-SWE of the breast has commonly assumed linear elastic isotropic tissue activated in the small strain range regime (Manduca et al., 2021) with exception to Sinkus et al. (2002), where they imaged isotropic nonlinearity proportional to an EX SEDF and Sinkus et al. (2005) where they measured breast anisotropy assuming linear elastic transverse isotropy.

Limitations: The main limitation with MR-SWE for breast mechanical properties imaging is the lack of hyperelastic reconstruction methods and the low strain range of the breast during testing. Both issues limit the ability to capture the large strain hyperelastic response of tissue.

7.2.3.1.5 Mechanical imaging

Mechanical imaging, originally coined "tactile imaging" by Parris Wellman measures the surface pressure response over a sensor pad as it is pressed and translated across the skin (Wellman, 1999; Wellman et al., 2001). With depth-sensitive reconstruction algorithms, tactile or *mechanical* imaging can measure the response of dense lesions and tissues beneath the skin and create either a 2D or 3D map of stiffness in the form of Young's and shear modulus, max pressure signal, and relative strain hardening measurements (Egorov et al., 2009; Wellman et al., 2001). This method also has the benefit of being able to measure the elastic modulus and thickness of skin during screening, which may be useful for biomechanical modeling discussed in Chapter 13 It is currently a low-cost option for developing countries introducing breast cancer screening protocols (Vaughan, 2019).

Limitations: The current limitations with mechanical imaging include its ability to reconstruct linear elastic mechanical properties, only assuming isotropic materials. It is depth-limited to about 3.5 cm, and thus is not suitable for imaging deep tumors properties (Egorov and Sarvazyan, 2008; Shih et al., 2015).

7.2.3.2 Measurement of mechanical properties of ex-vivo tissue specimens

A common acceptable approach of characterizing the mechanical properties of biological tissue, including breast tissue, is founded on mechanical testing of small ex-vivo tissue specimens acquired from surgeries. Preparing a tissue specimen for mechanical testing involves cutting off the specimen from the specific anatomical site of the body, leading to blood supply cut off, slight disturbance of tissue integrity, and ceasing all physiological activities. This approach assumes that, despite all these changes, ex-vivo tissue specimens can be regarded as good representatives of their respective in-vivo tissue, provided that the specimen is tested within only few hours of its excision, not fixed using chemical agents (such as formalin), and kept moist and warm at close to physiological temperature. Meeting these conditions is feasible with breast tissue specimens (Samani et al., 2003).

Mechanical testing of ex-vivo soft tissue specimens is challenging as conventional uniaxial testing is neither feasible nor reliable. It is not feasible as uniaxial testing requires a sufficiently long sample cut into a regular shape (e.g., cylinder) to minimize sample end effects and stress and strain nonuniformity throughout the sample volume. It is also notable that cutting the two ends of the specimen, such that they have smooth planar surfaces, is not possible due to the tissue high deformability. Loading tissue specimens using this approach is done through mechanical contact between surfaces of the two ends and loading plates of the testing device. Contact stresses between the specimen surfaces and plates are inevitably nonuniform due to the specimen's uneven surface. This contact stress nonuniformity leads to an initial oscillatory stress-strain relationship, which does not represent the tissue intrinsic properties. To minimize the effect of this issue, a sufficient amount of preloading is applied to establish full contact between the specimen's end surfaces and loading plates. Though somewhat effective to remove the oscillatory behavior, it should be noted that the tissue stress and strain nonuniformity will persist and, unless the tissue is close to linear elastic, this nonuniformity can lead to significant errors in the measured tissue mechanical stiffness parameters. Because of the above-mentioned issues, despite being a direct method whereby the slope of stress-strain plot is considered a good representative of the tissue Young's modulus, uniaxial testing has not been accepted as a reliable technique for measuring ex-vivo soft tissue specimens.

Alternative indirect mechanical testing methods have been developed for measuring the mechanical properties of ex-vivo tissue specimens. These methods are founded on loading the tissue specimen such that the stress and strain distribution are nonuniform, though the nature of uniformity can be modeled using analytical or computational models derived from continuum mechanics. Examples of such methods include indentation (Krouskop et al., 1998; Samani et al., 2003) and pressurization (Ghadarghadar et al., 2013; Nafo and Al-Mayah, 2019). In these techniques, the mechanical response (e.g., force vs. indenter displacement) is measured and converted to the tissue mechanical parameters using a model derived from continuum mechanics, which takes into account tissue specimen and indenter geometry, loading, and boundary conditions. The most popular technique among methods in the class of indirect techniques is the indentation technique.

7.2.3.2.1 Indentation

Indentation follows the indirect method of tissue mechanical parameter estimation, where only a small part of the specimen's surface is loaded by an indenter. The mechanical response of indentation test is characterized by the indentation force vs. displacement data. This data is used to infer the tissue mechanical parameters. In the context of breast tissue, indentation has been most commonly applied to resected adipose, fibroglandular, and tumor tissues (Skovoroda et al., 1995; Krouskop et al., 1998; Wellman et al., 1999; Samani et al., 2003; Samani and Plewes, 2004; Matsumura et al., 2009; Samani et al., 2007; O'Hagan and Samani, 2008, 2009; Umemoto et al., 2014; Dempsey et al., 2021), whereas using indentation for measuring in-vivo skin mechanics is a more recent development (Flynn et al., 2011).

Measurement of Young's modulus of small tissue specimen using indentation

In earlier works and in the simplest isotropic case, the relationship between force and displacement was based on linear elasticity theory with several assumptions on tissue geometry. This allowed the use of analytical equations derived from theory of elasticity to measure the Young's modulus of thick tissue specimens using Eq. (7.12):

$$E = \frac{2\left(1 - \nu^2\right)}{\pi a} S, \tag{7.12}$$

where a is the radius of the indenter, ν is the tissue Poisson's ratio, and S the measured force vs. displacement slope. To account for specimen limited thickness, Hayes et al. (1972) developed a semi-analytical technique, where the Young's modulus is estimated using Eq. (7.13):

$$E = \kappa_0 \frac{(1 - \nu^2)}{2a} S, \tag{7.13}$$

where κ_0 is a correction factor that is dependent on a/h, where h is the specimen's thickness. Eq. (7.13) is only valid where the size of indenter is substantially smaller than the in-plane size of the tissue specimen. For a circular tissue slice with a radius of R, computational models confirmed by experimental evidence indicate that Eq. (7.13) is valid when $R \geq 5a$. For small tissue specimens where this condition is not met, Samani et al. (2003) developed the following general equation in Eq. (7.14):

$$E = \kappa S, \tag{7.14}$$

where κ depends on the geometry of the tissue sample and indenter and boundary conditions. This parameter can be computed experimentally or using inverse FE technique (Samani et al., 2003). Early works showed that E was sensitive to precompression, which is anticipated due to the tissue nonlinearity. This observation has been confirmed by many other researchers, including in the latter works of (Matsumura et al., 2009; Umemoto et al., 2014).

Measurement of tissue specimen hyperelastic parameters using indentation

Because soft tissue is nonlinear in nature, the linear elastic assumption is not valid, except where the tissue undergoes infinitesimal deformation as a result of indentation. Though such small deformation may be used in tissue mechanical stimulation in elastography applications, many other applications involve breast tissue large deformation, where models based on linear elasticity are insufficient, and nonlinear models are required. The latter require measurement of the tissue hyperelastic parameters. There are two techniques for estimating tissue hyperelastic parameters from indentation data. The first uses an analytical model presented by (Wellman et al., 1999). An important limitation of this model is that it was derived from a single SEDF. A more computationally expensive option was presented by Samani and Plewes (2004), which allows for any form of SEDF to be considered to describe the tissue. In essence, it follows a similar style to FE-USE, by first generating a FE mesh of the tissue sample and modeling the indenter's influence on the tissue with appropriate boundary conditions. The model is then used to optimize the hyperelastic material by maximizing the similarity of force data measured by the indentation experiment and corresponding forces generated by the model. This is extremely valuable in concept, because it allows the user to determine an ideal model to describe the mechanics of any sample. This methodology was used to study the hyperelastic coefficients of pathological tissues in O'Hagan and Samani (2008, 2009), and recently healthy tissue samples by Dempsey et al. (2021).

Limitations: the indentation technique has general limitations and methodology-specific limitations. For general limitations, the tissue being quantified is assumed homogeneous, which may be an issue due to presence of the Cooper's ligaments or presence of a mixture of fibroglandular and adipose tissue within the sample. As previously mentioned, another issue, which is not limited to indentation, is that the gravity force induces initial stress in the tissue, which would impact recorded forces, hence the accuracy of measured parameters. The influence of gravity has not been investigated rigorously in the context of indentation testing. In other biomechanical applications, where breast models are involved, it has been shown that a gravity-loaded reference state impacts the strain and stress distribution significantly (Gamage et al., 2019).

For methodology-specific limitations, if an analytical equation is followed, the main limitations are assumptions made pertaining to the equation derivation. Eq. (7.12) assumes that the sample has very large dimensions in both indentation loading direction and in the plane perpendicular to loading direction, ideally approaching a semi-infinite medium. Eq. (7.13) does not set a constraint on the sample's thickness but has similar size requirement in the plane perpendicular to the loading direction. In practice, satisfying this constraint is not always possible.

Another source of error pertains to FE inversion techniques used to compute the tissue mechanical parameters. In these techniques, the specimen's FE model is considered as a forward model, where uncertainty exists in the assumption of boundary conditions; e.g., assuming frictionless contact with the end of indenter contacting the specimen. The geometry is also assumed based on the sample preparation and cutting procedure. This geometry cannot be followed fully, and these simplifications will impact the results. To address this issue, Samani et al. (2003) developed a cutting technique, which produces relatively accurate block-shaped tissue specimens. It involves using a custom-made mold, where the irregularly shaped sample embedded within gelatin is placed in cylindrical sample holder equipped with slits to allow sharp blade to cut through the tissue to obtain a block-shaped specimen. Lastly, the initialization geometry should match the rest state without gravity-induced deformation, which has not been implemented in present methods.

7.2.3.2.2 Whole breast biomechanical model based optimization

Biomechanical models, as shown for FE-USE and inverse FE-based indentation techniques, can be used within inverse problem frameworks to optimize mechanical properties if the loading and deformation are known. Fortunately, during many medical imaging procedures, both the loading (gravity, and/or mammographic compression plates) and deformation (from medical imaging) are known. This provides sufficient information to predict the mechanical properties, through constructing the breast FE model before computing the tissue mechanical parameters, using optimization, where the model/image agreement is maximized. The finer details of building a biomechanical whole breast model are provided in Chapter 13 of the book, but basic expectations from a mechanics perspective will be discussed here.

As mentioned in the introduction, the model should consider unique mechanics of the skin, adipose, and fibroglandular tissues based on how unique they are both functionally and anatomically. If a tumor or lesion is present, that should also be considered unique. The model should also consider how the tissues are connected and how the breast is connected to the chest wall (i.e., modeling the retromammary space). The SEDF's for the tissues should match or be capable of capturing what has proven to fit experimental data. Lastly, since the whole breast is under significant deformation during medical imaging (pendant during prone MRI, leaning on the lateral side of the body due to chest wall curvature in supine imaging, or compressed during mammography), the geometry of the model should be initialized in the breasts' rest state. Submerging the breasts in a water bath has shown some success to estimate the rest state given the similar densities of water and soft tissue (Carter et al., 2009). However, it should be noted that submerging the breast in water does not neutralize the tissue locally throughout the breast volume. There has also been success using optimization to predict the rest state via assumed mechanical properties obtained from the literature to then match an ideal image of a loaded breast state geometry (Eiben et al., 2016; Rajagopal et al., 2007; Vavourakis et al., 2016b). What to incorporate in breast models is ultimately left to the researcher, as a variety of models and simplifications are typically made. Once made, the mechanics can then be tuned between the deformation of two subsequent images to maximize image information. Examples include prone to supine matching (Hipwell et al., 2016) and prone MRI to mammogram compression (García et al., 2018).

Limitations: Limitations for this patient personalization biomechanical method pertain to the simplifications made when creating the model; this will be discussed in further details in Section 7.3. Briefly, tissue isotropy is assumed in most cases, whereas constitutive model choice has not reached a consensus. There are varying opinions on the importance of modeling the skin of the breast, the muscle tissues, and whether to consider adipose and fibroglandular tissues homogeneous or separate (Gamage et al., 2019). Finally, viscoelastic effects are also largely ignored.

7.2.4 Constitutive model coefficients

Though there are many constitutive models and methods to test for model coefficients, as discussed in Sections 7.2.2 and 7.2.3, direct comparison between models and modalities remains challenging due to lack of widely accepted strain range protocol in addition to the issue of hyperelastic parameters' nonuniqueness. In this section, the primary aim is presenting constitutive model parameter results obtained from mechanical testing with a similar strain range. The results include initial shear modulus (μ_0) of the tissue, which is a unique parameter that can be calculated analytically from linear elastic theory and many hyperelastic models. Also reported are hyperelastic parameters of the VW model, which has been used by researchers in breast modeling and elastography applications. Thus Table 7.2 contains comparable mechanical coefficient information from studies that are, in theory, suitable for general mechanics.

In this table, several early studies have been excluded based on their a-posteriori limited success in breast biomechanical models, as discussed in Section 7.3. For instance, the early indentation works of Krouskop et al. (1998) has led to parameters consistent with unrealistically high stiffness values, hence have not succeeded in rest state optimized biomechanical modeling. UCT results are also omitted given

Table 7.2 Linear and hyperelastic coefficients for various SEDF models of adipose and fibroglandular breast tissues from select studies of elastography, indentation, and biomechanical model optimization. Model acronyms are defined in Table 7.1.

Method	Reference	Model	Max Strain (Abs)	Mechanical Parameters
FE-USE	(Goenezen et al., 2012)	VW	>12%	Mixed (N = 20): $\mu_0 = N/A$, $\gamma = 7.3 \pm 6.3$
Mechanical Imaging	(Egorov and Sarvazyan, 2008)	LE	~5%	Mixed (N = 6): μ_0(kPa) $= 2.3 \pm 0.46$
Indentation	(Wellman et al., 1999)	EX	~10%	Adipose (N = 26): μ_0(kPa) $= 1.49 \pm 0.78$, m $= 7.4 \pm 4$ Fibroglandular (N = 7): μ_0(kPa) $= 5.06 \pm 2.25$, m $= 12.3 \pm 7.4$
	(Samani et al., 2007)	LE	<5%	Adipose (N = 71): μ_0(kPa) $= 1.08 \pm 0.30$ Fibroglandular (N = 26): μ_0(kPa) $= 1.08 \pm 0.20$
	(Stewart et al., 2011) evaluating (Wellman, 1999)	NH	~10%	Mixed (N = U): μ_0(kPa) $= 1.1$
	(Goenezen et al., 2012) evaluating (Krouskop et al., 1998)	VW	U	Mixed (N = U): $\mu_0 = N/A$, $\gamma \cong 1.3$
	(Dempsey et al., 2021)	VW	>15%	Mixed (N = 72): μ_0(kPa) $= 1.33$ [0.88, 1.67], $\gamma = 11.82$ [5.16, 14.29]
		OG (N = 3)	>15%	Mixed (N = 72): μ_0(kPa) $= 0.87$ [0.84, *]
Biomechanical Modeling Based	(Rajagopal et al., 2008)	NH	~20%	Mixed (N = 2): μ_0(kPa) $= 0.23 \pm 0.2$
	(Carter et al., 2009)			Adipose (N = 2): μ_0(kPa) $= 0.7$ Fibroglandular (N = 2): μ_0(kPa) $= 1.1$
	(Gamage et al., 2012)			Mixed (N = 1): μ_0(kPa) $= 0.1$
	(Eiben et al., 2016)			Mixed (N = 14): μ_0(kPa) $= 0.32 \pm 0.06$
	(Gamage et al., 2017a)			Mixed (N = 2): μ_0(kPa) $= 0.48 \pm 0.38$

continued on next page

Table 7.2 (*continued*)

Method	Reference	Model	Max Strain (Abs)	Mechanical Parameters
MR-SWE	(Hawley et al., 2017)	LE	U	Mostly Adipose (N = 22): μ_0(kPa) $= 0.83 \pm 0.11$ Mostly Fibroglandular (N = 22): μ_0(kPa) $= 0.92 \pm 0.13$
	(Sinkus et al., 2005)	$\perp_{iso}$ LE	U Patient1 > Patient2	Patient 1: $\mu_{0\perp}$(kPa) $= 1.1 \pm 0.5$ $\mu_{0\parallel}$(kPa) $= 0.9 \pm 0.5$ Patient 2: $\mu_{0\perp}$(kPa) $= 0.36 \pm 0.15$ $\mu_{0\parallel}$(kPa) $= 0.26 \pm 0.2$
	(van Houten et al., 2003)	LE	U	Adipose (N = 5): μ_0(kPa) $= 7.56 \pm 0.86$ Fibroglandular (N = 5): μ_0(kPa) $= 9.39 \pm 0.83$
	(McKnight et al., 2002)	LE	U	Mostly Adipose (N = 4): μ_0(kPa) $= 3.3 \pm 1.9$ Mostly Fibroglandular (N = 4): μ_0(kPa) $= 7.5 \pm 3.6$
	(Lorenzen et al., 2002)	LVE	U	Mostly Adipose (N = 25): μ_0(kPa) $= 2$ [0.4-4] Mostly Fibroglandular (N = 25): μ_0(kPa) $= 2.5$ [1, 7]
	(Sinkus et al., 2000)	LVE	U	Patient 1: Mixed μ_0(kPa) $= 0.75$ [0.5, 1]

*N/A = Not applicable due to relative reconstruction, U = Unknown, [#,#] = min-max range, * = artefact, ^ = assumed, $\perp_{iso}$ = transverse isotropic, $\parallel$ = parallel to anisotropic axis, LVE = linear viscoelastic.*

the extreme sensitivity to Poisson's ratio. For biomechanical modeling, the only coefficients presented are based on models that made efforts to predict the breast rest state geometry given the impact of pre-stress under gravity loading configurations. Elastography results that report relative stiffness were also omitted. Table 7.2 shows some interesting similarities between multiple imaging modalities. First, the nonlinear elastography results from Goenezen et al. (2012), for healthy tissue agrees well with the findings of Dempsey et al. (2021), in terms of the nonlinear coefficient of the VW model. The γ's reported are well in range of each other for healthy tissue from ex-vivo and in-vivo independent studies. This contradicts some disregard in the field on the utility of ex-vivo punch indentation and the hypothesis that blood flow pressure may alter the mechanics significantly. It is curious that the γ reported in Goenezen et al. (2012), by reevaluating Krouskop et al. (1998), is so low. We believe this is impacted by how the coefficients were fit. Goenezen et al. (2012) and Dempsey et al. (2021), fit coefficients using inverse FE methods. It's unclear how Goenezen et al. (2012), fit the data from Krouskop et al. (1998), including whether all data was fit or only the data reported in their Fig. 7. If the data was fit to stress-strain curves in a least squares measure, then similar issues exist that the strain and stresses are not uniform to generate that curve. The FE indentation method in Dempsey et al. (2021), completely removes that risk by matching pure reactive force to applied displacement by only changing the mechanics.

Overall, there is a trend over time from the late 1990's to 2020 that the shear modulus approaches values <1 kPa. We believe that this represents convergence of modality accuracy as each respective field advances. Studies that use analytical equations report stiffer shear moduli than studies that included optimization. It is also encouraging that the mechanical models developed to predict deformation using tissue mechanical parameters obtained from testing small sample studies are in near agreement with ones obtained from large sample works of MR-SWE (Hawley et al., 2017) and indentation (Dempsey et al., 2021; Samani et al., 2007), although the stiffness optimized using biomechanical modeling is generally lower than the stiffness of these modalities. We hypothesize that the breast model mechanics are softer due to consideration of the rest state geometry, which pertains to the model in the zero pre-strain (rest state) geometry. If the hypothesis is true, then the mechanical properties obtained from indentation show the impact of prestress on the material (although the nonlinear coefficient would not be impacted in the VW model significantly). It is possible that the nonzero prestress also impacts MR-SWE, whereas a second influence may be the viscoelastic effects and activation frequency used in the imaging protocol. Another influence is simply the choice of hyperelastic model. The benefit of the study by Dempsey et al. (2021), is that multiple models were fit, which allowed for multiple measurements of the initial shear modulus. The difference between the VW model and the 3rd-order Ogden model is about 0.46 kPa. If this error is an approximate estimate of SEDF error, then the shear moduli become even closer.

None of the biomechanical models presented in Table 7.2 considered the skin of the breast. This will be discussed in detail in Section 7.3. A study on the superviscous properties of the in-vivo brain showed that at the quasi-static limit using MR-SWE,

Table 7.3 Hyperelastic parameters obtained for the Yeoh, 5th-order 5-term polynomial, Ogden (OG), and Veronda-Westmann (VW) models of combined breast tissue samples.

Yeoh (Pa)			**5-term Polynomial (5P) (Pa)**				
C_{10}	C_{20}	C_{30}	C_{10}	C_{01}	C_{20}	C_{11}	C_{02}
439.27	21778.85	5818.08	331.48	17258.08	102.46	3389.92	1838.03
3rd-order Ogden (OG)						**Veronda-Westmann (VW)**	
μ_1 (Pa)	α_1	μ_2 (Pa)	α_2	μ_3 (Pa)	α_3	μ_0 (Pa)	β
-247525.54	2.32	199192.77	3.80	49200.66	−3.62	1334.28	11.82

the in-vivo human brain shear modulus measured 0.042 kPa. This extreme softness could not be reproduced using a FE model of similar stiffness without considering an encapsulating membrane (Herthum et al., 2021). Similar considerations may be necessary for measuring tissue stiffness parameters of the breast adipose and fibroglandular tissue, where other constituents that have major mechanical role need to be taken into account.

Among all the results reported in Table 7.2 that differentiate adipose from fibroglandular tissue, there are contrasting results in tissue mechanical similarity. This difference is best attributed to the strain range being applied to the tissue. As discussed in Azar et al. (2001), adipose and fibroglandular tissue are mechanically different at small strains but approach near identical behavior at large strains. In Azar, this was mathematically enforced, but agreements at large strains have been observed by Umemoto et al. (2014) and Dempsey et al. (2021). It is possible that the MR-SWE differences are due to measurements only being induced at small strains but, since these studies do not report amplitude of activator, it is not possible to make an educated strain approximation. The work of Wellman et al. (1999), may have been influenced by how thin the tissue sample was and the works of Samani et al. (2007), that show similarity at small strains may be due to the simplification of linear elasticity. In general, there is nothing fully conclusive from a mechanical properties perspective other than to support observations that at large strains, adipose and fibroglandular tissue are similar, and at small strains, they are different. This is very important in the context of the breast mechanical modeling, as it leads to an efficient model development process, where segmentation of adipose and fibroglandular tissue and separation of these tissue types in FE meshing becomes unnecessary. Generally, it is reasonable to consider a single type of tissue representing both adipose and fibroglandular in breast models where the breast undergoes large strains. Dempsey et al. (2021), showed that, at all strain values their tested samples underwent, there was no statistically significant difference among stress values obtained for the three tissue types of adipose, fibroglandular, and mixed. As such, they reported hyperelastic coefficients of "mean" class of all 72 tested breast tissue samples obtained for four commonly used hyperelastic models as given in Table 7.3.

Regarding expected transverse isotropy in the breast due to Cooper's ligaments, relevant evidence presented by Sinkus et al. (2005), is the only existing evidence pointing to that direction, hence in-depth discussion and comparison is not possible. As will be shown in the following section, there are breast biomechanical models that do consider transverse isotropy where the rest state is not considered, hence the mechanical properties cannot be directly compared. The recent work of Briot et al. (2020), shows that the nonlinear stiffness of Cooper's ligaments is orders of magnitude higher than the tissues tested here, and directional dependence of these dispersed ligaments impacts mechanics, as shown in Han et al. (2012).

Lastly, we address the mechanics on breast tumors that vary according to type of tumor. The works of O'Hagan and Samani (2008, 2009), show a comprehensive hyperelastic mechanical description of many pathologies based on inverse FE indentation. This, in conjunction with Tables 7.2 and 7.3 presented in Ramião et al. (2016), cover the tumor mechanical coefficients. We will not discuss further the mechanical parameters of breast tumors since their stiffness varies greatly with pathology, and the literature is too sparse to discuss trends.

7.3 A community's use of mechanics

This section is centered around how the community utilizes the mechanics knowledge presented in Section 7.2 for breast biomechanical modeling applications. Table 7.4 summarizes most breast biomechanical models presented in the literature. This table includes information on what we believe is important in terms of capturing the true mechanics of the breast. The table's columns are organized by the following:

1. Reference.
2. Tissue homogeneity: indicates whether adipose and fibroglandular tissues are modeled as a single homogeneous type of tissue where the default (-) is considered homogeneous.
3. SEDF: see Table 7.1 for acronyms.
4. Parameter assignment: indicates whether the parameters were assigned based on the literature or were optimized for the model. The default (-) is being assigned based on the literature, which may include determining new parameters using averaging or selecting them based on a range given in the literature.
5. Isotropic or anisotropic | Incompressible or compressible: indicates whether the tissue was considered isotropic or anisotropic and if the tissue was considered compressible. Here the default (- | -) is (isotropic | incompressible)
6. Rest state: indicates the assumed rest state of the breast; "gravity loaded" means no compensation was considered; "reverse gravity" means using new geometry based on applying opposite gravity compared to the imaging direction; "optimization" refers to computing or approximating the rest state discussed in Section 7.2.3.2.2; and "water bath" involves imaging the breasts in water as a neutral buoyancy configuration. In this case, the default (-) is "gravity loaded."
7. Rib boundary conditions: indicates the boundary conditions pertaining to the breast model connection to the chest wall. Here, "fixed" implies no degrees of

freedom; "prescribed" is based on image deformable image registration; "fixed⊥" means fixed to chest wall but the nodes can slide transverse to the chest wall normal; "free __-__" indicates in what global axis directions the tissue can move; and "fixed in plane" means a plane was chosen at the back of the breast, where full sliding in plane is allowed.

8. Skin: indicates whether the skin was modeled. Here the default (-) is not being modeled.
9. Lesion: indicates whether a breast lesion existed and if the lesion was modeled independently with different mechanical properties. The default (- | -) here is (no lesion | not modeled)

It is noteworthy that we have deliberately excluded mass spring models and any completely in-silico models or models based on phantom studies. Though computationally efficient, mass spring models currently cannot account for incompressibility, which is a major limitation for soft tissue organs, and in-silico and phantom studies have been omitted since the focus is on real model mechanics.

To expand on the data presented in Table 7.4 and provide more detailed information in the context of tissue mechanics, Table 7.5 is provided in addition to the discussion that follows. This table represents a breakdown of Table 7.4 and shows the % distribution of values presented in the table.

Beginning with Column 2, it is not anticipated to see any trends with whether a model is heterogeneous or homogeneous for adipose and fibroglandular tissue, as this depends on the strain range involved in the biomechanical model. Recall earlier when we suggested low strain should be heterogeneous and large strain can be homogeneous. Even at large strains, treating the tissue as heterogeneous may still be beneficial if the mechanical parameters are tuned.

Column 3 comes somewhat as a surprise considering the popularity of the NH model (60%). This is surprising, as no other methodology that measures coefficients described in Section 7.2 has used this model. A popular argument when choosing the NH model is that it is the simplest extension from linear elastic, and thus easily optimized for when tuning to a patient with only one variable. For computation, this reason makes sense, but no external studies confirm that it is acceptable beyond the model success for the chosen application. Some works have carried out multiple material studies, as shown in Table 7.4, to determine whether a NH model is sufficient. However, those studies contain other simplifications that limit the confidence in their conclusions. For instance, not optimizing the rest state has important implications, and ignoring the skin in the model may force other mechanics to be stiffer. To show an example of how two hyperelastic models with near identical initial shear moduli respond, we present Fig. 7.1. This figure illustrates the uniaxial extension of an incompressible NH model with initial shear modulus of 1 kPa, which is approximately the mean obtained from studies presented in Table 7.2, and the mean 3rd-order Ogden model from Dempsey et al. (2021), which has an initial shear modulus of 0.87 kPa but is described with 6 variables in total.

As seen in Fig. 7.1, whereas similar at small strains, the 3rd order Ogden model can capture more nonlinearity. It is also noteworthy that, with incompressible tissue

Table 7.4 Breakdown of the mechanical influences of 52 breast FE biomechanical models in literature.

<table>
<tr><th>Reference</th><th>Tissue homogeneity</th><th>SEDF</th><th>Parameter Assignment</th><th>(Isotropic or Anisotropic | Incompressible)</th><th>Rest state</th><th>Rib Boundary Conditions</th><th>Skin</th><th>Lesion (exists | modeled)</th></tr>
<tr><td>(Azar et al., 2001)</td><td>Heterogeneous</td><td>MM</td><td>-</td><td>- | -</td><td>-</td><td>-</td><td>Y</td><td>Y | -</td></tr>
<tr><td>(Azar et al., 2002)</td><td>Heterogeneous</td><td>MM</td><td>-</td><td>- | -</td><td>-</td><td>Fixed$^{\perp}$</td><td>Y</td><td>Y | -</td></tr>
<tr><td>(Yin et al., 2004)</td><td>Heterogeneous</td><td>M-R</td><td>-</td><td>- | -</td><td>-</td><td>Free sup-inf</td><td>-</td><td>Y | Y</td></tr>
<tr><td>(Tanner et al., 2006)</td><td>Heterogeneous</td><td>MM</td><td>MMs</td><td>- | MMs</td><td>-</td><td>Variable</td><td>Y</td><td>- | -</td></tr>
<tr><td>(Ruiter et al., 2006)</td><td>Both</td><td>MM</td><td>MM</td><td>- | -</td><td>-</td><td>Prescribed</td><td>Y</td><td>Y | -</td></tr>
<tr><td>(del Palomar et al., 2008)</td><td>-</td><td>NH</td><td>Optimized</td><td>- | -</td><td>-</td><td>-*</td><td>Y</td><td>- | -</td></tr>
<tr><td>(Chung et al., 2008)</td><td>-</td><td>NH</td><td>-</td><td>- | -</td><td>Water Bath</td><td>-</td><td>-</td><td>- | -</td></tr>
<tr><td>(Rajagopal et al., 2008)</td><td>-</td><td>NH</td><td>Optimized</td><td>- | -</td><td>Water Bath</td><td>-</td><td>-</td><td>- | -</td></tr>
<tr><td>(Tanner et al., 2009)</td><td>Heterogeneous</td><td>LE</td><td>MMs</td><td>Aniso | -</td><td>-</td><td>Prescribed</td><td>Y</td><td>Y | Y</td></tr>
<tr><td>(Carter et al., 2009)</td><td>Heterogeneous</td><td>NH</td><td>Optimized</td><td>- | -</td><td>Water Bath</td><td>Prescribed</td><td>Y</td><td>- | -</td></tr>
<tr><td>(Shih et al., 2010)</td><td>Heterogeneous</td><td>M-R</td><td>-</td><td>- | -</td><td>-</td><td>Fixed$^{\perp}$</td><td>-</td><td>- | -</td></tr>
<tr><td>(Tanner et al., 2011)</td><td>Heterogeneous</td><td>NH</td><td>Optimized</td><td>Aniso | -</td><td>-</td><td>Prescribed</td><td>Y</td><td>Y | Y</td></tr>
<tr><td>(Hsu et al., 2011)</td><td>Heterogeneous</td><td>LE</td><td>-</td><td>- | -</td><td>-</td><td>Fixed in plane</td><td>Y</td><td>- | -</td></tr>
<tr><td>(Lapuebla-Ferri et al., 2011)</td><td>-</td><td>NH</td><td>-</td><td>- | -</td><td>-</td><td>-</td><td>Y</td><td>- | -</td></tr>
<tr><td>(Stewart et al., 2011)</td><td>-</td><td>NH</td><td>-</td><td>- | -</td><td>-</td><td>-</td><td>-</td><td>Y | Y</td></tr>
<tr><td>(Gamage et al., 2012)</td><td>-</td><td>NH</td><td>Optimized</td><td>- | -</td><td>Optimized</td><td>-</td><td>-</td><td>- | -</td></tr>
<tr><td>(Mertzanidou et al., 2012)</td><td>-</td><td>NH</td><td>-</td><td>Aniso | -</td><td>-</td><td>Fixed in plane</td><td>-</td><td>Y | -</td></tr>
<tr><td>(Hopp et al., 2012)</td><td>-</td><td>NH</td><td>-</td><td>- | -</td><td>-</td><td>-</td><td>-</td><td>Y | -</td></tr>
<tr><td>(Lago et al., 2012)</td><td>Heterogeneous</td><td>5P</td><td>-</td><td>- | -</td><td>-</td><td>-</td><td>Y</td><td>- | -</td></tr>
<tr><td>(Han et al., 2012)</td><td>-</td><td>NH</td><td>Optimized</td><td>Aniso | Comp</td><td>-</td><td>Fixed$^{\perp}$</td><td>-</td><td>Y | Y</td></tr>
<tr><td>(Lee et al., 2013)</td><td>-</td><td>NH</td><td>-</td><td>- | -</td><td>Optimized</td><td>Fixed$^{\perp}$</td><td>-</td><td>Y | -</td></tr>
<tr><td>(Chen et al., 2013)</td><td>-</td><td>NH</td><td>-</td><td>- | -</td><td>-</td><td>Fixed to Chest Model</td><td>-</td><td>- | -</td></tr>
</table>

continued on next page

Table 7.4 *(continued)*

<table>
<tr><th>Reference</th><th>Tissue homogeneity</th><th>SEDF</th><th>Parameter Assignment</th><th>(Isotropic or Anisotropic | Incompressible)</th><th>Rest state</th><th>Rib Boundary Conditions</th><th>Skin</th><th>Lesion (exists | modeled)</th></tr>
<tr><td>(Dufaye et al., 2013)</td><td>Heterogeneous</td><td>NH</td><td>-</td><td>- | -</td><td>-</td><td>Fixed*</td><td>Y</td><td>- | -</td></tr>
<tr><td>(Eder et al., 2014)</td><td>-</td><td>MM</td><td>MMs</td><td>- | -</td><td>Optimized</td><td>-</td><td>-</td><td>- | -</td></tr>
<tr><td>(Mertzanidou et al., 2014)</td><td>-</td><td>NH</td><td>-</td><td>Aniso | -</td><td>-</td><td>Fixed in plane</td><td>-</td><td>Y | -</td></tr>
<tr><td>(Han et al., 2014)</td><td>Heterogeneous</td><td>NH</td><td>Optimized</td><td>Aniso | -</td><td>-</td><td>Fixed$^{\perp}$</td><td>Y</td><td>- | -</td></tr>
<tr><td>(Solves-Llorens et al., 2014)</td><td>Heterogeneous</td><td>NH</td><td>-</td><td>- | -</td><td>Reverse Gravity</td><td>Fixed$^{\perp}$</td><td>Y</td><td>- | -</td></tr>
<tr><td>(Conley et al., 2015)</td><td>-</td><td>LE</td><td>-</td><td>- | -</td><td>-</td><td>-</td><td>-</td><td>Y | -</td></tr>
<tr><td>(Pianigiani et al., 2015)</td><td>Heterogeneous</td><td>MM</td><td>-</td><td>- | -</td><td>-</td><td>-</td><td>-</td><td>- | Y</td></tr>
<tr><td>(Hopp et al., 2015)</td><td>Heterogeneous</td><td>MM</td><td>UCT</td><td>- | -</td><td>-</td><td>-</td><td>-</td><td>Y | ~Y</td></tr>
<tr><td>(Vavourakis et al., 2016a)</td><td>Heterogeneous</td><td>M-R</td><td>-</td><td>- | -</td><td>Optimized</td><td>-</td><td>Y</td><td>Y | -</td></tr>
<tr><td>(Eiben et al., 2016)</td><td>-</td><td>NH</td><td>Optimized</td><td>- | -</td><td>Optimized</td><td>Surface constrained</td><td>-</td><td>- | -</td></tr>
<tr><td>(Haddad et al., 2016)</td><td>-</td><td>OG</td><td>-</td><td>- | -</td><td>-</td><td>-</td><td>Y*</td><td>- | Y</td></tr>
<tr><td>(Sturgeon et al., 2016)</td><td>Heterogeneous</td><td>NH</td><td>-</td><td>- | -</td><td>Reverse Gravity</td><td>Fixed in plane</td><td>Y</td><td>- | -</td></tr>
<tr><td>(Gamage et al., 2017a)</td><td>-</td><td>NH</td><td>Optimized</td><td>- | -</td><td>Optimized</td><td>Prescribed</td><td>-</td><td>- | -</td></tr>
<tr><td>(Liu et al., 2017)</td><td>Heterogeneous</td><td>M-R</td><td>-</td><td>- | -</td><td>-*</td><td>-*</td><td>-</td><td>- | -</td></tr>
<tr><td>(Martínez-Martínez et al., 2017)</td><td>Heterogeneous</td><td>M-R</td><td>-</td><td>- | Comp</td><td>-</td><td>Fixed$^{\perp}$</td><td>Y</td><td>- | -</td></tr>
<tr><td>(Lapuebla-Ferri et al., 2017)</td><td>-</td><td>NH</td><td>-</td><td>- | -</td><td>-</td><td>-</td><td>Y</td><td>Y | Y</td></tr>
<tr><td>(Griesenauer et al., 2017)</td><td>Heterogeneous</td><td>LE</td><td>Optimized</td><td>- | Comp</td><td>-</td><td>Prescribed</td><td>-</td><td>- | -</td></tr>
<tr><td>(Griesenauer et al., 2018)</td><td>Heterogeneous</td><td>LE</td><td>Optimized</td><td>- | -</td><td>-</td><td>Prescribed</td><td>-</td><td>- | -</td></tr>
<tr><td>(Na et al., 2019)</td><td>Heterogeneous</td><td>NH</td><td>-</td><td>- | -</td><td>Reverse Gravity</td><td>-</td><td>-</td><td>- | -</td></tr>
</table>

continued on next page

Table 7.4 *(continued)*

<table>
<tr><th>Reference</th><th>Tissue homogeneity</th><th>SEDF</th><th>Parameter Assignment</th><th>(Isotropic or Anisotropic | Incompressible)</th><th>Rest state</th><th>Rib Boundary Conditions</th><th>Skin</th><th>Lesion (exists | modeled)</th></tr>
<tr><td>(García et al., 2019)</td><td>Heterogeneous</td><td>NH</td><td>Optimized</td><td>Aniso | -</td><td>-</td><td>Fixed$^{\perp}$</td><td>-</td><td>- | -</td></tr>
<tr><td>(Gamage et al., 2019)</td><td>Homogeneous</td><td>NH</td><td>-</td><td>- | -</td><td>Optimized</td><td>-</td><td>Y</td><td>- | -</td></tr>
<tr><td>(Chen et al., 2019)</td><td>Heterogeneous</td><td>NH</td><td>-</td><td>- | -</td><td>-</td><td>-</td><td>-</td><td>- | -</td></tr>
<tr><td>(Sun et al., 2019)</td><td>-</td><td>P5</td><td>Optimized</td><td>- | -</td><td>-</td><td>-</td><td>-</td><td>- | -</td></tr>
<tr><td>(Esslinger et al., 2019)</td><td>-</td><td>NH</td><td>-</td><td>- | -</td><td>Reverse gravity</td><td>-</td><td>-</td><td>- | Y</td></tr>
<tr><td>(Bessa et al., 2019)</td><td>Heterogeneous</td><td>M-R</td><td>-</td><td>- | -</td><td>(U)</td><td>-</td><td>Y</td><td>Y | -</td></tr>
<tr><td>(Green et al., 2020)</td><td>Heterogeneous</td><td>LE</td><td>-</td><td>- | -</td><td>-</td><td>Prescribed</td><td>-</td><td>Y | Y</td></tr>
<tr><td>(Al abdi et al., 2020)</td><td>Heterogeneous</td><td>LPE</td><td>-</td><td>- | ~-</td><td>-</td><td>Fixed$^{\perp}$</td><td>-</td><td>Y | Y</td></tr>
<tr><td>(Bessa et al., 2020)</td><td>Heterogeneous</td><td>M-R</td><td>-</td><td>- | -</td><td>(U)</td><td>-</td><td>Y</td><td>Y | -</td></tr>
<tr><td>(Y. Sun et al., 2021)</td><td>-</td><td>P5</td><td>-</td><td>- | -</td><td>Reverse Gravity</td><td>-</td><td>-</td><td>- | -</td></tr>
<tr><td>(Xue et al., 2021)</td><td>Heterogeneous</td><td>NH</td><td>Optimized</td><td>- | -</td><td>(U)</td><td>Fixed$^{\perp}$ ant-pos</td><td>Y</td><td>Y | -</td></tr>
</table>

Y = Yes, (U) = Unclear, ⊥ = normal, MM = Multiple Materials, MMs = Multi-Material study, ∗ = assumed, ∼ = effectively, LPE = Linear poroelastic.

Table 7.5 Important statistics from Table 7.4 based on 52 breast biomechanical studies.

Column Number	Percentage of Possible Options from Table 7.4
Column 2	Heterogeneous = 62% Homogeneous = 38%
Column 3	NH = 60% M-R = 13% LE = 16%
Column 4	Literature = 72% Optimized = 27% UCT = <1%
Column 5 (Isotropic)	Isotropic = 87% Anisotropic = 13%
Column 5 (Compressibility)	Incompressible = 94% Compressible = 6%
Column 6	Gravity Loaded = 65% Optimized = 7% Reverse Gravity = 10% Water Bath = 6% Unclear = 6%
Column 8	No Skin = 56% With Skin = 44%
Column 9	Lesion exists but not modeled = 23% Lesion exists and modeled = 17% No lesion = 60%

under higher strain values, the NH model is known to exhibit softening rather than the biological tissue common characteristic of stiffening. There is also the risk of poor model choice, which could potentially impact the conclusions of studies. Fig. 7.1 shows that Ogden model exhibits significantly larger strain stiffening than the NH model. In Han et al. (2012) and Tanner et al. (2011), it was shown that transverse isotropic tissues limited chest normal elongation in breast biomechanical models. Whereas the conclusions concerning the influence of Cooper's ligaments in this direction is likely, the other influence could be that the NH baseline isotropic model simply could not stiffen fast enough in that direction. Since these studies only used NH and LE models over large strains that cannot capture stiffening like the Ogden showed, the confidence of the NH model remains questionable. These deficiencies suggest that the NH model has serious limitations, hence its suitability should be carefully assessed in a given modeling application.

The M-R model popularity is reasonable, as it is described by 2, or 5 variables if using the 5P model, which naturally increases complexity. To date researchers have used results from the literature and no optimization was performed. The LE cases in Table 7.4 come from small deformation studies, where the LE assumption is reasonably satisfied. Because of the LE assumption, in many cases the tissue was also

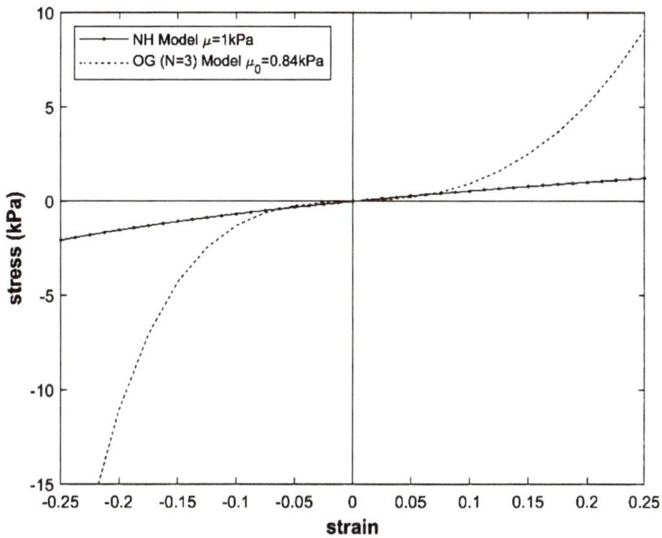

FIGURE 7.1

Plot of the uniaxial stress strain relationship for a neo-Hookean and 3rd-order Ogden model.

considered heterogeneous while optimization was carried out for parameter estimation.

Concerning parameter assignment in Column 4, 72% of real breast biomechanical models used mechanical properties from the literature. In general, there is no obvious trend that using optimization is becoming more popular. This could be due to several reasons. Ideally, the mechanical properties in the literature are an approximate global average, and using these parameters provides acceptable accuracy. However, this is not the case, because their choice of tissue mechanical model or SEDF has not been standardized and, as shown in mechanical studies, the properties from patient to patient vary greatly (see Fig. 6 of Dempsey et al., 2021). Another possible argument is that the estimated average mechanical parameters are expected to provide sufficient accuracy, which can be understandable; however, it loses sight of accurate modeling. Ideally, the mechanical parameters should be always optimized for given application to achieve desirable accuracy.

Column 5 indicates that 87% of the studies included in the table assume isotropic tissue. Since there are no recent measurement of parameters pertaining to nonlinear anisotropic models of the breast tissue, the field is presently limited from this perspective. There have been some studies that included transverse anisotropic models (13%), whereas studies that compared anisotropic vs. isotropic showed that modeling tissue anisotropy improved accuracy (Han et al., 2012; Tanner et al., 2011, 2009). We suspect the lack of adoption stems from the complexity of modeling anisotropy. However, considering the field of cardiac biomechanics, where full orthotropic material modeling is common with usually 4 plus variables to tune, this gap no longer exists (Avazmohammadi et al., 2019). The results on compressibility being incompressible 94% of the time make complete sense. An early MR-SWE study suggested that the breast is slightly compressible (van Houten et al., 2003). Though some breast models have considered the breast tissue to be compressible, the impact was shown to be minor as noted by (Tanner et al., 2006), nonetheless the best results were obtained by assuming incompressibility.

The rest state geometry in Column 6 is likely the largest point of contention from mechanics perspective. Without a proper rest state, the mechanical parameter initialization is incorrect and will lead to incorrect deformation. This seriously impacts all results from studies that do not consider the rest state, which may be why many studies opt for simple NH materials, which do not exhibit significant nonlinear variation, as shown in Fig. 7.1. Though reverse gravity is a step in the right direction, the rest geometry is then based on a model that is initially prestressed, hence it approaches a stress strain curve from the wrong side of expected stiffness. Ideally, the few research laboratories that have rest state optimization algorithms should share these codes using opensource hubs. Otherwise, application of all the breast tissue mechanical parameters will continue to suffer from a serious source of inaccuracy.

Results presented in Column 7 describe how the breast is in contact with the chest wall, where significant variations are seen. Clearly, there is no consensus, and given the difficulty of collecting detailed high-resolution information in that region from imaging, an accurate answer is not anticipated to be found soon. Since the retromammary space based on anatomy is loose, the boundary conditions of allowing sliding is reasonable. Some works involved using prescribed chest displacements based on imaging. The data was derived for instance from deformable image registration, surface-based constraints, and predicted sliding (Eiben et al., 2016; Gamage et al., 2019; Tanner et al., 2011).

Column 8, which presents information about skin inclusion in the respective models, is also clearly a strict point of contention with 56% modeling the skin, and 44% not. The difference in the mechanical properties between the skin and breast tissue is remarkable, and without a doubt, for clinical and other skin-related purposes, the skin should be accurately modeled. Much of the literature allows for the mechanical properties of the adipose and fibroglandular tissue to be stiffer to "capture" the effect of skin. This is not appropriate, however, as internal tissue displacement configuration accuracy is expected to suffer as a result of considering stiffer breast tissue. Associated errors can significantly impact registration and tumor tracking, as mentioned in many studies. We recognize that truly modeling anisotropic skin is not feasible given the inability to image the fiber directions, though adding a nonlinear isotropic membrane is better than none, and significant improvements in prediction accuracy have been seen by modeling even isotropic skin (Gamage et al., 2019). Since the skin has variable thickness, the best option to include skin is to select a thickness (in the range provided by literature), and then optimize the mechanics.

Lastly, Column 9 considers tumor modeling. As indicated earlier, in all instances of modeling a lesion, the mechanics should be personalized. However, only four out of nine studies that involved a tumor and modeled a tumor optimized mechanics. On top of that, 23% of studies involving lesions did not model them and opted to fill their respective volumes with healthy breast tissue. In the absence of rigorous computational and experimental models quantifying errors pertaining to model elements, such as tumor inclusion and rest state consideration, it is hard to assess their precise impact on breast model accuracy.

7.4 Final remarks and future directions

7.4.1 Final remarks

The mechanics of breast tissue impact many fields of research and accurate measurement of the mechanical parameters of breast tissue will only improve results and further the benefit of modeling in the clinic and industry. General trends observed based on the multiple modalities presented in Section 7.2 show that at least the initial shear modulus of breast adipose and fibroglandular tissue appears to be converging to about 0.8 kPa $\pm$ 0.46 kPa. Nonlinearity agreement exists between large sample studies and, in this range of strains, adipose and fibroglandular tissue can be considered mechanically comparable.

Concerning the application of mechanics in biomechanical modeling, there needs to be a strong push to advance the personalization and realistic modeling of breast models on many fronts. Optimization algorithms should be shared with the community, so the rest state optimization becomes common practice. All models should be pushed to optimize more than one variable NH models with emphasis that this simplification limits the ability to capture nonlinear behavior greatly. All unique tissues that compose the breast should be modeled, including the skin and lesions if present, whereas the presence of Cooper's ligaments can be modeled through utilizing a transverse anisotropic model for adipose and fibroglandular tissue as a start. Without the modeling considerations, while the accuracy of available techniques can still be acceptable, the field will continue to lag behind as the validity of measured parameters will be limited to conditions and scenario of the target model and are anticipated to provide only suboptimal accuracy for other applications/scenarios.

7.4.2 Future directions

Recently, techniques for the reconstruction of hyperelastic models using MRE has been presented by Capilnasiu et al. (2019). As this develops in the clinic, being able to measure other SEDF's with MRE in-vivo for breast tissues will provide a comparison to nonlinear models from ex-vivo indentation. Presently the nonlinear reconstruction has been tested on bovine tissue (Capilnasiu et al., 2020). Between nonlinear FE-USE (Goenezen et al., 2012; Goswami et al., 2020) and nonlinear SW-MRE, these measurements are expected to impact the clinic greatly for lesion diagnosis and identification (Rus et al., 2020).

In terms of modeling, recent work has highlighted the importance of both anisotropic mechanics and skin mechanics (Gamage et al., 2019), so introducing these mechanics into conventional breast models should improve accuracy of the desired biomechanical model goal.

References

Al abdi, R.M., Deng, B., Hijazi, H.H., Wu, M., Carp, S.A., 2020. Mechanical and hemodynamic responses of breast tissue under mammographic-like compression during functional dynamic optical imaging. Biomedical Optics Express 11 (10), 5425. https://doi.org/10.1364/boe.398110.

Ashikuzzaman, M., Gauthier, C.J., Rivaz, H., 2019. Global ultrasound elastography in spatial and temporal domains. IEEE Transactions on Ultrasonics, Ferroelectrics, and Frequency Control 66 (5), 876–887. https://doi.org/10.1109/TUFFC.2019.2903311.

Avazmohammadi, R., Soares, J.S., Li, D.S., Raut, S.S., Gorman, R.C., Sacks, M.S., 2019. A contemporary look at biomechanical models of myocardium. Annual Review of Biomedical Engineering 21 (1), 417–442. https://doi.org/10.1146/annurev-bioeng-062117-121129.

Azar, F.S., Metaxas, D.N., Schnall, M.D., 2001. A deformable finite element model of the breast for predicting mechanical deformations under external perturbations 1. Academic Radiology 8 (10), 965–975. https://doi.org/10.1016/S1076-6332(03)80640-2.

Azar, F.S., Metaxas, D.N., Schnall, M.D., 2002. Methods for modeling and predicting mechanical deformations of the breast under external perturbations. Medical Image Analysis 6, 1–27. https://doi.org/10.1016/S1361-8415(01)00053-6.

Barr, R.G., Zhang, Z., 2012. Effects of precompression on elasticity imaging of the breast. Journal of Ultrasound in Medicine 31 (6), 895–902. https://doi.org/10.7863/jum.2012.31.6.895.

Bessa, S., Carvalho, P.H., Oliveira, H.P., 2019. Registration of breast MRI and 3D scan data based on surface matching. In: 2019 IEEE 16th International Symposium on Biomedical Imaging (ISBI 2019), pp. 1074–1077.

Bessa, S., Gouveia, P.F., Carvalho, P.H., Rodrigues, C., Silva, N.L., Cardoso, F., Cardoso, J.S., Oliveira, H.P., Cardoso, M.J., 2020. 3D digital breast cancer models with multimodal fusion algorithms. Breast 49, 281–290. https://doi.org/10.1016/j.breast.2019.12.016.

Briot, N., Chagnon, G., Masri, C., Girard, E., Payan, Y., 2020. Ex-vivo mechanical characterisation of the breast Cooper's ligaments. Computer Methods in Biomechanics and Biomedical Engineering 23 (sup1), S49–S51. https://doi.org/10.1080/10255842.2020.1812155.

Cai, Y., Yu, W., Chen, L., 2016. Finite element modeling of bra fitting. In: Advances in Women's Intimate Apparel Technology. Elsevier Inc., pp. 147–168.

Capilnasiu, A., Bilston, L., Sinkus, R., Nordsletten, D., 2020. Nonlinear viscoelastic constitutive model for bovine liver tissue. Biomechanics and Modeling in Mechanobiology 19 (5), 1641–1662. https://doi.org/10.1007/s10237-020-01297-5.

Capilnasiu, A., Hadjicharalambous, M., Fovargue, D., Patel, D., Holub, O., Bilston, L., Screen, H., Sinkus, R., Nordsletten, D., 2019. Magnetic resonance elastography in nonlinear viscoelastic materials under load. Biomechanics and Modeling in Mechanobiology 18 (1), 111–135. https://doi.org/10.1007/s10237-018-1072-1.

Carter, T.J., Tanner, C., Hawkes, D.J., 2009. Determining material properties of the breast for image-guided surgery. In: Medical Imaging 2009: Visualization, Image-Guided Procedures, and Modeling, vol. 7261, p. 726124.

Chen, J.-H., Chan, S., Zhang, Y., Li, S., Chang, R.F., Su, M.Y., 2019. Evaluation of breast stiffness measured by ultrasound and breast density measured by MRI using a prone-supine deformation model. Biomarker Research 7 (1). https://doi.org/10.1186/s40364-019-0171-1.

Chen, L.-H., Ng, S.-P., Yu, W., Zhou, J., Wan, K.W.F., 2013. A study of breast motion using non-linear dynamic FE analysis. Ergonomics 56 (5), 868–878. https://doi.org/10.1080/00140139.2013.777798.

Chung, J.-H., Rajagopal, V., Nielsen, P.M.F., Nash, M.P., 2008. Modelling mammographic compression of the breast. In: Medical Image Computing and Computer-Assisted Intervention – MICCAI 2008.

Conley, R.H., Meszoely, I.M., Weis, J.A., Pheiffer, T.S., Arlinghaus, L.R., Yankeelov, T.E., Miga, M.I., 2015. Realization of a biomechanical model-assisted image guidance system for breast cancer surgery using supine MRI. International Journal of Computer Assisted Radiology and Surgery 10 (12), 1985–1996. https://doi.org/10.1007/s11548-015-1235-9.

Cruz, P., Hernandez, F.J., Zuñiga, M.L., Rodríguez, J.M., Figueroa, R., Vertiz, A., Pineda, Z., 2018. A biomimetic approach for designing a full external breast prosthesis: post-mastectomy. Applied Sciences 8 (3). https://doi.org/10.3390/app8030357.

Davis, L.C., Baumer, T.G., Bey, M.J., van Holsbeeck, M., 2019. Clinical utilization of shear wave elastography in the musculoskeletal system. Ultrasonography 38 (1), 2–12. https://doi.org/10.14366/usg.18039.

del Palomar, A.P., Calvo, B., Herrero, J., López, J., Doblaré, M., 2008. A finite element model to accurately predict real deformations of the breast. Medical Engineering & Physics 30 (9), 1089–1097. https://doi.org/10.1016/j.medengphy.2008.01.005.

Dempsey, S.C.H., O'Hagan, J.J., Samani, A., 2021. Measurement of the hyperelastic properties of 72 normal homogeneous and heterogeneous ex vivo breast tissue samples. Journal of the Mechanical Behavior of Biomedical Materials 124, 104794. https://doi.org/10.1016/j.jmbbm.2021.104794.

Dufaye, G., Cherouat, A., Bachmann, J.M., Borouchaki, H., 2013. Advanced finite element modelling for the prediction of 3D breast deformation. European Journal of Computational Mechanics 22 (2–4), 170–182. https://doi.org/10.1080/17797179.2013.820902.

Eder, M., Raith, S., Jalali, J., Volf, A., Settles, M., Machens, H.G., Kovacs, L., 2014. Comparison of different material models to simulate 3-D breast deformations using finite element analysis. Annals of Biomedical Engineering 42 (4), 843–857. https://doi.org/10.1007/s10439-013-0962-8.

Egorov, V., Kearney, T., Pollak, S.B., Rohatgi, C., Sarvazyan, N., Airapetian, S., Browning, S., Sarvazyan, A., 2009. Differentiation of benign and malignant breast lesions by mechanical imaging. Breast Cancer Research and Treatment 118 (1), 67–80. https://doi.org/10.1007/s10549-009-0369-2.

Egorov, V., Sarvazyan, A.P., 2008. Mechanical imaging of the breast. IEEE Transactions on Medical Imaging 27 (9), 1275–1287. https://doi.org/10.1109/TMI.2008.922192.

Eiben, B., Vavourakis, V., Hipwell, J.H., Kabus, S., Buelow, T., Lorenz, C., Mertzanidou, T., Reis, S., Williams, N.R., Keshtgar, M., Hawkes, D.J., 2016. Symmetric biomechanically guided prone-to-supine breast image registration. Annals of Biomedical Engineering 44 (1), 154–173. https://doi.org/10.1007/s10439-015-1496-z.

Esslinger, D., Rapp, P., Knödler, L., Preibsch, H., Tarín, C., Sawodny, O., Brucker, S.Y., Hahn, M., 2019. A novel finite element model–based navigation system–supported workflow for breast tumor excision. Medical & Biological Engineering & Computing 57 (7), 1537–1552. https://doi.org/10.1007/s11517-019-01977-0.

Flynn, C., Taberner, A., Nielsen, P., 2011. Measurement of the force-displacement response of in vivo human skin under a rich set of deformations. Medical Engineering & Physics 33 (5), 610–619. https://doi.org/10.1016/j.medengphy.2010.12.017.

Fovargue, D., Nordsletten, D., Sinkus, R., 2018. Stiffness reconstruction methods for MR elastography. NMR in Biomedicine 31 (10). https://doi.org/10.1002/nbm.3935.

Gamage, T.P.B., Baluwala, H.Y., Nash, M.P., Nielsen, P.M.F., 2017a. Registration of prone and supine breast MRI for breast cancer treatment planning. In: Wittek, A., Joldes, G., Nielsen, P.M.F., Doyle, B.J., Miller, K. (Eds.), Computational Biomechanics for Medicine. Springer International Publishing.

Gamage, T.P.B., Boyes, R., Rajagopal, V., Nielsen, P.M.F., Nash, M.P., 2012. Modelling prone to supine breast deformation under gravity loading using heterogeneous finite element models. In: Computational Biomechanics for Medicine: Deformation and Flow. Springer New York, pp. 29–38.

Gamage, T.P.B., Malcolm, D.T.K., Talou, G.M., Mîra, A., Doyle, A., Nielsen, P.M.F., Nash, M.P., 2019. An automated computational biomechanics workflow for improving breast cancer diagnosis and treatment. Interface Focus 9 (4). https://doi.org/10.1098/rsfs.2019.0034.

Gamage, T.P.B., Nielsen, P.M.F., Nash, M.P., 2017b. Clinical applications of breast biomechanics. In: Biomechanics of Living Organs: Hyperelastic Constitutive Laws for Finite Element Modeling. Elsevier Inc., pp. 215–242.

García, E., Diez, Y., Diaz, O., Lladó, X., Gubern-Mérida, A., Martí, R., Martí, J., Oliver, A., 2019. Breast MRI and X-ray mammography registration using gradient values. Medical Image Analysis 54, 76–87. https://doi.org/10.1016/j.media.2019.02.013.

García, E., Diez, Y., Diaz, O., Lladó, X., Martí, R., Martí, J., Oliver, A., 2018. A step-by-step review on patient-specific biomechanical finite element models for breast MRI to x-ray mammography registration. Medical Physics 45 (1), e6–e31. https://doi.org/10.1002/mp.12673.

Gemmeke, H., Hopp, T., Zapf, M., Kaiser, C., Ruiter, N.v., 2017. 3D ultrasound computer tomography: hardware setup, reconstruction methods and first clinical results. Nuclear Instruments & Methods in Physics Research. Section A, Accelerators, Spectrometers, Detectors and Associated Equipment 873, 59–65. https://doi.org/10.1016/j.nima.2017.07.019.

Ghadarghadar, N., Agrawal, S.K., Samani, A., Ladak, H.M., 2013. Estimation of the quasi-static Young's modulus of the eardrum using a pressurization technique. Computer Methods and Programs in Biomedicine 110 (3), 231–239. https://doi.org/10.1016/j.cmpb.2012.11.006.

Goenezen, S., Dord, J.-F., Sink, Z., Barbone, P.E., Jiang, J., Hall, T.J., Oberai, A.A., 2012. Linear and nonlinear elastic modulus imaging: an application to breast cancer diagnosis. IEEE Transactions on Medical Imaging 31 (8), 1628–1637. https://doi.org/10.1109/TMI.2012.2201497.

Gokhale, N.H., Barbone, P.E., Oberai, A.A., 2008. Solution of the nonlinear elasticity imaging inverse problem: the compressible case. Inverse Problems 24 (4). https://doi.org/10.1088/0266-5611/24/4/045010.

Goswami, S., Ahmed, R., Khan, S., Doyley, M.M., McAleavey, S.A., 2020. Shear induced non-linear elasticity imaging: elastography for compound deformations. IEEE Transactions on Medical Imaging 39 (11), 3559–3570. https://doi.org/10.1109/TMI.2020.2999439.

Green, C.A., Goodsitt, M.M., Roubidoux, M.A., Brock, K.K., Davis, C.L., Lau, J.H., Carson, P.L., 2020. Deformable mapping using biomechanical models to relate corresponding lesions in digital breast tomosynthesis and automated breast ultrasound images. Medical Image Analysis 60. https://doi.org/10.1016/j.media.2019.101599.

Griesenauer, R.H., Weis, J.A., Arlinghaus, L.R., Meszoely, I.M., Miga, M.I., 2017. Breast tissue stiffness estimation for surgical guidance using gravity-induced excitation. Physics in Medicine and Biology 62 (12), 4756–4776. https://doi.org/10.1088/1361-6560/aa700a.

Griesenauer, R.H., Weis, J.A., Arlinghaus, L.R., Meszoely, I.M., Miga, M.I., 2018. Toward quantitative quasistatic elastography with a gravity-induced deformation source for image-guided breast surgery. Journal of Medical Imaging 5 (01), 1. https://doi.org/10.1117/1.jmi.5.1.015003.

Haddad, S.M.H., Omidi, E., Flynn, L.E., Samani, A., 2016. Comparative biomechanical study of using decellularized human adipose tissues for post-mastectomy and post-lumpectomy breast reconstruction. Journal of the Mechanical Behavior of Biomedical Materials 57, 235–245. https://doi.org/10.1016/j.jmbbm.2015.12.005.

Han, L., Hipwell, J.H., Eiben, B., Barratt, D., Modat, M., Ourselin, S., Hawkes, D.J., 2014. A nonlinear biomechanical model based registration method for aligning proneand supine MR breast images. IEEE Transactions on Medical Imaging 33 (3), 682–694. https://doi.org/10.1109/TMI.2013.2294539.

Han, L., Hipwell, J.H., Tanner, C., Taylor, Z., Mertzanidou, T., Cardoso, J., Ourselin, S., Hawkes, D.J., 2012. Development of patient-specific biomechanical models for predicting large breast deformation. Physics in Medicine and Biology 57 (2), 455–472. https://doi.org/10.1088/0031-9155/57/2/455.

Hawley, J.R., Kalra, P., Mo, X., Raterman, B., Yee, L.D., Kolipaka, A., 2017. Quantification of breast stiffness using MR elastography at 3 Tesla with a soft sternal driver: a reproducibility study. Journal of Magnetic Resonance Imaging 45 (5), 1379–1384. https://doi.org/10.1002/jmri.25511.

Hayes, W.C., Keer, L.M., Herrmann, G., Mockros, L.F., 1972. A mathematical analysis for indentation tests of articular cartilage. Journal of Biomechanics 5 (5). https://doi.org/10.1016/0021-9290(72)90010-3.

Herthum, H., Dempsey, S.C.H., Samani, A., Schrank, F., Shahryari, M., Warmuth, C., Tzschätzsch, H., Braun, J., Sack, I., 2021. Superviscous properties of the in vivo brain at large scales. Acta Biomaterialia 121, 393–404. https://doi.org/10.1016/j.actbio.2020.12.027.

Hipwell, J.H., Vavourakis, V., Han, L., Mertzanidou, T., Eiben, B., Hawkes, D.J., 2016. A review of biomechanically informed breast image registration. Physics in Medicine and Biology 61 (2), R1–R31. https://doi.org/10.1088/0031-9155/61/2/R1.

Holzapfel, G., 2001. Nonlinear solid mechanics. A continuum approach for engineering, 2nd ed. John Wiley & Sons.

Hopp, T., Baltzer, P., Dietzel, M., Kaiser, W.A., Ruiter, N.v., 2012. 2D/3D image fusion of X-ray mammograms with breast MRI: visualizing dynamic contrast enhancement in mammograms. International Journal of Computer Assisted Radiology and Surgery 7 (3), 339–348. https://doi.org/10.1007/s11548-011-0623-z.

Hopp, T., Duric, N., Ruiter, N.v., 2015. Image fusion of Ultrasound Computer Tomography volumes with X-ray mammograms using a biomechanical model based 2D/3D registration. Computerized Medical Imaging and Graphics 40, 170–181. https://doi.org/10.1016/j.compmedimag.2014.10.005.

Hopp, T., Ruiter, N.v., 2021. Strain elastography with ultrasound computer tomography: a simulation study based on biomechanical models. In: Ruiter, N.v., Byram, B.C. (Eds.), Medical Imaging 2021: Ultrasonic Imaging and Tomography. SPIE.

Hsu, C.M.L., Palmeri, M.L., Segars, W.P., Veress, A.I., Dobbins, J.T., 2011. An analysis of the mechanical parameters used for finite element compression of a high-resolution 3D breast phantom. Medical Physics 38 (10), 5756–5770. https://doi.org/10.1118/1.3637500.

Johns, P.C., Yaffe, M.J., 1987. X-ray characterisation of normal and neoplastic breast tissues. Physics in Medicine and Biology 32 (6), 675–697. https://doi.org/10.1088/0031-9155/32/6/002.

Jor, J.W.Y., Parker, M.D., Taberner, A.J., Nash, M.P., Nielsen, P.M.F., 2013. Computational and experimental characterization of skin mechanics: identifying current challenges and future directions. Wiley Interdisciplinary Reviews: Systems Biology and Medicine 5 (5), 539–556. https://doi.org/10.1002/wsbm.1228.

Kheirkhah, N., Sadeghi-Naini, A., Samani, A., 2020. Analytical estimation of out-of-plane strain in ultrasound elastography to improve axial and lateral displacement fields*. In: 2020 42nd Annual International Conference of the IEEE Engineering in Medicine & Biology Society, pp. 2055–2058.

Krouskop, T.A., Wheeler, T.M., Kallel, F., Garra, B.S., Hall, T., 1998. Elastic moduli of breast and prostate tissues under compression. Ultrasonic Imaging 20 (4), 260–274. https://doi.org/10.1177/016173469802000403.

Kruse, S.A., Smith, J.A., Lawrence, A.J., Dresner, M.A., Manduca, A., Greenleaf, J.F., Ehman, R.L., 2000. Tissue characterization using magnetic resonance elastography: preliminary results Tissue characterization using magnetic resonance elastography: preliminary results*. Physics in Medicine and Biology 45, 1579–1590. https://doi.org/10.1088/0031-9155/45/6/313.

Lago, M.A., Martínez-Martínez, F., Rupérez, M.J., Monserrat, C., Alcañiz, M., 2012. Breast prone-to-supine deformation and registration using a time-of-flight camera. In: The Fourth IEEE RAS/EMBS International Conference on Biomedical Robotics and Biomechatronics, pp. 1161–1163.

Lapuebla-Ferri, A., Cegoñino-Banzo, J., Jiménez-Mocholí, A.-J., del Palomar, A.P., 2017. Towards an in-plane methodology to track breast lesions using mammograms and patient-specific finite-element simulations. Physics in Medicine and Biology 62 (22), 8720–8738. https://doi.org/10.1088/1361-6560/aa8d62.

Lapuebla-Ferri, A., del Palomar, A.P., Herrero, J., Jiménez-Mocholí, A.-J., 2011. A patient-specific FE-based methodology to simulate prosthesis insertion during an augmentation mammoplasty. Medical Engineering & Physics 33 (9), 1094–1102. https://doi.org/10.1016/j.medengphy.2011.04.014.

Lee, A.W.C., Rajagopal, V., Babarenda Gamage, T.P., Doyle, A.J., Nielsen, P.M.F., Nash, M.P., 2013. Breast lesion co-localisation between X-ray and MR images using finite element modelling. Medical Image Analysis 17 (8), 1256–1264. https://doi.org/10.1016/j.media.2013.05.011.

Limbert, G., 2019. Constitutive Modelling of Skin Mechanics. Studies in Mechanobiology, Tissue Engineering and Biomaterials, vol. 22. Springer, pp. 19–76.

Liu, Y.-L., Liu, P.-Y., Huang, M.-L., Hsu, J.-T., Han, R.-P., Wu, J., 2017. Simulation of breast compression in mammography using finite element analysis: a preliminary study. Radiation Physics and Chemistry 140, 295–299. https://doi.org/10.1016/j.radphyschem.2017.01.017.

Lorenzen, J., Sinkus, R., Biesterfeldt, M., Adam, G., 2003. Menstrual-cycle dependence of breast parenchyma elasticity: estimation with magnetic resonance elastography of breast tissue during the menstrual cycle. Investigative Radiology 38 (4), 236–240.

Lorenzen, J., Sinkus, R., Lorenzen, M., Dargatz, M., Leussler, C., Röschmann, P., Adam, G., 2002. MR elastography of the breast: preliminary clinical results. Rofo 174 (7), 830–834. https://doi.org/10.1055/s-2002-32690.

Manduca, A., Bayly, P.J., Ehman, R.L., Kolipaka, A., Royston, T.J., Sack, I., Sinkus, R., van Beers, B.E., 2021. MR elastography: principles, guidelines, and terminology. Magnetic Resonance in Medicine 85 (5), 2377–2390. https://doi.org/10.1002/mrm.28627.

Mariappan, Y.K., Glaser, K.J., Ehman, R.L., 2010. Magnetic resonance elastography: a review. Clinical Anatomy 23 (5), 497–511. https://doi.org/10.1002/ca.21006.

Martínez-Martínez, F., Rupérez-Moreno, M.J., Martínez-Sober, M., Solves-Llorens, J.A., Lorente, D., Serrano-López, A.J., Martínez-Sanchis, S., Monserrat, C., Martín-Guerrero, J.D., 2017. A finite element-based machine learning approach for modeling the mechanical behavior of the breast tissues under compression in real-time. Computers in Biology and Medicine 90, 116–124. https://doi.org/10.1016/j.compbiomed.2017.09.019.

Matsumura, T., Umemoto, T., Fujihara, Y., Ueno, E., Yamakawa, M., Shiina, T., Mitake, T., 2009. Measurement of elastic property of breast tissue for elasticity imaging. In: 2009 IEEE International Ultrasonics Symposium Proceedings, pp. 1451–1454.

McKnight, A.L., Kugel, J.L., Rossman, P.J., Manduca, A., Hartmann, L.C., Ehman, R.L., 2002. MR elastography of breast cancer: preliminary results. American Journal of Roentgenology 178 (6), 1411–1417. https://doi.org/10.2214/ajr.178.6.1781411.

Mertzanidou, T., Hipwell, J., Johnsen, S., Han, L., Eiben, B., Taylor, Z., Ourselin, S., Huisman, H., Mann, R., Bick, U., Karssemeijer, N., Hawkes, D., 2014. MRI to X-ray mammography intensity-based registration with simultaneous optimisation of pose and biomechanical transformation parameters. Medical Image Analysis 18 (4), 674–683. https://doi.org/10.1016/j.media.2014.03.003.

Mertzanidou, T., Hipwell, J.H., Han, L., Taylor, Z., Huisman, H., Bick, U., Karssemeijer, N., Hawkes, D.J., 2012. Intensity-based MRI to X-ray mammography registration with an integrated fast biomechanical transformation. In: Maidment, A.D.A., Bakic, P.R., Gavenonis, S. (Eds.), Breast Imaging. IWDM 2012. In: Lecture Notes in Computer Science, vol. 7361. Springer.

Mousavi, S.R., Rivaz, H., Sadeghi-Naini, A., Czarnota, G.J., Samani, A., 2017. Breast ultrasound elastography using full inversion-based elastic modulus reconstruction. IEEE Transactions on Computational Imaging 3 (4), 774–782. https://doi.org/10.1109/tci.2017.2741422.

Mousavi, S.R., Sadeghi-Naini, A., Czarnota, G.J., Samani, A., 2014. Towards clinical prostate ultrasound elastography using full inversion approach. Medical Physics 41 (3). https://doi.org/10.1118/1.4864476.

Myung, Y., Lee, J.G., Cho, M., Heo, C.Y., 2019. Finite element analysis of long-term changes of the breast after augmentation mammoplasty: implications for implant design. Archives of Plastic Surgery 46 (4), 386–389. https://doi.org/10.5999/aps.2019.00346.

Na, G.-Y., Yang, J., Cho, S., 2019. Development of a 3D breast shape generation and deformation system for breast implant fabrication. Journal of Mechanical Science and Technology 33 (3), 1293–1303. https://doi.org/10.1007/s12206-019-0230-4.

Nafo, W., Al-Mayah, A., 2019. Measuring hyperelastic properties of hydrogels using cavity expansion method. Experimental Mechanics 59 (7), 1047–1061. https://doi.org/10.1007/s11340-019-00504-4.

Oberai, A.A., Gokhale, N.H., Goenezen, S., Barbone, P.E., Hall, T.J., Sommer, A.M., Jiang, J., 2009. Linear and nonlinear elasticity imaging of soft tissue in vivo: demonstration of feasibility. Physics in Medicine and Biology 54 (5), 1191–1207. https://doi.org/10.1088/0031-9155/54/5/006.

Ogden, R.W., 1972. Large deformation isotropic elasticity on the correlation of theory and experiment for incompressible rubberlike solids. Proceedings of the Royal Society of London. Series A, Mathematical and Physical Sciences 326 (1567). https://doi.org/10.1098/rspa.1972.0026.

O'Hagan, J.J., Samani, A., 2008. Measurement of the hyperelastic properties of tissue slices with tumour inclusion. Physics in Medicine and Biology 53 (24), 7087–7106. https://doi.org/10.1088/0031-9155/53/24/006.

O'Hagan, J.J., Samani, A., 2009. Measurement of the hyperelastic properties of 44 pathological ex vivo breast tissue samples. Physics in Medicine and Biology 54 (8), 2557–2569. https://doi.org/10.1088/0031-9155/54/8/020.

Oliver, X., Agelet de Saracibar, C., 2017. Continuum Mechanics for Engineers. Theory and Problems, 2nd ed.

Oomens, C.W.J., van Vijven, M., Peters, G.W.M., 2017. Skin mechanics. In: Biomechanics of Living Organs: Hyperelastic Constitutive Laws for Finite Element Modeling. Elsevier Inc., pp. 347–357.

Pianigiani, S., Ruggiero, L., Innocenti, B., 2015. An anthropometric-based subject-specific finite element model of the human breast for predicting large deformations. Frontiers in Bioengineering and Biotechnology 3. https://doi.org/10.3389/fbioe.2015.00201.

Rajagopal, V., Chung, J.H., Bullivant, D., Nielsen, P.M.F., Nash, M.P., 2007. Determining the finite elasticity reference state from a loaded configuration. International Journal for Numerical Methods in Engineering 72 (12), 1434–1451. https://doi.org/10.1002/nme.2045.

Rajagopal, V., Lee, A., Chung, J.-H., Warren, R., Highnam, R.P., Nash, M.P., Nielsen, P.M.F., 2008. Creating individual-specific biomechanical models of the breast for medical image analysis. Academic Radiology 15 (11), 1425–1436. https://doi.org/10.1016/j.acra.2008.07.017.

Ramião, N.G., Martins, P.S., Rynkevic, R., Fernandes, A.A., Barroso, M., Santos, D.C., 2016. Biomechanical properties of breast tissue, a state-of-the-art review. Biomechanics and Modeling in Mechanobiology 15 (5), 1307–1323. https://doi.org/10.1007/s10237-016-0763-8.

Rivlin, R.S., Saunders, D.W., 1951. Large elastic deformations of isotropic materials VII. Experiments on the deformation of rubber. Philosophical Transactions of the Royal Society of London, Series A: Mathematical and Physical Sciences 243 (865). https://doi.org/10.1098/rsta.1951.0004.

Ruiter, N.v., Stotzka, R., Müller, T.O., Gemmeke, H., Reichenbach, J.R., Kaiser, W.A., 2006. Model-based registration of X-ray mammograms and MR images of the female breast. IEEE Transactions on Nuclear Science 53 (1), 204–211. https://doi.org/10.1109/TNS.2005.862983.

Rus, G., Faris, I.H., Torres, J., Callejas, A., Melchor, J., 2020. Why are viscosity and nonlinearity bound to make an impact in clinical elastographic diagnosis? Sensors (Switzerland) 20 (8). https://doi.org/10.3390/s20082379.

Samani, A., Bishop, J., Luginbuhl, C., Plewes, D.B., 2003. Measuring the elastic modulus of ex vivo small tissue samples. Physics in Medicine and Biology 48 (14), 2183–2198. https://doi.org/10.1088/0031-9155/48/14/310.

Samani, A., Plewes, D., 2004. A method to measure the hyperelastic parameters of ex vivo breast tissue samples. Physics in Medicine and Biology 49 (18), 4395–4405. https://doi.org/10.1088/0031-9155/49/18/014.

Samani, A., Zubovits, J., Plewes, D., 2007. Elastic moduli of normal and pathological human breast tissues: an inversion-technique-based investigation of 169 samples. Physics in Medicine and Biology 52 (6), 1565–1576. https://doi.org/10.1088/0031-9155/52/6/002.

Sandby-Møller, J., Poulsen, T., Wulf, H.C., 2003. Epidermal thickness at different body sites: relationship to age, gender, pigmentation, blood content, skin type and smoking habits. Acta Dermato-Venereologica 83 (6), 410–413. https://doi.org/10.1080/00015550310015419.

Shih, T.-C., Chen, J.H., Liu, D., Nie, K., Sun, L., Lin, M., Chang, D., Nalcioglu, O., Su, M.-Y., 2010. Computational simulation of breast compression based on segmented breast and fibroglandular tissues on magnetic resonance images. Physics in Medicine and Biology 55 (14), 4153–4168. https://doi.org/10.1088/0031-9155/55/14/013.

Shih, W.Y., Shih, W.-H., Xu, X., 2015. Depth measurement in tissue using piezoelectric sensors having different probe sizes. Patent No. US20160331300. United States Patent and Trademark Office.

Sigrist, R.M.S., Liau, J., Kaffas, A. el, Chammas, M.C., Willmann, J.K., 2017. Ultrasound elastography: review of techniques and clinical applications. Theranostics 7 (5), 1303–1329. https://doi.org/10.7150/thno.18650.

Sinkus, R., Lorenzen, J., Schrader, D., Lorenzen, M., Dargatz, M., Holz, D., 2000. High-resolution tensor MR elastography for breast tumour detection. Physics in Medicine and Biology 45, 1649–1664. https://doi.org/10.1088/0031-9155/45/6/317.

Sinkus, R., Tanter, M., Catheline, S., Lorenzen, J., Kuhl, C., Sondermann, E., Fink, M., 2005. Imaging anisotropic and viscous properties of breast tissue by magnetic resonance-elastography. Magnetic Resonance in Medicine 53 (2), 372–387. https://doi.org/10.1002/mrm.20355.

Sinkus, R., Weiss, S., Wiggert, E., Lorenzen, J., Dargatz, M., Kuhl, C., 2002. Non-linear elastic tissue properties of the breast measured by MR-elastography-initial in-vitro and in-vivo results. In: Proceedings ISMRM 10th Annual Meeting.

Skovoroda, A.R., Klishko, A.N., Gusakyan, D.A., Mayevskii, Ye.I., Yermilova, V.D., Oranskaya, G.A., Sarvazyan, A.P., 1995. Quantitative analysis of the mechanical characteristics of pathologically changed soft biological tissues. Biophysics 40 (6), 1359–1364. https://www.researchgate.net/publication/281044249.

Solanki, M., Raja, V., 2010. VRCAI 2010: proceedings: ACM SIGGRAPH conference on virtual-reality continuum and its applications in industry: Seoul, South Korea December 12-13, 2010. In: VRCAI '10: Proceedings of the 9th ACM SIGGRAPH Conference on Virtual-Reality Continuum and Its Applications in Industry, pp. 221–224.

Solves-Llorens, J.A., Rupérez, M.J., Monserrat, C., Feliu, E., García, M., Lloret, M., 2014. A complete software application for automatic registration of x-ray mammography and magnetic resonance images. Medical Physics 41 (8). https://doi.org/10.1118/1.4885957.

Stewart, M.L., Smith, L.M., Hall, N., 2011. A numerical investigation of breast compression: a computer-aided design approach for prescribing boundary conditions. IEEE Transactions on Biomedical Engineering 58 (10 PART 1), 2876–2884. https://doi.org/10.1109/TBME.2011.2162063.

Sturgeon, G.M., Kiarashi, N., Lo, J.Y., Samei, E., Segars, W.P., 2016. Finite-element modeling of compression and gravity on a population of breast phantoms for multimodality imaging simulation. Medical Physics 43 (5), 2207–2217. https://doi.org/10.1118/1.4945275.

Sun, Y., Chen, L., Yick, K. lun, Yu, W., Lau, N., Jiao, W., 2019. Optimization method for the determination of Mooney-Rivlin material coefficients of the human breasts in-vivo using static and dynamic finite element models. Journal of the Mechanical Behavior of Biomedical Materials 90, 615–625. https://doi.org/10.1016/j.jmbbm.2018.11.016.

Sun, Y., Yick, K. lun, Cai, Y., Yu, W., Chen, L., Lau, N., Zhang, S., 2021. Finite element analysis on contact pressure and 3D breast deformation for application in women's bras. Fibers and Polymers. https://doi.org/10.1007/s12221-021-0878-0.

Sun, Z., Lee, S.H., Gepner, B.D., Rigby, J., Hallman, J.J., Kerrigan, J.R., 2021. Comparison of porcine and human adipose tissue loading responses under dynamic compression and shear: a pilot study. Journal of the Mechanical Behavior of Biomedical Materials 113. https://doi.org/10.1016/j.jmbbm.2020.104112.

Sung, H., Ferlay, J., Siegel, R.L., Laversanne, M., Soerjomataram, I., Jemal, A., Bray, F., 2021. Global cancer statistics 2020: GLOBOCAN estimates of incidence and mortality worldwide for 36 cancers in 185 countries. CA: A Cancer Journal for Clinicians 71 (3), 209–249. https://doi.org/10.3322/caac.21660.

Tanner, C., Schnabel, J.A., Hill, D.L.G., Hawkes, D.J., Leach, M.O., Hose, D.R., 2006. Factors influencing the accuracy of biomechanical breast models. Medical Physics 33 (6), 1758–1769. https://doi.org/10.1118/1.2198315.

Tanner, C., White, M., Guarino, S., Hall-Craggs, M.A., Douek, M., Hawkes, D.J., 2009. Anisotropic behaviour of breast tissue for large compressions. In: 2009 IEEE International Symposium on Biomedical Imaging: From Nano to Macro, pp. 1223–1226.

Tanner, C., White, M., Guarino, S., Hall-Craggs, M.A., Douek, M., Hawkes, D.J., 2011. Large breast compressions: observations and evaluation of simulations. Medical Physics 38 (2), 682–690. https://doi.org/10.1118/1.3525837.

Umemoto, T., Ueno, E., Matsumura, T., Yamakawa, M., Bando, H., Mitake, T., Shiina, T., 2014. Ex vivo and in vivo assessment of the non-linearity of elasticity properties of breast tissues for quantitative strain elastography. Ultrasound in Medicine & Biology 40 (8), 1755–1768. https://doi.org/10.1016/j.ultrasmedbio.2014.02.005.

van Houten, E.E.W., Doyley, M.M., Kennedy, F.E., Weaver, J.B., Paulsen, K.D., 2003. Initial in vivo experience with steady-state subzone-based MR elastography of the human breast. Journal of Magnetic Resonance Imaging 17 (1), 72–85. https://doi.org/10.1002/jmri.10232.

Vaughan, C.L., 2019. Novel imaging approaches to screen for breast cancer: recent advances and future prospects. Medical Engineering & Physics 72, 27–37. https://doi.org/10.1016/j.medengphy.2019.09.001.

Vavourakis, V., Eiben, B., Hipwell, J.H., Williams, N.R., Keshtgar, M., Hawkes, D.J., 2016a. Multiscale mechano-biological finite element modelling of oncoplastic breast surgery - numerical study towards surgical planning and cosmetic outcome prediction. PLoS ONE 11 (7). https://doi.org/10.1371/journal.pone.015976.

Vavourakis, V., Hipwell, J.H., Hawkes, D.J., 2016b. An inverse finite element u/p-formulation to predict the unloaded state of in vivo biological soft tissues. Annals of Biomedical Engineering 44 (1), 187–201. https://doi.org/10.1007/s10439-015-1405-5.

Veronda, D.R., Westmann, R.A., 1970. Mechanical characterization of skin—finite deformations. Journal of Biomechanics 3 (1). https://doi.org/10.1016/0021-9290(70)90055-2.

Wang, Y., Bayer, M., Jiang, J., Hall, T.J., 2020. An improved region-growing motion tracking method using more prior information for 3-D ultrasound elastography. IEEE Transactions on Ultrasonics, Ferroelectrics, and Frequency Control 67 (3), 580–597. https://doi.org/10.1109/TUFFC.2019.2948984.

Wellman, P., Howe, R., Dalton, E., Kern, K.A., 1999. Breast Tissue Stiffness in Compression is Correlated to Histological Diagnosis.

Wellman, P.S., 1999. Tactile Sensing.

Wellman, P.S., Dalton, E.P., Krag, D., Kern, K.A., Howe, R.D., 2001. Tactile imaging of breast masses first clinical report. Archives of Surgery 136, 204–208. https://doi.org/10.1001/archsurg.136.2.204.

Xue, C., Tang, F.H., Lai, C.W.K., Grimm, L.J., Lo, J.Y., 2021. Multimodal patient-specific registration for breast imaging using biomechanical modeling with reference to AI evaluation of breast tumor change. Life 11 (8). https://doi.org/10.3390/life11080747.

Yin, H.M., Sun, L.Z., Wang, G., Yamada, T., Wang, J., Vannier, M.W., 2004. ImageParser: a tool for finite element generation from three-dimensional medical images. BioMedical Engineering OnLine 3 (31). https://doi.org/10.1186/1475-925X-3-31.

Zhang, C., Hilts, M., Batchelar, D., Orlando, N., Gardi, L., Fenster, A., Crook, J., 2021. Characterization and registration of 3D ultrasound for use in permanent breast seed implant brachytherapy treatment planning. Brachytherapy 20 (1), 248–256. https://doi.org/10.1016/j.brachy.2020.07.010.

Zhang, S., Yick, K. lun, Chen, L., Yu, W., Lau, N., Sun, Y., 2020. Finite-element modelling of elastic woven tapes for bra design applications. Journal of the Textile Institute 111 (10), 1470–1480. https://doi.org/10.1080/00405000.2019.1704962.

CHAPTER

Evolution of mechanical properties with pathology & aging: application to pelvic tissues?

8

Steven Abramowitch, Liam Martin, and Madeline Hackett
Translational Research Laboratory in Urogynecology, Pittsburgh, PA, United States

8.1 Introduction

The study of pelvic tissues in humans is particularly challenging due to accessibility, relevance (e.g., cadaveric tissues are accessible but are not always relevant to the demographic or pathology of interest), and a lack of anatomically similar animal models. When those challenges are coupled with a lack of research funding related to matters of women's pelvic health, the result is often a dearth of information available in the literature on the mechanical properties of female pelvic tissues, especially information that identifies changes in their behaviors in the context of pathology and aging. Of the data that is available, it is often plagued with sourcing issues and/or differences in mechanical testing protocols such that large variations exist for reported values that, in some cases, lead to vastly different conclusions. Though this may be discouraging, it is important to remember that these are still early days for modern engineering and engineering approaches to be applied in this space. The good news is that there are now more and more engineers entering this space as their primary research focus. As a result, the last few years have seen a substantial growth in both the quantity and quality of studies aimed at mechanically characterizing female pelvic tissues. Such momentum is critical for delineating the impact of specific female conditions, developing new innovations, and ultimately identifying treatment and prevention strategies that improve the quality of life of women. It is also critical to gain the attention of funding agencies and investors.

This chapter will review the available literature on the impact of pathology and aging on the mechanical properties of female pelvic tissues. It will first focus on the major tissue constituents, their contributions to the mechanical properties of tissues, and the impact of pathologies and aging that are specific to these constituents. Though we will incorporate examples of tissues in the female pelvis as much as possible; where gaps remain, we will look to other organ systems for insights, while simultaneously reflecting on the known similarities and differences between these systems of those of the female pelvis. Indeed, female pelvic tissues are often unique

Biomechanics of the Female Reproductive System: Breast and Pelvic Organs. https://doi.org/10.1016/B978-0-12-823403-7.00020-8

in terms of their form and function; thus seemingly universal theories about what contributes to tissue mechanical behavior need to be thought about carefully and ultimately tested in this space. Nevertheless, looking at other organ systems for insights can be critical in forming hypotheses about the mechanics of female pelvic tissues.

The next major focus of this chapter will be pathologies that are specific to pelvic tissues and that predominately affect women, namely pelvic floor disorders, including stress urinary incontinence and pelvic organ prolapse. These pathologies are mechanical in nature and largely multifactorial in etiology. Whereas the impact of constituent specific pathologies and aging can contribute to pelvic floor disorders (as will be highlighted), the complex nature of these disorders requires a more overarching discussion that includes how the association of these disorders with tissue mechanical properties. Finally, pregnancy, while certainly not a pathology, is often associated with tremendous changes in the mechanical properties of pelvic tissues that are likely consequential in the prevention of injury during birthing and the development of future pathologies. In addition, premature softening of the cervix prior to 37 weeks of gestation has many important connections between tissue composition, organization, and mechanical function that are necessary to highlight in the context of this chapter. Though a review of premature rupture of fetal membranes also deserves significant attention, this falls outside of the current scope of this chapter, which is focused on female pelvic tissues. Instead, readers are referred to the work of Owen and others for more information related specifically to the mechanics of fetal membranes (Lindsay Wood et al., 2020; Stock et al., 2015; Graham and Bakaysa, 2019).

It is hoped that the information provided here, while certainly not all-encompassing, lends some background and context for the reader on the impact of pathology and aging on the mechanical properties of female pelvic tissues. There is still much work to be done in this space, so it is hoped that researchers, including those with background in other organ systems, will use this to develop hypotheses that are worthy of investigation related to female pelvic tissues and become motivated to contribute more knowledge to this space. It is also important to note that, though this chapter focuses on the science, the impact of disorders related to pelvic tissues cannot be overstated for women. The number of those with one or more pathologies related to pelvic tissues, many of whom are suffering in silence, is far greater than public perception; the range is from the young gymnast in competition to the grandmother who is simply looking to be more active. Continued research and innovation in this space has the potential to improve the quality of life for millions.

8.2 Mechanical properties of tissues

Female pelvic tissues are responsible for many significant and critical functions, ranging from structural support to allowing for the propagation of our species. Elimination of waste, sexual function, reproduction, and anatomical support are all processes that are primarily mechanical or include significant mechanical components. Yet, they are also affected by pathology and aging, with the potential for serious short and long-term consequences. As such, understanding the mechanical properties of pelvic tissues and how they are affected by pathologies and aging is critical to the health of women (Easley et al., 2017).

Though the study of the mechanical properties of tissues is a mature discipline, much of the work on this topic has been applied to fields of medicine other than obstetrics and urogynecology. Nevertheless, this textbook has already highlighted some of the work related to characterizing the mechanical properties of pelvic tissues (see Chapter 6) and has provided some fundamental concepts and definitions. This chapter will try to be readable without an intimate knowledge of earlier chapters in this textbook. However, the reader is asked to review Chapter 6 if any of the details below on mechanics is not clear. For the rest of this chapter, the term "mechanical properties" will refer to the relationship between stress and strain. Some refer to this relationship as "material properties," which is also considered acceptable. In a broad sense, "material properties" are inclusive of "mechanical properties," but also include properties such as density and thermal conductivity, etc. However, for the purposes of this textbook "mechanical properties" and "material properties" are both referring to the relationship between stress and strain.

It should further be noted that many biologic tissues, in addition to responding to external stresses imposed upon them, can generate internal forces as well. For this reason, it is often important to distinguish between two subclasses of mechanical properties: passive and active.

a. Passive: The stress-strain relationship is not influenced by a biological process that requires the consumption of energy to do mechanical work, e.g., muscle contraction. Instead, it is driven by the noncontractile response of structural tissue constituents and fluids to the application of an external force.

Many studies that focus on the mechanical properties of tissues characterize passive mechanical properties, meaning that experiments are conducted on excised tissues from cadavers or animals that are no longer biologically viable or whose ability to intrinsically generate force has been disabled (typically via chemical modulation). These samples can be fresh, fixed, or frozen, and experimental techniques can include failure testing in one (uniaxial) or two (biaxial) directions. Alternatively, nondestructive viscoelastic characterization, as discussed in Chapter 6, can also be performed. Specific parameters that are used to describe the stress-strain relationship (e.g., value of the peak stress or slope of the stress-strain curve) can be used to compare tissues. Constitutive models can also be fitted to these data with specific parameters that can be useful for comparisons between tissues, depending on the model being used and the context. Passive mechanical properties generally dominate the literature due to ethical and technical challenges related to testing viable tissue in the laboratory and performing *in vivo* testing.

b. Active: The stress strain relationship dictated by biological processes that require the consumption of energy to do mechanical work, e.g., muscle contraction. These are generally driven by actin-myosin interactions that are fueled by chemical energy in the form of ATP.

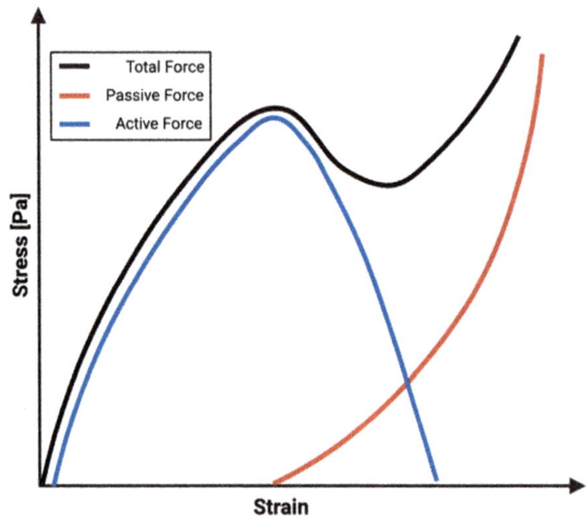

FIGURE 8.1

Example of passive and active mechanical properties.

For most tissues in which active mechanical properties are relevant, they must include a contractile cell phenotype. Examples of these tissues include skeletal, smooth, and cardiac muscle. However, it should be noted that cell motility and division are also active mechanical processes, for which quantification of active mechanical properties may be relevant. These processes generally occur at longer time scales than muscle contraction and will not be highlighted in this chapter. However, that does not mean that the mechanical significance of these processes is not important.

One challenge in quantifying the active mechanical properties of a tissue is being able to separate that response from that tissue's passive mechanical properties. Typically, this requires repeated tests of the same sample with the contractile behavior activated and again with it passivated, assuming no interaction between the constituents providing passive mechanical properties from those contributing to the active mechanical properties. The results of these two tests can be subtracted to determine the tissue-specific active mechanical properties (Fig. 8.1). Typically, this is controlled using chemical or electrical modulation. Depending on the tissue and the experimental conditions, it may be difficult to distinguish between the active and passive responses, so it is important for the reader to understand the context and what is being referred to when the term "mechanical properties" or "material properties" is being used.

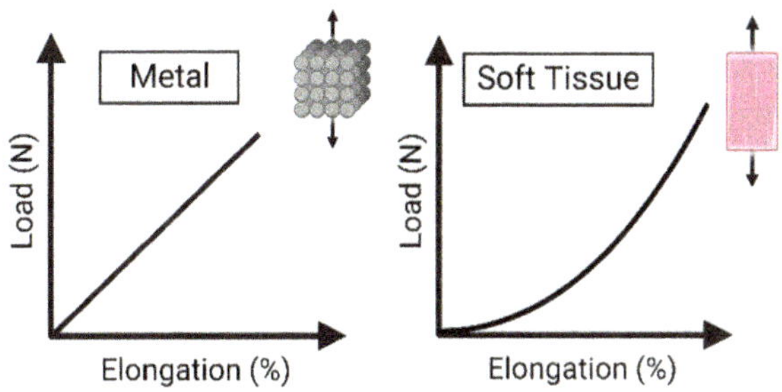

FIGURE 8.2

Mechanical properties of traditional engineering materials vs. soft tissue.

8.3 The mechanical importance of tissue composition and microstructure

Many engineering materials, such as steel, behave linearly when a modest tensile or compressive stress (i.e., one that does not cause permanent change) is applied to them. As an increasing tensile stress (force divided by the beam's cross-sectional area) is applied to a steel beam, the strain (change in length divided by its initial length) will change proportionately until it reaches a strain that results in a permeant change to the internal crystalline structure of the steel. Until this point, the curve resulting from plotting stress versus strain is linear. This stress-strain relationship is due in large part to the microstructure of the material. Steel's microstructure is a three-dimensional lattice of iron atoms with some other alloying materials replacing iron in select locations. Because of this nearly perfect arrangement of atoms, any force passed into the material is spread evenly across the lattice structure. Though the mechanical properties of soft biologic materials, such as tendons, ligaments, organs, and muscles, are nonlinear due to a much different composition and microstructure, the analogy is important because their microstructure, like that of steel, largely dictates the way these materials behave mechanically (Fig. 8.2).

Compared to the typically more regular microstructure of engineering materials, the microstructure of soft tissues is much less regular. Soft tissues are composed of many different microscopic structures that all contribute to their mechanical behavior (Subit et al., 2008). Collagen, glycosaminoglycans (GAGs), and elastin form a complex network of fibrous and nonfibrous extracellular matrix proteins, which are often specifically tailored to a tissue and its mechanical role within the body (Hay, 1981). As the ratio of these constituents, their different subtypes, and arrangements vary from tissue type to tissue type, so do the mechanical properties. Thus when considering the impact of pathology and aging on the mechanical properties of tissues, a reasonable place to start is to consider the mechanical contributions of each of these individual constituents and pathologies or aging mechanisms that are specific to these constituents. Though this reductionistic approach is logical, the reader should be cautioned that changes to tissues generally do not occur in isolation. For example, a genetic defect that specifically affects one constituent often causes changes to the mechanics within adjoining or adjacent tissues, which can trigger cells to respond differently when producing other matrix constituents. Thus understanding cause and effect can be challenging. Nevertheless, this approach allows for the identification of some associations that, when coupled with other findings, can allow us to formulate some general hypotheses about how composition and structure relate to the mechanics of tissues in the context of pathology and aging.

8.3.1 Collagen

Collagen makes up approximately 30% of protein dry weight in humans and is present in most body tissues, comprising up to 90% of the dry weight in ligaments, which are known for their load bearing capacity in the musculoskeletal system (Ricard-Blum, 2011; Fujii et al., 1994). It comes in many subtypes and can undergo different degrees of crosslinking (with mineralization in the case of bone) that makes it incredibly versatile in providing varying degrees of stiffness, strength, and mechanical energy absorption across different tissues (Fujii et al., 1994; Pal, 2009; Burr, 2002). Most of the load bearing collagens within pelvic tissues consist of collagen types I and III, with generally a much smaller percentage of type V (Gong and Xia, 2019). Though there is not a simple relationship between the ratios of these collagens– or even the total amount of collagen– to the mechanical properties of a specific tissue, some general trends have been observed. For example, tissues that are higher in type I tend to be stiffer with all other factors being equal, whereas those with higher percentages of type III and V tend to be less stiff. However, the reader is cautioned about making gross generalizations in this regard, because factors such as crosslinking, the spatial arrangement of the collagen relative to the direction of applied forces, and the interaction of collagen with other tissue constituents can result in dramatically different properties between two tissues, even if the relative amounts of collagen and the ratios of their subtypes are equal. Collagen is thought to be the primary contributor to the overall shape of a tissue's stress-strain relationship, which is generally highly nonlinear. This means that most soft tissues are very compliant (easy to change shape) in response to very low stresses but will stiffen dramatically (becoming more difficult to change shape) in response to high stresses. This is mechanically very advantageous, because it reduces the risk of tissue failure and injury in response to sudden changes in stress. In fact, the "flexibility" that is built into tissues is analogous to bridges being purposely built to be able to bend and twist based on the stresses it is experiencing. Without the ability to do this, repeated small changes in stress would be more likely to contribute to failure of the entire system. The nonlinearity in the stress-strain relationship arises, in part, from two related phenomena associated with collagen: uncrimping and recruitment. When no forces are applied to a tissue, collagen often appears "crimped" or wavy. Like slack in a rope, it is quite easy to displace the two ends of collagen away from one another when it has this configuration. However, like the rope, once that "slack"—or in this case "crimp"—is removed and the fiber becomes taut, it is much more difficult to displace the two ends to be further apart. The stress at which a single fiber becomes "taut" is not uniform throughout the tissue. Thus the tissue progressively stiffens with increasing stress as more and more collagen fibers uncrimp and are recruited to contribute to the stiffening behavior of the tissue. When all of the collagen is recruited, the stress-strain relationship becomes more linear.

8.3.1.1 Pathologies specific to collagen

Collagen-associated soft tissues disorders are a large classification of disorders that have systemic effects on the mechanics of many tissues throughout the body. Some of the most common disorders are specific types of Ehlers–Danlos syndrome (EDS), osteogenesis imperfecta (OI), and osteoporosis. Though the latter are thought to be disorders involving bone and cartilage, all have also been associated with pelvic floor disorder (Lammers et al., 2012; Carley and Schaffer, 2000). As a primary example of a disorder that is specific to collagen, this chapter will focus on EDS.

EDS is an inherited group of connective tissue disorders that fit into a category of disorders described as hypermobility spectrum disorders. They are generally characterized by abnormal collagen synthesis and can affect all collagenous tissue in the body (Parapia and Jackson, 2008). One of the most visually recognized symptoms of Ehlers–Danlos is extremely stretchy skin. In addition to skin hyperextensibility, patients with EDS can experience joint hypermobility and increased vascular fragility (Shirley et al., 2012). It can also take longer for their bodies to heal following injury (Babak et al., 2003; Trudgian and Trotman, 2011). EDS is classified into six different subcategories that describe the region and/or tissues that it affects. However, the end result is an increase in tissue compliance for the affected tissues. An increase in laxity of the connective tissues that support the pelvic organs would result in increased anatomical motion, resulting from increases of intraabdominal pressures or sudden acceleration/deceleration events. Thus it would be anticipated that these patients would have an increased risk of developing pelvic organ prolapse and stress urinary incontinence. However, this condition may also come with some advantages, namely the likelihood for some tissues to fail would be reduced, because it would be difficult to achieve the required strains to result in injury. Studies of the mechanical properties of tissues with EDS have provided some interesting results. In a mouse model that simulates EDS, the authors found that mice without collagen V had significantly thinner and stiffer corneas compared to wild type mice (Kling et al., 2021). These corneal tissues also failed at a lower ultimate stress and strain. The authors of this study hypothesized that this lower stress and strain failure contributed to the increased likelihood of arterial rupture in human patients with EDS (Kling et al., 2021; Royer and Han, 2022). The stress-strain relationship for tissues with EDS is flatter and longer compared to corresponding nonpathologic tissues (Royer and Han, 2022). This is likely due to the altered collagen content that is common with many forms of EDS (Royer and Han, 2022).

Interestingly, the effects of EDS on childbirth are well established with diagnosed patients being at a higher risk of complications during pregnancy and birth (Volkov et al., 2006). Overall, patients are more likely to have preterm births, antepartum hemorrhage, and premature rupture of membranes within the body (Spiegel et al., 2020). Preterm births are most likely due to cervical insufficiency, as patients with EDS have up to three times the risk of cervical insufficiency compared to the normal population (Spiegel et al., 2020; Makatsariya et al., 2020; Ploeckinger et al., 1997). Complications appear to be more severe in patients with the vascular subtype, and they run increased risk of aorta or uterine rupture (Shirley et al., 2012; Volkov et al., 2006). Unfortunately, the pregnancy-related mortality rate for these patients is reported to be as high as 50% (Elanie Epin et al., 2000; Rudd et al., 1983), but there has been recent research that suggests that safe birth is possible (Palmquist et al., 2009). For example, there is a case report of a patient that had their child birthed via cesarean section with no complications (Palmquist et al., 2009). However, the authors recommend that any patients with EDS be cared for by a multidisciplinary team with sufficient patient counseling about the risks of childbirth with EDS.

In terms of urogynecological conditions, there is a well-established association between EDS and other hypermobility spectrum disorders and several pelvic floor disorders (Gilliam et al., 2020). Rates of urinary incontinence, pelvic organ prolapse, and rectal prolapse are all higher in this population (Lammers et al., 2012; Li et al., 2020; Olafsdottir et al., 2020). Though the greatest risk factor for the development of these conditions in healthy women is vaginal childbirth (Kurt et al., 2018), patients with EDS, and hypermobility spectrum disorders, in general, often can develop these conditions independent of pregnancy and birthing (Norton et al., 1995; Mastoroudes et al., 2013). This might cause one to conclude that the support provided to the pelvic organs via the connective tissues, as opposed to the levator ani and other urogenital muscles, is the most critical in the development of pelvic floor disorders. Indeed, a recent genetics study has demonstrated that up to 40% of cases of pelvic organ prolapse can be explained by a woman's genetic phenotype (Olafsdottir et al., 2020). In addition, the phenotypes at greatest risk all had changes in genes that were primarily related to connective tissues. However, muscle attachment to bone is established through collagenous connective tissues (Zelzer et al., 2014). Furthermore, the extracellular matrix of muscle tissue that helps to form the connections between muscle fascicles is also primarily composed of collagen (Csapo et al., 2020). Thus EDS and other hypermobility spectrum disorders are also likely to negatively impact the form and function muscles that provide active support to the pelvic organs.

8.3.1.2 Impact of aging on collagen

Tendons and ligaments are collagen-rich structures with up to 90% of their dry weight being made up of type I collagen (Derby and Akhtar, 2015). Older populations are at a higher risk for tendinopathy and tendon injury, which suggests a deterioration or alteration in mechanical properties (Dunkman et al., 2013; Ackerman et al., 2017). Some studies have found decreased stiffness in aged tendons, whereas others have found no age-related differences in tendon passive stiffness (Ackerman et al., 2017; e Lima et al., 2018; Lindemann et al., 2020; McCrum et al., 2018; Svensson et al., 2021). However, a study of the viscoelastic properties of the patellar tendon in female mice found that aged tendons are less stiff, demonstrated more viscous behavior, and ultimately displayed a decreased ability to transmit force (Dunkman et al., 2013). This suggests that while the quasi-static mechanical properties (stress-strain behavior independent of time) may or may not undergo significant changes with aging, viscoelastic analyses also should be performed to determine if these tissues are dissipating more energy with increased age. In general, however, trends in the mechanical properties of aging tendons and ligaments are less consistent than those of aging skeletal muscles, especially in physically active populations.

Osteoporosis is another collagen-associated disorder, which is common in older patients and primarily affects the microstructure of bone (Rachner et al., 2011). Osteoporosis is a condition that is often associated with "normal" aging processes characterized by the loss of bone density. This condition is becoming more common as the median age of women continues to climb in countries with reasonably good access to health care. Though potentially treatable with bone density preserving medications, nutrition, and physical activity, studies have shown that there is an increased risk of urinary incontinence in patients who are undergoing an evaluation for osteoporosis (Meyer et al., 2020; Sran, 2009). However, it is difficult to tell if osteoporosis is causal or if it is simply because both conditions are common in women of advancing age (Sran, 2009; Khosla and Hofbauer, 2017; Johnson et al., 2000). Since tissues that support the pelvic organs are anchored to bone, it is possible that changes in bone density could allow for increased distensibility of these tissues at their attachment to bone; however, more work needs to be done to confirm if there is a causal relationship here or if these are two common, but otherwise unrelated conditions.

8.3.2 Elastin

As the name suggests, elastin provides elasticity to the tissue, enabling them to return to their initial shape after deformation (Kielty et al., 2002). Despite many associating the term "elasticity" with an ability to stretch, the true definition does not depend on the amount of stretch that a tissue undergoes. Instead, it is based on that tissue's ability to store energy with deformation and return that energy as the tissue returns to its initial shape. Thus elasticity in a tissue is typically measured by its ability to instantaneously "recoil" after a force that caused a deformation (change in its shape) is removed. The more that recoil is instantaneous and complete, the more elastic that tissue is behaving. An excellent visual example of elastic behavior in tissue is the external visible part of the ear, also known as the pinnae. Here the tissue is composed of elastic cartilage, where bundles of elastin fibers that are composed of elastin are distributed throughout the cartilage in the ear, allowing the ear to maintain its shape. Even in response to large deformations, where the ear is folded on itself, removal of the force results in a nearly instantaneous recoil of that tissue back to its initial shape.

Elastin is present throughout the connective tissues of the body and the majority of your body's elastin is generally thought to be formed during development and never again. Due to its long half-life *in vivo* (estimated to be about 70 years), it is suggested that the mechanical breakdown of tissues as we age is in part due to a progressive breakdown in elastin (Shin et al., 2019). Since elastin is intimately coupled with smooth muscle in a number of tissues, a loss of elastin is also thought to be partially to blame for the loss in smooth muscle as we age. These effects are generally most consequential in vascular tissues, where the content of elastin and smooth muscle are high and necessary for the function of these tissues. Interestingly, some feel that tissues within the female reproductive tract are an exception to this rule. In other words, events such as pregnancy and birthing may stimulate these tissues to turnover elastin to some degree. In its mature form, the fibers of elastin typically appear in a disordered conformation when a tissue is unloaded, which is related to the function of elastin. It prefers that arrangement, and thus is energetically driven to return to that conformation as long as no forces on the tissue are dictating otherwise. A loss of elastin through enzymic degradation, injury, or disease is often associated with significant changes in mechanical behavior.

8.3.2.1 Pathologies specific to elastin

Pathologies related to elastin include mutations to the ELN gene that encode one of the two components that are necessary for the body to produce elastic fibers. Two conditions that are associated with these gene mutations are cutis laxa and supravalvular aortic stenosis. Marfan syndrome, which is caused by a mutation in the FBM1 gene, the other necessary component for the formation of elastic fibers is another condition that directly impacts the form and function of connective tissues. To serve as an example of elastic fiber deficiency, this chapter will highlight cutis laxa and its relation to pelvic floor disorders.

Cutis laxa (CL), which results in significantly reduced elastin content in affected connective tissues, can lead to dysfunctional connective tissue mechanical behavior. Although rare, a recent case study has identified uterovaginal prolapse and rectal prolapse associated with CL in a patient as young as two months old (Choudhary et al., 2011). In another case, study of a forty-eight-year-old CL patient with vaginal prolapse, biopsies of the uterine ligaments revealed increased collagen content and decreased elastin content when compared to normal levels (Paladini et al., 2007). Other publications also cite that certain manifestations of this disease are associated with urogenital prolapse (Salter et al., 2006; Eser et al., 2015). Although still anecdotal, these reports suggest that when pelvic tissues are affected by CL, it can result in a significant loss of structural support to the pelvic organs. These findings are congruent with findings in women experiencing uterovaginal prolapse without CL. It was found that the elastin content was lower than in women with prolapse and the elastin metabolism had been altered relative to controls (Zong et al., 2010). Animal models seemed to confirm the importance of elastin in connective tissue development and function. There exist knockout mice for lysyl oxidase-like 1 (LOX-L1) and fibulin-5, which both play essential roles in the maturation of elastic fibers (although their roles also extend beyond elastin). Nevertheless, these models are known for the sometimes-spontaneous loss of pelvic organ support independent of birthing or that develops following birthing (Ali Borazjani et al., 2019; Lee et al., 2008; Drewes et al., 2007). This has led to some researchers to examine LOX-L1 and fibulin-5 expression in the uterosacral ligaments of human patients undergoing abdominal hysterectomy (Klutke et al., 2008; Jung et al., 2009). Post-menopausal women were found to have less expression than premenopausal women in one study (Zhao and Zhou, 2012). In another, patients with advanced pelvic organ prolapse were found to have decreased mRNA expression relative to controls (Brizzolara et al., 2009). Still other studies found an increase in expression of fibulin-5 with increased severity of prolapse and with patients that had cervical elongation (Wang et al., 2019). Thus no direct conclusions can be made as to whether elastin metabolism is responsible for the development of prolapse in the majority of women. However, it certainly seems to play a critical role for women that have pathologies specifically related to elastin.

8.3.2.2 Impact of aging on elastin

In aging skin, elastin undergoes fragmentation and thinning due to intrinsic aging or external factors, such as UV light exposure and air pollution. Due to their low turnover rate, elastic fibers accumulate such damage over time. Sufficient elastin damage can lead to reduced tissue elasticity or stiffening and failure of elastin-rich organs in the cardiorespiratory system. Aging is also marked by low-grade inflammation which may be caused, in part, by the products of elastin degradation. Inflammation increases the risk for development of pathologies such as arthrosclerosis, COPD, cancer, and neurodegenerative diseases (Heinz, 2021; Fulop et al., 2012). The reorganization and degradation of elastin may contribute to the advancement of cerebrovascular disease. Aged cranial arteries displayed a disorganized elastin network, stiffening, and lowered compliance when compared to younger cerebral arteries, while the area fraction remained the same. This suggests that the change in mechanical properties associated with elastin is in part due to elastin degradation and reorganization (Fonck et al., 2009). The same mechanisms of elastin degradation and reorganization and their consequences on stiffness and compliance have been noted in major human arteries, including the aorta (Duca et al., 2016).

Presumably, these same aging mechanisms are at work on female pelvic tissues and may be partially responsible for the age-related increase in pelvic floor disorders. Hormonal changes associated with menopause and hormone therapy may have a role in either accelerating or decelerating these processes. However, this remains an active area of investigation, because it is often difficult to perform the controlled, prospective studies that are necessary to delineate cause and effect.

8.3.3 Glycosaminoglycans

GAGs are polar polysaccharide molecules that can play several roles in the body. By attracting water into tissues, they can aid in lubrication between constituents at smaller length scales, e.g., collagen fibril sliding, or tissue surfaces at larger scales, e.g., cartilage surfaces within a joint. When these molecules are contained within an outer enclosure that consists of collagen, the water attracted to the interior causes an increase in volume, which then tensions the surrounding collagen enclosure. This creates an internal pressure within the tissue, which enables it to be used for shock absorption. This is commonly seen in cartilage and the intervertebral discs of the spine

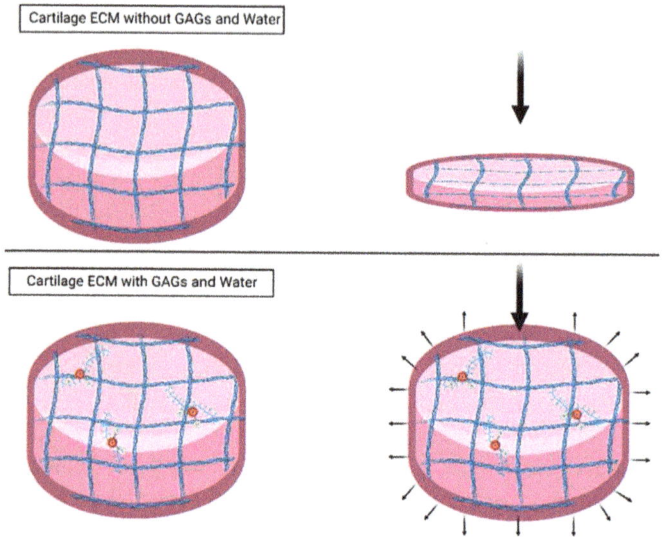

FIGURE 8.3

Glycosaminoglycans pressurize cartilage sample under compression.

(Pal, 2009; Urban et al., 1979). Collagen cannot resist compressive loads by itself. However, when sounding a region that is high in GAGs and water, any compressive loads applied to the tissue cause an increase in internal pressure that results in the collagen being loaded in tension, which is what collagen is designed to resist. Thus GAGs play a critical role in allowing compressive loads to be resisted by collagen in tension. Without GAGs attracting water to the interior of the tissues, there would be no internal pressure and the collagen matrix would simply collapse in response to compression (Fig. 8.3).

There are several different GAGs that each have distinct roles within tissues. In addition to load bearing, GAGs play roles in cell binding, diffusion of proteins; they can coat cells to form a barrier against infection, and bind/store growth factors within the extracellular matrix for later use. Often, they can be bound in a chain or multiple chains to a core protein that forms a proteoglycan. This core protein can maintain the spatial localization of many GAG chains that would normally repel one another due to like charges. Proteoglycans are often found in connective tissues and help to form the ground substance, which is the nonfibrous part of the extracellular matrix. They can contribute to the stiffness and viscoelastic behavior of the tissue by binding directly to the collagen network. Depending on the size of the proteoglycan, its core protein, and its attached GAG chains, proteoglycans can serve a wide variety of mechanical roles.

8.3.3.1 Pathologies specific to glycosaminoglycans

Pathologies related to GAGs and proteoglycans can result from over accumulation, accelerated degradation, and specific mutations. Since GAGs play so many roles within the body, the symptoms of any specific pathology will depend on the molecules involved, their location, and their functional role at that location (Mizumoto and Yamada, 2021). Thus it can often be difficult to determine if mechanical changes measured in a tissue are the direct result of a corresponding change to GAGs/proteoglycans or a secondary event that is more downstream, e.g., a change in cell signaling that eventually manifests in changes to the collagen or elastin matrix.

Many congenital disorders are believed to be caused by defects in the biosynthesis and/or attachment of GAGs to extracellular proteins (Mizumoto and Yamada, 2021; Mizumoto et al., 2020, 2015). Genetic sequencing of patients with some well-known disorders have attributed them to defective anabolism and catabolism of GAG molecules. Defective catabolism of GAG molecules has been directly tied to significant anatomical changes and/or changes in cognition, demonstrating how essential and ubiquitous GAGs are throughout the body. Some of these conditions are some subtypes of Ehlers–Danlos syndrome, osteochondrodysplasia, brachydactyly, and some forms of dwarfism (Mizumoto and Yamada, 2021). Unfortunately, there is still ongoing research into the exact molecular pathways that are directly responsible for the development of these congenital disorders (Mizumoto and Yamada, 2021). Mucopolysaccharidoses, a collection of inherited diseases which are characterized by multi-organ accumulation of GAGs caused by impaired GAG catabolism, has many significant systemic effects. Though there is not much data in the literature on how these conditions impact urogenital tissues, there is a case series of pregnancy and birthing in women with this disorder (Stewart et al., 2016). Due to their condition, 75% of the deliveries studied were a planned cesarean section, compared to the 31% in patients without this condition (Stewart et al., 2016). This increase was in part due to short material stature with normal-sized babies leading to cephalopelvic disproportion (Stewart et al., 2016). Though up to 75% of this case series experienced complications, most were not related to the pelvic organs or pelvic organ support.

Studies have attempted to link GAG content to development of pelvic floor disorders (Mizumoto and Yamada, 2021; Suzme et al., 2007; Kati et al., 2010). These studies do not agree on whether there is a relationship between changes in GAG content and pelvic organ prolapse or other disorders. Some studies do show that there is a change in the concentration of sulfated GAGs in some patients, but these differences were unable to show a causal relationship (Mizumoto and Yamada, 2021; Suzme et al., 2007). Future studies will have to be aimed at directly quantifying if such changes exist and if they meaningfully contribute to the mechanics of pelvic tissues. With that said, GAGs do appear to be important in regulating the mechanical behavior of the cervix. During pregnancy GAG content rises significantly, peaking on the onset of labor, and then falls dramatically after delivery (Osmers et al., 1993). This coincides with patterns in cervical stiffness that are observed in pregnancy and will be discussed later in the chapter.

8.3.3.2 Impact of aging on glycosaminoglycans

Tissues tend to become stiffer with advancing age. The loss of water in soft tissues due to a lack of or improperly functioning GAGs could contribute to tissue stiffening. In the patellar tendon of rats, it was found that tendon stiffness increased, and GAG content decreased (Vailas et al., 1985). GAGs have also been identified in chondroid-like bodies of the equine superficial digital flexor tendon. Loss of GAGs in chondroid-like bodies with age may contribute to tendon stiffening (Ali et al., 2021). Aging human skin displays laxity, thinning, and wrinkling. It has been hypothesized that these changes are primarily due to aged collagen, but increased collagen turnover is insufficient to rectify these issues. Instead, it appears that GAGs play a more significant role in skin aging by binding to more proteins over time, leaving less space for association with water. Lack of water in the skin may be the culprit behind skin wrinkling, thinning, and laxity (Lee et al., 2016).

In terms of pelvic tissues, aging and menopause may play a significant role in the quantity and types of sulfated GAGs that are present in periurethral tissues. Postmenopausal women had fewer sulfated GAGs in total, with dermatan sulfate and chondroitin sulfate showing significant reductions. The mechanical significance of this remains unclear (Feldner et al., 2008). However, it is possible that this may be related to tissue atrophy following menopause.

8.3.4 Smooth muscle

Smooth muscles are involuntary, nonstriated muscles within the body, which are involved with many functions, including the control of blood pressure and digestion. Due to their wide use within the body, they come in many phenotypic forms. In general, the two ends of the phenotypic continuum are contractile and synthetic smooth muscles. Contractile smooth muscle cells appear as elongated, spindle-shaped cells, which primarily regulate the shape of tissues by contracting (Sugita et al., 2019; Rensen et al., 2007; Gomez and Owens, 2012). Synthetic smooth muscles exhibit much higher growth rates and migratory activity and produce more proteins than their contractile counterpart (Sugita et al., 2019; Rensen et al., 2007). In terms of vascular tissues, each type of smooth muscle cell phenotype has a distinct purpose within the vessel they surround, and each vessel has a distinct ratio of smooth muscle cells phenotypes and percentage of those phenotypes to other matrix constituents, which, when disturbed, can result in the development of pathologies (Rensen et al., 2007). For example, a higher ratio of synthetic smooth muscle cells to contractile muscle cells has been found in patients with atherosclerosis and aneurisms (Gomez and Owens, 2012; Petsophonsakul et al., 2019).

The role of smooth muscle within some of the organs of the pelvis is immediately clear, e.g., bladder, rectum, uterus, urethra. In the case of the bladder and rectum, its relaxation allows for waste storage, whereas its contraction allows for waste elimination. One thing that is somewhat remarkable is the coordination of smooth muscle within spatially localized and even connected tissues. For example, during micturition (elimination of urine), the smooth muscle of the bladder contracts to increase pressure within the bladder, whereas the smooth muscle of the urethra must relax such that its tone does not prevent the flow of urine (Turner and Brading, 1997). A similar coordination must occur between the rectum and smooth muscle within the anal sphincter. For the uterus, smooth muscle contractions are responsible for the cramping that occurs with menstruation to aid in the monthly sheading of the inner lining of the uterus (Solomon et al., 2005). It is also critical for generating the pressures necessary for birthing and to restore the shape of the uterus following birth (Riemer and Heymann, 1998). Interestingly, smooth muscle is also found in the vagina and the supporting ligaments of the pelvic organs. Presumably, this is to enable these tissues to accommodate significant changes in shape in response to an applied force and the ability to recover that shape after the force is removed. Though elastin is normally thought to play this role, smooth muscle is often collocated with elastin and likely enable the tissue to achieve changes in shape that extend beyond that which elastin could achieve alone. Smooth muscle has also been found to play a protective role for the collagen within the tissues by allowing the collagen fibers to stay relaxed longer, effectively delaying collagen recruitment (Danso et al., 2020).

8.3.4.1 Pathologies specific to smooth muscle

Smooth muscle dysfunction syndrome (SMDS) is a condition characterized by aortic and cerebrovascular disease, including lesions, stenosis, aneurysm, stroke, and seizures. SMDS is caused by a mutation in the ACTA2 gene (Regalado et al., 2018). In the case of a twenty-six-year-old woman with SMDS, her cesarean section was complicated by severe uterine bleeding caused by the lack of uterine tone. It can be inferred that SMDS affected the smooth muscle of the uterus causing atony and uterine hemorrhage. Other conditions have also been linked to smooth muscle dysfunction; one of the most prevalent is bladder instability. Bladder instability is caused by the smooth muscle that surrounds the bladder not being able to function normally in response to filling and emptying (Turner and Brading, 1997). In nonpathological patients, the smooth muscle that surrounds the bladder must be compliant enough to allow for filling, but strong enough to force the bladder to keep its shape (Turner and Brading, 1997). In cases where either of these fail, patients can develop urinary incontinence (Turner and Brading, 1997; Delancey and Ashton-Miller, 2004). Loss of smooth muscle has also been linked to decreased airway responsiveness and viscoelasticity, which in turn may lead to alveolar collapse (Südy et al., 2020, 2021). Finally, loss of smooth muscle sphincter tone can sometimes result in fecal incontinence, which some studies have attempted to treat by introducing myogenic progenitor cells to the sphincter region (Thurner et al., 2020). Once these cells are introduced, they can induce muscle regeneration and restoration of fecal continence. Currently this has only been shown in a mouse model, but this may develop into a potential human treatment of fecal incontinence (Thurner et al., 2020).

8.3.4.2 Impact of aging on smooth muscle

In large arteries, vascular smooth muscle cells (mVSMCs) may modulate arterial stiffness by facilitating contraction, mechanotransduction, and distribution of tensile forces between elastin and collagen (Lacolley et al., 2017). In one study, vessels were grown *in vitro* from porcine VSMCs derived from infant or adult subjects to compare their composition. In vessels derived from infant cells, there were less collagen crosslinks observed and higher burst pressures (Solan and Age, 2005). This suggests that VSMCs in older populations lead to stiffer tissues with reduced strength. In another study comparing younger and older human thoracic aortic samples, it was similarly found that older samples displayed an increased collagen content and subsequent stiffness and decreased elastin function. The authors determined that in aged samples, mechanically sensitive smooth muscle cells increase collagen production to compensate for increased tension and return their mechanical environment closer to normal (Taghizadeh and Tafazzoli-Shadpour, 2017). These findings were supported in another study, where it was posed that aging results in an increased mechanical load on the vessel wall caused by inflammation and change in the stiffness gradient of the tissue. In response, VSMCs increase contractility and produce more collagen (Lacolley et al., 2018). Smooth muscle contractile function is known to decrease with age. The most prominent location of this decline is in the vascular smooth muscle that surrounds arteries (Seawright et al., 2018). As this decrease in muscle tone occurs, patients are more likely to develop vascular smooth muscle complications, such as atherosclerosis (Monk and George, 2015).

Smooth muscle may play a role in normal vaginal support and function. In prolapsed vaginal tissue, smooth muscle organization, composition, and tone are decreased. Smooth muscle contributed significantly to the biaxial mechanical properties of the murine vagina, with increased muscle tone being related to decreased vaginal length and decreased material stiffness. Similar results pertaining to material stiffness were found in the murine artery (Clark et al., 2019). Clearly, there is a relationship between SMCs and collagen and elastin, where increased muscle tone partially unloads collagen and elastin, decreasing material stiffness, and decreased muscle tone causes the load to be transferred to collagen and elastin and increases material stiffness. The role of smooth muscle in the development and treatment of pelvic organ prolapse is likely underappreciated and deserves further investigation.

8.3.5 Skeletal muscle

Skeletal muscles are the most common muscle type within the human body. It is the only muscle type that is striated and voluntary, which makes them vital for locomotion (Dave et al., 2021). The muscles that help to support the pelvic organs are the iliococcygeus, coccygeus, puboviseral, and puborectalis (Raizada and Mittal, 2009). Interestingly, these muscles, unlike their musculoskeletal counterparts, are typically under continuous action that is mostly autonomic. This ensures that the genital hiatus remain closed and the pelvic organs remain in place (Raizada and Mittal, 2009; Kepenekci et al., 2011). Typically, one has to consciously relax these muscles to urinate or defecate. If the continuous action of these muscles is interrupted permanently (e.g., resulting from muscle injury), the pelvic organs can become more mobile, thereby increasing the risk for pelvic floor prolapse (Dunivan et al., 2016).

Though increasing genital hiatal measurements are not a perfect predictor of prolapse, it has been linked to increasing severity (Dunivan et al., 2016). On the other hand, if the autonomic contractions of these muscles are too excessive or if these muscles are unable to relax, this can lead to conditions such as levator ani syndrome (Bharucha and Trabuco, 2008) and dyspareunia (Tayyeb and Gupta, 2022). Levator ani syndrome is characterized by frequent dull anorectal pain (Bharucha and Trabuco, 2008). It is hypothesized to be due to increased muscle activity around the anus, which is supported by increased EMG activity in individuals with the condition (Grimaud et al., 1991). However, the exact pathophysiology of the condition is unknown. Dyspareunia is defined as pain before, during, or after intercourse. Once again, the exact cause of this condition is unknown, but it is hypothesized that increased levator muscle activity around the time of intercourse may increase the likelihood of occurrence (Tayyeb and Gupta, 2022).

8.3.5.1 Pathologies specific to skeletal muscle

Injury and disease in the skeletal muscles can range from tears to rare neurological diseases, such as amyotrophic lateral sclerosis, to disorders of the cellular mechanisms that are responsible for generating forceful contractions. For example, similar to the effects of aging on skeletal muscle structure and function, muscular dystrophy causes increased amounts of collagen, increased collagen crosslinks, and subsequent muscle stiffening (Virgilio et al., 2015). Injury changes the mechanical environment in the muscle and can alter the behavior of mechanosensitive constituents leading to atrophy and scar formation. This can manifest as limitations in strength, mobility, muscle stiffness, and increases the risk for repeated injury. For example, in rotator cuff tears, it has been observed that individuals frequently will not recover their full muscle mass (Gibbons et al., 2018). In the hamstring muscle, patients who have a prior injury have higher strains compared to those who do not have a prior injury (Silder et al., 2010a,b). However, it is possible to compensate for a local muscle injury by increasing the functional capability of the noninjured portion of the muscle if the original injury was minimal to moderate (Silder et al., 2010b). The only difference this group noted was an increased biceps femoris tendon size on the injured leg (Silder et al., 2010b).

In the female pelvis muscle, injuries during delivery are thought to manifest in one of four ways: 1) a stretch injury that results in a partial or complete tear; 2) a compression injury, where the muscle gets pinched between the fetal head and the maternal pelvis; 3) a nerve crush or nerve stretch injury that prevents functional signals from getting to the muscle; or 4) an ischemic injury, where the blood supply that feeds the muscle is cut off for a long enough period of time that the muscle dies (Catanzarite et al., 2018; Kearney et al., 2006). Of course, a combination of one or more of these injuries is also possible. Whereas there is imaging that supports the possibility of all of these existing in women, the most common presentation of muscle injury via imaging appears to be a stretch injury that results in a complete or partial avulsion of the puboviseral muscle attachment from its insertion on the pubic bone (Catanzarite et al., 2018; Kearney et al., 2006). It is common to see this type of injury in women with prolapse; however, it does not explain all, or even the majority, of prolapse cases. With that said, imaging has limitations, and there is a significant need for validated functional muscle assessments.

8.3.5.2 Impact of aging on skeletal muscle

In the aging musculoskeletal system, skeletal muscles tend to exhibit higher muscle tone and stiffness and a decrease in elasticity and force generating potential (Bajpai et al., 2021; Kragstrup et al., 2011; Niyomchan et al., 2019; Pavan et al., 2020; Hasson and Caldwell, 2012). Aging is marked by sarcopenia, or loss of muscle mass, which is a major contributor to overall muscle weakness (Derby and Akhtar, 2015). The mechanical behavior of soft tissues is viscoelastic, or time dependent. After repeated loading that comes with increasing age, tissues can begin to behave more viscous than elastic as a result of elastin damage due to repeated loading and stiffer as a result of collagen cross linking stiffening the extracellular matrix (mECM). There is an ongoing debate as to whether skeletal muscle changes are a function of aging exclusively, or rather a consequence of a more sedentary lifestyle that comes with advancing age. Several studies have demonstrated that physical activity may prevent changes in mechanical properties due to aging. In older animals, there was a decrease in collagen cross links and muscle stiffness for those who underwent a physical training program (Kragstrup et al., 2011). For physically active elderly people enrolled in a skiing intervention study, there were no age-related decreases found in muscle strength or muscle architecture in the vastus lateralis of male or female subjects. It appears that the ECM is the main driver of muscle mechanics when compared to individual muscle fibers. In studies of the mechanical properties of single muscle fibers, investigators found no difference in passive force and an increased viscoelastic function in aged human muscle fibers (Pavan et al., 2020; Lim et al., 2019).

The passive mechanical properties of PFMs, including the coccygeus, iliococcygeus, and pubovisceralis were compared between biopsies of older and younger female cadavers. It was found that older PFMs were dramatically stiffer than younger PFMs, supporting that age-related changes in collagen crosslinking and muscle stiffness also occur in PFMs (Burnett et al., 2020). Similar to skeletal muscle, it was found that aging resulted in decreased cross-sectional area and increased collagen content with aging in the coccygeus, iliococcygeus, and pubovisceralis muscles (Alperin et al., 2016).

In one study, the obturator internus muscle was compared among younger, middle aged, and older female cadavers. It was assumed that the function of the obturator internus affected the function of the levator ani based on physical therapy techniques, in which obturator internus strengthening programs were used to treat levator ani dysfunction. It was found that the obturator internus displayed a decreased cross-sectional area and increased collagen content in older subjects when compared to younger and middle-aged subjects. These muscle properties were highly correlated with decreased force-generating capacity in the levator ani muscle (Cook et al., 2017).

Though it is compelling to think of PFMs as skeletal muscles, the comparison may not be so straightforward. In a study of the passive mechanical properties of

PFMs compared to appendicular muscles in older and younger female cadavers, it was found that PFM stiffness increased dramatically with age, while the stiffness of the appendicular muscles remained unchanged. This could suggest that female PFMs undergo sex-specific remodeling based on changing biochemistry with age (Burnett et al., 2020).

8.3.6 Pathologies directly affecting multiple aspects of pelvic tissue mechanics

Despite many of the aforementioned pathologies being broadly systemic, meaning that they affect multiple parts of the body or the entire body, nevertheless, their mechanism is associated with a singular constituent within tissues. However, there are other pathologies that impact the mechanics of tissues, for which the mechanism is not specific and has the potential to directly affect multiple tissue constituents simultaneously. For example, obesity is a multifactorial pathology that can have direct consequences on pelvic tissues based on their mechanical burden (or lack thereof if the subject is sedentary). However, obesity is also often associated with changes in inflammation pathways and hormones that can also directly impact the mechanics of pelvic tissues. Obesity alone is a common risk factor for pelvic floor disorders and deserves much more basic science attention within obstetrics and urogynecology.

Still other pathologies affect how tissue constituents interconnect. One such pathology, which is often part of the obesity spectrum of disorders and is very common in the population, is diabetes. This is a challenging pathology, because it is broadly systemic, and it affects accumulate with age (Morley, 2008). Diabetes is among the top ten causes of death, affects at least 451 million people in the world (2017 and is projected to effect 693 million people by 2045). It costs the US healthcare system 327 billion dollars annually (Yang et al., 2018; Lin et al., 2020). With this trend, only starting to slow in some countries, understanding the effects of diabetes on the human body is imperative (Lin et al., 2020). Diabetes-associated complications include blindness, increased fall risk, end stage renal disease, lower extremity amputation, increased risk of stroke, among others (Ahmed and Thornalley, 2007; Suji and Sivakami, 2004; Morley, 2013). However, diabetes also induces multifaceted changes in mechanical properties for most of the tissues within the body. Studies have shown that there is an increase in bone mineral density due to diabetes (Patel et al., 2008; Vestergaard, 2007). Others have shown that diabetes also rearranges collagen fibers through a process known as nonenzymatic glycation (Brownlee et al., 1984).

To understand the impact of diabetes, it is important to understand that there are two types of crosslinks that occur to collagen: enzymatic and nonenzymatic (Snedeker and Gautieri, 2014; Kerkhof et al., 2009). Enzymatic crosslinks occur naturally during collagen maturation and are responsible for forming stable intermolecular collagen crosslinks, which are mediated by lysyl oxidase (Snedeker and Gautieri, 2014; Kerkhof et al., 2009; Bailey, 2001). Without these strong intermolecular bonds, the collagen fibrils are much weaker, and this can have a major impact on tissue function (Kerkhof et al., 2009; Bailey, 2001; Vaidya et al., 2022). In mature, healthy tissue, the formation of these crosslinks plateaus once the tissue has reached its required stiffness (Snedeker and Gautieri, 2014). However, connective tissue stiffness has been shown to continue to increase in tissues due to increased age and diabetes due to the continued formation of crosslinks (Snedeker and Gautieri, 2014; Iel and Oxlund, 1996; Bank et al., 1998; Abate et al., 2010; Wang et al., 2002; Sell and Monnier, 2012; Vashishth et al., 2001). Nonenzymatic glycosylation is the process where a glucose molecule attaches itself to a protein without the involvement of an enzyme such as lysyl oxidase (Brownlee et al., 1984). Studies have shown that this process is not localized, and that increased stiffness occurs throughout the body in ligaments, tendons, skin, arteries, and bone (Sell and Monnier, 2012; Vashishth et al., 2001; Karim et al., 2018; Kim et al., 2017). This process also eventually produces advanced glycation end-products (AGEs), which are increasingly problematic (Snedeker and Gautieri, 2014; Vaidya et al., 2022; Avery and Bailey, 2006, 2005; Chen et al., 2015).

AGEs have been shown to have a major effect on both the biomechanical behavior of tissue and its biologic activity (Snedeker and Gautieri, 2014). AGEs react with cell surface receptors, which can lead to many downstream effects (Abate et al., 2010). For example, they can directly upregulate pro-inflammatory mediators, upregulate adhesion molecules, and may lead to a dysfunctional cell phenotypes (Snedeker and Gautieri, 2014; Kerkhof et al., 2009; Abate et al., 2010). Additionally, AGE production has been shown to be the major cause of substantial dysfunction of tissue that contain collagen and elastin, being partially responsible for soft tissue complications with age (Kerkhof et al., 2009). In terms of AGEs' specific effects on the biomechanical properties of tissues, however, it is not an easy issue to understand. Increased crosslinks are known to increase the failure load of single fibrils, while also decreasing the overall tissue's failure properties (Snedeker and Gautieri, 2014; Bank et al., 1998; Abate et al., 2010; Wang et al., 2002; Goldin et al., 2006). This seemingly contradictory relationship has a simple explanation that goes back to the stiff bridge analogy at the beginning of this chapter. The crosslinks that are formed do not only stiffen the material; they also increase the material's brittleness (Kent et al., 1985; Reddy, 2004). A brittle material is a material that fractures with little deformation applied to it. They can absorb relatively less energy before failure, even in materials with a high strength. In tissues such as tendons and ligaments, this becomes increasingly problematic, as the very nature of the tissue is to absorb high amounts of energy without failure. Unfortunately, this increased brittleness can lead to many injuries due to premature failure of connective tissues. This phenomenon is not isolated to ligaments and tendons. It also occurs in the bones of patients with diabetes, who are at an increased risk of developing hip fractures from falling (Morley, 2008, 2013).

Due to the systematic effects that diabetes can have on connective tissues, it is unsurprising that the pelvic organs and their supportive tissue would be affected by diabetes. Pelvic floor disorders result from a lack of support to pelvic organs. That can occur via direct tissue injury and/or a process that results in a change in tissue mechanics that is independent of or secondary to injury (Eric Jelovsek et al., 2007).

In either case, diabetes could play a role. For example, diabetic patients often have more difficulty healing after injury. Thus an obstetric injury that would not result in the manifestation of a future pelvic floor disorder in a nondiabetic patient might put a diabetic patient at significant risk. In addition, the changes in the stiffness biologic tissues that is common with diabetes likely places pelvic tissues, especially during activities that involve significant deformations, at a greater likelihood for injury. That could be the result of an acute stretch event or the additive accumulation of a number of microtraumas that simply do not heal adequately between subsequent micro-traumatic events. Any of these mechanisms might explain why diabetes is associated with an increased risk of developing a pelvic floor disorder, such as stress urinary incontinence (SUI) (Micussi et al., 2015; Lawrence et al., 2007). However, it is not just about the connective tissues. Other research has shown that there is a decrease in muscle tone of the pelvic floor muscles in women with diabetes (Chen et al., 2015; Micussi et al., 2015). Again, it is difficult to appreciate the cause and effect for studies that establish these associations, because diabetes is associated with so many other co-morbidities, e.g., obesity. However, studies are trying to move beyond associative comparisons. For example, AGEs were present at a higher concentration in the vaginal wall of diabetic women when compared to controls (Derby and Akhtar, 2015; Chen et al., 2015). This study does not claim that there is a causal relationship between AGEs and development of pelvic organ prolapse. Nevertheless, it is a step towards being more mechanistic about the relationship between diabetes and pelvic floor disorders. Certainly, more work needs to be done in this area.

8.3.7 Pathologies and aging of pelvic tissues

Thus far, this chapter has examined pathologies that are not necessarily pelvic tissue-specific to gain insight into factors that influence the mechanics of pelvic tissues and how they might be related to disorders such as pelvic organ prolapse. However, disorders such as pelvic organ prolapse, stress urinary incontinence, rectal prolapse, etc., can be pathologies independent of any other comorbidity. Thus it is important to also examine what we know about these pathologies from a tissue mechanics perspective.

Smooth muscle, collagen, and elastin are abundant in all of the major pelvic organs, and we can roughly understand the behavior of some pelvic tissues based on the relative ratios of these constituents. For example, the bladder is made up of more smooth muscle and elastin, whereas the vagina and rectum contain relatively more collagen. Thus one would expect the bladder to be more compliant and undergo more significant changes in shape with contraction and relaxation of its smooth muscle component than the vagina and rectum. On the other hand, the collagen of the vagina and rectum would make it tougher than the bladder (Chantereau et al., 2014). Indeed, in a study of the mechanical properties of the uterosacral, broad, and round ligaments, bladder, vaginal wall, and rectum of older and younger female cadavers, the bladder was found to be more compliant than the more the more collagenous ligaments, vaginal wall, and rectum. Younger tissues exhibited different mechanical behavior than older tissues, with older tissues being stiffer and stronger than younger tissues (Chantereau et al., 2014). These findings were similar to another study that showed that the vagina was stiffer than the rectum, which was stiffer than the bladder in female cadavers (Rubod et al., 2012).

When examining pelvic organ support, it is important to consider the role of pelvic ligaments, which serve to transmit forces. Unlike ligaments in the musculoskeletal system, ligaments in the pelvis undergo much larger strains and are composed of more elastin and smooth muscle (note: the latter is often not a component of ligaments in the musculoskeletal system). Nevertheless, much like their musculoskeletal counterparts, they are stiffer along the direction of loading (Becker and de Vita, 2015). In humans, when comparing the broad, round, and utero-sacral ligaments from cadavers of similar age, the uterosacral ligament was the stiffest, followed by the round ligament, and then the broad ligament (Riemer and Heymann, 1998). The uterosacral, broad, and round ligaments have also been compared between younger and older cadavers. Except for the broad ligament, it was found that aging resulted in an increased stiffness and elongation in the uterine ligaments (Chantereau et al., 2014). Though the same effect was not observed in the broad ligament, the small sample size combined with the greater compliance of the broad ligament may have made it more difficult to detect age related changes in this structure.

8.3.8 Pelvic floor disorders

Pelvic floor disorders are a collection of wide-ranging conditions affecting the tissues and organs of the pelvis. As with the earlier sections of this chapter, highlighting all of relevant conditions in this class of disorders is beyond the scope of this chapter. Instead, the focus will be on the two most common: stress urinary incontinence and pelvic organ prolapse.

8.3.8.1 Stress urinary incontinence

Stress urinary incontinence is one of the most researched pathologies within the field of obstetrics and urogynecology and is one of the more common subtypes of urinary incontinence (Botlero et al., 2011). Stress urinary incontinence has been shown to be associated with decreased collagen content in ligaments, including the uterosacral, cardinal, and paraurethral tissues (Han et al., 2014). Specifically, patients with SUI and pelvic organ prolapse were found to have a 36%, 42%, and 31% decrease in collagen type I content of the cardinal, uterosacral, and paraurethral tissues, respectively, compared to controls (Fujii et al., 1994; Asgari et al., 2017). Changes in collagen type III showed similar trends with a 48% decrease in the cardinal ligament, 39% decrease in the uterosacral ligament, and 35% decrease in the paraurethral tissues (Han et al., 2014). Presumably, the increased laxity that would be expected to accompany this decrease would increase urethral mobility, which likely leads to increased symptom severity in some patients. This concept has been supported by finite element studies that showed that decreases in stiffness of tissues supporting the urethra were associated with increased bladder neck mobility (Brandão et al., 2015). Other finite element studies have also associated the mechanical properties of the urethra, perineal membrane, bladder, and paraurethral connective tissues with urethral motion and shape, which in turn were associated with symptomatology (Routzong et al., 2021). Though these finite element studies are unable to explain why these tissues are less stiff, it is conceivably caused by decreases in collagen content. However, based on the previous sections in this chapter, focusing solely on one constituent has the potential to lead to misinterpretations. Thus it is necessary to get a more complete picture of the time course of tissue remodeling that takes place leading up to the onset of symptoms. Because of the number of patients that would have to be tracked for a significant duration of time, such studies are generally not possible, which is why finite element studies and animal models are important (Jiang and Damaser, 2011). They can help us identify the potential key features and time course of changes so researchers can be more informed when designing studies in humans.

One of the challenges of performing rigorous computational studies in this space is the lack of quality mechanical data that serve as inputs to simulations. For example, a literature search for mechanical properties of the urethra in patients with and without SUI reveals exceptionally large variations in reported values. Of the few studies available, the properties range between 5 KPa and 10 MPa (Müller et al., 2008; Spirka et al., 2013), i.e., differences of 3 orders of magnitude. Of the computational models that have been developed to study the urethra, all of those studies utilized values from one or two sources (Fung and Biomechanics, 1981; Freeman, 1971), which are from the 1970s and early 1980s. That is not to say that early work was not of excellent quality. The point here is that the broad ranges of values in the literature are likely not representative, and more studies on the experimental mechanics of the human urethra are sorely needed, especially since it is a complicated multilayered, composite structure. Such work need not be done using traditional mechanical testing approaches to obtain meaningful information. For example, one study quantified a metric that they described as "urethral elastance," which was defined as the internal luminal pressure of the urethra divided by its cross-sectional area while the bladder was empty and the patient was at rest (Lose, 1989; Lose and Colstrup, 1990). Though not a rigorous measure of the mechanical properties of the urethra, this allowed for a comparison between pathologic and nonpathologic tissues. The authors that used this method showed that there was a decrease in "urethral elastance" in patients with SUI (Lose and Colstrup, 1990). On average, there was a 50% decrease at the level of the bladder neck; that finding correlated with the decreases in collagen and elastin. When interpreting these data, one needs to be cognizant of what might be contributing to the urethral pressures and whether they are solely reflective of urethral function. In addition, differences in the smooth muscle component's contractile function and organization may result in changes to pressure measurements that do not correspond with changes in cross-sectional areas. Nevertheless, this study highlights an outside-of-the-box approach that obtains some mechanical information from human subjects that would be difficult to obtain otherwise.

Based on the correlations of SUI with decreased collagen content that was highlighted in the earlier two paragraphs, it would seem that preventing collagen degradation in tissues that support the urethra should be the focus of prevention and treatment strategies. Thus it may be paradoxical to learn that a nonsurgical treatment option for mild to moderate stress urinary incontinence aims to denature collagen using radiofrequency treatment. Though the results of the first prospective, randomized, sham controlled, double blind study on this treatment is still ongoing at the time of writing this chapter (Pope et al., 2022), preliminary findings have shown success in some patients (Appell et al., 2007). The theory is that, though radiofrequency energy can be used to break down collagen, it also causes targeted tissues to shrink and tighten. If the tissue can heal, while maintaining that "tightened" configuration, it may lead to increased urethral support (Appell et al., 2007; Dillon and Dmochowski, 2009; Ross et al., 2002). One concern is that, although the tissue may be shorter, it may have inferior mechanical properties, i.e., increased compliance, which negate the reduction in length. Regardless of the outcomes, this will be an interesting clinical trial to watch moving forwards.

8.3.8.2 Pelvic organ prolapse

Another common pelvic floor disorder is pelvic organ prolapse. Much like stress urinary incontinence, many theories claim that pelvic organ prolapse is due to a lack of supportive capacity in the tissues whose role is to support the pelvic organs. Thus as seems to be the case for SUI, it would appear logical that there should be a corresponding decrease in collagen content in the connective tissues that support the pelvic organs. However, there are contradictory results in the literature thus far (Gong and Xia, 2019; Han et al., 2014; Liu et al., 2017; Vulic et al., 2011; Yucel et al., 2013; Sun et al., 2016). Some studies have shown that there is indeed a decrease in both collagen types I and III in the uterosacral, cardinal ligaments, and paraurethral tissues in women with pelvic organ prolapse (Han et al., 2014; Liu et al., 2017; Vulic et al., 2011). In these studies, they observed a decrease in the type I collagen content of 26, 29, and 24 percent in the uterosacral ligament, cardinal ligament, and paraurethral tissues, respectively (Han et al., 2014; Yucel et al., 2013). Other studies have shown trends of a decrease in collagen type I, but contradict previous work to show an increase in collagen type III (Yucel et al., 2013; Sun et al., 2016), a response that is consistent with a "scar-type" healing response in musculoskeletal tissues (Xue and Jackson, 2015). These studies found that there was a 5.2% (Yucel et al., 2013) and 35% (Sun et al., 2016) decrease in collagen type I within the uterosacral ligament, respectively. However, neither of these differences were significant between groups (Yucel et al., 2013; Sun et al., 2016). According to these two studies, there was a concomitant increase in the amount of type III collagen in the uterosacral ligament (7.9% (Yucel et al., 2013) and 135% (Sun et al., 2016)). A primary theory for differences found in these studies is that the biopsy site along the length of the ligament could be different (Liu et al., 2017). However, at this point it is difficult to draw any definitive conclusions. First, it is unclear if these changes are causative or develop secondary to prolapse. Assuming that it is causative, it is likely that similar presentations of prolapse may result from different mechanistic pathways. For example, an injury response might look very different from a degenerative response. In addition, the supportive tissues involved may not always be the same or become differentially involved, depending on the stage of prolapse. Finally, there is not a direct correlation between a tissue's mechanical properties and its collagen content (including ratios of subtypes). As mentioned for SUI, a more complete picture of the time course of remodeling in terms of all mechanically contributing constituents is needed, requiring a combination of basic science studies in humans, animal studies, computational simulations, and creative *in vivo* measurements to obtain such knowledge.

To this end, some data do exist on how the mechanical properties of pelvic tissues change in pathologic patients compared to healthy controls (Mangera et al., 2012). When comparing the mechanical properties of the female paravaginal tissues in patients with and without prolapse, samples were found to fail at lower values of stress and strain in patients who presented with POP (Mangera et al., 2012). Pathologic tissues were also reported to have a higher modulus, and these findings correlated with increased symptoms (Mangera et al., 2012). Some studies compared the mechanical properties of nonpathologic and pathologic tissue. The ultimate tensile strength of vaginal tissue from prolapse patients ranged between 0.27 MPa and 0.60 MPa, whereas healthy tissues ranged between 0.42 MPa and 0.79 MPa. A similar trend was reported with ultimate strain, where vaginal tissues from prolapsed patients failed at strains ranging from 0.14 to 0.5, whereas healthy tissue failed at strains between 0.37 and 0.6. Additionally, the moduli of vaginal tissue also had a different range comparing pathologic and nonpathologic tissue. Prolapsed tissue ranged between 9.45 MPa and 12.1 MPa, whereas healthy tissue ranged between 6.65 MPa and 10.26 MPa. In general, most literature agrees that there is an association between vaginal stiffness and POP severity (Goh, 2002; Lei et al., 2007; Martins et al., 2013; Goh, 2003; Rubod et al., 2008). Patients with POP were associated with higher stiffness and stress at failure for the anterior wall when compared to the posterior vaginal wall (Martins et al., 2013). Some have hypothesized that these differences are not causal and are instead due to remodeling of the wall in response to prolapse (Hsu et al., 2005; Alperin and Moalli, 2006). Based on what we know about the loss of smooth muscle with prolapse and how that loss is typically associated with an increase in mechanical properties, this could be correct. Again, much work remains in terms of answering these questions and understanding the mechanisms of prolapse.

8.3.9 Pregnancy and childbirth

Pregnancy and childbirth are two of the most physically demanding physiological processes that the human body undergoes. Though not pathological processes, pregnancy has been closely correlated with changes in the mechanical properties of the cervix (Myers et al., 2015; Osol and Cipolla, 1993; Liao et al., 2014; Akins et al., 2011). If these changes occur prematurely, normal processes of pregnancy can become a pathological condition. During pregnancy the cervix must support the weight of the growing fetus within the uterus; then, during childbirth, must soften to allow the fetus to pass. Human cervical tissues have been shown to become 5 to 10 times more compliant during pregnancy (Bauer et al., 2009; Cabrol, 1991). Ultimately, it returns to approximately its original shape and stiffness afterwards. This process must be carefully orchestrated, as early softening may lead to a pre-term birth that can be fatal for the fetus or result in a lifetime of complications (Myers et al., 2015). Patients that experience premature softening of the cervix may also be at risk for subclinical chorioamnionitis or decidual hemorrhage (Weiner et al., 2005; Gonen, 2006). Additionally, a weakened cervix may increase the likelihood of infection (Banicevic et al., 2014).

Studies on pregnancy in a mouse model have answered many questions about how the body begins to prepare for birth. The changes in cervical extracellular matrix (ECM) content begin relatively early in gestation, which leads to a progressive decrease in stiffness of the cervix, which peaks at birth. Throughout pregnancy mature cervical collagen, with strong intermolecular crosslinks, sees a reduction in those crosslinks. Collagen with fewer crosslinks is more compliant due to fewer intermolecular bonds. There is much more to this story, with details that are continually emerging. However, a major challenge in this field is that methodologies for assessing cervical integrity during pregnancy are not standardized (House et al., 2009). Future work to increase the standardization of assessment of mechanical properties using fluorescence, tonometry, and/or ultrasound are necessary (House et al., 2009; Bauer et al., 2007; Schlembach et al., 2009). These methodologies, which are in preclinical testing, will hopefully allow for a more direct quantification of the human cervix during pregnancy. Once these data become available more directed hypotheses surrounding cervical material properties can be tested.

8.4 Conclusion

It is an unfortunate truth that pathologies and aging are part of our lives. For those of us that have or have had the misfortune of experiencing serious challenges related to urinary or defecatory dysfunction, it becomes immediately obvious how much we all take these activities for granted. Suddenly, a significant portion of your mental attention is focused on the effects of these disorders. At the same time, society tells us to keep these conditions private, necessitating that the person affected bears a considerable psychological burden in addition to their physical symptoms. For these reasons, pelvic floor disorders can often result in an inferior quality of life for those afflicted, and those afflicted are disproportionally female.

Though pelvic floor disorders often result from injury related to childbirth, the types of injury and how those injuries manifest over time are still largely unknown. However, the risk of developing pelvic floor disorders also has a lot to do with the mechanical integrity of pelvic tissues, and there are many pathologies that can influence them. Though not inclusive of all pathologies that can contribute to an increased risk in pelvic floor disorders, this chapter has demonstrated, through a few examples, that the mechanical properties of pelvic tissues are necessary for pelvic organ support, and that all the constituents that contribute to the composition of tissue play a direct or indirect role in a tissue's mechanical function. By taking a reductionistic approach to help identify the individual contributions of each of these constituents, it became apparent that biology is not so simple. Pathologies, and the effects of aging, do not occur in isolation. It is often that a change in one constituent leads to changes in others. Another take-home message is that pelvic floor disorders are multifactorial. Different sets of processes and events can lead to relatively similar outcomes. When this fact is coupled with patients often present with comorbidities that can be intimately connected with the processes of aging, finding a one-size-fits-all solution to these pelvic floor disorders is probably unlikely. Instead, it will be important to identify common themes and fundamental principles that underly each of these various pathways.

One concept that has been shown to be important to biological systems is that they often exist in a state of mechanical homeostasis. In other words, tissues are specifically tuned to the mechanical demands that exist in their immediate environment. In addition, they have the ability to adapt to changes in their mechanical environment provided that the changes are not sudden and extreme. There is often significant redundancy built into these systems such that the system can respond to a number of different loading conditions. If one component is insufficient, another is there to compensate, that is, to the degree to which it is able, minimizing the risk of catastrophic failures. For small disruptions in the mechanical environment, there is often an attempt to restore homeostasis within the system. Much of this is mediated by the cells that maintain these tissues responding to mechanical cues. The so-called field of mechanobiology aims to bridge the gap between the mechanical properties of tissues and their biology to help us answer questions about what regulates the composition and organization of tissues and how a specific pathology is mechanistically linked to changes in mechanical function. Until those details are known, it is often useful to think about these tissues in the context of mechanical homeostasis, especially when the reductionistic approach to answering these questions breaks down.

Ultimately, identifying preventative and treatment strategies for pelvic floor disorders will require significantly more science. Since many of these disorders are caused by changes to the mechanical function of pelvic tissues, the science absolutely needs to include fundamental biomechanics. As stated in the introduction, more engineers are entering this space in recent years, and that is leading to an increase in the quality and quantity of biomechanical research related to female pelvic tissues. Thus it is hoped that the next decade will see exciting progress towards reducing the incidence of disorders related to pelvic tissues and positively impacting the quality of life of women who are living with these disorders. This is a pivotal time to be a researcher in this space. Indeed, few areas remain in medicine where some basic experimentation has the potential for a huge translational impact. However, the silver lining stemming from centuries of ignoring these conditions in women is that such opportunities currently exist in the fields of obstetrics and urogynecology. In addition, we are now armed with deep knowledge from research in other areas of medicine that can help to fast-track our efforts. There are still going to be challenges that are pelvic tissue-specific, but those challenges present incredible opportunities for innovation; in this regard the future looks promising.

References

Abate, M., Schiavone, C., Pelottf, P., Salinp, V., 2010. Limited joint mobility in diabetes and ageing: recent advances in pathogenesis and therapy. International Journal of Immunopathology and Pharmacology 23.

Ackerman, J.E., Bah, I., Jonason, J.H., Buckley, M.R., Loiselle, A.E., 2017. Aging does not alter tendon mechanical properties during homeostasis, but does impair flexor tendon healing. Journal of Orthopaedic Research 35 (12), 2716–2724.

Ahmed, N., Thornalley, P.J., 2007. Advanced glycation endproducts: what is their relevance to diabetic complications? Diabetes, Obesity and Metabolism 9, 233–245.

Akins, M.L., Luby-Phelps, K., Bank, R.A., Mahendroo, M., 2011. Cervical softening during pregnancy: regulated changes in collagen cross-linking and composition of matricellular proteins in the mouse. Biology of Reproduction 84 (5), 1053–1062.

Ali Borazjani, A., Couri, B.M., Kuang, M., Balog, B.M., Damaser, M.S., 2019. Role of lysyl oxidase like 1 in regulation of postpartum connective tissue metabolism in the mouse vagina. Biology of Reproduction 101 (5), 916–927. Available from: https://academic.oup.com/biolreprod/advance-article-abstract/doi/10.1093/biolre/ioz148/5545484.

Ali, O.J., Ehrle, A., Comerford, E.J., Canty-Laird, E.G., Mead, A., Clegg, P.D., et al., 2021. Intrafascicular chondroid-like bodies in the ageing equine superficial digital flexor tendon comprise glycosaminoglycans and type II collagen. Journal of Orthopaedic Research 39 (12), 2755–2766.

Alperin, M., Cook, M., Tuttle, L.J., Esparza, M.C., Lieber, R.L., 2016. Impact of vaginal parity and aging on the architectural design of pelvic floor muscles. American Journal of Obstetrics and Gynecology.

Alperin, M., Moalli, P.A., 2006. Remodeling of vaginal connective tissue in patients with prolapse. Urogynecology 18, 544–550.

Appell, R.A., Singh, G., Klimberg, I.W., Graham, C., Juma, S., Wells, W.G., et al., 2007. Nonsurgical, radiofrequency collagen denaturation for stress urinary incontinence: retrospective 3-year evaluation. Expert Review of Medical Devices 4 (4), 455–461.

Asgari, M., Latifi, N., Heris, H.K., Vali, H., Mongeau, L., 2017. In vitro fibrillogenesis of tropocollagen type III in collagen type i affects its relative fibrillar topology and mechanics. Scientific Reports 7 (1).

Avery, N.C., Bailey, A.J., 2005. Enzymic and non-enzymic cross-linking mechanisms in relation to turnover of collagen: relevance to aging and exercise. Scandinavian Journal of Medicine & Science in Sports, 231–240.

Avery, N.C., Bailey, A.J., 2006. The effects of the Maillard reaction on the physical properties and cell interactions of collagen. Pathologie Et Biologie 54 (7), 387–395.

Babak, A., Thayer, D., Glat, P.M., 2003. Two Cases of Traumatic Wounds in Patients with Ehlers Danlos Syndrome Successfully Treated with a Bioengineered Skin Equivalent, vol. 15.

Bailey, A.J., 2001. Molecular mechanisms of ageing in connective tissues [Internet]. Mechanisms of Ageing and Development 122. Available from: http://www.elsevier.com/locate/mechagedev*.

Bajpai, A., Li, R., Chen, W., 2021. The Cellular Mechanobiology of Aging: from Biology to Mechanics. Annals of the New York Academy of Sciences, vol. 1491. Blackwell Publishing Inc., pp. 3–24.

Banicevic, A., Popovic, M., Ceric, A., 2014. Cervical length measured by transvaginal ultrasonography and cervicovaginal infection as predictor of preterm birth risk. Acta Informatica Medica 22 (2), 139–141.

Bank, R.A., Bayliss, M.T., Lafeber, F.P.J.G., Maroudas, A., Tekoppele, J.M., 1998. Ageing and zonal variation in post-translational modification of collagen in normal human articular cartilage the age-related increase in non-enzymatic glycation affects biomechanical properties of cartilage. Biochemical Journal 330.

Bauer, M., Mazza, E., Jabareen, M., Sultan, L., Bajka, M., Lang, U., et al., 2009. Assessment of the in vivo biomechanical properties of the human uterine cervix in pregnancy using the aspiration test A feasibility study. European Journal of Obstetrics & Gynecology and Reproductive Biology 144 (SUPPL 1).

Bauer, M., Mazza, E., Nava, A., Zeck, W., Eder, M., Bajka, M., Cacho, F., Lang, U., Holzapfel, G., et al., 2007. In vivo characterization of the mechanics of human uterine cervices. Annals of the New York Academy of Sciences 1101.

Becker, W.R., de Vita, R., 2015. Biaxial mechanical properties of swine uterosacral and cardinal ligaments. Biomechanics and Modeling in Mechanobiology 14 (3), 549–560.

Bharucha, A.E., Trabuco, E., 2008. Functional and chronic anorectal and pelvic pain disorders. Gastroenterology Clinics of North America 37, 685–696.

Botlero, R., Davis, S.R., Urquhart, D.M., Bell, R.J., 2011. Incidence and resolution rates of different types of urinary incontinence in women: findings from a cohort study. The Journal of Urology 185 (4), 1331–1337.

Brandão, S., Parente, M., Mascarenhas, T., da Silva, A.R.G., Ramos, I., Jorge, R.N., 2015. Biomechanical study on the bladder neck and urethral positions: simulation of impairment of the pelvic ligaments. Journal of Biomechanics 48 (2), 217–223.

Brizzolara, S.S., Killeen, J., Urschitz, J., 2009. Gene expression profile in pelvic organ prolapse. Molecular Human Reproduction 15 (1), 59–67.

Brownlee, M., Vlassara, H., Cerami, A., York, N, 1984. Nonenzymatic glycosylation and the pathogenesis of diabetic complications. Annals of Internal Medicine 101, 527–537. Available from: http://annals.org/.

Burnett, L.A., Cook, M., Shah, S., Michelle Wong, M., Kado, D.M., Alperin, M., 2020. Age-associated changes in the mechanical properties of human cadaveric pelvic floor muscles. Journal of Biomechanics 98, 109436. https://doi.org/10.1016/j.jbiomech.2019.109436.

Burr, D.B., 2002. The contribution of the organic matrix to bone's material properties. Bone 31 (1), 8–11.

Cabrol, D., 1991. Cervical distensibility changes in pregnancy, term, and preterm labor. Seminars in Perinatology 15 (2), 133–139.

Carley, M.E., Schaffer, J., 2000. Urinary incontinence and pelvic organ prolapse in women with Marfan or Ehlers-Danlos syndrome. American Journal of Obstetrics and Gynecology 182 (5), 1021–1023.

Catanzarite, T., Bremner, S., Barlow, C.L., Bou-Malham, L., O'Connor, S., Alperin, M., 2018. Pelvic muscles' mechanical response to strains in the absence and presence of pregnancy-induced adaptations in a rat model. American Journal of Obstetrics and Gynecology 218 (5), 512.e1–512.e9.

Chantereau, P., Brieu, M., Kammal, M., Farthmann, J., Gabriel, B., Cosson, M., 2014. Mechanical properties of pelvic soft tissue of young women and impact of aging. International Urogynecology Journal 25 (11), 1547–1553.

Chen, Y., Huang, J., Hu, C., Hua, K., 2015. Relationship of advanced glycation end products and their receptor to pelvic organ prolapse [Internet]. International Journal of Clinical and Experimental Pathology 8. Available from: http://www.ijcep.com/.

Choudhary, S., Bisati, S., Koley, S., 2011. Congenital cutis laxa with rectal and uterovaginal prolapse. Indian Journal of Dermatology, Venereology and Leprology 77 (3), 321–324.

Clark, G.L., Pokutta-Paskaleva, A.P., Lawrence, D.J., Lindsey, S.H., Desrosiers, L., Knoepp, L.R., et al., 2019. Smooth muscle regional contribution to vaginal wall function. Interface Focus 9 (4).

Cook, M.S., Bou-Malham, L., Esparza, M.C., Alperin, M., 2017. Age-related alterations in female obturator internus muscle. International Urogynecology Journal 28 (5), 729–734.

Csapo, R., Gumpenberger, M., Wessner, B., 2020. Skeletal muscle extracellular matrix – what do we know about its composition, regulation, and physiological roles? A narrative review. Frontiers in Physiology 11.

Danso, E.K., Schuster, J.D., Johnson, I., Harville, E.W., Buckner, L.R., Desrosiers, L., et al., 2020. Comparison of biaxial biomechanical properties of post-menopausal human prolapsed and non-prolapsed uterosacral ligament. Scientific Reports 10 (1).

Delancey, J.O.L., Ashton-Miller, J.A., 2004. Pathophysiology of adult urinary incontinence. Gastroenterology 126 (1), S23–S32.

Derby, B., Akhtar, R., 2015. Mechanical properties of aging soft tissues [Internet], vol. 1. Available from: http://www.springer.com/series/4604.

Dillon, B., Dmochowski, R., 2009. Radiofrequency for the treatment of stress urinary incontinence in women. Female Urology, 369–374.

Drewes, P.G., Yanagisawa, H., Starcher, B., Hornstra, I., Csiszar, K., Marinis, S.I., et al., 2007. Pelvic organ prolapse in fibulin-5 knockout mice: pregnancy-induced changes in elastic fiber homeostasis in mouse vagina. The American Journal of Pathology 170 (2), 578–589.

Duca, L., Blaise, S., Romier, B., Laffargue, M., Gayral, S., el Btaouri, H., et al., 2016. Matrix aging and vascular impacts: focus on elastin fragmentation. Cardiovascular Research 110 (3), 298–308.

Dunivan, G.C., Lyons, K.E., Jeppson, P.C., Ninivaggio, C.S., Komesu, Y.M., Alba, F.M., et al., 2016. Pelvic organ prolapse stage and the relationship to genital hiatus and perineal body measurements. Female Pelvic Medicine and Reconstructive Surgery 22 (6), 497–500.

Dunkman, A.A., Buckley, M.R., Mienaltowski, M.J., Adams, S.M., Thomas, S.J., Satchell, L., et al., 2013. Decorin expression is important for age-related changes in tendon structure and mechanical properties. Matrix Biology 32 (1), 3–13.

e Lima, K.M.M., Costa Júnior, J.F.S., Pereira, W.C. de A., de Oliveira, L.F., 2018. Assessment of the mechanical properties of the muscle-tendon unit by supersonic shear wave imaging elastography: a review. Ultrasonography. Korean Society of Ultrasound in Medicine 37, 3–15.

Easley, D.C., Abramowitch, S.D., Moalli, P.A., 2017. Female pelvic floor biomechanics: bridging the gap. Current Opinion in Urology 27, 262–267.

Elanie Epin, M.P., Lrike Chwarze, U.S., Ndrea Uperti-f Urga, A.S., Eter, B., Yers, P.H., 2000. Clinical and genetic features of Ehlers-Danlos syndrome type IV, the vascular type. The New England Journal of Medicine 342, 673.

Eric Jelovsek, J., Maher, C., Barber, M.D., Jelovsek, J.E., Barber, M.D., 2007. Pelvic organ prolapse. The Lancet 369. Available from: http://www.thelancet.com.

Eser, A., Unlubilgin, E., Hizli, F., Acar, M., Kamalak, Z., Kosus, A., et al., 2015. Is there a relationship between pelvic organ prolapse and tissue fibrillin-1 levels? International Neurourology Journal 19 (3), 164–170.

Feldner, P.C., Sartori, M.G.F., Nader, H.B., Dietrich, C.P., Rodrigues de Lima, G., Baracat, E.C., et al., 2008. Sulfated glycosaminoglycans of periurethral tissue in pre- and postmenopausal women. European Journal of Obstetrics & Gynecology and Reproductive Biology 139 (2), 252–255.

Fonck, E., Feigl, G.G., Fasel, J., Sage, D., Unser, M., Rüfenacht, D.A., et al., 2009. Effect of aging on elastin functionality in human cerebral arteries. Stroke 40 (7), 2552–2556.

Freeman, M.A.R., 1971. Strength of biological materials. Journal of Bone and Joint Surgery. British Volume.
Fujii, K., Yamagishi, T., Nagafuchi, T., Tsuji, M., Kuboki, Y., 1994. Biochemical properties of collagen from ligaments and periarticular tendons of the human knee. In: Experimental Investigations. Knee Surgery, Sports Traumatology, Arthroscopy 2.
Fulop, T., Khalil, A., Larbi, A., 2012. The role of elastin peptides in modulating the immune response in aging and age-related diseases. Pathologie Et Biologie 60 (1), 28–33.
Fung, Y.C., Skalak, R., 1981. Biomechanics mechanical properties of living tissues. Journal of Biomechanical Engineering.
Gibbons, M.C., Singh, A., Engler, A.J., Ward, S.R., 2018. The role of mechanobiology in progression of rotator cuff muscle atrophy and degeneration. Journal of Orthopaedic Research 36 (2), 546–556.
Gilliam, E., Hoffman, J.D., Yeh, G., 2020. Urogenital and pelvic complications in the Ehlers-Danlos syndromes and associated hypermobility spectrum disorders: a scoping review. Clinical Genetics 97, 168–178.
Goh, J.T.W., 2002. Biomechanical properties of prolapsed vaginal tissue in pre-and post-menopausal women. International Urogynecology Journal 13, 76–79.
Goh, J.T.W., 2003. Biomechanical and biochemical assessments for pelvic organ prolapse. Urogynecology 15, 391–394.
Goldin, A., Beckman, J.A., Schmidt, A.M., Creager, M.A., 2006. Advanced glycation end products: sparking the development of diabetic vascular injury. Circulation 114, 597–605.
Gomez, D., Owens, G.K., 2012. Smooth muscle cell phenotypic switching in atherosclerosis. Cardiovascular Research 95, 156–164.
Gonen, R., 2006. A short cervix in women with preterm labor and intactmembranes: a risk factor for microbial invasionof the amniotic cavity. American Journal of Obstetrics and Gynecology 194, 901.
Gong, R., Xia, Z., 2019. Collagen changes in pelvic support tissues in women with pelvic organ prolapse. European Journal of Obstetrics & Gynecology and Reproductive Biology 234, 185–189.
Graham, G., Bakaysa, S., 2019. Preterm premature rupture of membranes (PPROM). Evidence Based Obstetrics and Gynecology 1.
Grimaud, J.C., Bouvier, M., Naudy, B., Guien, C., Salducci, J., 1991. Manometric and radiologic investigations and biofeedback treatment of chronic idiopathic anal pain. Diseases of the Colon & Rectum 34 (8), 690–695.
Han, L., Wang, L., Wang, Q., Li, H., Zang, H., 2014. Association between pelvic organ prolapse and stress urinary incontinence with collagen. Experimental and Therapeutic Medicine 7 (5), 1337–1341.
Hasson, C.J., Caldwell, G.E., 2012. Effects of age on mechanical properties of dorsiflexor and plantarflexor muscles. Annals of Biomedical Engineering 40 (5), 1088–1101.
Hay, E.D., 1981. Extracellular matrix. The Journal of Cell Biology 91 (3). Available from: http://rupress.org/jcb/article-pdf/91/3/205s/1075485/205s.pdf.
Heeransh, D. Dave, Shook, Micah, Varacallo, Matthew, 2021. Anatomy, Skeletal Muscle.
Heinz, A., 2021. Elastic fibers during aging and disease. Ageing Research Reviews 66.
House, M., Kaplan, D.L., Socrate, S., 2009. Relationships between mechanical properties and extracellular matrix constituents of the cervical stroma during pregnancy. Seminars in Perinatology 33, 300–307.
Hsu, Y., Chen, L., Delancey, J.O.L., Ashton-Miller, J.A., 2005. Vaginal thickness, cross-sectional area, and perimeter in women with and those without prolapse. Obstetrics and Gynecology 105 (5), 1012–1017.

Iel, A.B., Oxlund, H., 1996. Changes in biomechanical properties, composition of collagen and elastin, and advanced glycation endproducts of the rat aorta in relation to age. Atherosclerosis 127.

Jiang, H.H., Damaser, M.S., 2011. Animal Models of Stress Urinary Incontinence. Handbook of Experimental Pharmacology, vol. 202, pp. 45–67.

Johnson, T.M., Kincade, J.E., Bernard, S.L., Busby-Whitehead, J., DeFriese, G.H., 2000. Self-care practices used by older men and women to manage urinary incontinence: results from the national follow-up survey on self-care and aging. Journal of the American Geriatrics Society 48 (8), 894–902.

Jung, H.J., Jeon, M.J., Yim, G.W., Kim, S.K., Choi, J.R., Bai, S.W., 2009. Changes in expression of fibulin-5 and lysyl oxidase-like 1 associated with pelvic organ prolapse. European Journal of Obstetrics & Gynecology and Reproductive Biology 145 (1), 117–122.

Karim, L., Moulton, J., van Vliet, M., Velie, K., Robbins, A., Malekipour, F., et al., 2018. Bone microarchitecture, biomechanical properties, and advanced glycation end-products in the proximal femur of adults with type 2 diabetes. Bone 114, 32–39.

Kati, L.M., Feldner, P.C., Aquino De Castro, R., Kobayashi, E.Y., Gracio, M., Sartori, F., et al., 2010. Analysis of glycosaminoglycans in the parametrium and vaginal apex of women with and without uterine prolapse. Journal of Women's Health 19 (7).

Kearney, R., Miller, J.M., Ashton-Miller, J.A., Delancey, J.O.L., 2006. Obstetric factors associated with levator ani muscle injury after vaginal birth LEVEL OF EVIDENCE: II-3. Obstetrics and Gynecology 107.

Kent, M.J.C., Light, N.D., Bailey, A.J., 1985. Evidence for glucose-mediated covalent cross-linking of collagen after glycosylation in vitro. Biochemical Journal 225.

Kepenekci, I., Keskinkilic, B., Akinsu, F., Cakir, P., Elhan, A.H., Erkek, A.B., et al., 2011. Prevalence of pelvic floor disorders in the female population and the impact of age, mode of delivery, and parity. Diseases of the Colon & Rectum 54 (1), 85–94.

Kerkhof, M.H., Hendriks, L., Brölmann, H.A.M., 2009. Changes in connective tissue in patients with pelvic organ prolapse - a review of the current literature. International Urogynecology Journal 20, 461–474.

Khosla, S., Hofbauer, L.C., 2017. Osteoporosis treatment: recent developments and ongoing challenges. The Lancet Diabetes & Endocrinology 5, 898–907.

Kielty, C., Sherratt, M., Shittleworth, A., 2002. Elastic fibres. Journal of Cell Science 115 (14), 2817–2828.

Kim, C.S., Park, S., Kim, J., 2017. The role of glycation in the pathogenesis of aging and its prevention through herbal products and physical exercise. Journal of Exercise Nutrition & Biochemistry 21 (3), 55–61.

Kling, S., Torres-Netto, E.A., Abdshahzadeh, H., Espana, E.M., Hafezi, F., 2021. Collagen V insufficiency in a mouse model for Ehlers Danlos-syndrome affects viscoelastic biomechanical properties explaining thin and brittle corneas. Scientific Reports 11 (1).

Klutke, J., Ji, Q., Campeau, J., Starcher, B., Felix, J.C., Stanczyk, F.Z., et al., 2008. Decreased endopelvic fascia elastin content in uterine prolapse. Acta Obstetricia et Gynecologica Scandinavica 87 (1), 111–115.

Kragstrup, T.W., Kjaer, M., Mackey, A.L., 2011. Structural biochemical, cellular, and functional changes in skeletal muscle extracellular matrix with aging. Scandinavian Journal of Medicine & Science in Sports 21, 749–757.

Kurt, S., Canda, M.T., Bal, M., Tasyurt, A., 2018. Are there any preventable risk factors for women who had surgery for pelvic organ prolapse and stress urinary incontinence? Pakistan Journal of Medical Sciences 34 (4), 874–878.

Lacolley, P., Regnault, V., Avolio, A.P., 2018. Smooth muscle cell and arterial aging: basic and clinical aspects. Cardiovascular Research 114, 513–528.

Lacolley, P., Regnault, V., Segers, P., Laurent, S., 2017. Vascular smooth muscle cells and arterial stiffening: relevance in development, aging, and disease. Physiological Reviews 97 (4), 1555–1617.

Lammers, K., Lince, S.L., Spath, M.A., VanKempen, L.C.L.T., Hendriks, J.C.M., Vierhout, M.E., et al., 2012. Pelvic organ prolapse and collagen-associated disorders. International Urogynecology Journal 23 (3), 313–319.

Lawrence, J.M., Lukacz, E.S., Liu, I.L.A., Nager, C.W., Luber, K.M., 2007. Pelvic floor disorders, diabetes, and obesity in women: findings from the Kaiser Permanente continence associated risk epidemiology study. Diabetes Care 30 (10), 2536–2541.

Lee, D.H., Oh, J.H., Chung, J.H., 2016. Glycosaminoglycan and proteoglycan in skin aging. Journal of Dermatological Science 83, 174–181.

Lee, U.J., Marcus Gustilo-Ashby, A., Daneshgari, F., Kuang, M., Vurbic, D., Lin, D.L., et al., 2008. Lower urogenital tract anatomical and functional phenotype in lysyl oxidase like-1 knockout mice resembles female pelvic floor dysfunction in humans. American Journal of Physiology. Renal Physiology 295, 545–555. Available from: http://www.ajprenal.orgF545.

Lei, L., Song, Y., Chen, R.Q., 2007. Biomechanical properties of prolapsed vaginal tissue in pre- and postmenopausal women. International Urogynecology Journal 18 (6), 603–607.

Li, L., Sun, Z., Chen, J., Zhang, Y., Shi, H., Zhu, L., 2020. Genetic polymorphisms in collagen-related genes are associated with pelvic organ prolapse. Menopause 27 (2), 223–229.

Liao, D., Hee, L., Sandager, P., Uldbjerg, N., Gregersen, H., 2014. Identification of biomechanical properties in vivo in human uterine cervix. Journal of the Mechanical Behavior of Biomedical Materials 39, 27–37.

Lim, J.Y., Choi, S.J., Widrick, J.J., Phillips, E.M., Frontera, W.R., 2019. Passive force and viscoelastic properties of single fibers in human aging muscles. European Journal of Applied Physiology 119 (10), 2339–2348.

Lin, X., Xu, Y., Pan, X., Xu, J., Ding, Y., Sun, X., et al., 2020. Global, regional, and national burden and trend of diabetes in 195 countries and territories: an analysis from 1990 to 2025. Scientific Reports 10 (1).

Lindemann, I., Coombes, B.K., Tucker, K., Hug, F., Dick, T.JM., 2020. Age-related differences in gastrocnemii muscles and Achilles tendon mechanical properties in vivo. Journal of Biomechanics 112.

Lindsay Wood, S., Szychowski, J.M., Owen, J., Williams, B.N., 2020. The effect of intramuscular 17α-hydroxyprogesterone in women screened for shortened cervical length. American Journal of Perinatology 37 (7), 659–665.

Liu, C., Wang, Y., Li, B.S., Yang, Q., Tang, J.M., Min, J., et al., 2017. Role of transforming growth factor β-1 in the pathogenesis of pelvic organ prolapse: a potential therapeutic target. International Journal of Molecular Medicine 40 (2), 347–356.

Lose, G., 1989. Mechanical properties of the urethra in females with genuine stress incontinence: static measurements in the resting urethra. Neurourology and Urodynamics 8.

Lose, G., Colstrup, H., 1990. Mechanical properties of the urethra in healthy and stress incontinent females: dynamic measurements in the resting urethra. The Journal of Urology 144 (5), 1258–1262.

Makatsariya, A., Radetskaya, L., Bitsadze, V., Khizroeva, J., Khamani, N., Makatsariya, N., 2020. Prenatal care and labor in patients with mesenchimal dysplasias (Marfan syndrome, Ehlers–Danlos syndrome, hereditary hemorrhagic telangiectasia). Journal of Maternal-Fetal & Neonatal Medicine 33 (3), 373–379.

Mangera, A., Bullock, A.J., Chapple, C.R., MacNeil, S., 2012. Are biomechanical properties predictive of the success of prostheses used in stress urinary incontinence and pelvic organ prolapse? A systematic review. Neurourology and Urodynamics 31, 13–21.

Martins, P., Lopes Silva-Filho, A., Rodrigues Maciel Da Fonseca, A.M., Santos, A., Santos, L., Mascarenhas, T., et al., 2013. Biomechanical properties of vaginal tissue in women with pelvic organ prolapse. Gynecologic and Obstetric Investigation 75 (2), 85–92.

Mastoroudes, H., Giarenis, I., Cardozo, L., Srikrishna, S., Vella, M., Robinson, D., et al., 2013. Prolapse and sexual function in women with benign joint hypermobility syndrome. BJOG 120 (2), 187–192.

McCrum, C., Leow, P., Epro, G., König, M., Meijer, K., Karamanidis, K., 2018. Alterations in leg extensor muscle-tendon unit biomechanical properties with ageing and mechanical loading. Frontiers in Physiology 9.

Meyer, I., Morgan, S.L., Markland, A.D., Szychowski, J.M., Richter, H.E., 2020. Pelvic floor disorder symptoms and bone strength in postmenopausal women. International Urogynecology Journal 31, 1777–1784. https://doi.org/10.1007/s00192-020-04254-z.

Micussi, M.T., Freitas, R.P., Angelo, P.H., Soares, E.M., Lemos, T.M., Maranhão, T.M., 2015. Evaluation of the relationship between the pelvic floor muscles and insulin resistance. Diabetes, Metabolic Syndrome and Obesity 8, 409–413.

Mizumoto, S., Janecke, A.R., Sadeghpour, A., Povysil, G., McDonald, M.T., Unger, S., et al., 2020. CSGALNACT1-congenital disorder of glycosylation: a mild skeletal dysplasia with advanced bone age. Human Mutation 41 (3), 655–667.

Mizumoto, S., Yamada, S., 2021. Congenital disorders of deficiency in glycosaminoglycan biosynthesis. Frontiers in Genetics 12.

Mizumoto, S., Yamada, S., Sugahara, K., 2015. Mutations in biosynthetic enzymes for the protein linker region of chondroitin/dermatan/heparan sulfate cause skeletal and skin dysplasias. BioMed Research International 2015.

Monk, B.A., George, S.J., 2015. The effect of ageing on vascular smooth muscle cell behaviour - a mini-review. Gerontology 61, 416–426.

Morley, J.E., 2008. Diabetes and aging: epidemiologic overview. Clinics in Geriatric Medicine 24, 395–405.

Morley, J.E., 2013. Frailty, falls, and fractures. Journal of the American Medical Directors Association 14, 149–151.

Müller, B., Ratia Garcia, J., Marti, F., Leippold, T., 2008. Mechanical properties of urethral tissue. Journal of Biomechanics 41 (July), S61.

Myers, K.M., Feltovich, H., Mazza, E., Vink, J., Bajka, M., Wapner, R.J., et al., 2015. The mechanical role of the cervix in pregnancy. Journal of Biomechanics 48 (9), 1511–1523.

Niyomchan, A., Panichareon, B., Siriphorn, A., Wongtawatchai, T., 2019. Age-related structural-mechanical property changes in human peroneus longus muscle. Folia Morphologica (Poland) 78 (2), 401–407.

Norton, P.A., Baker, J., Sharp, H.C., Waxenski, J.C., 1995. Genitourinary prolapse and joint hypermobility in women. Obstetrics and Gynecology 85 (2), 225–228.

Olafsdottir, T., Thorleifsson, G., Sulem, P., Stefansson, O.A., Medek, H., Olafsson, K., et al., 2020. Genome-wide association identifies seven loci for pelvic organ prolapse in Iceland and the UK Biobank. Communications Biology 3 (1).

Osmers, R., Rath, W., Pflanz, M.A., Kuhn, W., Stuhlsatz, H.W., Szeverenyi, M., 1993. Glycosaminoglycans in cervical connective tissue during pregnancy and parturition. Obstetrics and Gynecology 81 (1), 88–92.

Osol, G., Cipolla, M., 1993. Pregnancy-induced changes in the three-dimensional mechanical properties of pressurized rat uteroplacental (radial) arteries. American Journal of Obstetrics and Gynecology 168 (1), 268–274.

Pal, L., 2009. Pelvic organ prolapse and relationship with skeletal integrity. Women's Health 5 (3), 325–333.

Paladini, D., di Spiezio Sardo, A., Mandato, V.D., Guerra, G., Bifulco, G., Mauriello, S., et al., 2007. Association of cutis laxa and genital prolapse: a case report. International Urogynecology Journal 18 (11), 1367–1370.

Palmquist, M., Pappas, J.G., Petrikovsky, B., Blakemore, K., Roshan, D., 2009. Successful pregnancy outcome in Ehlers-Danlos syndrome, vascular type. Journal of Maternal-Fetal & Neonatal Medicine 22 (10), 924–927.

Parapia, L.A., Jackson, C., 2008. Ehlers-Danlos syndrome – a historical review. British Journal of Haematology 141 (1), 32–35.

Patel, S., Hyer, S., Tweed, K., Kerry, S., Allan, K., Rodin, A., et al., 2008. Risk factors for fractures and falls in older women with type 2 diabetes mellitus. Calcified Tissue International 82 (2), 87–91.

Pavan, P., Monti, E., Bondí, M., Fan, C., Stecco, C., Narici, M., et al., 2020. Alterations of extracellular matrix mechanical properties contribute to age-related functional impairment of human skeletal muscles. International Journal of Molecular Sciences 21 (11).

Petsophonsakul, P., Furmanik, M., Forsythe, R., Dweck, M., Schurink, G.W., Natour, E., et al., 2019. Role of vascular smooth muscle cell phenotypic switching and calcification in aortic aneurysm formation involvement of Vitamin K-dependent processes. Arteriosclerosis, Thrombosis, and Vascular Biology 39, 1351–1368.

Ploeckinger, B., Ulm, M.R., Chalubinski, K., 1997. Ehlers-Danlos syndrome type II in pregnancy, vol. 14. American Journal of Perinatology.

Pope, C., Cotterill, N., Drake, M.J., Fitzgerald, B., Greenwell, T., Jha, S., et al., 2022. The importance of overcoming the challenges in delivering the Proper Understanding of Recurrent Stress Urinary Incontinence Treatment (PURSUIT) study. Continence 1, 100020.

Rachner, T.D., Khosla, S., Hofb, L.C., 2011. Osteoporosis: now and the future. The Lancet 377, 1276–1287.

Raizada, Varuna, Mittal, R.K., 2009. Pelvic floor anatomy and applied physiology. Gastroenterology Clinics of North America 37 (3).

Reddy, G.K., 2004. Cross-linking in collagen by nonenzymatic glycation increases the matrix stiffness in rabbit Achilles tendon. Experimental Diabesity Research 5 (2), 143–153.

Regalado, E.S., Mellor-Crummey, L., de Backer, J., Braverman, A.C., Ades, L., Benedict, S., et al., 2018. Clinical history and management recommendations of the smooth muscle dysfunction syndrome due to ACTA2 arginine 179 alterations. Genetics in Medicine 20 (10), 1206–1215.

Rensen, S., Doevendans, P., van Eys, G., 2007. Regulation and characteristics of vascular smooth muscle cell phenotypic diversity. Netherlands Heart Journal 15 (3), 100–108.

Ricard-Blum, S., 2011. The collagen family. Cold Spring Harbor Perspectives in Biology 3 (1), 1–19.

Riemer, K., Heymann, M., 1998. Regulation of uterine smooth muscle function during gestation. Pediatric Research 44, 615–627.

Ross, J.W., Galen, D.I., Abbott, K., Albala, D., Presthus, J., Su-Ou, C., et al., 2002. A prospec tive multisite study ofRadiofrequency bipolar energy for treatment of genuine stress incontinence. The Journal of the American Association of Gynecologic Laparoscopists 9 (4), 493–499.

Routzong, M.R., Martin, L.C., Rostaminia, G., Abramowitch, S., 2021. Urethral support in female urinary continence part 2: a computational, biomechanical analysis of Valsalva. International Urogynecology Journal. Available from: http://link.springer.com/10.1007/s00192-021-04694-1.

Royer, S.P., Han, S.J., 2022. Mechanobiology in the comorbidities of Ehlers Danlos syndrome. Frontiers in Cell and Developmental Biology 10.

Rubod, C., Boukerrou, M., Brieu, M., Jean-Charles, C., Dubois, P., Cosson, M., 2008. Biomechanical properties of vaginal tissue: preliminary results. International Urogynecology Journal 19 (6), 811–816.

Rubod, C., Brieu, M., Cosson, M., Rivaux, G., Clay, J.C., de Landsheere, L., et al., 2012. Biomechanical properties of human pelvic organs. Urology.

Rudd, N.L., Holbrook, K.A., Nimrod, C., Byers, P.H., 1983. Pregnancy complications in type IV Ehlers-Danlos syndrome. The Lancet, 50–53.

Salter, S.A., Batra, R.S., Rohrer, T.E., Kohli, N., Kimball, A.B., 2006. Striae and pelvic relaxation: two disorders of connective tissue with a strong association. Journal of Investigative Dermatology 126 (8), 1745–1748.

Schlembach, D., Mackay, L., Shi, L., Maner, W.L., Garfield, R.E., Maul, H., 2009. Cervical ripening and insufficiency: from biochemical and molecular studies to in vivo clinical examination. European Journal of Obstetrics & Gynecology and Reproductive Biology 144 (SUPPL 1).

Seawright, J.W., Sreenivasappa, H., Gibbs, H.C., Padgham, S., Shin, S.Y., Chaponnier, C., et al., 2018. Vascular smooth muscle contractile function declines with age in skeletal muscle feed arteries. Frontiers in Physiology 9.

Sell, D.R., Monnier, V.M., 2012. Molecular basis of arterial stiffening: role of glycation-a mini-review. Gerontology 58, 227–237.

Shin, J.W., Kwon, S.H., Choi, J.Y., Na, J.I., Huh, C.H., Choi, H.R., et al., 2019. Molecular mechanisms of dermal aging and antiaging approaches. International Journal of Molecular Sciences 20 (9).

Shirley, E.D., DeMaio, M., Bodurtha, J., 2012. Ehlers-Danlos syndrome in orthopaedics: etiology, diagnosis, and treatment implications. Sports Health 4 (5), 394–403.

Silder, A., Reeder, S.B., Thelen, D.G., 2010a. The influence of prior hamstring injury on lengthening muscle tissue mechanics. Journal of Biomechanics 43 (12), 2254–2260.

Silder, A., Thelen, D.G., Heiderscheit, B.C., 2010b. Effects of prior hamstring strain injury on strength, flexibility, and running mechanics. Clinical Biomechanics 25 (7), 681–686.

Snedeker, J.G., Gautieri, A., 2014. The role of collagen crosslinks in ageing and diabetes-the good, the bad, and the ugly. Ligaments and Tendons Journal 4.

Solan, A., Niklason, L., 2005. Age effects on vascular smooth muscle: an engineered tissue approach. Cell Transplant 14, 481–488. Available from: http://www.cognizantcommunication.com.

Solomon, L.A., Schimp, V.L., Ali-Fehmi, R., Diamond, M.P., Munkarah, A.R., 2005. Clinical update of smooth muscle tumors of the uterus. Journal of Minimally Invasive Gynecology 12, 401–408.

Spiegel, E., Nicholls-Dempsey, L., Czuzoj-Shulman, N., Abenhaim, H.A., 2020. Pregnancy outcomes in women with Ehlers-Danlos Syndrome. Journal of Maternal-Fetal & Neonatal Medicine, 1–7.

Spirka, T., Kenton, K., Brubaker, L., Damaser, M.S., 2013. Effect of material properties on predicted vesical pressure during a cough in a simplified computational model of the bladder and urethra. Annals of Biomedical Engineering 41 (1), 185–194.

Sran, M.M., 2009. Prevalence of urinary incontinence in women with osteoporosis. Journal of Obstetrics and Gynaecology Canada 31 (5), 434–439.
Stewart, F.J., Bentley, A., Burton, B.K., Guffon, N., Hale, S.L., Harmatz, P.R., et al., 2016. Pregnancy in patients with mucopolysaccharidosis: a case series. Molecular Genetics and Metabolism Reports 8, 111–115.
Stock, O., Gordon, L., Kapoor, J., Walker, S.P., Whitehead, C., Kaitu'U-Lino, T.J., et al., 2015. Chorioamnionitis occurring in women with preterm rupture of the fetal membranes is associated with a dynamic increase in mRNAs coding cytokines in the maternal circulation. Reproductive Sciences 22 (7), 852–859.
Subit, D., Masson, C., Brunet, C., Chabrand, P., 2008. Microstructure of the ligament-to-bone attachment complex in the human knee joint. Journal of the Mechanical Behavior of Biomedical Materials 1 (4), 360–367.
Südy, R., Peták, F., Kiss, L., Balogh, Á.L., Fodor, G.H., Korsós, A., et al., 2021. Obesity and diabetes: similar respiratory mechanical but different gas exchange defects. The American Journal of Physiology 320 (3), L368–L376.
Südy, R., Schranc, Á., Fodor, G.H., Tolnai, J., Babik, B., Peták, F., 2020. Lung volume dependence of respiratory function in rodent models of diabetes mellitus. Respiratory Research 21 (1).
Sugita, S., Mizutani, E., Hozaki, M., Nakamura, M., Matsumoto, T., 2019. Photoelasticity-based evaluation of cellular contractile force for phenotypic discrimination of vascular smooth muscle cells. Scientific Reports 9 (1).
Suji, G., Sivakami, S., 2004. Glucose, glycation and aging. Biogerontology 5, 365–373.
Sun, M.J., Cheng, Y.S., Sun, R., Cheng, W.L., Liu, C. san, 2016. Changes in mitochondrial DNA copy number and extracellular matrix (ECM) proteins in the uterosacral ligaments of premenopausal women with pelvic organ prolapse. Taiwanese Journal of Obstetrics and Gynecology 55 (1), 9–15.
Suzme, R., Yalcin, O., Gurdol, F., Gungor, F., Bilir, A., 2007. Connective tissue alterations in women with pelvic organ prolapse and urinary incontinence. Acta Obstetricia et Gynecologica Scandinavica 86 (7), 882–888.
Svensson, R.B., Eriksen, C.S., Tran, P.H.T., Kjaer, M., Magnusson, S.P., 2021. Mechanical properties of human patellar tendon collagen fibrils. An exploratory study of aging and sex. Journal of the Mechanical Behavior of Biomedical Materials 124.
Taghizadeh, H., Tafazzoli-Shadpour, M., 2017. Characterization of mechanical properties of lamellar structure of the aortic wall: effect of aging. Journal of the Mechanical Behavior of Biomedical Materials 65, 20–28.
Tayyeb, M., Gupta, V., 2022. Dyspareunia.
Thurner, M., Thurner, M., Deutsch, M., Janke, K., Messner, F., Kreutzer, C., et al., 2020. Generation of myogenic progenitor cell-derived smooth muscle cells for sphincter regeneration. Stem Cell Research & Therapy 11 (1).
Trudgian, J., Trotman, S., 2011. Ehlers-Danlos syndrome and wound healing: injury in a collagen disorder. British Journal of Nursing 20 (Sup2), S10–S20.
Turner, W., Brading, A., 1997. Smooth muscle of the bladder in the normal and the diseased state: pathophysiology, diagnosis and treatment. Pharmacology 75 (2), 77–110.
Urban, J.P.G., Maroudas, A., Bayliss, M.T., Dillon, J., 1979. Swelling pressures of proteoglycans at the concentrations found in cartilaginous tissues. Biorheology 16 (6), 447–464.
Vaidya, R., Lake, S.P., Zellers, J.A., 2022. Effect of diabetes on tendon structure and function: not limited to collagen crosslinking. Journal of Diabetes Science and Technology, 193229682211008. Available from: http://journals.sagepub.com/doi/10.1177/19322968221100842.

Vailas, A.C., Pedrini, V.A., Pedrini-Mille, A., Holloszy, J.O., 1985. Patellar tendon matrix changes associated with aging and voluntary exercise. American Physiological Society, 1572–1576. Available from: http://www.physiology.org/journal/jappl.

Vashishth, D., Gibson, G.J., Khoury, J.I., Schaffler, M.B., Kimura, J., Fyhrie, D.P., 2001. Influence of Nonenzymatic Glycation on Biomechanical Properties of Cortical Bone.

Vestergaard, P., 2007. Discrepancies in bone mineral density and fracture risk in patients with type 1 and type 2 diabetes - a meta-analysis. Osteoporosis International 18 (4), 427–444.

Virgilio, K.M., Martin, K.S., Peirce, S.M., Blemker, S.S., 2015. Multiscale models of skeletal muscle reveal the complex effects of muscular dystrophy on tissue mechanics and damage susceptibility. Interface Focus 5 (2).

Volkov, N., Nisenblat, V., Ohel, G., Gonen, R., 2006. Ehlers-Danlos syndreome: insights on obstetric aspects. Obstetrical & Gynecological Survey 62 (1), 51–57.

Vulic, M., Strinic, T., Tomic, S., Capkun, V., Jakus, I.A., Ivica, S., 2011. Difference in expression of collagen type i and matrix metalloproteinase-1 in uterosacral ligaments of women with and without pelvic organ prolapse. European Journal of Obstetrics & Gynecology and Reproductive Biology 155 (2), 225–228.

Wang, H., Kira, Y., Hamuro, A., Takase, A., Tachibana, D., Koyama, M., 2019. Differential gene expression of extracellular-matrix-related proteins in the vaginal apical compartment of women with pelvic organ prolapse. International Urogynecology Journal 30 (3), 439–446.

Wang, X., Shen, X., Li, X., Mauli Agrawal, C., 2002. Age-Related Changes in the Collagen Network and Toughness of Bone.

Weiner, C.P., Lee, K.Y., Buhimschi, C.S., Christner, R., Buhimschi, I.A., 2005. Proteomic biomarkers that predict the clinical success of rescue cerclage. American Journal of Obstetrics and Gynecology, 710–718.

Xue, M., Jackson, C.J., 2015. Extracellular matrix reorganization during wound healing and its impact on abnormal scarring. Advances in Wound Care (New Rochelle) 4 (3), 119–136.

Yang, W., Dall, T.M., Beronjia, K., Lin, J., Semilla, A.P., Chakrabarti, R., et al., 2018. Economic costs of diabetes in the U.S. in 2017. Diabetes Care 41 (5), 917–928.

Yucel, N., Usta, A., Guzin, K., Kanter, M., Bilgic, E., Ozel, N.O., et al., 2013. Immunohistochemical analysis of connective tissue in patients with pelvic organ prolapse. Journal of Molecular Histology 44 (1), 97–102.

Zelzer, E., Blitz, E., Killian, M.L., Thomopoulos, S., 2014. Tendon-to-bone attachment: from development to maturity. Birth Defects Research Part C Embryo Today Reviews 102 (1), 101–112.

Zhao, B.H., Zhou, J.H., 2012. Decreased expression of elastin, fibulin-5 and lysyl oxidase-like 1 in the uterosacral ligaments of postmenopausal women with pelvic organ prolapse. Journal of Obstetrics and Gynaecology Research 38 (6), 925–931.

Zong, W., Stein, S.E., Starcher, B., Meyn, L.A., Moalli, P.A., 2010. Alteration of vaginal elastin metabolism in women with pelvic organ prolapse. Obstetrics and Gynecology 115, 953–961. Available from: http://links.lww.com/AOG/A172.

CHAPTER

9 Mechanical properties of pelvic implants: interaction between implants and tissue

Barbara Röhrnbauer

Zurich University of Applied Sciences, Institute of Mechanical Systems, Biomechanical Engineering, Winterthur, Switzerland

9.1 Motivation

9.1.1 Current issues with prosthetic meshes for pelvic organ prolapse repair

Vaginal meshes, hazard or cure? This is the topic of an ongoing discussion in the scientific community, e.g. in Barber (2015); Mancuso et al. (2020); Mironska et al. (2019). In fact, during the last decade, severe and frequent complications, such as erosions, pain, and dyspareunia, in particular related to transvaginal placement of meshes led to a re-classification of transvaginal meshes into class III products in the US, thousands of lawsuits and in some countries, such as New Zealand, Australia, England, and Scotland, even to a ban of synthetic meshes (Barber, 2015; Campbell et al., 2018; Mancuso et al., 2020). But what are the causes for this escalation?

Synthetic meshes are applied to treat severe cases of pelvic organ prolapse (POP). Though not being a life-threatening disease, POP is known to strongly impair the quality of life. And the prevalence is high: A number as high as 50% of the female population is affected by this condition, the main risk factor being parity (Brandt et al., 2018; Maher et al., 2016; Olsen et al., 1997). Of those, up to 20% need surgical intervention (Chapple et al., 2017; Løwenstein et al., 2014; Olsen et al., 1997). In fact, this already high number can be expected to increase in the future as women strive for a more active lifestyle (Slieker-ten Hove et al., 2009). Applying traditional native tissue repair, recurrence rates are reported up to 30% (Maher and Baessler, 2006; Moore and Miklos, 2009). Driven by the success of hernia meshes and encouraged by simplified FDA approval through 510(k), between 2002 and 2008, the market was flushed by a large variety of meshes and easy-to-use mesh kits for pelvic floor repair. And in fact, mesh application showed a better anatomical outcome compared to native tissue repair (Maher et al., 2016; Moore and Miklos, 2009). But this initial success was soon overshadowed by an increasing number of severe complications, bringing to light the

Biomechanics of the Female Reproductive System: Breast and Pelvic Organs. https://doi.org/10.1016/B978-0-12-823403-7.00021-X

consequences of this rushed market introduction without sufficient evidence of safety and efficacy for pelvic floor applications. Lacking alternatives, international gynecological organizations have addressed this situation (Chapple et al., 2017). There is a consensus, that next to a critical review of indications for mesh implantation and experience of the surgeon, implant materials with carefully tailored mechanical properties will improve safety and performance of POP repair using meshes (Mancuso et al., 2020; Mangera et al., 2012; Paul et al., 2020; Roman et al., 2016). And in fact, there are positive indications in that direction: a recent European study carried out by specialists in the field and using a modern ultra-lightweight mesh in transvaginal technique reported good objective outcome with a low number of mesh-related problems (Brandt et al., 2019, 2018). Although more data are needed, this study might better reflect the current situation and provide motivation to deeper investigate the processes on anatomical and tissue level, particularly also in regard to mechanical interactions between implant and tissue.

9.1.2 Why mechanics matter

Prosthetic meshes have two major functions: (i) to reconstruct the anatomy by suspending organs in the pelvic floor, (ii) to support tissues to recreate tissue integrity. Both functions are mechanical in character. The corresponding outcome therefore depends on the mechanical behavior of the mesh. Moreover, these two functions are effective at different length scales: the structural length scale of the anatomy and the material length scale of the tissue. Requirements for the mechanical behavior are very different at these length scales. Structurally, the mesh should provide enough support to maintain long-term anatomical integrity, and it should be compliant to mimic the motion of the soft tissues in the pelvic floor. At the length scale of the underlying tissues, the mesh should favor tissue ingrowth and limit adverse local reactions, such as injuries and inflammations.

It is well known that next to the raw material and the topology of an implant, the mechanical properties influence the biological response. Cells react to mechanical stimuli (Stevens and George, 2005), which determine their proliferation, migration, and differentiation (Kennedy et al., 2017; Vogel, 2006). By building focal adhesions, cells attach to surfaces, e.g., to the filaments of a mesh. Through complex mechanisms of mechanosensing and mechanotransduction, changes in environmental stiffness are recognized by the cells and trigger certain behavior. The implant communicates with cells via its mechanical properties. Detailed knowledge of the mechanical implant behavior might allow for correlations with the specific cell response, which would be a valuable contribution towards bottom-up derived design criteria for implants. Mechanical properties influence cell response, and cell ingrowth, in turn, also influences the mechanical response of the implant. Tissue integrated between the filaments of a knitted mesh restricts their relative movement or the collapse of pores, changing its overall mechanical response.

Thus there are different mechanical functions of a prosthetic mesh and different mechanisms of mechanobiological or biomechanical interactions, which determine the in vivo processes following mesh implantation. Only by better understanding the interplay of mechanics and biology at the different length scales, the root causes for mesh-related complications can be identified, paving the way to a systematic design of new and mechanically biocompatible pelvic floor materials.

9.1.3 Hypotheses

Specifically, the following is hypothesized: (i) a lack of structural mechanical biocompatibility leads to an unfavorable anatomical outcome; (ii) a lack of mechanical biocompatibility at the length scales of the underlying tissues leads to relative movement between mesh and tissues, which might cause local inflammations, erosions, or wrinkle formation. None of these aspects can be examined in isolation as they strongly depend on each other. In addition, other factors, such as the patients' clinical condition, the surgeons' experience, or the implantation technique play an important role as these set the boundary conditions, also for the mechanical performance (Mangir et al., 2019).

9.1.4 Outline of this chapter

In this chapter, vaginal meshes and their potential interplay with biology shall be reviewed from a mechanical perspective. In the first section, meshes shall be introduced as a type of textile, featuring a hierarchical structure. The current state of the art regarding mechanical testing of pelvic floor and hernia meshes shall be reviewed in the following section. It shall be discussed with respect to physiological relevance, meaningful mechanical parameters, and the consideration of different length scales. Conclusions shall be sketched based on potential interactions with native tissue and on an ideal mesh design. A discussion on how numerical models could contribute to an efficient mechanical characterization of meshes and, ultimately, to an improved mesh design is presented in the subsequent section. Finally, the focus shall be shifted to alternative materials, which seem promising, also from a mechanical perspective.

Remark: Being aware that there is a range of materials available for pelvic floor implants, the reviews in this chapter (except for the last section) are mainly restricted to textile prosthetic meshes, knitted from polypropylene filaments.

9.2 Mechanical and physical properties of prosthetic meshes

9.2.1 Textile meshes as hierarchical structures

In the previous section, the multi-scale character of prosthetic meshes has been indicated, differentiating between the structural length scale, relating to the anatomy and the length scale of the underlying tissues. For textile prosthetic meshes, the latter can be subdivided into further hierarchy levels (Fig. 9.1), which are a result of the manufacturing process, e.g., knitting. At the macro-scale, the textile sheet is

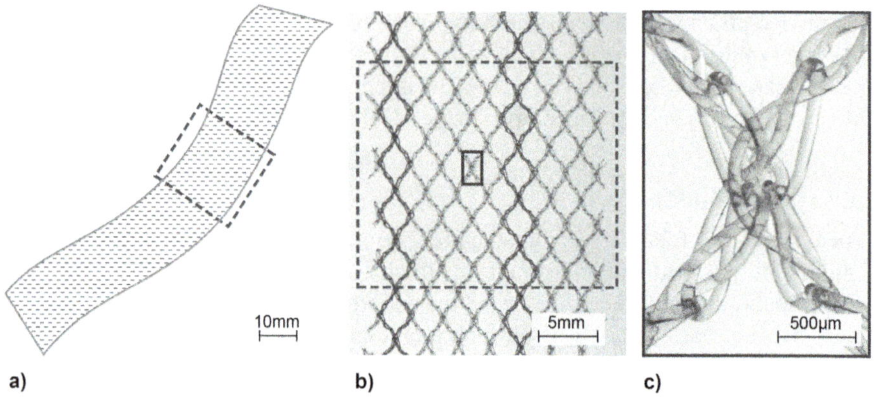

FIGURE 9.1

Hierarchical structure of a prosthetic mesh (exemplarily shown for Ultrapro, Ethicon Johnson&Johnson) a) Macro-scale: membrane-like structure, b) Meso-scale: two-dimensional periodic pattern of pores, c) Micro-scale: three-dimensional inter-looping of filaments.

a two-dimensional structure, flexible in bending and with soft in-plane properties (Fig. 9.1a). At the meso-scale, a two-dimensional periodic pattern of pores determines the mechanical behavior by mechanism-like collapse and expansion of these pores (Fig. 9.1b). And finally, at the microscale, the mechanical behavior is determined by the complex, three-dimensional inter-looping of filaments, which are highly-curved, beam-like structures in frictional contact with each other (Fig. 9.1c). This hierarchical structure is the basis for understanding many of the intrinsic mechanical phenomena of prosthetic meshes in uniaxial and biaxial stress loading conditions reported in Röhrnbauer and Mazza (2014) and summarized in what follows.

9.2.2 Non-linear stress-strain relationship

A non-linear stress-strain relationship in uniaxial stress loading conditions is typical for most soft biological tissues subject to large deformations (Fung, 1993). Such behavior is also observed for knitted textiles (Fig. 9.2), which is why they seem to be ideal implant materials for soft tissue support. The non-linear stress-strain behavior of knits can be attributed to three major mechanisms; the collapse of the pores at the meso-scale leads to a geometry-based non-linear stress response. Initially, low stresses are required to rotate the strands until the pores are collapsed. Once the strands are aligned, the loops and knots formed by the filaments at the micro-scale are tightened. For this, higher stresses are required as the bending resistance of and the friction between the filaments must be overcome. Applying even higher stress levels, the material itself, e.g., the polypropylene is strained.

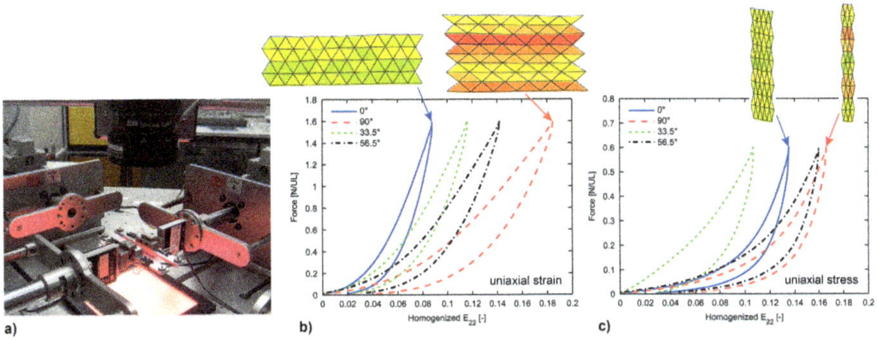

FIGURE 9.2

Mechanical behavior of a prosthetic mesh *From Röhrnbauer, B. & Mazza, E. (2014). Uniaxial and biaxial mechanical characterization of a prosthetic mesh at different length scales. Journal of the Mechanical Behavior of Biomedical Materials, 29, 7–19. https://doi.org/10.1016/j.jmbbm.2013.07.021.* a) Uniaxial test setup for prosthetic meshes showing a uniaxial strain, i.e. biaxial stress test, b-c) Macro-scale stress-strain responses in uniaxial strain (b) and uniaxial stress (c) loading conditions. Prosthetic meshes show a non-linear stress response, anisotropy and hysteresis. The curves show the average stress-strain relationship for different directions of load application (0°, 90°, 33.5°, 56.5°). The contour plots of the samples schematically indicate the deformed shape of the sample and the local strain distribution at the level of the pores.

9.2.3 Anisotropy

Due to the manufacturing technique, the knitting pattern features a certain directionality, shape, and orientation of pores. The material, or more precisely, the structure, is not symmetric with respect to any arbitrary direction, and neither are the material properties symmetric with respect to any arbitrary loading direction (Fig. 9.2). Depending on the manufacturing technique, immense differences are observed in the mechanical properties in the two principal in-plane directions (0°- and 90°-directions in Fig. 9.2). In addition, non-obvious effects might be observed for arbitrary off-axis directions, e.g., an unproportionate stiff stress response due to locking effects (33.5°-direction in Fig. 9.2c)), which will be further discussed later in this chapter (see also Fig. 9.4).

9.2.4 Viscoelasticity

The inter-looping of filaments and the frictional contacts between them, lead to dissipative behavior, i.e., hysteresis, observed in macroscopic mechanical tests (Fig. 9.2). Time-dependent effects, such as creep and relaxation can be associated with both alignment and re-arranging of filaments and visco-elastic properties of the polypropylene filaments themselves.

9.2.5 Preconditioning

Preconditioning in soft tissues and materials can be described as a right-shift of the stress-strain curve in cyclic loading conditions (Fung, 1993). At the macro-scale, preconditioning is typically experienced as softening, as the initial tangent stiffness is reduced during cyclic load. However, for knitted textiles, preconditioning effects could even be experienced as stiffening. The underlying mechanisms can be explained at the micro- and the meso-scale. Due to friction, the tightening of the loops and knots and, therefore, also, the collapse of the pores is only partly reversible. This leads to an elongation of the textile, which, in a mechanical test, is seen as a strip sagging between the clamps of the test machine. This results in a vanishing initial stiffness. However, once the textile strip is flattened, the stress response is significantly stiffer compared to the original configuration, as the filaments are already aligned, and the knots are already tightened. Therefore the mechanical behavior of textiles strongly depends on their strain history. It is important to realize that textiles subject to cyclic load, such as pelvic floor meshes, might show significantly different mechanical behavior compared to the original material without any strain history.

All these mechanical phenomena contribute to the mechanical behavior of textiles. They need to be considered when using textiles as implant materials. Correlating these with clinical experience might help to derive design criteria for improved materials, although the textile manufacturing process might only allow for limited adjustment options.

9.3 Experimental mechanical characterization of prosthetic meshes

9.3.1 Physiologically relevant test conditions

Mechanical testing has become an essential part when characterizing the performance of prosthetic meshes. Numerous studies applying uniaxial (Afonso et al., 2008; Ai et al., 2020; Barone et al., 2016; Bellón et al., 2009; Boukerrou et al., 2007; Deeken et al., 2011; Dietz et al., 2003; Edwards et al., 2013; Feola et al., 2013; Hernández-Gascón et al., 2011; Hilger et al., 2006; Hollinsky et al., 2008; Horbach et al., 2017; Jones et al., 2009; Krause et al., 2008; Morch et al., 2017; Ozog et al., 2011; Shepherd et al., 2012; Trabuco et al., 2007) or more recently also biaxial loading conditions (Deeken et al., 2011; Edwards et al., 2013; Eliason et al., 2011; Feola et al., 2013; Horan et al., 2009; Konerding et al., 2011; Lu et al., 2017; Ulrich et al., 2016) have been reported. Biaxial loading conditions can be obtained by (equi-)biaxial tension, inflation, or uniaxial strain tests (Fig. 9.3). Note: Uniaxial strain testing is often referred to as pure shear or strip biaxial testing. In the case of meshes, where the thickness is assumed to remain constant, a state of uniaxial strain is created.

Addressing different micro-structural mechanisms, the biaxial stress response of textiles, as for most soft tissues, differs significantly from the uniaxial stress response. Biaxial stiffnesses of prosthetic meshes have been observed to be by a factor of 2–4

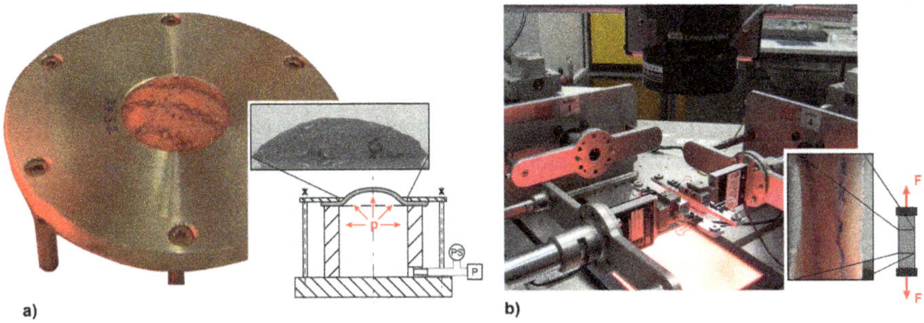

FIGURE 9.3

Different test setups for dry meshes and explants. a) Inflation test setup: Samples are inflated by pressure, creating a biaxial state of stress, shown for a mesh explant. b) Uniaxial test setup: The large photo shows a uniaxial strain test on a dry mesh, creating a biaxial state of stress. The smaller photo shows a uniaxial stress test on a mesh explant.

higher compared to uniaxial stiffnesses (Maurer et al., 2014; Röhrnbauer and Mazza, 2014). Moreover, anisotropic and preconditioning effects are more pronounced in uniaxial as compared to biaxial loading conditions. And finally, certain effects, such as locking, might only be seen in uniaxial but not in biaxial loading conditions. Thus the mechanical behavior strongly depends on the loading conditions. Only by applying physiologically relevant loading conditions, conclusions can be drawn regarding the mechanical biocompatibility of textile implant materials.

Reported test specimens are either dry meshes (Afonso et al., 2008; Barone et al., 2016; Deeken et al., 2011; Dietz et al., 2003; Edwards et al., 2013; Eliason et al., 2011; Feola et al., 2013, 2014; Hernández-Gascón et al., 2011; Hollinsky et al., 2008; Horbach et al., 2017; Jones et al., 2009; Krause et al., 2008; Lu et al., 2017; Röhrnbauer and Mazza, 2014; Shepherd et al., 2012; Ulrich et al., 2016), meshes embedded in an elastomeric matrix (Morch et al., 2019; Röhrnbauer and Mazza, 2013) or explants, i.e., the original implant and the surrounding ingrown host tissue, derived from animal models (Ai et al., 2020; Bellón et al., 2009; Boukerrou et al., 2007; Hilger et al., 2006; Horan et al., 2009; Konerding et al., 2011; Morch et al., 2017, 2021; Ozog et al., 2011; Röhrnbauer et al., 2013; Trabuco et al., 2007; Ulrich et al., 2016) (Fig. 9.3). Dry meshes represent the initial state of the mesh after implantation. They allow to study the kinematics and stress response of the mesh, which the underlying native tissue is exposed to directly after implantation. In contrast, embedded meshes or explants account for a "matrix" component, which blocks the pores and restricts their collapse. This condition mimics the ingrown in vivo state. Explants also take into account the in vivo immune reaction to the implant in the selected animal model. One major drawback of explants, however, is the typically small sample size and the large scatter introduced by the tissue component (Boukerrou et al., 2007; Ozog et al., 2011), which often impairs the informative value of a mechanical characterization. Addressing this situation, it has been shown that meshes embedded in an elastomeric matrix can effectively predict the stiffness response in biaxial inflation loading conditions (Röhrnbauer and Mazza, 2013). Such model systems, whenever applicable, allow to reduce the number of animal studies and at the same time increase the reliability of mechanical data. In conclusion, the test conditions, dry, embedded, or ingrown, influence the micro- and the mesostructure of the prosthetic mesh and, thus, determine the outcome of the mechanical test. For a meaningful interpretation of the outcome, the physiologically relevant condition for the intended study goals needs to be mimicked.

9.3.2 Mechanical behavior: a broad term

As one of the first groups, mechanically characterizing prosthetic meshes for hernia repair, Schumpelick et al. suggested a broad range of physical and mechanical properties, mainly inspired by the textile industry (Schumpelick et al., 1999). Based on safety considerations, rupture-related properties, such as material strength, seam tear-out force or maximum strain were focused on. Along the same line, many of the early experimental studies on prosthetic meshes, also for prolapse repair, concentrated on measuring material strength (Bellón et al., 2009; Deeken et al., 2011; Dietz et al., 2003; Eliason et al., 2011; Hilger et al., 2006; Hollinsky et al., 2008; Horan et al., 2009; Krause et al., 2008; Trabuco et al., 2007). These early tests and experiences showed that the material strength of synthetic meshes is by orders of magnitude higher compared to the expected physiological load range. In addition to providing enough strength, prosthetic meshes need to prevent stress shielding (Liang et al., 2016; Ozog et al., 2014). Therefore, more recently, it has been understood that the deformation behavior, i.e., the shape of the non-linear stress-strain curve is even more relevant for the performance of the meshes. In particular, the stiffness in the physiological load range has become a relevant parameter for mesh characterization (Afonso et al., 2008; Ai et al., 2020; Barone et al., 2016; Edwards et al., 2013; Feola et al., 2013, 2014; Hernández-Gascón et al., 2011; Jones et al., 2009; Konerding et al., 2011; Ozog et al., 2011; Shepherd et al., 2012). Alternatively, hyperelastic material model formulations have been fitted to experimental data, and the corresponding parameters have been used to quantify the mechanical behavior of explants (Morch et al., 2017, 2021). Ideally, the extracted values match the corresponding values of the native tissue. To make this comparison and thus evaluate this aspect of mechanical biocompatibility of the prosthetic mesh, a good estimate of the mechanical behavior of the native tissue in physiologically relevant loading conditions is needed.

As outlined in the previous section, in addition to the (uniaxial) elastic behavior, knitted polypropylene meshes feature a large range of other clinically relevant mechanical phenomena, such as strong preconditioning effects (Edwards et al., 2013; Jones et al., 2009; Krause et al., 2008; Liang et al., 2016; Shepherd et al., 2012; Ulrich et al., 2016), and anisotropy (Barone et al., 2019, 2016; Feola et al., 2013), which are less often characterized as standardized experimental protocols and parameters are missing. Moreover, their clinical impact is rarely discussed. Preconditioning causes irreversible deformations, which after a few cycles of load lead to a significant change of the structure and of the force response (Morch et al., 2021; Röhrnbauer and Mazza,

2014). In vivo, an initially perfectly positioned mesh could soon lose a significant part of its support function, and thus impair anatomic reconstruction. Anisotropy, if not considered, might lead to unexpected mechanical behavior in multi-axial loading conditions (Barone et al., 2019). In particular, pre-cut mesh systems could deform inhomogeneously, and thus in combination with irreversible preconditioning effects, could be re-positioned unfavorably. In Maurer et al. (2014), a comprehensive test protocol and a corresponding set of parameters for prosthetic meshes is suggested, which account for the different mechanical phenomena described above. Moreover, local deformation patterns are considered, which are discussed in more detail in the next section of this chapter.

In conclusion, for a mechanical characterization of prosthetic meshes, one strength or stiffness value, even if physiological (loading) conditions (dry or embedded, uniaxial or biaxial loading) are considered, is not sufficient. Other effects, such as preconditioning, and anisotropy are similarly important to more comprehensively judge the mechanical behavior of prosthetic meshes in vivo.

9.3.3 Mechanical behavior at the lower length scales

In the previous sections, the macro-scale mechanical properties have been discussed, i.e., the mechanical phenomena of the mesh as a membrane in different loading conditions. However, at lower length scales, the mesh is not a homogeneous membrane, and the observed mechanical behavior might differ significantly from what is seen at the macro-scale. Deformation patterns at the level of a pore or within a knot have been reported to be non-affine compared to the far-field macro-scale deformation (Maurer et al., 2014). Optically evaluated regions of local delamination between a prosthetic mesh and an elastomeric matrix have been proposed to describe the deformation mismatch between mesh and matrix. Going deeper into details, it has been shown that in uniaxial stress loading conditions, the collapse of pores is accompanied by large rotations of stitches (Röhrnbauer and Mazza, 2014) (Fig. 9.4a). Similarly, in Barone et al. (2019) and Horbach et al. (2017), it has been reported that macro-porous meshes might turn into unfavorable micro- or even non-porous structures as the pores collapse in uniaxial stress loading conditions. Then again, there are loading directions, where perpendicularly oriented stitches lock the transverse contraction, keeping pores open and local deformations restricted (Barone et al., 2016; Röhrnbauer and Mazza, 2014) (Fig. 9.4b).

These local, mechanism-like deformation patterns are supposed to cause relative movements between mesh and underlying tissue, even if the macro-scale deformation between mesh and tissue might match well. Relative movements can impair the ingrowth of cells and can lead to injuries or local inflammation. Moreover, mesh erosions might stem from these locally mismatching deformation behaviors. Currently, only a few studies consider the meso- and micro-scale mechanical behavior of meshes. However, whereas macro-scale deformations interact with the anatomy and the tissue as a continuum, lower scale deformations interact with the tissue microstructure at the cell level and directly impact the cell response. Therefore the

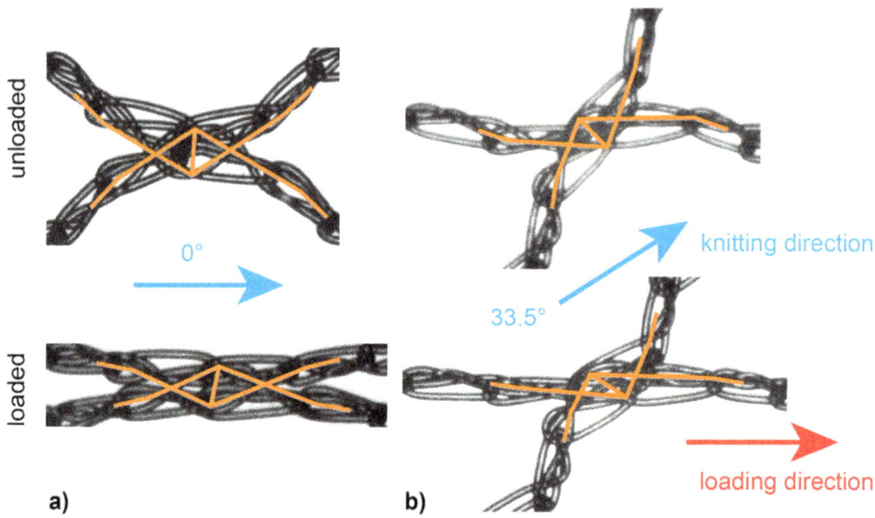

FIGURE 9.4

Local deformations of a prosthetic mesh (exemplarily shown for Ultrapro, Ethicon Johnson&Johnson) *Modified from Röhrnbauer, B., Kress, G. & Mazza, E. (2014). A physically based structural model for a textile prosthetic mesh. International Journal of Solids and Structures, 51(3–4), 633–646. https://doi.org/10.1016/j.ijsolstr.2013.10.030.* a) Rotation of stitches and collapse of pores in uniaxial stress loading conditions along the knitting direction of the mesh (0°-direction), b) Constrained transverse contraction by stitches oriented perpendicular to the loading direction for uniaxial stress loading conditions under 33.5° to the knitting direction.

concept of mechanical biocompatibility of pelvic floor implants must as well comprise these lower length scales.

9.3.4 Structural mechanical behavior

The previously described mechanical tests are suitable to characterize prosthetic meshes as micro-structure-based membranes. In contrast, structural tests characterize the mechanical response of the prosthetic mesh at the anatomical length scale. Structural tests consider the final configuration of the mesh, e.g., a mesh kit or a pre-cut strip in a spatial orientation representative for the implanted state in physiologically relevant loading and boundary conditions. In contrast to the previously described membrane tests, structural tests often lead to inhomogeneous states of deformation, and can bring to light unexpected phenomena, e.g., local reductions of pore size and increased material burden (Barone et al., 2019). And yet, structural tests are rarely performed, as test setups and the application of boundary conditions are complex to realize. Moreover, various implantation techniques and patient-specific load and boundary conditions make it difficult to define representative standards. Numerical simulations offer more flexibility, in terms of accounting for this variability, but also in terms of application of load and boundary conditions, which are difficult to realize experimentally. This topic is further elaborated on later in this chapter.

9.3.5 Local mesh-tissue interaction

The interaction between the local mechanical behavior of meshes and cells is rarely investigated explicitly. Animal studies and corresponding explant tests, as described before, account for the tissue component. However, so far, most studies focus on the evaluation of the macro-scale mechanical behavior and the overall host tissue response, and no explicit correlation is made between meso- or micro-scale mechanical behavior and cell response. From such investigations and collected patient outcomes, rough design guidelines for mesh porosity and weight, as described in the following sections, have been derived. On the other hand, tissue engineering approaches for pelvic floor implant materials, as outlined in the final section of this chapter, mainly grow cells in static conditions. There are only a few examples of dynamically stimulated cell cultivation and in these cases, the mechanical properties of the substrate material is not looked at in detail (Vashaghian et al., 2019). Thus, so far, mesh-tissue interaction is mainly limited to the cell response to the static mesh or its global mechanical behavior and detailed investigations of mesh mechanobiology are missing. Such investigations, however, might be interesting to formulate more fundamental and bottom-up justified criteria for mechanical biocompatibility.

9.3.6 Experimental observations: what can we learn for the mechanical evaluation and design of meshes?

From the large amount of experimental data available, the question arises: What can we learn from these experimental observations regarding mechanical evaluation and design of meshes? Without claiming completeness, the following points represent a summary of main findings:

- The non-linear stress-strain relationship requires the definition of relevant load cases and load levels for parameter evaluation. The mechanical behavior cannot be extrapolated or superposed from different loading conditions.
- Different loading conditions incite different mechanisms at the lower length scales, resulting in different macro-scale behavior. The biaxial initial stiffness of meshes is significantly higher than the uniaxial. Thus for a meaningful comparison of stiffness values, the tested loading conditions need to match the physiological ones.
- Due to the manufacturing process, textile meshes feature distinct anisotropy, which needs to be considered for the design of mesh implants and kits to ensure the anatomical support function. Structural tests might help to identify unexpected mechanical phenomena.
- Cyclic loading leads to inelastic deformations (preconditioning), changing the structure of the mesh, in particular the size of its pores, drastically. Optimally, the

mesh design aims for a stable mesh configuration. Alternatively, the instability of shape might be considered by suitable procedures before mesh implantation.

- Although the macro-scale mechanical behavior of the mesh and the underlying tissue might match, non-affine meso- and micro-scale deformations might cause local deformation mismatches, which might impair tissue ingrowth. Thus mechanical biocompatibility needs to be evaluated at different length scales.

9.3.7 What would an ideal mesh look like?

From the previously summarized mechanical aspects, it can be argued which mechanical behavior or phenomenon might cause problems, but it is much more challenging to state what the ideal implant material should look like. Nevertheless, attempts can be made to derive design criteria that have the potential to render a mesh mechanically biocompatible. In the past, meshes have not been designed to match the specific needs of pelvic floor tissues and structures. They have mainly been copied from meshes used for hernia repair. To derive design guidelines, the concept of mechanical biocompatibility has been coined, and comprehensively outlined in Mazza and Ehret (2015). It demands similar mechanical behavior of implant materials compared to the host tissue. During the last decade, a common understanding has developed that monofilament, large porous (> 1500 μm), ultra-light weight (18–42 g/m^2) and high-porous (> 55%) meshes are preferable. These are known to yield a favorable host tissue response in terms of tissue ingrowth, vascularization, and reduced inflammation (Chapple et al., 2017; Liang et al., 2016; Lu et al., 2017; Mangir et al., 2018). However, they are often also correlated with low tissue-like stiffness (Feola et al., 2013; Ozog et al., 2011), which is not generally the case, as has been shown in Maurer et al. (2015). In this study, nine different prosthetic meshes used for hernia and/or pelvic floor repair have been mechanically tested aiming at a comprehensive set of mechanical and physical parameters and corresponding criteria describing their mechanical biocompatibility. Different independent and uncorrelated aspects, such as uniaxial and biaxial stiffness, anisotropy, the influence of prior deformation history (preconditioning), local mismatch of deformation mechanisms, and changes in mechanical properties when embedded in a homogeneous matrix have been addressed. According to these criteria, potentially favorable meshes could be identified, but clinical evidence is needed to confirm the proposed hypotheses.

Whereas the design of textile meshes is limited by the manufacturing technique, current trends to improve the critical situation in prolapse repair follow different directions: On the one hand, available meshes are modified by adding resorbable components, an embedding matrix, or coatings (Dias et al., 2015; Edwards et al., 2013). These measures shall improve biocompatibility and the host immune response by reducing material burden or by delivering substances. On the other hand, alternative manufacturing techniques, such as 3D printing or electrospinning have been proposed offering greater freedom of design (see final section of this chapter). Though initial animal studies with these new materials report promising outcomes in terms of tissue integration and host immune response, detailed mechanical investigations, in particular covering meso- and micro-scale behavior are still missing. The development towards alternative materials is further reviewed in the last section of this chapter. Regarding textile meshes, the question is justified whether there can be an ideal mesh design at all or whether there are other materials which intrinsically feature a better mechanical biocompatibility.

9.4 Numerical modeling of pelvic implants

Generally, the value of numerical simulations, in particular, in combination with experimental studies is known, but is rarely applied for prosthetic mesh design. Numerical models allow to investigate local and three-dimensional quantities, such as stresses and strains, which cannot be easily assessed experimentally. Moreover, they allow for efficient design optimization through virtual prototyping. As described in the previous sections, prosthetic meshes are multi-scale materials, which must be considered by the modeling approach. In the following paragraphs, different modeling techniques are reviewed and discussed with respect to the gained insight into the mechanical behavior of prosthetic meshes and their potential interaction with anatomy and tissues.

9.4.1 Structural simulations

Structural simulations of the whole pelvic floor have been performed to study the pathogenesis of POP and to quantify the influence of parameters, such as abdominal pressure, stiffness of ligaments, and muscles (Babayi et al., 2019; Brandão et al., 2017; Gordon et al., 2019; Jeanditgautier et al., 2016; Mayeur et al., 2016; Silva et al., 2017). Moreover, they are used to analyze surgical procedures (Bhattarai et al., 2018; Brandão et al., 2017). On the one hand, creating a representative geometry that allows for studying general mechanisms valid for a large population is a challenge. On the other hand, patient-specific geometry models require cost and time-efficient procedures (Jiang et al., 2019). The patient-specific surgery planning seems to be a future perspective, including the choice of an appropriate or even patient-specific mesh (Bhattarai et al., 2018; Brandão et al., 2017; Jiang et al., 2019; Silva et al., 2017). When modeling prosthetic meshes in the implanted state, it is common practice to use hyperelastic continuum model formulations, e.g., according to Veronda–Westman (Brandão et al., 2017), Mooney–Rivlin, Neo-Hooke, or Yeoh (Silva et al., 2017; Mayeur et al., 2016; Jeanditgautier et al., 2016). The material constants are typically derived from the macro-scale elastic force response in uniaxial loading conditions. The biaxial force response or the kinematic response are not necessarily captured by this approach. Moreover, such phenomenological continuum models do not provide any link to the microstructure of the mesh. Therefore the described approach is useful to investigate, e.g., the elastic suspension function of implanted mesh strips in sacrocolpopexy. However, simulating multi-axial loading conditions, as seen for mesh sheets, would require parameters fitted to a broader experimental basis. Moreover, phenomenological, structural simulations do not allow for drawing conclusions regarding mesh-tissue or even mesh-cell interactions. Such investigations would demand more detailed model formulations, e.g., microstructure inspired continuum approaches or multi-scale approaches.

9.4.2 Meso-/microstructure-inspired modeling approaches

In Feola et al. (2014) and Barone et al. (2019), different PP meshes were modeled as continua with cut-out holes (i.e., pores) and varying mechanical properties for knot and fiber regions. Thus a mesostructural geometry was used with homogenized material properties of the filament loops. This modeling approach allowed performing structural simulations of whole mesh kits, quantifying experimentally observed local effects, such as pore collapse or increased material burden in multiaxial loading conditions. Chanda et al. used a similar approach, to investigate local effects of suturing (Chanda et al., 2018). Alternatively, continuum model formulations for macrostructural geometries, i.e., membranes, have been suggested for dry meshes and explants accounting for meso- and microstructural properties in a homogenized manner. Horbach et al. (2017) suggested a hyperelastic continuum approach, linking to porosity as a local effect described by the ratio of cross-sectional area occupied by the mesh as a whole and by the fibers, respectively. Morch et al. (2019) proposed a model for mesh tissue composites based on a microsphere approach, originally developed for micro-mechanically based rubber-like materials (Miehe et al., 2004). Similar models have been developed and applied for fibrous soft tissues (e.g., Kuhl et al., 2005). Thus anisotropies and changing fiber orientation can be considered in addition to the non-linear elastic stress-strain response. Such meso- and microstructure-inspired model formulations are of significant value, as they allow correlating meso- and micro-structural effects with the macro-scale and even structural behavior efficiently. Computational efforts and procedures of implementation are still moderate, although considerably higher compared to standard hyperelastic model formulations. However, these modeling approaches are not bottom-up approaches and abstract certain properties, such as porosity or fiber alignment in a homogenized manner. Therefore detailed investigations of mesh-cell interactions are not possible.

9.4.3 Meso-/microstructural and multi-scale models

In textile research, micro- and mesostructural modeling techniques based on bottom-up unit cell approaches are used to study local effects, such as crimp exchange, shear locking, and yarn slip. They are mainly applied for weaves (Assidi et al., 2011; Ben Boubaker et al., 2007; Carvelli, 2009; Grujicic et al., 2009; King et al., 2005; Nadler et al., 2006a; Nayfeh and Kress, 1997; Xiao et al., 2011). The basic idea behind such models is to build a simplified but physical micro- or meso-scale structure to capture mechanical phenomena at the higher length scales. Ideally, the mechanical properties of the micro- or meso-scale elements are directly measured in corresponding experiments or deduced from manufacturer specifications. This approach has also been applied for prosthetic meshes (Röhrnbauer et al., 2014), although it is rarely used for knitted textiles. The unit cell geometry for knits is generally more complex

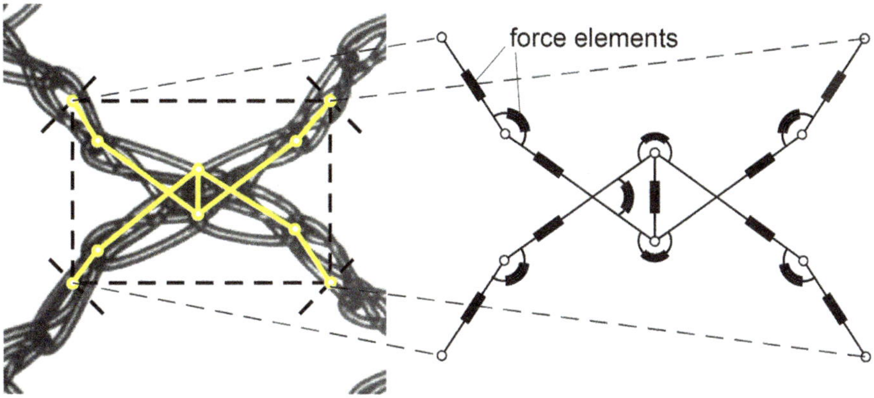

FIGURE 9.5

Structural unit cell model of a knitted prosthetic mesh *Modified from Röhrnbauer, B., Kress, G. & Mazza, E. (2014). A physically based structural model for a textile prosthetic mesh. International Journal of Solids and Structures, 51(3–4), 633–646. https://doi.org/10.1016/j.ijsolstr.2013.10.030.* The stitches and knots are abstracted by tension/compression and rotational force elements.

than for weaves and the mechanical properties of the meso-structural components, e.g., the strands of filament loops themselves feature non-linear mechanical behavior (Fig. 9.5).

Structural unit cell models show excellent descriptive capabilities. They map the global non-linear force response due to geometric and material non-linearities and the local kinematics, e.g., large Poisson ratios determined by the collapse of pores (Assidi et al., 2011; Ben Boubaker et al., 2007; King et al., 2005; Nadler et al., 2006a; Nayfeh and Kress, 1997). Moreover, they intrinsically account for anisotropy through the unit cell topology (Assidi et al., 2011; King, 2006; Nadler et al., 2006a). In fact, these models are typically also predictive for new loading conditions. Micro- and meso-structural models can be used to investigate the interaction between the mesh and the underlying tissue or cells. The amount of relative motion between mesh and tissue or stress concentrations due to restricted relative motion, which might be associated with local complications such as mesh erosion, can be investigated. Thus even quantitative and conclusive criteria for the local mechanical biocompatibility of a textile implant, e.g., based on the amount of non-affine deformations can be derived (Röhrnbauer et al., 2014). Moreover, these models represent excellent tools for virtual prototyping and design optimization. To use these models for structural simulations, which can link the global structural behavior to the meso- or micro-scale design, multi-scale homogenization techniques are needed. Homogenized constitutive model formulations have to be derived and implemented, e.g., in finite element codes (Antonietti et al., 2011; Grujicic et al., 2009; King, 2006; Nadler et al., 2006b). However, this step is not straightforward, requiring mathematical techniques and computational power. Such multi-scale constitutive model formulations would allow drawing conclusions from

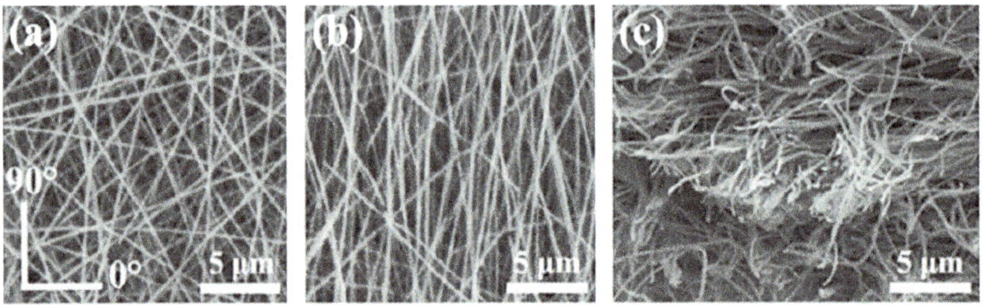

FIGURE 9.6

Microstructure of electrospun networks *From Tonsomboon, K. & Oyen, M. (2013). Composite electrospun gelatin fiber-alginate gel scaffolds for mechanically robust tissue engineered cornea. Journal of the Mechanical Behavior of Biomedical Materials, 21, 185–194. https://doi.org/10.1016/j.jmbbm.2013.03.001*. SEM of electrospun gelatin fibers obtained from a) a static collector, b) a mandrel at 3100 RPM, and c) SEM of the cross-section of randomly oriented gelatin fibers.

macroscopic, structural simulations, e.g., representing different implantation techniques or even patient-specific anatomies for an improved micro-/meso-scale mesh design.

In conclusion, numerical models can help to investigate the interaction of prosthetic meshes with the anatomy, the underlying tissues or even cells. The selection of the modeling approach needs to consider the mechanical phenomena and the length scales, which shall be accounted for and the corresponding efforts in terms of implementation and computational power.

9.5 Alternative materials: electrospun networks

As indicated earlier in this chapter, alternative manufacturing technologies might lead to more mechanically biocompatible structures compared to textile meshes. Due to the urgent need for less risk-bearing pelvic floor materials, in recent years different biomaterials have been suggested to replace conventional meshes (Dias et al., 2015). Additive manufacturing techniques offer new dimensions of tailoring the topology, and thus also the mechanical properties. Though 3D-printed materials are still in their infancy (Farmer et al., 2020, 2021; Paul et al., 2020), in recent years, different leading research groups in the field started investigating electrospun networks (ESNs) as alternative materials for pelvic floor repair (Glindtvad et al., 2018; Hansen et al., 2020; Hympanova et al., 2018, 2017; Mori da Cunha et al., 2020, 2019; Paul et al., 2019; Roman et al., 2016; Vashaghian et al., 2018, 2019, 2017).

ESNs are created by spinning thin fibers from a polymeric solution driven by electrostatic forces (Fig. 9.6). They offer enormous design freedom by varying (i) raw material; (ii) diameter, density, and orientation of the fibers; and (iii) shape, size, and distribution of pores. Due to their nanofibrous structure, ESNs were observed to excellently mimic extracellular matrix, and thus favor cell ingrowth and vascularization (Mauck et al., 2009; Stella et al., 2010a). For these reasons, ESNs have become common scaffold materials for tissue engineering, e.g., for pulmonary valve leaflets (Stella et al., 2010a), and have also been suggested as implant materials for hernia repair (Mazza and Ehret, 2015). First in vitro (Vashaghian et al., 2019, 2017) and in vivo animal studies using ESNs in an abdominal wall (Glindtvad et al., 2018; Hansen et al., 2020; Hympanova et al., 2018, 2017; Mori da Cunha et al., 2020, 2019; Paul et al., 2019; Roman et al., 2016) or a vaginal model (Hympánová et al., 2020) have been reported, evaluating the suitability of ESNs also for POP repair. Applying different base materials and different microstructures, ESNs have been observed to lead to good tissue integration in vitro and in vivo and to a good to moderate inflammatory immune response in vivo. Moreover, functionalization of the ESNs and/or tissue engineering approaches have been applied with promising outcomes (Glindtvad et al., 2018; Mori da Cunha et al., 2020; Paul et al., 2019). Detailed mechanical investigations, however, or approaches to specifically tailor mechanical properties for pelvic floor applications have not been reported yet.

In general, ESN tailoring and mechanical characterization has been studied extensively (Mauck et al., 2009) and is still a topic of current research. Given the enormous freedom of design, criteria are needed to tune the process parameters towards ESNs with the desired physical and mechanical properties. Comparable to textile meshes, the complex hierarchical structure of ESNs needs to be considered, requiring knowledge of the deformation mechanisms at different length scales. Very fundamental studies have been performed by Stella et al. (2010a,b, 2008) quantifying the microstructural kinematics of ESNs and their interaction with ingrown cells. Similar to textile meshes, also for ESNs, non-affine deformation patterns have been identified, leading to differences between the global deformation and local deformations as experienced by the cells. Along the same line, Zündel et al. (2019) introduced a multi-scale stiffness factor derived by numerical simulation of ESNs, quantifying the differences in stiffness at the macro-scale and local stiffnesses at the length scale of the cells. Taking into account such multi-scale mechanical considerations to design ESNs for pelvic floor repair will result in bottom-up tailored implant materials with improved implant-cell interaction and improved mechanical biocompatibility.

For current synthetic meshes, early tests and experience showed that static or dynamic rupture, i.e., fatigue, are not the main modes of failure. It is rather a mismatch of deformation patterns that is assumed to cause complications, such as erosion, wrinkling, or pain. However, emerging into new manufacturing technologies and alternative materials, all potential modes of failure, including rupture and a lack of mechanical biocompatibility, need to be considered. Thus a mechanical failure mode analysis accompanied by extensive mechanical testing is required to mechanically qualify new materials for pelvic floor repair.

9.6 What are the needs?

From a mechanical perspective, pelvic organ prolapse is a multi-scale problem, and the same is true for the treatment options, such as mesh implants. Therefore, an understanding of the mechanical behavior of the pelvic floor and according implants at all length scales is crucial and at the same time complex to achieve. At the structural level, more research is required to understand the patient-specific pelvic floor disorders and the required mechanical support structures to restore the anatomy. Moreover, the patients' subjective perception of symptoms is rarely looked at in engineering disciplines as it is difficult to objectify. However, it might be related to the mechanical behavior of the implant and might lead to paths beyond a pure reconstruction of the anatomy. At the material macro-scale level, the objective to create implants with a similar mechanical response compared to native tissue, requires good knowledge of the targeted tissue. This might vary with patient-specific implantation sites. Its mechanical behavior, which is subject to in vivo loading conditions, is still a major challenge to measure. Sophisticated ex vivo or even in vivo measurement techniques are needed to capture the physiologically relevant mechanical response of the native tissue of interest. Finally, at the meso- or micro-scale level, the length scale of the cells, more knowledge is needed about the complex mechanobiology between implant material and cells. Such insights could explain the causes for the experienced severe complications and allow for improved material designs.

Whereas these topics are part of current research activities, there is a need for pragmatic guidelines and criteria to better judge current prosthetic meshes also from a mechanical perspective. For most medical devices featuring mechanical functionality, such as orthopedic or dental implants, standardized procedures exist to prove mechanical safety and efficacy. However, such standards are still missing for prosthetic meshes (Maurer et al., 2014; Whitehead-Clarke et al., 2022). Whitehead-Clarke et al. (2022) just recently, reviewed the state of the art regarding in vivo testing of hernia meshes and found significant inconsistency in methods and comparators. Clear guidelines on minimum requirements regarding the analysis of mechanical biocompatibility of prosthetic meshes is supposed to help industry to develop safer and more effective implant materials, while research aims at alternative design and manufacturing techniques for the implant materials of the future.

References

Afonso, J.S., Martins, P.A.L.S., Girao, M.J.B.C., Natal Jorge, R.M., Ferreira, A.J.M., Mascarenhas, T., Fernandes, A.A., Bernardes, J., Baracat, E.C., Rodrigues de Lima, G., Patricio, B., 2008. Mechanical properties of polypropylene mesh used in pelvic floor repair. International Urogynecology Journal 19 (3), 375–380. https://doi.org/10.1007/s00192-007-0446-1.

Ai, F.F., Mao, M., Zhang, Y., Kang, J., Zhu, L., 2020. Experimental study of a new original mesh developed for pelvic floor reconstructive surgery. International Urogynecology Journal 31 (1), 79–89. https://doi.org/10.1007/s00192-019-03947-4.

Antonietti, P.F., Biscari, P., Tavakoli, A., Verani, M., Vianello, M., 2011. Theoretical study and numerical simulation of textiles. Applied Mathematical Modelling 35 (6), 2669–2681. https://doi.org/10.1016/j.apm.2010.11.062.

Assidi, M., Ben Boubaker, B., Ganghoffer, J.F., 2011. Equivalent properties of monolayer fabric from mesoscopic modelling strategies. International Journal of Solids and Structures 48 (20), 2920–2930. https://doi.org/10.1016/j.ijsolstr.2011.06.010.

Babayi, M., Azghani, M.R., Hajebrahimi, S., Berghmans, B., 2019. Three-dimensional finite element analysis of the pelvic organ prolapse: a parametric biomechanical modeling. Neurourology and Urodynamics 38 (2), 591–598. https://doi.org/10.1002/nau.23885.

Barber, M.D., 2015. Mesh use in surgery for pelvic organ prolapse. BMJ. British Medical Journal 350 (June), 1–2. https://doi.org/10.1136/bmj.h2910.

Barone, W.R., Knight, K.M., Moalli, P.A., Abramowitch, S.D., 2019. Deformation of transvaginal mesh in response to multiaxial loading. Journal of Biomechanical Engineering 141 (2), 1–8. https://doi.org/10.1115/1.4041743.

Barone, W.R., Moalli, P.A., Abramowitch, S.D., 2016. Textile properties of synthetic prolapse mesh in response to uniaxial loading. American Journal of Obstetrics and Gynecology September, 326.e1–326.e9. https://doi.org/10.1097/01.numa.0000435373.80608.40.

Bellón, J.M., Rodríguez, M., García-Honduvilla, N., Gómez-Gil, V., Pascual, G., Buján, J., 2009. Comparing the behavior of different polypropylene meshes (heavy and lightweight) in an experimental model of ventral hernia repair. Journal of Biomedical Materials Research. Part B, Applied Biomaterials 89 (2), 448–455. https://doi.org/10.1002/jbm.b.31234.

Ben Boubaker, B., Haussy, B., Ganghoffer, J.F., 2007. Discrete woven structure model: yarn-on-yarn friction. Comptes Rendus. Mécanique 335 (3), 150–158. https://doi.org/10.1016/j.crme.2007.02.006.

Bhattarai, A., Jabbari, M., Anding, R., Staat, M., 2018. Surgical treatment of vaginal vault prolapse using different prosthetic mesh implants: a finite element analysis. Technisches Messen 85 (5), 331–342.

Boukerrou, M., Boulanger, L., Rubod, C., Lambaudie, E., Dubois, P., Cosson, M., 2007. Study of the biomechanical properties of synthetic mesh implanted in vivo. European Journal of Obstetrics & Gynecology and Reproductive Biology 134 (2), 262–267. https://doi.org/10.1016/j.ejogrb.2007.02.023.

Brandao, S., Parente, M., Da Roza, T.H., Silva, E., Ramos, I.M., Mascarenhas, T., Jorge, R.M.N., 2017. On the stiffness of the mesh and urethral mobility: a finite element analysis. Journal of Biomechanical Engineering 139 (8), 1–9. https://doi.org/10.1115/1.4036606.

Brandt, A., Kuszka, A., Niesel, A., Lutz, H., Fünfgeld, C., Mengel, M., Ulrich, D., 2019. 1-year outcome after treatment of uterovaginal prolapse with a 6-point fixation mesh. Neurourology and Urodynamics 38 (4), 1129–1134. https://doi.org/10.1002/nau.23968.

Brandt, A., Ulrich, D., Kuszka, A., Niesel, A., Lutz, H., Fünfgeld, C., Kieback, D.G., 2018. Treatment of pelvic organ prolapse using a lightweight modified HexaPro mesh. Pelviperineology 37, 39–41. https://www.pelviperineology.org/june-2018/pdf/pelvic-organ-prolapse-treatment-lightweight-hexapro-mesh.pdf.

Campbell, P., Jha, S., Cutner, A., 2018. Vaginal mesh in prolapse surgery. The Obstetrician & Gynaecologist 20 (1), 49–56. https://doi.org/10.1111/tog.12454.

Carvelli, V., 2009. Monofilament technical textiles: an analytical model for the prediction of the mechanical behaviour. Mechanics Research Communications 36 (5), 573–580. https://doi.org/10.1016/j.mechrescom.2009.02.006.

Chanda, A., Ruchti, T., Upchurch, W., 2018. Biomechanical modeling of prosthetic mesh and human tissue surrogate interaction. Biomimetics 3 (3), 1–16. https://doi.org/10.3390/biomimetics3030027.

Chapple, C., Cruz, F., Deffieux, X., Milani, A.L., Arlandis, S., Artibani, W., Bauer, R.M., Brukhard, F., Cardozo, L., Castro-Diaz, D., Cornu, J.N., Deprest, J., Gunneman, A., Gyhagen, M., Heesakkers, J., Koelbl, H., MacNeil, S., Naumann, G., Roovers, J.-P.W.R., et al., 2017. Re: consensus statement of the European urology association and the European urogynaecological association on the use of implanted materials for treating pelvic organ prolapse and stress urinary incontinence. European Urology, 424–431. https://doi.org/10.1016/j.juro.2018.03.068.

Deeken, C.R., Abdo, M.S., Frisella, M.M., Matthews, B.D., 2011. Physicomechanical evaluation of polypropylene, polyester, and polytetrafluoroethylene meshes for inguinal hernia repair. Journal of the American College of Surgeons 212 (1), 68–79. https://doi.org/10.1016/j.jamcollsurg.2010.09.012.

Dias, F.G.F., Fernandes Dias, P.H.G., Prudente, A., Riccetto, C., 2015. New strategies to improve results of mesh surgeries for vaginal prolapses repair - an update. International Brazilian Journal of Urology 41 (4), 623–634. https://doi.org/10.1590/S1677-5538.IBJU.2014.0163.

Dietz, H.P., Vancaillie, P., Svehla, M., Walsh, W., Steensma, A.B., Vancaillie, T.G., 2003. Mechanical properties of urogynecologic implant materials. International Urogynecology Journal 14 (4), 239–243. https://doi.org/10.1007/s00192-003-1041-8.

Edwards, S.L., Werkmeister, J.A., Rosamilia, A., Ramshaw, J.A.M., White, J.F., Gargett, C.E., 2013. Characterisation of clinical and newly fabricated meshes for pelvic organ prolapse repair. Journal of the Mechanical Behavior of Biomedical Materials 23, 53–61. https://doi.org/10.1016/j.jmbbm.2013.04.002.

Eliason, B.J., Frisella, M.M., Matthews, B.D., Deeken, C.R., 2011. Effect of repetitive loading on the mechanical properties of synthetic hernia repair materials. Journal of the American College of Surgeons 213 (3), 430–435. https://doi.org/10.1016/j.jamcollsurg.2011.05.018.

Farmer, Z.L., Domínguez-Robles, J., Mancinelli, C., Larrañeta, E., Lamprou, D.A., 2020. Urogynecological surgical mesh implants: new trends in materials, manufacturing and therapeutic approaches. International Journal of Pharmaceutics 585 (March), 119512. https://doi.org/10.1016/j.ijpharm.2020.119512.

Farmer, Z.L., Utomo, E., Domínguez-Robles, J., Mancinelli, C., Mathew, E., Larrañeta, E., Lamprou, D.A., 2021. 3D printed estradiol-eluting urogynecological mesh implants: influence of material and mesh geometry on their mechanical properties. International Journal of Pharmaceutics 593, 120145. https://doi.org/10.1016/j.ijpharm.2020.120145.

Feola, A., Barone, W., Moalli, P., Abramowitch, S., 2013. Characterizing the ex vivo textile and structural properties of synthetic prolapse mesh products. International Urogynecology Journal and Pelvic Floor Dysfunction 24 (4), 559–564. https://doi.org/10.1007/s00192-012-1901-1.

Feola, A., Pal, S., Moalli, P., Maiti, S., Abramowitch, S., 2014. Varying degrees of nonlinear mechanical behavior arising from geometric differences of urogynecological meshes. Journal of Biomechanics 47 (11), 2584–2589. https://doi.org/10.1016/j.jbiomech.2014.05.027.

Fung, Y.C., 1993. Biomechanics - Mechanical Properties of Living Tissues, 2nd ed. Springer.

Glindtvad, C., Chen, M., Vinge Nygaard, J., Wogensen, L., Forman, A., Danielsen, C.C., Taskin, M.B., Andersson, K.E., Axelsen, S.M., 2018. Electrospun biodegradable microfibers induce new collagen formation in a rat abdominal wall defect model: a possible treatment for pelvic floor repair? Journal of Biomedical Materials Research. Part B, Applied Biomaterials 106 (2), 680–688. https://doi.org/10.1002/jbm.b.33875.

Gordon, M.T., DeLancey, J.O.L., Renfroe, A., Battles, A., Chen, L., 2019. Development of anatomically based customizable three-dimensional finite-element model of pelvic floor support system: POP-Sim1.0. Interface Focus 9 (4). https://doi.org/10.1098/rsfs.2019.0022.

Grujicic, M., Bell, W.C., Arakere, G., He, T., Cheeseman, B.A., 2009. A meso-scale unit-cell based material model for the single-ply flexible-fabric armor. Materials & Design 30 (9), 3690–3704. https://doi.org/10.1016/j.matdes.2009.02.008.

Hansen, S.G., Taskin, M.B., Chen, M., Wogensen, L., Vinge Nygaard, J., Axelsen, S.M., 2020. Electrospun nanofiber mesh with fibroblast growth factor and stem cells for pelvic floor repair. Journal of Biomedical Materials Research. Part B, Applied Biomaterials 108 (1), 48–55. https://doi.org/10.1002/jbm.b.34364.

Hernández-Gascón, B., Peña, E., Melero, H., Pascual, G., Doblaré, M., Ginebra, M.P., Bellón, J.M., Calvo, B., 2011. Mechanical behaviour of synthetic surgical meshes: finite element simulation of the herniated abdominal wall. Acta Biomaterialia 7 (11), 3905–3913. https://doi.org/10.1016/j.actbio.2011.06.033.

Hilger, W.S., Walter, A., Zobitz, M.E., Leslie, K.O., Magtibay, P., Cornella, J., 2006. Histological and biomechanical evaluation of implanted graft materials in a rabbit vaginal and abdominal model. American Journal of Obstetrics and Gynecology 195 (6), 1826–1831. https://doi.org/10.1016/j.ajog.2006.07.006.

Hollinsky, C., Sandberg, S., Koch, T., Seidler, S., 2008. Biomechanical properties of lightweight versus heavyweight meshes for laparoscopic inguinal hernia repair and their impact on recurrence rates. Surgical Endoscopy and Other Interventional Techniques 22 (12), 2679–2685. https://doi.org/10.1007/s00464-008-9936-6.

Horan, R.L., Bramono, D.S., Stanley, J.R.L., Simmons, Q., Chen, J., Boepple, H.E., Altman, G.H., 2009. Biological and biomechanical assessment of a long-term bioresorbable silk-derived surgical mesh in an abdominal body wall defect model. Hernia 13 (2), 189–199. https://doi.org/10.1007/s10029-008-0459-9.

Horbach, A.J., Duong, M.T., Staat, M., 2017. Modelling of compressible and orthotropic surgical mesh implants based on optical deformation measurement. Journal of the Mechanical Behavior of Biomedical Materials 74 (March), 400–410. https://doi.org/10.1016/j.jmbbm.2017.06.012.

Hympanova, L., Mori da Cunha, M.G.M.C., Rynkevic, R., Wach, R.A., Olejnik, A.K., Dankers, P.Y.W., Arts, B., Mes, T., Bosman, A.W., Albersen, M., Deprest, J., 2018. Experimental reconstruction of an abdominal wall defect with electrospun polycaprolactone-ureidopyrimidinone mesh conserves compliance yet may have insufficient strength. Journal of the Mechanical Behavior of Biomedical Materials 88 (August), 431–441. https://doi.org/10.1016/j.jmbbm.2018.08.026.

Hympanova, L., Mori da Cunha, M.G.M.C., Rynkevic, R., Zündel, M., Gallego, M.R., Vange, J., Callewaert, G., Urbankova, I., Van der Aa, F., Mazza, E., Deprest, J., 2017. Physiologic musculofascial compliance following reinforcement with electrospun polycaprolactone-ureidopyrimidinone mesh in a rat model. Journal of the Mechanical Behavior of Biomedical Materials 74 (June), 349–357. https://doi.org/10.1016/j.jmbbm.2017.06.032.

Hympánová, L., Rynkevic, R., Román, S., Mori da Cunha, M.G.M.C., Mazza, E., Zündel, M., Urbánková, I., Gallego, M.R., Vange, J., Callewaert, G., Chapple, C., MacNeil, S., Deprest, J., 2020. Assessment of electrospun and ultra-lightweight polypropylene meshes in the sheep model for vaginal surgery. European Urology Focus 6 (1), 190–198. https://doi.org/10.1016/j.euf.2018.07.024.

Jeanditgautier, F., Mayeur, O., Brieu, M., Lamblin, G., Rubod, C., Cosson, M., 2016. Mobility and stress analysis of different surgical simulations during a sacral colpopexy, using a finite element model of the pelvic system. International Urogynecology Journal 27 (6), 951–957. https://doi.org/10.1007/s00192-015-2917-0.

Jiang, Z., Mayeur, O., Witz, J.F., Lecomte-Grosbras, P., Dequidt, J., Cosson, M., Duriez, C., Brieu, M., 2019. Virtual image correlation of magnetic resonance images for 3D geometric modelling of pelvic organs. Strain 55 (3), 1–17. https://doi.org/10.1111/str.12305.

Jones, K.A., Feola, A., Meyn, L., Abramowitch, S.D., Moalli, P.A., 2009. Tensile properties of commonly used prolapse meshes. International Urogynecology Journal 20 (7), 847–853. https://doi.org/10.1007/s00192-008-0781-x.

Kennedy, K.M., Bhaw-Luximon, A., Jhurry, D., 2017. Cell-matrix mechanical interaction in electrospun polymeric scaffolds for tissue engineering: implications for scaffold design and performance. Acta Biomaterialia 50, 41–55. https://doi.org/10.1016/j.actbio.2016.12.034.

King, M.J., 2006. A continuum constitutive model for the mechanical behavior of woven fabrics including slip and failure. PhD Thesis. https://doi.org/10.1016/j.ijsolstr.2004.10.030.

King, M.J., Jearanaisilawong, P., Socrate, S., 2005. A continuum constitutive model for the mechanical behavior of woven fabrics. International Journal of Solids and Structures 42 (13), 3867–3896. https://doi.org/10.1016/j.ijsolstr.2004.10.030.

Konerding, M.A., Bohn, M., Wolloscheck, T., Batke, B., Holste, J.L., Wohlert, S., Trzewik, J., Förstemann, T., Hartung, C., 2011. Maximum forces acting on the abdominal wall: experimental validation of a theoretical modeling in a human cadaver study. Medical Engineering & Physics 33 (6), 789–792. https://doi.org/10.1016/j.medengphy.2011.01.010.

Krause, H., Bennett, M., Forwood, M., Goh, J., 2008. Biomechanical properties of raw meshes used in pelvic floor reconstruction. International Urogynecology Journal 19 (12), 1677–1681. https://doi.org/10.1007/s00192-008-0711-y.

Kuhl, E., Garikipati, K., Arruda, E.M., Grosh, K., 2005. Remodeling of biological tissue: mechanically induced reorientation of a transversely isotropic chain network. Journal of the Mechanics and Physics of Solids 53 (7), 1552–1573. https://doi.org/10.1016/j.jmps.2005.03.002.

Liang, R., Knight, K., Abramowitch, S., Moalli, P., 2016. Exploring the basic science of prolapse meshes. Current Opinion in Obstetrics and Gynegology 28 (5), 413–419. https://doi.org/10.1016/j.physbeh.2017.03.040.

Løwenstein, E., Ottesen, B., Gimbel, H., 2014. Incidence and lifetime risk of pelvic organ prolapse surgery in Denmark from 1977 to 2009. International Urogynecology Journal and Pelvic Floor Dysfunction 26 (1), 49–55. https://doi.org/10.1007/s00192-014-2413-y.

Lu, Y., Dong, S.N., Chen, Y.C., Zhang, P.H., Tong, X.W., 2017. Fabrication, characterization and application of polypropylene macroporous mesh for repairing pelvic floor defects. Textile Research Journal 87 (7), 878–888. https://doi.org/10.1177/0040517516639833.

Maher, C., Baessler, K., 2006. Surgical management of anterior vaginal wall prolapse: an evidence based literature review. International Urogynecology Journal 17 (2), 195–201. https://doi.org/10.1007/s00192-005-1296-3.

Maher, C., Feiner, B., Baessler, K., Christmann-Schmid, C., Haya, N., Marjoribanks, J., 2016. Transvaginal mesh or grafts compared with native tissue repair for vaginal prolapse. Cochrane Database of Systematic Reviews 2016 (2). https://doi.org/10.1002/14651858.CD012079.

Mancuso, E., Downey, C., Doxford-Hook, E., Bryant, M.G., Culmer, P., 2020. The use of polymeric meshes for pelvic organ prolapse: current concepts, challenges, and future perspectives. Journal of Biomedical Materials Research. Part B, Applied Biomaterials 108 (3), 771–789. https://doi.org/10.1002/jbm.b.34432.

Mangera, A., Bullock, A.J., Chapple, C.R., MacNeil, S., 2012. Are biomechanical properties predictive of the success of prostheses used in stress urinary incontinence and pelvic organ prolapse? A systematic review. Neurourology and Urodynamics 31, 13–21. https://doi.org/10.1002/nau.

Mangir, N., Chapple, C.R., MacNeil, S., 2018. Synthetic materials used in the surgical treatment of pelvic organ prolapse: problems of currently used material and designing the ideal material. In: Pelvic Floor Disorders, pp. 35–55. Chapter 4.

Mangir, N., Dikici, B.A., Chapple, C.R., MacNeil, S., 2019. Landmarks in vaginal mesh development: polypropylene mesh for treatment of SUI and POP. Nature Reviews Urology 16 (11), 675–689. https://doi.org/10.1038/s41585-019-0230-2.

Mauck, R.L., Baker, B.M., Nerurkar, N.L., Burdick, J.A., Li, W.J., Tuan, R.S., Elliott, D.M., 2009. Engineering on the straight and narrow: the mechanics of nanofibrous assemblies for fiber-reinforced tissue regeneration. Tissue Engineering. Part B: Reviews 15 (2), 171–193. https://doi.org/10.1089/ten.teb.2008.0652.

Maurer, M.M., Röhrnbauer, B., Feola, A., Deprest, J., Mazza, E., 2014. Mechanical biocompatibility of prosthetic meshes: a comprehensive protocol for mechanical characterization. Journal of the Mechanical Behavior of Biomedical Materials 40, 42–58. https://doi.org/10.1016/j.jmbbm.2014.08.005.

Maurer, Manfred M., Röhrnbauer, B., Feola, A., Deprest, J., Mazza, E., 2015. Prosthetic meshes for repair of hernia and pelvic organ prolapse: comparison of biomechanical properties. Materials 8 (5), 2794–2808. https://doi.org/10.3390/ma8052794.

Mayeur, O., Witz, J.F., Lecomte, P., Brieu, M., Cosson, M., Miller, K., 2016. Influence of geometry and mechanical properties on the accuracy of patient-specific simulation of women pelvic floor. Annals of Biomedical Engineering 44 (1), 202–212. https://doi.org/10.1007/s10439-015-1401-9.

Mazza, E., Ehret, A.E., 2015. Mechanical biocompatibility of highly deformable biomedical materials. Journal of the Mechanical Behavior of Biomedical Materials 48, 100–124. https://doi.org/10.1016/j.jmbbm.2015.03.023.

Miehe, C., Göktepe, S., Lulei, F., 2004. A micro-macro approach to rubber-like materials - Part I: the non-affine micro-sphere model of rubber elasticity. Journal of the Mechanics and Physics of Solids 52 (11), 2617–2660. https://doi.org/10.1016/j.jmps.2004.03.011.

Mironska, E., Chapple, C.R., MacNeil, S., 2019. Recent advances in pelvic floor repair. F1000Research 8, 1–11. https://doi.org/10.12688/f1000research.18900.1.

Moore, R.D., Miklos, J.R., 2009. Vaginal mesh kits for pelvic organ prolapse, friend or foe: a comprehensive review. TheScientificWorldJournal 9, 163–189. https://doi.org/10.1100/tsw.2009.19.

Morch, A., Pouseele, B., Doucède, G., Witz, J.F., Lesaffre, F., Lecomte-Grosbras, P., Brieu, M., Cosson, M., Rubod, C., 2017. Experimental study of the mechanical behavior of an explanted mesh: the influence of healing. Journal of the Mechanical Behavior of Biomedical Materials 65, 190–199. https://doi.org/10.1016/j.jmbbm.2016.07.033.

Morch, Annie, Astruc, L., Witz, J.F., Lesaffre, F., Lecomte-Grosbras, P., Soulat, D., Brieu, M., 2019. Modeling of anisotropic hyperelastic heterogeneous knitted fabric reinforced composites. Journal of the Mechanics and Physics of Solids 127, 47–61. https://doi.org/10.1016/j.jmps.2019.03.006.

Morch, Annie, Doucède, G., Lecomte-Grosbras, P., Brieu, M., Rubod, C., Cosson, M., 2021. Pelvic organ prolapse meshes: can they preserve the physiological behavior? Journal of the Mechanical Behavior of Biomedical Materials 120, 104569. https://doi.org/10.1016/j.jmbbm.2021.104569.

Mori da Cunha, M.G.M.C., Arts, B., Hympanova, L., Rynkevic, R., Mackova, K., Bosman, A.W., Dankers, P.Y.W., Deprest, J., 2020. Functional supramolecular bioactivated electrospun mesh improves tissue ingrowth in experimental abdominal wall reconstruction in rats. Acta Biomaterialia 106, 82–91. https://doi.org/10.1016/j.actbio.2020.01.041.

Mori da Cunha, M.G.M.C., Hympanova, L., Rynkevic, R., Mes, T., Bosman, A.W., Deprest, J., 2019. Biomechanical behaviour and biocompatibility of ureidopyrimidinone-polycarbonate electrospun and polypropylene meshes in a hernia repair in rabbits. Materials 12 (7), 1–14. https://doi.org/10.3390/ma12071174.

Nadler, B., Papadopoulos, P., Steigmann, D.J., 2006a. Convexity of the strain-energy function in a two-scale model of ideal fabrics. Journal of Elasticity 84 (3), 223–244. https://doi.org/10.1007/s10659-006-9063-3.

Nadler, Ben, Papadopoulos, P., Steigmann, D.J., 2006b. Multiscale constitutive modeling and numerical simulation of fabric material. International Journal of Solids and Structures 43 (2), 206–221. https://doi.org/10.1016/j.ijsolstr.2005.05.020.

Nayfeh, A.H., Kress, G.R., 1997. Non-linear constitutive model for plain-weave composites. Composites. Part B, Engineering 28 (5–6), 627–634.

Olsen, A.L., Smith, V.J., Bergstrom, J.O., Colling, J.C., Clark, A.L., 1997. Epidemiology of surgically managed pelvic organ prolapse and urinary incontinence. Obstetrics and Gynecology 89 (4), 501–506.

Ozog, Y., Deprest, J., Haest, K., Claus, F., De Ridder, D., Mazza, E., 2014. Calculation of membrane tension in selected sections of the pelvic floor. International Urogynecology Journal 25 (4), 499–506. https://doi.org/10.1007/s00192-013-2253-1.

Ozog, Y., Konstantinovic, M.L., Werbrouck, E., De Ridder, D., Edoardo, M., Deprest, J., 2011. Shrinkage and biomechanical evaluation of lightweight synthetics in a rabbit model for primary fascial repair. International Urogynecology Journal 22 (9), 1099–1108. https://doi.org/10.1007/s00192-011-1440-1.

Paul, K., Darzi, S., McPhee, G., Del Borgo, M.P., Werkmeister, J.A., Gargett, C.E., Mukherjee, S., 2019. 3D bioprinted endometrial stem cells on melt electrospun poly ε-caprolactone mesh for pelvic floor application promote anti-inflammatory responses in mice. Acta Biomaterialia 97, 162–176. https://doi.org/10.1016/j.actbio.2019.08.003.

Paul, K., Darzi, S., Werkmeister, J.A., Gargett, C.E., Mukherjee, S., 2020. Emerging nano/micro-structured degradable polymeric meshes for pelvic floor reconstruction. Nanomaterials 10 (6). https://doi.org/10.3390/nano10061120.

Röhrnbauer, B., Kress, G., Mazza, E., 2014. A physically based structural model for a textile prosthetic mesh. International Journal of Solids and Structures 51 (3–4), 633–646. https://doi.org/10.1016/j.ijsolstr.2013.10.030.

Röhrnbauer, B., Mazza, E., 2013. A non-biological model system to simulate the in vivo mechanical behavior of prosthetic meshes. Journal of the Mechanical Behavior of Biomedical Materials 20, 305–315. https://doi.org/10.1016/j.jmbbm.2013.01.029.

Röhrnbauer, B., Mazza, E., 2014. Uniaxial and biaxial mechanical characterization of a prosthetic mesh at different length scales. Journal of the Mechanical Behavior of Biomedical Materials 29, 7–19. https://doi.org/10.1016/j.jmbbm.2013.07.021.

Röhrnbauer, B., Ozog, Y., Egger, J., Werbrouck, E., Deprest, J., Mazza, E., 2013. Combined biaxial and uniaxial mechanical characterization of prosthetic meshes in a rabbit model. Journal of Biomechanics 46 (10), 1626–1632. https://doi.org/10.1016/j.jbiomech.2013.04.015.

Roman, S., Urbánková, I., Callewaert, G., Lesage, F., Hillary, C., Osman, N.I., Chapple, C.R., Deprest, J., MacNeil, S., 2016. Evaluating alternative materials for the treatment of stress urinary incontinence and pelvic organ prolapse: a comparison of the in vivo response to meshes implanted in rabbits. The Journal of Urology 196 (1), 261–269. https://doi.org/10.1016/j.juro.2016.02.067.

Schumpelick, V., Klinge, U., Welty, G., Klosterhalfen, B., 1999. Meshes in der Bauchwand. Der Chirurg 70 (8), 876–887. https://doi.org/10.1007/s001040050737.

Shepherd, J.P., Feola, A.J., Abramowitch, S.D., Moalli, P.A., 2012. Uniaxial biomechanical properties of seven different vaginally implanted meshes for pelvic organ prolapse. International Urogynecology Journal 23 (5), 613–620. https://doi.org/10.1007/s00192-011-1616-8.

Silva, M.E.T., Brandão, S., Parente, M.P.L., Mascarenhas, T., Natal Jorge, R.M., 2017. The influence of pelvic organ prolapse on the passive biomechanical properties of pelvic floor muscles. Journal of Mechanics in Medicine and Biology 17 (6), 1–19. https://doi.org/10.1142/S0219519417500907.

Slieker-ten Hove, M.C.P., Pool-Goudzwaard, A.L., Eijkemans, M.J.C., Steegers-Theunissen, R.P.M., Burger, C.W., Vierhout, M.E., 2009. Symptomatic pelvic organ prolapse and possible risk factors in a general population. American Journal of Obstetrics and Gynecology 200 (2), 184.e1–184.e7. https://doi.org/10.1016/j.ajog.2008.08.070.

Stella, J.A., D'Amore, A., Wagner, W.R., Sacks, M.S., 2010a. On the biomechanical function of scaffolds for engineering load-bearing soft tissues. Acta Biomaterialia 6 (7), 2365–2381. https://doi.org/10.1016/j.actbio.2010.01.001.

Stella, J.A., Liao, J., Hong, Y., David Merryman, W., Wagner, W.R., Sacks, M.S., 2008. Tissue-to-cellular level deformation coupling in cell micro-integrated elastomeric scaffolds. Biomaterials 29 (22), 3228–3236. https://doi.org/10.1016/j.biomaterials.2008.04.029.

Stella, J.A., Wagner, W.R., Sacks, M.S., 2010b. Scale-dependent fiber kinematics of elastomeric electrospun scaffolds for soft tissue engineering. Journal of Biomedical Materials Research. Part A 93 (3), 1032–1042. https://doi.org/10.1002/jbm.a.32593.

Stevens, M.M., George, J.H., 2005. Exploring and engineering the cell surface interface. Science 310 (5751), 1135–1138. https://doi.org/10.1126/science.1106587.

Trabuco, E.C., Zobitz, M.E., Klingele, C.J., Gebhart, J.B., 2007. Effect of host response (incorporation, encapsulation, mixed incorporation and encapsulation, or resorption) on the tensile strength of graft-reinforced repair in the rat ventral hernia model. American Journal of Obstetrics and Gynecology 197 (6), 638.e1–638.e6. https://doi.org/10.1016/j.ajog.2007.08.049.

Ulrich, D., Edwards, S.L., Alexander, D.L.J., Rosamilia, A., Werkmeister, J.A., Gargett, C.E., Letouzey, V., 2016. Changes in pelvic organ prolapse mesh mechanical properties following implantation in rats. American Journal of Obstetrics and Gynecology 214, 260.e1–260.e8.

Vashaghian, M., Diedrich, C.M., Zandieh-Doulabi, B., Werner, A., Smit, T.H., Roovers, J.P., 2019. Gentle cyclic straining of human fibroblasts on electrospun scaffolds enhances their regenerative potential. Acta Biomaterialia 84, 159–168. https://doi.org/10.1016/j.actbio.2018.11.034.

Vashaghian, M., Ruiz-Zapata, A.M., Kerkhof, M.H., Zandieh-Doulabi, B., Werner, A., Roovers, J.P., Smit, T.H., 2017. Toward a new generation of pelvic floor implants with electrospun nanofibrous matrices: a feasibility study. Neurourology and Urodynamics 36, 565–573. https://doi.org/10.1002/nau.

Vashaghian, M., Zaat, S.J., Smit, T.H., Roovers, J.P., 2018. Biomimetic implants for pelvic floor repair. Neurourology and Urodynamics 37 (2), 566–580. https://doi.org/10.1002/nau.23367.

Vogel, V., 2006. Mechanotransduction involving multimodular proteins: converting force into biochemical signals. Annual Review of Biophysics and Biomolecular Structure 35, 459–488. https://doi.org/10.1146/annurev.biophys.35.040405.102013.

Whitehead-Clarke, T., Karanjia, R., Banks, J., Beynon, V., Parker, S., Sanders, D., Mudera, V., Windsor, A., Kureshi, A., 2022. The experimental methodology and comparators used for in vivo hernia mesh testing: a 10-year scoping review. Hernia 26 (1), 297–307. https://doi.org/10.1007/s10029-020-02360-x.

Xiao, M., Geng, Z., Liao, K., 2011. An improved model of rigid bodies for plain-weave fabrics based on the dynamics of multibody systems. Textile Research Journal 81 (13), 1381–1394. https://doi.org/10.1177/0040517511402128.

Zündel, M., Ehret, A.E., Mazza, E., 2019. The multiscale stiffness of electrospun substrates and aspects of their mechanical biocompatibility. Acta Biomaterialia 84, 146–158. https://doi.org/10.1016/j.actbio.2018.11.012.

CHAPTER

Constitutive models of soft connective tissues under large strain: application to pelvic tissue?

10

Annie Morch[a], Laure Astruc[b,c], Jean-François Witz[a], Pauline Lecomte-Grosbras[a], and Mathias Brieu[d]

[a]*Univ. Lille, CNRS, Centrale Lille, UMR 9013-LaMcube – Laboratoire de Mécanique, Multiphysique, Multiéchelle, Lille, France*

[b]*Univ. Lille, ENSAIT, ULR2461-GEMTEX-Génie et Matériaux Textiles, Lille, France*

[c]*Junia, Equipe robotique-mécanique, Lille, France*

[d]*College of Engineering, Computer Science and Technology, Dept. of Mechanical Engineering, California State University, Los Angeles, CA United States*

10.1 Introduction

The numerical modeling of a mechanical system has a major requirement: accurate constitutive models for every single material considered. Mastering the definition of the constitutive models is an absolute prerequisite to any desire to use the digital tools to analyze and understand physiological or pathological situations and ultimately evaluate, or even predict, the care's performances.

Soft tissues, as opposed to stiff tissues, such as bones or cartilages, are known to have non-linear elastic behavior in large deformation, also known as hyperelastic (Bergel, 1961; Bäck and Gasser, 2013; O'Hagan and Samani, 2009; Brunon et al., 2010; Calvo et al., 2014; Egorov et al., 2002).

As for the soft tissues studied in the works cited above, the pelvic system constitutive models represent, from that perspective, a great and exciting challenge that requires to introduce constitutive models under large deformations (Scholten et al., 1988; Fung, 1993; Ettema et al., 1998; Cosson et al., 2004). This constitutes a double challenge.

It is a challenge on the one hand, because the mechanical behavior of the constituent tissues is very diverse in nature (pseudo-linear or non-linear, isotropic or anisotropic, pseudo-elastic or visco-elastic, ...). The literature is full of models to account for this type of behavior. We highly recommend the remarkable synthesis work done by Chagnon et al. (2015, 2017).

Biomechanics of the Female Reproductive System: Breast and Pelvic Organs. https://doi.org/10.1016/B978-0-12-823403-7.00022-1

On the other hand, it is also challenging since the mechanical characterization of soft tissues is extremely sensitive: they are very flexible and pre-stressed in their natural *in vivo* state. There are moreover no standards yet, defining the conditions of soft biological materials testing. Before selecting a model and identifying it, the researcher must ask himself a set of questions: Are the experimental test conditions representative of *in vivo* conditions? Are the tests reproducible? Are the measurements and information collected accurate and sufficient for the selected model? All these questions are not trivial and remain mostly open.

In these conditions, the question that arises in the first place is not, in fact, to carry out the ideal mechanical test, nor to know which constitutive model is the most representative of the studied materials. Indeed, identification and experimentation are the two dimensions of a single question: Which mechanical behavior of the considered tissue am I able to identify? In other words, which model can we select, taking into account the nature and quality of the experimental information at our disposal?

Though the authors are well aware that it is sometimes complex, if not impossible, to conduct all the studies at once to master not only the experimental data, but also the modeling and numerical simulation, they nonetheless believe it is crucial to act consciously.

For such a reason, this chapter (and its authors) aims at providing a comprehensive *vade mecum* for anyone involved in the mechanical study of soft biological tissue. This chapter therefore contains the basics of what everyone needs to know to model the behavior of soft tissues in the context of hyperelasticity: from the mechanical formulation to the choice of an adequate model with respect to the working hypotheses and experimental conditions.

10.2 Mechanical formulation

Soft biological tissues exhibit a nonlinear hyperelastic behavior in large deformation. Their mechanical response is usually compared to rubber-like materials. This first section introduces the usual but mandatory theoretical framework and common hypotheses to study non-linear hyperelastic behavior under large strain, and thus soft biological tissues.

10.2.1 Theoretical framework: large deformation

Let Ω_0 be the initial, undeformed configuration of a material $\mathcal{M}$ in the euclidean space $\mathcal{E}$ with the orthonormal basis $(\underline{\boldsymbol{e}}_1, \underline{\boldsymbol{e}}_2, \underline{\boldsymbol{e}}_3)$. Let Ω_t be the current, deformed configuration of the said material. They are, respectively, called the Lagrangian and Eulerian configurations and illustrated in Fig. 10.1.

Let $\underline{\boldsymbol{u}}$ be the displacement vector of a material point, located at M_0 in the undeformed configuration and M in the deformed one: $\underline{\boldsymbol{u}} = \underline{MM_0}$. The coordinates of M

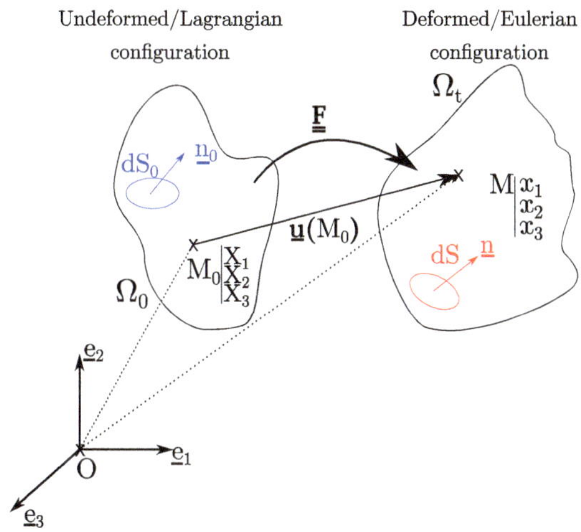

FIGURE 10.1

Material $\mathcal{M}$ undergoes deformation from the reference Ω_0 state to the deformed Ω_t configuration.

are written as

$$\begin{aligned} \underline{\boldsymbol{OM}} &= \underline{\boldsymbol{OM_0}} + \underline{\boldsymbol{u}}(M_0), \\ x_i &= X_i + u_i(X_1, X_2, X_3). \end{aligned} \tag{10.1}$$

The deformation gradient $\underline{\underline{\boldsymbol{F}}}$ is then expressed as

$$\begin{aligned} \underline{\underline{\boldsymbol{F}}} &= \underline{\underline{\boldsymbol{1}}} + \nabla_{\underline{X}} \underline{\boldsymbol{u}}, \\ F_{IJ} &= \delta_{IJ} + \frac{\partial u_i}{\partial X_j}, \end{aligned} \tag{10.2}$$

where $\nabla_{\underline{X}}$ is the gradient operator with respect to the Lagrangian coordinates. The transformation of an infinitesimal vector $\underline{\boldsymbol{dX}}$ and an infinitesimal area dS_0 are written

$$\begin{aligned} \underline{\boldsymbol{dx}} &= \underline{\underline{\boldsymbol{F}}}.\underline{\boldsymbol{dX}}., \\ dS\underline{\boldsymbol{n}} &= J\underline{\underline{\boldsymbol{F}}}^{-1}\underline{\boldsymbol{n}}_0 dS_0. \end{aligned} \tag{10.3}$$

The deformation gradient conveys the deformation of the material $\mathcal{M}$ undergoing a given loading, *i.e.*, how the material goes from an initial state Ω_0 to a deformed one Ω. But it contains simultaneously rigid-body motions and local deformation of the continuum. The right, $\underline{\underline{\boldsymbol{C}}}$, and left, $\underline{\underline{\boldsymbol{B}}}$, Cauchy–Green tensors are built to remove the contribution of the rigid-body rotations. Both contain only deformation information:

$$\begin{aligned} \underline{\underline{\boldsymbol{B}}} &= \underline{\underline{\boldsymbol{F}}}.\underline{\underline{\boldsymbol{F}}}^T, \\ \underline{\underline{\boldsymbol{C}}} &= \underline{\underline{\boldsymbol{F}}}^T.\underline{\underline{\boldsymbol{F}}}. \end{aligned} \tag{10.4}$$

However, when the material does not deform or translates, both measures are not zeroed. The Green–Lagrange deformation tensor is built from the deformation gradient to ensure that zero strain appears in the case of rigid-body motion and a stress-free state as

$$\begin{aligned}\underline{\underline{E}} &= \frac{1}{2}(\underline{\underline{F}}^T.\underline{\underline{F}} - \underline{\underline{1}}) \\ &= \frac{1}{2}(\nabla_{\underline{X}}\underline{u} + \nabla_{\underline{X}}\underline{u}^T + \nabla_{\underline{X}}\underline{u}^T.\nabla_{\underline{X}}\underline{u}).\end{aligned} \tag{10.5}$$

It must be mentioned that in the infinitesimal strain theory, the quadratic term in the expression of the Green–Lagrange strain is neglected. In large transformation, one can observe the non-linearity of the deformation with respect to the deformation gradient, which is not the case under small strain.

Though the Lagrangian and Eulerian configurations superimpose under small strain, this assumption no longer holds when large strains are considered. One has therefore to be especially careful about the configuration used to express the stress. In the Eulerian configuration, the force $\underline{f}$ applied on the surface element ds, in Fig. 10.1, is

$$\underline{f} = \underline{\underline{\sigma}}.\underline{n}ds \tag{10.6}$$

where $\underline{n}$ is the normal vector to the surface element ds, and $\underline{\underline{\sigma}}$ is the Cauchy stress.

The second Piola-Kirchhoff stress $\underline{\underline{T}}$ is related to the force $\underline{f_0}$ on the surface element dS by $\underline{f_0} = \underline{\underline{T}}.\underline{n_0}dS$. One can see that the Cauchy and second Piola-Kirchhoff (PK2) relate the force to the surface, respectively, in the current and undeformed configurations.

If both Lagrangian and Eulerian representation are widely used in computational approaches, both measures remain difficult to grasp fully from an experimenter's point of view. The Cauchy stress, also called true stress, describes the current stress state within the material. It requires measuring the force and cross-section in the deformed state. Though the force measurement is straightforward, the cross-section evaluation is easier to assess in the undeformed configuration. The Cauchy stress is difficult to get from an experimental point of view but provides a meaningful description of the material. The PK2 lacks physical meaning: the measure of the force applied on a sample only happens in the eulerian configuration. A third stress measure is preferred: the first Piola-Kirchhoff stress (PK1) $\underline{\underline{\tau}}$. It is a two-point tensor, similarly to the deformation gradient, it sits astride both configurations and links the force in the deformed configuration to areas in the reference one: $\underline{f} = \underline{\underline{\tau}}.\underline{n_0}dS$.

With Nanson's formula, Eq. (10.3) and push-forward and pull-back operations, one goes from one representation to the other:

$$\begin{aligned}\underline{\underline{\tau}} &= J\underline{\underline{\sigma}}.\underline{\underline{F}}^{-T}, \\ \underline{\underline{T}} &= J\underline{\underline{F}}^{-1}.\underline{\underline{\sigma}}.\underline{\underline{F}}^{-T}, \\ \underline{\underline{\sigma}} &= \frac{1}{J}\underline{\underline{F}}.\underline{\underline{T}}.\underline{\underline{F}}^T.\end{aligned} \tag{10.7}$$

The Cauchy stress is a symmetric tensor, so also is the PK2 stress. Since the deformation gradient is not symmetric, as a result of its two-point configuration, the PK1 stress, as well, is generally not symmetric.

10.2.2 Hyperelasticity and non-linear elastic modeling

Hyperelasticity is a subclass of elasticity. The behavior is assumed reversible without dissipative phenomena (damage, plastic deformation, viscous effects, ...). The material is supposed homogeneous: the deformation gradient tensor or stress tensor are uniform at the observation scale. A material is called hyperelastic if the stress derives from a potential depending on the deformation. The first and second laws of thermodynamics lead to the Clausius–Duhem inequality. If the transformation is isotherm, adiabatic, and non-dissipative, the Clausius–Duhem inequality imposes that the PK1 stress derives from a potential:

$$\underline{\underline{\boldsymbol{\tau}}} = \frac{\partial W}{\partial \underline{\underline{\boldsymbol{F}}}}. \tag{10.8}$$

The Lagrangian formulation of the Clausius–Duhem inequality leads to an equivalent expression of the PK2 stress:

$$\underline{\underline{\boldsymbol{T}}} = \frac{\partial W}{\partial \underline{\underline{\boldsymbol{E}}}} = 2\frac{\partial W}{\partial \underline{\underline{\boldsymbol{C}}}}. \tag{10.9}$$

With the push-forward operation of Eq. (10.7), the Cauchy stress is given by $\underline{\underline{\boldsymbol{\sigma}}} = \frac{1}{J}\underline{\underline{\boldsymbol{F}}}.\frac{\partial W}{\partial \underline{\underline{\boldsymbol{E}}}}.\underline{\underline{\boldsymbol{F}}}^{-T} = \frac{1}{J}\frac{\partial W}{\partial \underline{\underline{\boldsymbol{F}}}}.\underline{\underline{\boldsymbol{F}}}^{T}$.

The potential W is a scalar quantity called a strain energy density (SED). In brief, a strain energy density consists basically of a scalar function of an adequate deformation tensor and constants. The constants, also called mechanical parameters, are either dimensionless or equivalent to rigidity modules expressed in MPa. The SED contains geometrical and material non-linearities. The former comes from the deformation tensor since the quadratic term is no longer neglected. The latter concern the non-linear stress response. To consider both, highly non-linear functions of the deformation are used in the SED.

A few conditions are imposed on the strain energy density. The strain energy remains positive and increases with deformation. In the undeformed configuration, *i.e.*, $\underline{\underline{\boldsymbol{F}}} = \underline{\underline{\boldsymbol{1}}}$, no energy is stored:

$$\begin{aligned} &W(\underline{\underline{\boldsymbol{1}}}) = 0, \\ &W(\underline{\underline{\boldsymbol{F}}}) \geq 0,\ \forall \underline{\underline{\boldsymbol{F}}}. \end{aligned} \tag{10.10}$$

In mechanics, the direct optimization problem consists in minimizing the gap between the functionals of internal and external energies, namely

$$\underset{\underline{\underline{F}}}{\operatorname{argmin}}\, f(\underline{\underline{F}}) = \underset{\underline{\underline{F}}}{\operatorname{argmin}}\, W_{int}(\underline{\underline{F}}) - W_{ext}(\underline{\underline{F}}). \tag{10.11}$$

Rubber-like materials present non-linear stress vs. strain response and large deformation. They are traditionally studied in the non-linear hyperelasticity framework. Rubber-like materials are moreover considered incompressible: $J = \det \underline{\underline{F}} = 1$. This adds a constraint: $\mathrm{r}(\underline{\underline{F}}) = \det \underline{\underline{F}} - 1 = 0$. The Lagrange multiplier p is used to solve constrained optimization problems and imposes that the gradient of the functional f and the constraint r are proportional. In the minimization problem of Eq. (10.11), this imposes that equilibrium is respected when

$$\underset{\underline{\underline{F}}}{\operatorname{argmin}}\, W_{int}(\underline{\underline{F}}) - p\mathrm{r}(\underline{\underline{F}}) - W_{ext}(\underline{\underline{F}}). \tag{10.12}$$

The PK1 stress within the material is then expressed from Eq. (10.8), which transforms then directly from Eq. (10.12) into

$$\underline{\underline{\tau}} = \frac{\partial W(\underline{\underline{F}})}{\partial \underline{\underline{F}}} - p\underline{\underline{F}}^{-T}. \tag{10.13}$$

The Lagrange multiplier p is also called the hydrostatic pressure. It is calculated with respect to the boundary conditions and equilibrium equations.

10.2.3 Characterizing soft biological tissues: a challenging task

Human tissues split up into two main categories from a mechanical perspective: hard and soft tissues. Hard tissues, such as bones, are commonly modeled within the small strain theory framework: the behavior is said to be linear elastic with respect to the linearized Green–Lagrange deformation tensor. Abdominal and organ walls, ligaments, muscles, and skin belong to the second category: they allow for much larger (non-linear) deformation than hard tissues to ensure physiological and postural functionalities. They also present a non-linear stress-strain response in large deformation, as shown in Fig. 10.2. Every known women pelvic tissues (vagina, bladder, and rectum walls, ligaments, fascias) belong in this last category.

This behavior compares well with a rubber-like material, which makes the non-linear hyperelasticity framework particularly suitable for studying these unusual materials. However, one has to be cautious when applying this theory to soft biological tissues. Unlike elastomers, their testing and mechanical identification disclose challenges.

10.2.3.1 Initial state: impact on the mechanical parameters

Most of the literature refers to *ex vivo* mechanical testing (see among many others Moalli et al., 2005; Rubod et al., 2007; Peña et al., 2010, 2011; Rubod et al.,

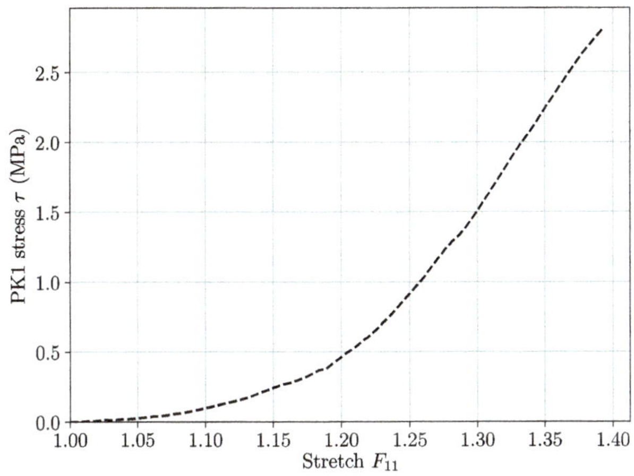

FIGURE 10.2

Stress vs. stretch response of a human vaginal sample in uniaxial monotonic tension test. Adapted from Jean-Charles et al. (2010).

2012; Feola et al., 2013). *Ex vivo* experimental characterization indeed looks easier to perform than *in vivo* for obvious ethical reasons but also for experimental ones. *Ex vivo* characterization ensures that only a single tissue is tested, whereas *in vivo* tests cannot allow the surrounding tissues or mechanical loads to be isolated.

When they are dissected from the body, retraction frequently happens (Fung, 1967, 1973; Hernández et al., 2011). Tissues in the body are often pre-stressed in their physiological loading conditions to ensure their specific role. Regardless of the wide variety of testing protocols, it remains challenging to gather a statistically representative population: soft tissues are rare, sometimes small, and complicated to obtain and test. To make the most of the samples collected, characterization protocols can favor mechanical tests, which are not physiologically representative but provide statistically significant numbers of samples. The inability to measure the level of prestress or to reproduce *in vivo* boundary conditions impact the experimental measure of force versus deformation.

One can observe an initial plateau with almost zero force measure while the sample stretches. Fig. 10.3 shows such plateau in the case of the uniaxial testing of a rat abdominal wall sample. This plateau is most of the time not representative of a mechanical response of the material, rather of a structural reorganization of the fibers within the tissue. It would be meaningless to model it with a strain energy density. It thus has to be removed before the identification of the SED.

To our knowledge, there is no standard procedure to determine the end of the initial plateau. A previous study (Morch et al., 2020) shows the impact of this initial plateau on the identified mechanical properties: depending on the choice of the modeler, mechanical parameters are strongly impacted. To compare mechanical properties (*i.e.*, SED choices), one has to get access to the raw data and be able to apply the

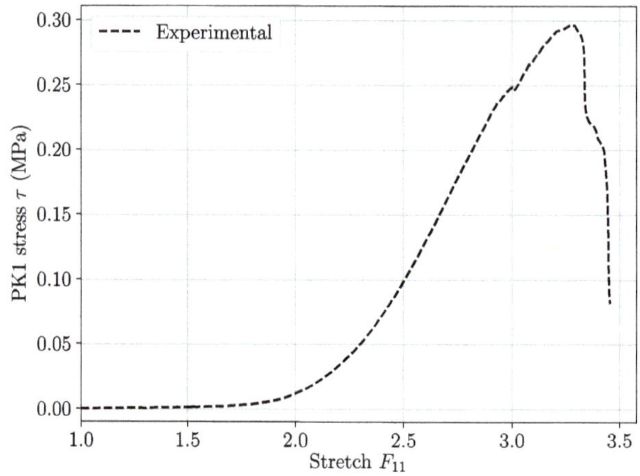

FIGURE 10.3

Raw experimental stress versus stretch response of a rat abdominal wall sample (Morch et al., 2017).

procedure of his choice to remove the plateau. Soft biological tissue studies should as much as possible provide the experimental raw data in addition to identified mechanical parameters.

The post-processing of experimental measures falls under the experimenter's or modeler's responsibility. As experimental do's and don'ts are not the main subject of this chapter, let us assume that this aspect is already set.

10.2.3.2 Incompressibility assumption

Soft tissues are widely accepted as incompressible in experimental works due to their high water content (Gao et al., 2010; Brieu et al., 2016). However, this hypothesis has rarely been documented and, more importantly, remains hardly verifiable experimentally. Compressible or nearly-incompressible approaches are often used as well (Nolan et al., 2014; Doll and Schweizerhof, 2000). The incompressible constraint in Eq. (10.13) is replaced by introducing a volumetric potential in the SED. Among others (Doll and Schweizerhof, 2000), we cite the following:

$$\begin{aligned} W_{vol} &= \frac{k(J-1)^2}{2}, \\ W_{vol} &= k\frac{\ln(J)^2}{2}, \\ W_{vol} &= k\left(\frac{J^2-1}{2} - \ln J\right), \end{aligned} \tag{10.14}$$

where $J = \det \underline{\underline{F}}$, and k is a constant in MPa. To enforce incompressibility, the coefficient k takes a much greater value than 1. In a simulation, removing this strong constraint using compressible or quasi-incompressible approaches helps to speed up the computation convergence. However, using the incompressibility assumption helps to remove the unknowns of the tensorial equation, Eq. (10.13), and thus makes the identification of mechanical properties of soft tissues easier.

It has to be noted that the strain energy density is in that case often written with decoupled volumetric and deviatoric parts:

$$W(\underline{\underline{F}}) = W_{dev}(\underline{\underline{\bar{F}}}) + W_{vol}(J), \tag{10.15}$$

where $\underline{\underline{\bar{F}}} = J^{-\frac{1}{3}}\underline{\underline{F}}$ is the deviatoric part of the deformation gradient.

The PK2 stress is given by $\underline{\underline{T}} = \dfrac{\partial W_{dev}}{\partial \underline{\underline{\bar{E}}}} : J^{-\frac{2}{3}}\left(\mathbb{I} - \dfrac{1}{3}\underline{\underline{C}} \times \underline{\underline{C}}^{-T}\right) - J\dfrac{\partial W_{vol}}{\partial J}\underline{\underline{F}}^{-T}$, with the fourth-order tensor defined as $\mathbb{I}_{ijkl} = \frac{\delta_{ik}\delta_{jl}+\delta_{il}\delta_{jk}}{2}$. With Eq. (10.7), one can then easily find the PK1 or Cauchy stress.

Depending on the context, *i.e.*, use of the SED, one should wisely choose between incompressible, nearly incompressible, and compressible approaches.

10.2.3.3 Homogeneous deformation

In the theoretical canvas, the last hypothesis lies in the homogeneity of the deformations. Similar to the incompressibility of soft tissues, this assumption is often admitted in the literature. Deformation measure takes mainly three forms: global measures with the displacement of the motor, local measures with video-extensometry, or deformation field measures with DIC or stereo-correlation. The two former are widely represented in the literature but do not provide any information about the deformation's homogeneity. Global measures only give information about the axial deformation, and they are easily impacted by slack in the testing equipment. The deformation is assumed to remain the same whether in the center sample or close to the grips. Videoextensometry technique commonly follows four contrasted spots on the center front of the sample during the test. This method already assumes the deformation's homogeneity, but provides a more accurate evaluation of the strain within the sample compared to global measures. Field methods provide richer information than the former. However, deformations fields often appear heterogeneous (Brunon et al., 2011; Mallett and Arruda, 2017; Jayyosi et al., 2017). Taking advantage of those broad measures raises new problems: use of FEMU approaches for identifying mechanical properties, increasing computation time, etc.

Models are used in two contexts: numerical simulations from a mechanical test to a complex system (surgery, physiological situation) and the identification of mechanical properties from experimental data. Regarding the latter, the model identification requires the knowledge of the experimental gradient deformation tensor $\underline{\underline{F}}$, which takes the general form

$$\underline{\underline{F}} = \begin{pmatrix} F_{11} & F_{12} & F_{13} \\ F_{21} & F_{22} & F_{23} \\ F_{31} & F_{32} & F_{33} \end{pmatrix}. \tag{10.16}$$

Given an in-plane loading $(O, \underline{e}_1, \underline{e}_2)$, and assuming local strain measurements means (video-extenso, DIC) are used, one can access the longitudinal and transverse stretch, F_{11} and F_{22}, and the in-plane shear components F_{12}, F_{21}. It requires much more complex techniques and tools to evaluate the out of plane stretches (F_{33}) and shears. For example, F_{32} and F_{23} are out of reach with traditional methods: X-Ray tomography measures alone could fill in these tensor components. In short, the general gradient deformation tensor takes intricate forms that can raise additional unknowns to determine in addition to the mechanical parameters.

Previous statements about commonly accepted working hypotheses may seem somewhat crushing. It does not mean that the hyperelastic modeling of soft biological tissues is impossible. However, we believe it is fundamental to keep in mind the necessary pre-requisites for the use of models and the inherent limitations related to the study of biological materials. The main objective of this study is not another review about the SED used for soft tissues. Chagnon et al. (2015) and Wex et al. (2015) both proposed comprehensive reviews about the strain energy densities, either isotropic and anisotropic. We instead aim to propose a few guidelines for the experimental or numerical researcher to strike a balance between the experimental or numerical study of soft tissues and the choice of an adequate SED and relevant work hypotheses. In what follow, we will always assume that the deformation field is homogeneous and that the initial state of the sample is fully known.

10.3 Isotropic modeling

10.3.1 Isotropic models: simplification of the SED and stress computation

Symmetric positive definite tensors of $\mathbb{R}^3$ can be diagonalized. Therefore Cauchy and PK2 stress, and left and right Cauchy–Green, Green–Lagrange tensor are diagonalizable. Their spectral decomposition leads to the identification of their respective eigenvalues and eigenvectors. They are called principal stresses for a stress tensor and principal stretches for dilatation tensor.

In the isotropic case, the principal stress and stretch directions coincide. That helps to determine some unknowns of the deformation gradient tensor. Let $(\underline{e}_I, \underline{e}_{II}, \underline{e}_{III})$ be the principal stress directions of the Cauchy stress. The deformation gradient now reads as follows:

$$
\begin{aligned}
\underline{\underline{\sigma}} &= \begin{pmatrix} \sigma_I & 0 & 0 \\ 0 & \sigma_{II} & 0 \\ 0 & 0 & \sigma_{III} \end{pmatrix}_{(\underline{e}_I, \underline{e}_{II}, \underline{e}_{III})} = \operatorname{diag}(\sigma_I, \sigma_{II}, \sigma_{III})_{(\underline{e}_I, \underline{e}_{II}, \underline{e}_{III})}, \\
\underline{\underline{F}} &= \begin{pmatrix} F_{11} & 0 & 0 \\ 0 & F_{22} & 0 \\ 0 & 0 & F_{33} \end{pmatrix}_{(\underline{e}_I, \underline{e}_{II}, \underline{e}_{III})} = \operatorname{diag}(F_{11}, F_{22}, F_{33})_{(\underline{e}_I, \underline{e}_{II}, \underline{e}_{III})}.
\end{aligned}
\tag{10.17}
$$

Table 10.1 Isotropic strain energy densities depending on the strain invariants.

Reference	Strain energy density
Neo-Hookean	$C_0(I_1-3)$
Mooney–Rivlin	$C_1(I_1-3)+C_2(I_2-3)$
Yeoh	$C_{10}(I_1-3)+C_{20}(I_1-3)^2+C_{30}(I_1-3)^3$
Generalized Rivlin	$\sum_{i,j} C_{ij}(I_1-3)^i(I_2-3)^j$
Gent and Thomas	$C_1(I_1-3)+C_2\ln(\frac{I_2}{3})$
Hart–Smith	$C_1\int \exp C_3(I_1-3)^2 dI_1 + C_2\ln(\frac{I_2}{3})$
Ogden	$W=\sum_i \frac{\mu_i}{\alpha_i}(\lambda_1^{\alpha_i}+\lambda_2^{\alpha_i}+\lambda_3^{\alpha_i}-3)$

One can show that isotropic SEDs only depend on the invariants of the Cauchy–Green right tensor $\underline{\underline{C}}$.[1] The SED is written

$$W(\underline{\underline{F}}) = W(I_1, I_2) + W_{vol}(J), \tag{10.18}$$

where I_1, I_2 are the first two invariants of $\underline{\underline{C}}$. The invariants are defined as

$$I_1 = Tr(\underline{\underline{C}}),\ I_2 = \frac{1}{2}(Tr(\underline{\underline{C}})^2 - Tr(\underline{\underline{C}}^2)),\ I_3 = \det \underline{\underline{C}} = J^2. \tag{10.19}$$

The first PK1 is obtained from the differentiation of W with respect to $\underline{\underline{F}}$, Eq. (10.13) now gives

$$\underline{\underline{T}} = 2\left(\frac{\partial W}{\partial I_1}\underline{\underline{F}} + \frac{\partial W}{\partial I_2}(I_1\underline{\underline{F}} - \underline{\underline{F}}.\underline{\underline{C}})\right) + \frac{\partial W_{vol}}{\partial J}\underline{\underline{F}}^{-T} \tag{10.20}$$

knowing the following partial derivatives:

$$\begin{aligned} \frac{\partial I_1}{\partial \underline{\underline{F}}} &= 2\underline{\underline{F}}, \\ \frac{\partial I_2}{\partial \underline{\underline{F}}} &= 2(I_1\underline{\underline{F}} - \underline{\underline{F}}.\underline{\underline{C}}), \\ \frac{\partial J}{\partial \underline{\underline{F}}} &= 2J\underline{\underline{F}}^{-T}. \end{aligned} \tag{10.21}$$

Numerous strain energy densities in I_1 and I_2 exist in the literature. Table 10.1 introduces a few of the most famous ones.

[1] This equally true with the invariants of the Cauchy–Green left tensor.

The Ogden model lies in using the eigenvalues, also called principal stretches of the right Cauchy–Green tensor. The principal stretches are also invariant and can easily be expressed as a function of the stretch invariants.

Except for the Neo-Hookean, all of the models of Table 10.1 are phenomenological. Physically-based models were also developed. They are based on the microstructural description introduced for the elastomers that remains valid for the biological soft tissues. Elastomers are basically a network of entangled macromolecules. Each of them is assumed to contribute to the global mechanical behavior of the material. These models are attractive in biological tissue modeling, as there is a direct analogy between elastomer and soft tissues. Instead of the macromolecules, soft tissues are made of entangled collagen and elastin fibers, whose density and distribution are responsible for their mechanical response.

The global strain energy density is then often described as the continuous or discrete integral over the unit sphere surface of the macromolecule elementary density (Treloar and Riding, 1979; Miehe et al., 2004; Alastrué et al., 2009; Merckel et al., 2011). We propose to recall the continuous form (Treloar and Riding, 1979):

$$W(\underline{\underline{F}}) = \iint_S w(\nu(\underline{u}))n(\underline{u})d^2\Omega \tag{10.22}$$

$$\text{with } \nu(\underline{u}) = \underline{u}.\underline{\underline{F}}.\underline{u}, \tag{10.23}$$

where $\underline{u}$ is a material direction defined by two angles in spherical coordinates $\underline{u} = (\cos\theta \sin\phi, \sin\theta \sin\phi, \cos\phi)_{(\underline{e}_1,\underline{e}_2,\underline{e}_3)}$, *i.e.*, a unit vector over the sphere surface S, $n(\underline{u})d^2\Omega$ the chain density in direction $\underline{u}$, and w the elementary density. This SED mainly works with Gaussian (Kuhn, 1936; Miehe et al., 2004) and non-Gaussian (Kuhn and Grün, 1942) statistical distributions:

$$\begin{aligned} w &= \frac{3}{2}nkT\nu^2, \\ w &= nkT\left(\frac{\nu}{\sqrt{N}} + \ln\frac{\beta}{\sinh\beta}\right) \text{ with } \beta = \mathcal{L}^{-1}\left(\frac{\nu}{\sqrt{N}}\right), \end{aligned} \tag{10.24}$$

where n is the chain density per volume unit, k the Boltzmann constant, T the absolute temperature, and N is the chain limit of extensibility. $\mathcal{L}^{-1}$ denotes the inverse of the Langevin function $\mathcal{L}(x) = \coth(x) - \frac{1}{x}$.

Table 10.2 sums up a few references pertaining to a richer literature using physically-based models.

10.3.2 Quick application

Let us consider an isotropic material under uniaxial tension along $\underline{e}_1$. The material only sees a longitudinal stress along $\underline{e}_1$, which is noted $\tau = \frac{f}{S_0}$, the ratio between the measured force and the initial cross-section. From Eq. (10.17), the deformation gradient and PK1 stress tensor are also diagonal in the basis $(\underline{e}_1, \underline{e}_2, \underline{e}_3)$:

$$\underline{\underline{\tau}} = diag(\tau, 0, 0),$$

Table 10.2 A few examples of isotropic strain energy densities and stresses in directional approaches.

Reference	Model	Expression
Treloar (1944)	Neo-hookean	$W = nkT(I_1 - 3)$
Kuhn and Grün (1942)	3-chains	$\forall i \in \{1,2,3\}$ $\sigma_i = \frac{nkT}{3}\frac{\lambda}{\sqrt{N}}\lambda_i \mathcal{L}^{-1}(\frac{\lambda_i}{\sqrt{N}})$ with $\lambda = \frac{\sqrt{I_1}}{\sqrt{3}}$
Arruda and Boyce (1993)	8-chains	$\forall i \in [1,8]$ $\sigma_i = \frac{nkT}{3}\frac{\lambda}{\sqrt{N}}\lambda_i^2 \mathcal{L}^{-1}(\frac{\lambda_i}{\sqrt{N}})$ with $\lambda = \frac{\sqrt{I_1}}{\sqrt{3}}$
Miehe et al. (2004); Gillibert et al. (2010)	non-affine microsphere model	$W = \sum_{i=0}^{n} \omega_i w(\nu_i)$

$$\underline{\underline{F}} = diag(F_{11}, F_{22}, F_{33}).$$

Since we assume isotropy, the two transverse directions undergo the same stretches: $F_{22} = F_{33}$. Experimentally, F_{11} and F_{22} are estimated as the length ratios in the axial (*i.e.*, the direction of the loading) and transverse (*i.e.*, the in-plane transverse direction of the loading) directions, respectively.

With the incompressibility assumption, one can draw a relation between the deformation gradient components:

$$F_{11}F_{22}F_{33} = 1. \tag{10.25}$$

We assume a SED depending only on I_1 (Fukahori and Seki, 1992). Combining Eq. (10.13) and (10.21) gives the expression of the PK1:

$$\underline{\underline{\tau}} = 2\frac{\partial W}{\partial I_1}\underline{\underline{F}} - p\underline{\underline{F}}^{-T}.$$

Knowing that $\tau_{22} = \tau_{33} = 0$, the hydrostatic pressure p is estimated, and one can find

$$\tau = 2\left(F_{11} - \frac{1}{F_{11}^2}\right)\frac{\partial W}{\partial I_1}. \tag{10.26}$$

Fig. 10.4 shows a fit of the experimental data of Fig. 10.2 with a second-order Yeoh model, *i.e.*, $C_{30} = 0$. The uniaxial stress expression is given by

$$\tau = 2\left(F_{11} - \frac{1}{F_{11}^2}\right)\left(C_{10} + 2C_{20}(F_{11}^2 + \frac{2}{F_{11}} - 3)\right).$$

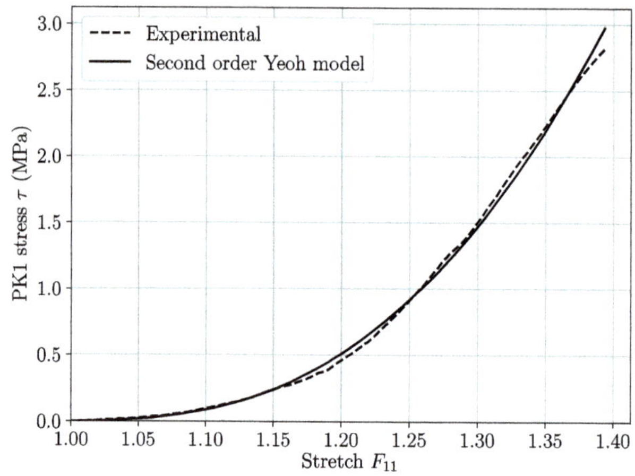

FIGURE 10.4

Experimental and modeled stress versus stretch response of a human prolapsed vaginal tissue sample.

Table 10.3 Examples of the PK1 stress tensor, corresponding deformation gradient tensor, and hydrostatic pressure of an isotropic material with an invariant-based SED tested in uniaxial tension (UT) in $\underline{e}_1$ or equibiaxial tension along $\underline{e}_1$ and $\underline{e}_2$.

	Loading case **UT along $\underline{e}_1$**	**EBT along $\underline{e}_1$ and $\underline{e}_2$**
PK1 stress	$\underline{\underline{\tau}} = \text{diag}(\tau, 0, 0)$	$\underline{\underline{\tau}} = \text{diag}(\tau, \tau, 0)$
Deformation gradient	$\underline{\underline{F}} = \text{diag}(F_{11}, F_{22}, F_{22})$	$\underline{\underline{F}} = \text{diag}(F_{11}, F_{11}, F_{33})$
Incompressibility assumption	$F_{22} = \frac{1}{\sqrt{F_{11}}}$ $p = \frac{2}{F_{11}^2}\left(\frac{\partial W}{\partial I_1} + \frac{\partial W}{\partial I_2}(F_{11}^2 + \frac{1}{F_{11}})\right)$	$F_{33} = \frac{1}{F_{11}^2}$ $p = \frac{2}{F_{11}^4}(\frac{\partial W}{\partial I_1} + 2\frac{\partial W}{\partial I_2}F_{11}^2)$

Here, the least-square minimization provides $C_{10} = 0.04$ MPa, and $C_{20} = 2.20$ MPa.

The same method is applied to an equibiaxial loading and leads to Table 10.3 results. In Table 10.3, the first two rows recall the expression of the PK1 stress and corresponding deformation gradient tensor in the case of a uniaxial tension test and equibiaxial tension test. The third one includes the computation of the hydrostatic pressure, considering an isotropic incompressible material provided with a I_1- and I_2-based SED.

10.3.3 The elephant in the room: why and when to use an isotropic SED?

Ideally, isotropic modeling should only concern perfectly isotropic material. The testing of elastomer materials (natural or synthetic rubber-like material) enables to check that the material is perfectly isotropic. Experimental characterization is easier: tests are repeatable, material quantity is sufficient to perform tests in multiple directions and/or in multiple loading conditions.

It is unfortunately not accurate for soft biological samples. We are first confronted with the rarity of the raw materials. Dissection does not always provide large enough samples to perform a test in multiaxial conditions. Moreover, by nature, biological tissues present a large inter- and intra-individual dispersion in terms of dimensions, composition, and mechanical properties. To obtain statistically significant observations of isotropy, one would have to perform a large number of tests in multiaxial loading conditions or multiple directions in the case of uniaxial tension. Given the rarity of the material, both options are often not feasible.

Uniaxial tension seems to be one of the most documented and used testing approach, as it requires smaller samples and straightforward post-processing methods. However, most soft tissues, excepted ligaments or tendons, undergo multiaxial loadings in the body. Multiaxial testing is also used in the literature (biaxial testing, inflation (Lyons et al., 2014; Cooney et al., 2015; Whitford et al., 2015)), but such method often requires the use of constitutive law implemented in FE code (UMat subroutine for instance) and simulation work. That additional step could make the mechanical parameters identification more complex.

Using an isotropic modeling approach to determine a soft biological tissue's mechanical properties can be relevant under a few conditions:

- For example, if the samples are characterized in uniaxial tension in one direction, no information is gathered about the other directions' mechanical response. It is impossible to use an anisotropic model, because the off-axis properties are left unrestricted and can therefore take an infinite number of values. Isotropic modeling seems to be the only valid option to extrapolate the mechanical response outside the measurement range.
- Deformation measurement on biological tissues may prove difficult. Sometimes local measurements are not acquired, which impairs the determination of the experimental deformation gradient's components. The isotropy and the incompressibility assumption helps to simplify the unknowns: stress and deformation gradient are diagonal on the same basis. One can deduce some components of the deformation gradient depending on the loading type (see Table 10.3 for application in uniaxial and biaxial case).
- Regarding numerical studies, one could prefer using isotropic laws. Depending on the system's complexity to simulate and boundary conditions, applying isotropic models might favor and speed up the computation convergence. Applying anisotropic behavioral models in a complex system would not necessarily

provide any better results or information in simulation but could slow down the computation.

In reality, soft biological tissues are more likely anisotropic. They are made from the entanglement of collagen and elastin fibers. Each tissue results from a specific distribution (orientation and density) of elastin and collagen fibers that depends on its function in the body and varies between individuals. The microstructure, *i.e.*, distribution of the tissue's fibers, seems responsible for soft tissues' anisotropic response. Given an adequate number of testing directions or rich measurements, anisotropic modeling remains accessible.

10.4 Anisotropic modeling

In biomechanics, the description of the tissue structure, regardless of the observation scale, often sets the starting point for anisotropic modeling strategies of the mechanical behavior. Earliest methods, based on Fung's work (Fung et al., 1979; Chuong and Fung, 1983) rely on a fourth-order description of the elasticity tensor. The general strain energy density is written as a function of the Green–Lagrange tensor $\mathbf{E}$ and a fourth-order tensor $\mathbb{A}$ (Humphrey, 1995):

$$W = k\left(\exp(A_{ijkl}E_{ij}E_{kl}) - 1\right). \tag{10.27}$$

However, alternative methods are often preferred in the literature. The first two sections showed in broad outline that an isotropic SED consists in a non-linear function f of the deformation gradient coupled with a constant parameter equivalent to a rigidity modulus c:

$$W = cf(\underline{\underline{C}}) = cf(\underline{\underline{F}}). \tag{10.28}$$

The same formalism also holds for the modeling of anisotropic non-linear hyperelastic behaviors. Fig. 10.5 presents three material types. Isotropic soft tissues are constituted of fibers with the same mechanical parameters in all directions. One can then visualize the anisotropy as a difference of material, *i.e.*, fiber, mechanical properties in specific directions, Fig. 10.5-B, or as the result of the organization of fibers sharing the same mechanical properties in preferred directions, see Fig. 10.5-C.

According to this simplified representation of a soft tissue, anisotropic modeling strategies part into two main categories: one is affecting the mechanical parameters of the fibers, and the other relies only on the structural organization of the fiber. The former, also the prevailing one, is a result of the stress with respect to the fiber's directions, meaning that the mechanical properties of the fiber change according to its orientation. The latter involves a strain penalty: the mechanical parameters of the fibers remain stable but their orientation impacts the mechanical behavior. The next subsections detail both modeling strategies.

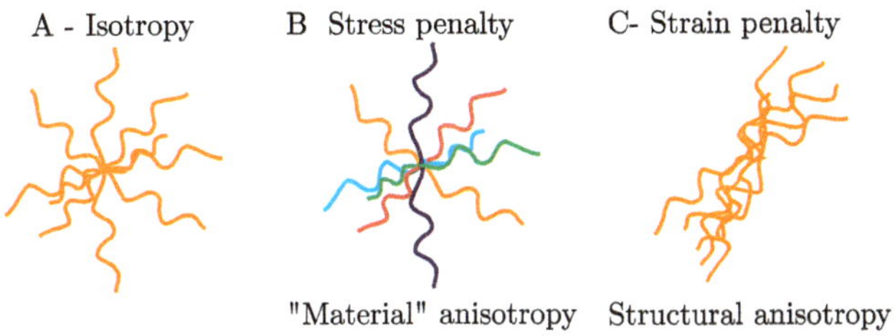

FIGURE 10.5

Schematic representation of isotropic and anisotropic soft tissues. A) Isotropic tissue: all the fibers have the same mechanical parameters and are homogeneously distributed. B) Anisotropic tissue: fibers are homogeneously distributed in space, but each fiber presents its own mechanical parameters. C) Anisotropic tissue: fibers, showing separately the same mechanical properties, are strongly oriented along a preferred direction.

10.4.1 Stress penalty

The first option considers that the tissues are made of an isotropic matrix reinforced with collagen fibers in specific directions (Lanir, 1983). If the structure of the tissues is fully known, or the fibers inducing the anisotropic behavior are determined in terms of orientation, the SED is decomposed with an isotropic and anisotropic part:

$$W(\underline{\underline{F}}) = W_{iso}(I_1, I_2, I_3) + W_{aniso}(\underline{\underline{F}}). \tag{10.29}$$

The fibers and their orientations should rule the anisotropic part of the SED.

To express their contribution in the SED, new invariants are introduced that depend on specific directions of fibers oriented along the unit vector $\underline{u}^i$ in spherical coordinates:

$$\underline{u}_i = (\cos(\theta_i)\sin(\phi_i), \sin(\theta_i)\sin(\phi_i), \cos(\phi_i)). \tag{10.30}$$

Most popular ones are I_4 and I_5:

$$\begin{aligned} I_4^{(i)} &= \underline{u}^i.\underline{\underline{C}}.\underline{u}^i, \\ I_5^{(i)} &= \underline{u}^i.\underline{\underline{C}}^2.\underline{u}^i, \end{aligned} \tag{10.31}$$

where i represents the ith identified fibers direction.

Similarly to the isotropic modeling, for one family of fibers, the SED can take polynomial, exponential, power development forms as summed up in Table 10.4.

The general form of the SED, with i directions of anisotropy will read

$$W = W_{iso} + \sum_{i=1}^{k} W_{aniso}\left(I_4^{(i)}, I_5^{(i)}\right). \tag{10.32}$$

This method is rather natural and follows the isotropy framework, recalled in Eq. (10.29). The anisotropic response is attained by adding specific directions $\underline{n}^i$

Table 10.4 Examples of I_4-, I_5-based anisotropic SED. For further examples, the reader is referred to the comprehensive work of Chagnon et al. (2015).

Reference	Type	SED
Kaliske (2000); Park and Youn (1998)	Polynomial	$W_{aniso} = \sum_{k=2}^{n} c_k (I_4 - 1)^k$
		$W_{aniso} = c_1(I_4 - 1) + c_2(I_5 - 1)$
		$W_{aniso} = c(I_5 - 1)^2$
Alastrué et al. (2010)	Polynomial	$W_{aniso} = c(\sqrt{I_4} - 1)^2$
Holzapfel et al. (2000)	Exponential	$\frac{c_1}{2c_2}\left(\exp\left(c_2(I_4 - 1)^2\right) - 1\right)$

and corresponding moduli c^i. This method is particularly adapted to a discrete representation of the microstructure, or a few directions of anisotropy (Holzapfel et al., 2000; Cai et al., 2015). A general stress penalty SED of a fiber family in direction θ_i would come from Eq. (10.29) to

$$W_{aniso} = c(\underline{u}^i) f(\underline{\underline{F}}, \underline{u}^i). \tag{10.33}$$

The SED is now a non-linear function of the deformation gradient and a fiber's direction $\underline{n}^i$. The rigidity modulus now also depends on the orientation $\underline{n}^i$.

Microsphere representation was historically used for the modeling of isotropic elastomers affected by the Mullins effect, and thus induced anisotropy (Gillibert et al., 2010; Göktepe and Miehe, 2005). The modeling canvas can indeed easily modify to introduce an anisotropic response. With a discrete integration scheme over the sphere surface, the global strain energy density is written

$$W = \sum_{i=1}^{n} \omega_i w(\nu_i), \tag{10.34}$$

where ω_i are the integration weights over the sphere surface. The elementary SED w can take several forms: statistical for a physically-based representation, polynomial, exponential, as recalled in Table 10.4 for phenomenological approaches (Alastrué et al., 2009, 2010; Holzapfel et al., 2015). Noting that ν_i is equivalent to I_4^i, it appears possible to change the apparent modulus c^i in specific directions.

If the directions of anisotropy are known and discrete, one could indeed modify the mechanical parameters of certain directions over the sphere surface (Morch et al., 2019). If the microstructure is more chaotic, this remains feasible within the micro-sphere framework. An orientation distribution function $\rho(\theta, \phi)$ (ODF) can be defined to describe the continuous 3D orientation of the fibers observed with an imaging technique (e.g., confocal microscopy, CT, microCT, small-angle X-ray scattering, ...). The global SED is still obtained with the integration over sphere of an elementary density. The network is still assumed to be a sphere, but the anisotropy is dealt with the introduction of the ODF in the SED expression. The ODF assigns to each direction distributed over the sphere surface an orientation concentration factor (Sáez et al., 2016). The elementary energy density is thus weighted by the ODF (Alastrué et al., 2009, 2010):

$$W = \sum_{i=1}^{n} \omega_i \rho(\theta_i, \phi_i) w(v_i). \tag{10.35}$$

The ODF only impacts the rigidity modulus of each elementary density. One can therefore note a penalization of the stresses with respect to the fiber direction. The general expression of the SED matches with Eq. (10.33), which transforms into

$$W = \sum_{i=1}^{n} \omega_i c_f(\underline{\boldsymbol{u}}_i) f(\underline{\underline{\boldsymbol{F}}}, \underline{\boldsymbol{u}}_i), \tag{10.36}$$

where c_f is a modulus depending on the orientation of the fiber and f a non-linear function of the deformation gradient and the unit vector corresponding to the fiber direction. The parameter introducing the anisotropy in the behavior is mostly the modulus, which changes with the fiber's direction, as in Fig. 10.5-B.

10.4.2 Strain penalty

Soft tissues are considered as polymers resulting from the entanglement of collagen and elastin fibers, whose mechanical properties depend on the arrangement of these fiber's network. The soft tissue mechanical response results from a material (collagen and elastin mechanical properties) and structural (arrangement of the fibers within the tissue) parameters. In the strain penalty modeling, one considers that the material properties are intrinsic, *i.e.*, collagen and elastin mechanical properties are independent from the tissue, specimen, or even individual, whereas the microstructure can vary between tissues, specimen, and individuals. The anisotropy fully relies on the microstructure of the tissue.

Astruc et al. (2019) developed another modeling strategy based on the observation of the collagen network in abdominal rectus sheath tissues. The structure tensor method is applied to 3D images of collagen and elastin to analyze their texture. It provides the distribution of the orientations observed in an anisotropic sample. This analysis helps to build an ellipsoid that depicts the anisotropy of the material: the main direction of the ellipsoid corresponds to the main direction of the fibers, the small axis characterize their distribution. In this approach, the unit sphere is replaced by an ellipsoid with a unit major axis and minor axes smaller than one.

The strain penalty approach was developed within the microsphere framework. The integration is however no longer performed on the unit sphere but on the ellipsoid that carries information about the microstructure and the anisotropy. Going back to the microsphere general expression, Eq. (10.22) and (10.35), it means that the unit vector $\underline{\boldsymbol{u}}_i$ over the unit sphere deforms into a non-unit vector $\underline{\boldsymbol{r}}_i$ over the ellipsoid

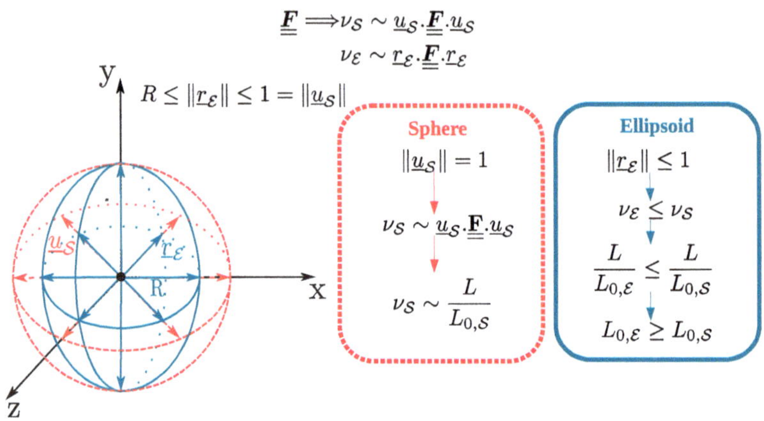

FIGURE 10.6

Interpretation of the effect of the ellipsoid on the mechanical quantities.

surface:

$$||\underline{\boldsymbol{u}}_i|| = 1 > ||\underline{\boldsymbol{r}}_i||. \tag{10.37}$$

Therefore the ellipsoid impacts now the projected deformation $\nu_i = \underline{\boldsymbol{r}}_i.\underline{\underline{\boldsymbol{F}}}.\underline{\boldsymbol{r}}_i$, Eq. (10.23) rather than the modulus. Fig. 10.6 provides a physical interpretation of the integration and projection over an ellipsoid surface in the case of transverse isotropy. The ellipsoid is a spheroid with the two minor axes of radius $R < 1$. The projected deformation ν_i is equivalent to a length ratio between the actual and initial length of the fiber. In the sphere representation, all fibers have the same initial length. In the ellipsoid one, the initial length varies according to the orientation of the fiber and is greater or equal to the initial length over the sphere surface.

The model representation induces a change in the initial length of the fibers, but not in their stiffness. The mechanical parameters of the elementary density can remain unchanged, considered as intrinsic material parameters. A general formulation of the SED in the discrete microsphere framework would be written from Eq. (10.36):

$$W = \sum_{i=1}^{n} \omega_i c_f f(\underline{\underline{\boldsymbol{F}}}, \underline{\boldsymbol{r}}_i), \tag{10.38}$$

where ω_i are the integration weights over the ellipsoid surface, c_f is the modulus of the fiber, independent of its orientation, and $\underline{\boldsymbol{r}}_i = (R\cos(\theta_i)\sin(\phi_i), R\sin(\theta_i)\sin(\phi_i), \cos(\phi_i))$ depends this time on the fibers distribution R.

In stress penalty approach, the distribution of the fibers was described with an ODF that gives to each direction of the unit sphere a penalty factor on the stiffness moduli: this method assigns locally different mechanical parameters to the fibers. The anisotropy of the distribution does not appear inside the non-linear function f. With strain penalty, the stiffness moduli, and all material parameters of the elementary density are intrinsic: each fiber have the same mechanical parameters, the anisotropy is dealt with changes in the fibers network. In short, collagen or elastin properties are universal, whereas their orientation and distribution are tissue- and even patient-specific.

10.4.3 Anisotropic representation: how to carefully open the Pandora's box?

Identifying anisotropic mechanical parameters requires, just as in the isotropic case, some specific attention. In the previous section, isotropic models were used to identify mechanical properties in specific conditions: when the material is isotropic of course, but more frequently with limited experimental measures that lack information about the general behavior of the sample. Fitting an anisotropic law on one uniaxial tensile test would be pointless: mechanical parameters would not be consistent in directions differing from the one used for the identification. With richer experimental measures, such as multiaxial testing, testing in several directions, local strain measures, anisotropic mechanical properties become accessible. It however requires some additional efforts.

We indeed recalled in the previous section that the principal stresses and stretches coincide in the isotropic case. This consideration is no longer true in the anisotropic case: the principal stresses and stretches referential do not align. In the principal stresses basis $(\underline{e}_I, \underline{e}_{II}, \underline{e}_{III})$, the right Cauchy–Green tensor, therefore the deformation gradient, is not diagonal:

$$\underline{\underline{\sigma}} = \begin{pmatrix} \sigma_I & 0 & 0 \\ 0 & \sigma_{II} & 0 \\ 0 & 0 & \sigma_{III} \end{pmatrix}_{(\underline{e}_I, \underline{e}_{II}, \underline{e}_{III})} \quad \text{and} \quad \underline{\underline{F}} = \begin{pmatrix} F_{11} & F_{12} & F_{13} \\ F_{21} & F_{22} & F_{23} \\ F_{31} & F_{32} & F_{33} \end{pmatrix}_{(\underline{e}_I, \underline{e}_{II}, \underline{e}_{III})} . \tag{10.39}$$

As already mentioned, it is impossible to experimentally measure all the components of the deformation gradient. In the loading plane, one can only compute the in-plane components: for example, in a uniaxial test in direction $\underline{e}_1$, the in-plane components of F_{11}, F_{22} and shears F_{12}, F_{21} can be measured as well as the transverse one in the plane $(O, \underline{e}_1, \underline{e}_3)$. However in the plane $(O, \underline{e}_2, \underline{e}_3)$, *i.e.*, where the sample is clamped, the shear components F_{23} and F_{32} cannot be easily measured; the only way would be tomography. This would be out of question for practical reasons, in the case of soft tissues characterization. In experimentally-based work, even with richer data (loading direction, multiaxial testing, field measures), this issue is often neglected.

With the isotropy and incompressibility assumption, we can take advantage of simplifications to fill in the deformation gradient. In the anisotropic case, these simplification no longer apply. It is mandatory to extrapolate the components of the deformation gradient. The minimization process takes both the direct (identification of the deformation gradient) and indirect (identification of the mechanical parameters of the modeling law) problems. This is often performed with an inverse method coupled with an FE simulation. If FE simulation is not used, this means that the optimization of the mechanical properties has to go with the resolution of the deformation gradient tensor (Astruc et al., 2019).

10.5 Conclusion

As mentioned by Chagnon et al. (2017) the selection of a constitutive model "strongly depend on the chosen experimental data." This comment is extremely true. It is indeed sometimes extremely difficult, if not impossible, to develop the experimental means according to the model we try to identify, it is thus easier to adapt the model to the experimental data that we are able to obtain.

According to James (1911) "The true, to put it very briefly, is only the expedient in the way of our thinking, just as the right is only the expedient in the way of our behaving." From a mechanical and experimental perspective, this only means that selecting and identifying a constitutive model for an application to pelvic tissues requires to be pragmatic and to analyze with objectivity the data and models at our disposal and to find the best compromise between accuracy, relevance, and realism.

Instead of presenting THE accurate models for the pelvic tissues, this chapter aimed at providing some elements of analysis and evaluation that help to consciously develop a better and more consistent strategy of identification of constitutive models with respect to the available experimental data and the expected degree of complexity.

References

Alastrué, V., Martínez, M.A., Doblaré, M., Menzel, A., 2009. Anisotropic micro-sphere-based finite elasticity applied to blood vessel modelling. Journal of the Mechanics and Physics of Solids (ISSN 0022-5096) 57 (1), 178–203. https://doi.org/10.1016/j.jmps.2008.09.005.

Alastrué, V., Sáez, P., Martínez, M.A., Doblaré, M., 2010. On the use of the Bingham statistical distribution in microsphere-based constitutive models for arterial tissue. Mechanics Research Communications (ISSN 0093-6413) 37 (8), 700–706. https://doi.org/10.1016/j.mechrescom.2010.10.001.

Arruda, E.M., Boyce, M.C., 1993. A three-dimensional constitutive model for the large stretch behavior of rubber elastic materials. Journal of the Mechanics and Physics of Solids (ISSN 0022-5096) 41 (2), 389–412. https://doi.org/10.1016/0022-5096(93)90013-6.

Astruc, L., Morch, A., Witz, J.-F., Novacek, V., Turquier, F., Hoc, T., Brieu, M., 2019. An anisotropic micro-ellipsoid constitutive model based on a microstructural description of fibrous soft tissues. Journal of the Mechanics and Physics of Solids 131, 56–73. https://doi.org/10.1016/j.jmps.2019.06.019.

Bäck, M., Gasser, T., 2013. Biomechanical factors in the biology of aortic wall and aortic valves diseases. Cardiovascular Research 99, 232–241.

Bergel, D., 1961. The static elastic properties of arterial wall. Journal of Physiology 156, 445–457.

Brieu, M., Chantereau, P., Gillibert, J., de Landsheere, L., Lecomte, P., Cosson, M., 2016. A nonlinear-elastic constitutive model for soft connective tissue based on a histologic description: application to female pelvic soft tissue. Journal of the Mechanical Behavior of Biomedical Materials (ISSN 1878-0180) 58, 65–74. https://doi.org/10.1016/j.jmbbm.2015.09.023.

Brunon, A., Bruyere-Garnier, K., Coret, M., 2010. Mechanical characterization of liver capsule through uniaxial quasi-static tensile tests until failure. Journal of Biomechanics 11 (43), 2221–2227.

Brunon, A., Bruyère-Garnier, K., Coret, M., 2011. Characterization of the nonlinear behaviour and the failure of human liver capsule through inflation tests. Journal of the Mechanical Behavior of Biomedical Materials (ISSN 1751-6161) 4 (8), 1572–1581. https://doi.org/10.1016/j.jmbbm.2010.12.016.

Cai, R., Holweck, F., Feng, Z.-Q., Peyraut, F., 2015. A new hyperelastic model for anisotropic hyperelastic materials with one fiber family. International Journal of Solids and Structures (ISSN 0020-7683) 84, 1–16. https://doi.org/10.1016/j.ijsolstr.2015.11.008.

Calvo, B., Sierra, M., Grasa, J., Muñoz, M., Peña, E., 2014. Determination of passive viscoelastic response of the abdominal muscle and related constitutive modelling. Journal of the Mechanical Behavior of Biomedical Materials 36, 47–58.

Chagnon, G., Ohayon, J., Martiel, J.-L., Favire, D., 2017. Chapter 1: hyperlasticity modeling for incompressible passive biological tissues. In: Biomechanics of Living Organs. In: Academic Press Series in Biomedical Engineering, London.

Chagnon, G., Rebouah, M., Favier, D., 2015. Hyperelastic energy densities for soft biological tissues: a review. Journal of Elasticity (ISSN 0374-3535) 120 (2), 129–160. https://doi.org/10.1007/s10659-014-9508-z.

Chuong, C.J., Fung, Y.C., 1983. Three-dimensional stress distribution in arteries. Journal of Biomechanical Engineering (ISSN 0148-0731) 105 (3), 268–274. https://doi.org/10.1115/1.3138417.

Cooney, G.M., Moerman, K.M., Takaza, M., Winter, D.C., Simms, C.K., 2015. Uniaxial and biaxial mechanical properties of porcine linea alba. Journal of the Mechanical Behavior of Biomedical Materials (ISSN 1878-0180) 41, 68–82. https://doi.org/10.1016/j.jmbbm.2014.09.026.

Cosson, M., Lambaudie, E., Boukerrou, M., Lobry, P., Crépin, G., Ego, A., 2004. A biomechanical study of the strength of vaginal tissues, results on 16 post-menopausal patienties presenting genital prolapse. European Journal of Obstetrics & Gynecology and Reproductive Biology 112, 201–205.

Doll, S., Schweizerhof, K., 2000. On the development of volumetric strain energy functions. Journal of Applied Mechanics 67, 17–21.

Egorov, V., Schastlivtsev, I., Prut, E., Baranov, A., Turusov, R., 2002. Mechanical properties of the human gastrointestinal tract. Journal of Biomechanics 10 (35), 1417–1425.

Ettema, G., Goh, J., Forwood, M., 1998. A new method to measure elastic properties of plastic-viscoelastic connective tissue. Medical Engineering & Physics 20, 308–314.

Feola, A., Duerr, R., Moalli, P., Abramowitch, S., 2013. Changes in the rheological behavior of the vagina in women with pelvic organ prolapse. International Urogynecology Journal (ISSN 0937-3462) 24 (7), 1221–1227. https://doi.org/10.1007/s00192-012-2002-x.

Fukahori, Y., Seki, W., 1992. Molecular behaviour of elastomeric materials under large deformation: 1. Re-evaluation of the Mooney-Rivlin plot. Polymer (ISSN 0032-3861) 33 (3), 502–508. https://doi.org/10.1016/0032-3861(92)90726-D.

Fung, Y., 1967. Elasticity of soft tissues in simple elongation. American Journal of Physiology-Legacy Content (ISSN 0002-9513) 213 (6), 1532–1544. https://doi.org/10.1152/ajplegacy.1967.213.6.1532.

Fung, Y., 1973. Biorheology of soft tissues. Biorheology 10, 139–155.

Fung, Y., 1993. Biomechanics. Mechanical Properties of Living Tissue, 2nd ed. Springer, New York, NY.

Fung, Y., Fronek, K., Patitucci, P., 1979. Pseudoelasticity of arteries and the choice of its mathematical expression. The American Journal of Physiology (ISSN 0002-9513) 237 (5), H620–H631. https://doi.org/10.1152/ajpheart.1979.237.5.h620.

Gao, Z., Lister, K., Desail, J., 2010. Constitutive modeling of liver tissue: experiment and theory. Annals of Biomedical Engineering 38 (2), 505–516. https://doi.org/10.1007/s10439-009-9812-0.

Gillibert, J., Brieu, M., Diani, J., 2010. Anisotropy of direction-based constitutive models for rubber-like materials. International Journal of Solids and Structures (ISSN 0020-7683) 47 (5), 640–646. https://doi.org/10.1016/j.ijsolstr.2009.11.002.

Göktepe, S., Miehe, C., 2005. A micro-macro approach to rubber-like materials. Part III: the micro-sphere model of anisotropic Mullins-type damage. Journal of the Mechanics and Physics of Solids (ISSN 0022-5096) 53 (10), 2259–2283. https://doi.org/10.1016/j.jmps.2005.04.010.

Hernández, B., Peña, E., Pascual, G., Rodríguez, M., Calvo, B., Doblaré, M., Bellón, J.M., 2011. Mechanical and histological characterization of the abdominal muscle. A previous step to modelling hernia surgery. Journal of the Mechanical Behavior of Biomedical Materials (ISSN 1751-6161) 4 (3), 392–404. https://doi.org/10.1016/j.jmbbm.2010.11.012.

Holzapfel, G.A., Gasser, T.C., Ogden, R.W., 2000. A new constitutive framework for arterial wall mechanics and a comparative study of material models. Journal of Elasticity 61 (1), 1–48.

Holzapfel, G.A., Niestrawska, J.A., Ogden, R.W., Reinisch, A.J., Schriefl, A.J., 2015. Modelling non-symmetric collagen fibre dispersion in arterial walls. Journal of the Royal Society Interface. https://doi.org/10.1098/rsif.2005.0073.

Humphrey, J., 1995. Mechanics of the arterial wall: review and directions. Critical Reviews in Biomedical Engineering (ISSN 0278-940X) 23 (1–2), 1–162. https://doi.org/10.1615/critrevbiomedeng.v23.i1-2.10.

James, W., 1911. The Meaning of Truth. Longman Green and Co., New York, NY.

Jayyosi, C., Bruyère-Garnier, K., Coret, M., 2017. Geometry of an inflated membrane in elliptic bulge tests: evaluation of an ellipsoidal shape approximation by stereoscopic digital image correlation measurements. Medical Engineering & Physics (ISSN 1350-4533), 1–8. https://doi.org/10.1016/j.medengphy.2017.06.020.

Jean-Charles, C., Rubod, C., Brieu, M., Boukerrou, M., Fasel, J., Cosson, M., 2010. Biomechanical properties of prolapsed or non-prolapsed vaginal tissue: impact on genital prolapse surgery. International Urogynecology Journal (ISSN 1433-3023) 21 (12), 1535–1538. https://doi.org/10.1007/s00192-010-1208-z.

Kaliske, M., 2000. A formulation of elasticity and viscoelasticity for fibre reinforced material at small and finite strains. Computer Methods in Applied Mechanics and Engineering 185, 225–243.

Kuhn, W., 1936. Beziehungen zwischen Molekülgrösse, statistischer Molekülgestalt und elastischen Eigensehaften hochpolymerer Stoffe. Kolloid-Zeitschrift, 258–271.

Kuhn, W., Grün, F., 1942. Beziehungen zwischen elastischen Konstanten und Dehnungsdoppelbrechung hochelastischer Stoffe. Kolloid-Zeitschrift (ISSN 0303-402X) 101 (3), 248–271. https://doi.org/10.1007/BF01793684.

Lanir, Y., 1983. Constitutive equations for fibrous connective tissues. Journal of Biomechanics 16 (1), 1–12.

Lyons, M., Winter, D.C., Simms, C.K., 2014. Mechanical characterisation of porcine rectus sheath under uniaxial and biaxial tension. Journal of Biomechanics (ISSN 0021-9290) 47 (8), 1876–1884. https://doi.org/10.1016/j.jbiomech.2014.03.009.

Mallett, K., Arruda, E.M., 2017. Digital image correlation-aided mechanical characterization of the anteromedial and posterolateral bundles of the anterior cruciate ligament. In: Gradient Biomaterials Special Issue. Acta Biomaterialia (ISSN 1742-7061) 56, 44–57. https://doi.org/10.1016/j.actbio.2017.03.045.

Merckel, Y., Diani, J., Roux, S., Brieu, M., 2011. A simple framework for full-network hyperelasticity and anisotropic damage. Journal of the Mechanics and Physics of Solids (ISSN 0022-5096) 59 (1), 75–88. https://doi.org/10.1016/j.jmps.2010.09.010.

Miehe, C., Göktepe, S., Lulei, F., 2004. A micro-macro approach to rubber-like materials - Part I: the non-affine micro-sphere model of rubber elasticity. Journal of the Mechanics and Physics of Solids (ISSN 0022-5096) 52 (11), 2617–2660. https://doi.org/10.1016/j.jmps.2004.03.011.

Moalli, P., Howden, N.S., Lowder, J.L., Navarro, J., Debes, K.M., Abramowitch, S.D., Woo, S.L.-Y., 2005. A rat model to study the structural properties of the vagina and its supportive tissues. American Journal of Obstetrics and Gynecology (ISSN 0002-9378) 192 (1), 80–88. https://doi.org/10.1016/j.ajog.2004.07.008.

Morch, A., Astruc, L., Lecomte-grosbras, P., Soulat, D., Brieu, M., 2019. Modeling of anisotropic hyperelastic heterogeneous knitted fabric reinforced composites. Journal of the Mechanics and Physics of Solids 127, 47–61. https://doi.org/10.1016/j.jmps.2019.03.006.

Morch, A., Astruc, L., Mayeur, O., Witz, J.-F., Lecomte-Grosbras, P., Brieu, M., 2020. Is there any objective and independent characterization and modeling of soft biological tissues? Journal of the Mechanical Behavior of Biomedical Materials (ISSN 1751-6161) 110, 103915. https://doi.org/10.1016/j.jmbbm.2020.103915.

Morch, A., Pouseele, B., Doucède, G., Witz, J.-F., Lesaffre, F., Lecomte-Grosbras, P., Brieu, M., Cosson, M., Rubod, C., 2017. Experimental study of the mechanical behavior of an explanted mesh: the influence of healing. Journal of the Mechanical Behavior of Biomedical Materials 65, 190–199.

Nolan, D.R., Gower, A.L., Destrade, M., Ogden, R.W., McGarry, J.P., 2014. A robust anisotropic hyperelastic formulation for the modelling of soft tissue. Journal of the Mechanical Behavior of Biomedical Materials (ISSN 1751-6161) 39, 48–60. https://doi.org/10.1016/j.jmbbm.2014.06.016.

O'Hagan, J., Samani, A., 2009. Measurement of the hyperelastic properties of 44 pathological ex vivo breast tissue samples. Medical Physics 10 (39), 5835–5848.

Park, H.C., Youn, S.-k., 1998. Finite element analysis and constitutive modelling of anisotropic nonlinear hyperelastic bodies with convected frames. Computer Methods in Applied Mechanics and Engineering 7825 (97).

Peña, E., Calvo, B., Martínez, M.A., Martins, P., Mascarenhas, T., Jorge, R.M.N., Ferreira, A., Doblaré, M., 2010. Experimental study and constitutive modeling of the viscoelastic mechanical properties of the human prolapsed vaginal tissue. Biomechanics and Modeling in Mechanobiology (ISSN 1617-7959) 9 (1), 35–44. https://doi.org/10.1007/s10237-009-0157-2.

Peña, E., Martins, P., Mascarenhas, T., Jorge, R.M.N., Ferreira, A., Doblaré, M., Calvo, B., 2011. Mechanical characterization of the softening behavior of human vaginal tissue. Journal of the Mechanical Behavior of Biomedical Materials (ISSN 1751-6161) 4 (3), 275–283. https://doi.org/10.1016/j.jmbbm.2010.10.006.

Rubod, C., Boukerrou, M., Brieu, M., Dubois, P., Cosson, M., 2007. Biomechanical properties of vaginal tissue. Part 1: new experimental protocol. Journal of Urology (ISSN 0022-5347) 178 (1), 320–325. https://doi.org/10.1016/j.juro.2007.03.040.

Rubod, C., Brieu, M., Cosson, M., Rivaux, G., Clay, J.-C., de Landsheere, L., Gabriel, B., 2012. Biomechanical properties of human pelvic organs. Urology (ISSN 0090-4295) 79 (4), 968.e17–968.e22. https://doi.org/10.1016/j.urology.2011.11.010.

Sáez, P., García, A., Peña, E., Gasser, T.C., Martínez, M.A., 2016. Microstructural quantification of collagen fiber orientations and its integration in constitutive modeling of the porcine carotid artery. Acta Biomaterialia 33, 183–193. https://doi.org/10.1016/j.actbio.2016.01.030.

Scholten, P., Schultz, A., Luchies, C., Miller, J., 1988. Motions and loads with the human pelvis: a biomechanical model study. Journal of Orthopaedic Research 6 (6), 840–850.

Treloar, L.R.G., 1944. Stress-strain data for vulcanised rubber under various types of deformation. Transactions of the Faraday Society 40.

Treloar, L.R.G., Riding, G., 1979. A non-Gaussian theory for rubber in biaxial strain. I. Mechanical properties. Proceedings of the Royal Society of London. Series A 369, 261–280.

Wex, C., Arndt, S., Stoll, A., Bruns, C., Kupriyanova, Y., 2015. Isotropic incompressible hyperelastic models for modelling the mechanical behaviour of biological tissues: a review. Biomedical Engineering (Biomedizinische Technik) 60 (6), 577–592. https://doi.org/10.1515/bmt-2014-0146.

Whitford, C., Studer, H., Boote, C., Meek, K.M., Elsheikh, A., 2015. Biomechanical model of the human cornea: considering shear stiffness and regional variation of collagen anisotropy and density. Journal of the Mechanical Behavior of Biomedical Materials (ISSN 1878-0180) 42, 76–87. https://doi.org/10.1007/978-94-007-4059-4_12.

PART

3

Clinical imaging, investigations tools, and characterization

CHAPTER

11 Medical imaging and patient-specific modeling of women pelvic system: application to magnetic resonance images

Zhifan Jiang[a], Olivier Mayeur[b], Jean-François Witz[b], Pauline Lecomte-Grosbras[b], Michel Cosson[c], and Mathias Brieu[d]

[a]*Sheikh Zayed Institute for Pediatric Surgical Innovation, Children's National Hospital, Washington, DC, United States*

[b]*Univ. Lille, CNRS, Centrale Lille, UMR 9013-LaMcube – Laboratoire de Mécanique, Multiphysique, Multiéchelle, Lille, France*

[c]*CHU Lille, Service de Chirurgie Gynécologique, Lille, France*

[d]*College of Engineering, Computer Science and Technology, Dept. of Mechanical Engineering, California State University, Los Angeles, CA, United States*

11.1 Introduction

The dysfunctions in the female pelvic system, such as the genital prolapse, have a large impact on the female population (Samuelsson et al., 1999; Swift, 2000). As medical imaging, especially the magnetic resonance imaging (MRI), has become an accurate source of information for pelvic floor disorder diagnoses and treatment planning (Pannu et al., 2000), numerical finite element (FE) simulation is the essential tool for studying the biomechanics of female pelvic system and better understanding its physio-pathology, highly correlated with pelvic mobility (Bump et al., 1996; Dell'oro et al., 2013; Lecomte-Grosbras et al., 2015; Meuleman et al., 2009). Medical images provide spatial and temporal information on anatomical structures for computing simulation inputs; reciprocally, the simulation can enrich the images by providing objective and quantitative analysis that cannot be retrieved directly from images. This chapter will focus efforts on the missing part that connects MRI and FE simulation: how to build from MRI a patient-specific and simulation-compatible 3D geometric model for the pelvic system.

The female pelvic system is complex, including principal pelvic organs (bladder, vagina, rectum), pelvis (bone structures), suspension devices (*e.g.*, ligaments and fascia), and support system (muscle: pelvic floor). To study hypermobility and large deformation presented in the pelvic system, particular interest has been paid to

Biomechanics of the Female Reproductive System: Breast and Pelvic Organs. https://doi.org/10.1016/B978-0-12-823403-7.00024-5

modeling soft tissues: pelvic organs, supporting structures, and muscles (Lecomte-Grosbras et al., 2013; Mayeur et al., 2017, 2014; Rao et al., 2010; Vallet et al., 2011). The reconstructed 3D geometric models are required to be compatible with simulation models, in terms of regularity and continuity and be consistent with patient anatomy. However, the geometric modeling of the pelvic system from medical images (MRI) has been found to be a complex operator-dependent and time-consuming process, not adapted to patient-specific applications.

The classic way to obtain appropriate geometric models is to apply a mesh generation algorithm to segmented volumes, and then use directly or repair the surface meshes for simulation. Many applications have been developed to automatically segment the targeted volumes or to improve the precision (He et al., 2019; Heimann and Meinzer, 2009; Hoyte et al., 2011; Lei et al., 2020; Paragios et al., 2006). However, little attention has been paid to the geometric model itself. Without a flexible and compatible geometric support, the potential use of meshes will be very limited due to its undesirable quality and its deviation from MRI after manual repairing and fairing. Moreover, it can hardly be changed to be advantageous to meet other simulation requirements, such as remeshing.

Due to their flexibility, precision, and versatility, non-uniform rational B-splines (NURBS) models arc widely used for shapes modeling in mechanics. Although it contains complex mathematical concepts and methods, advanced computer-aided design (CAD) software, which requires extensive training, can help create NURBS geometries for various purposes. Hence, creating parametric surfaces (NURBS) from segmented volume or points cloud is another way to model the pelvic system. This procedure is nevertheless time-consuming and requires a lot of experience of the operator, for example, how to merge a surface patch onto a points cloud is usually based on subjective design experience.

Virtual image correlation (VIC) method was introduced and developed by Réthoré and François (2014) and Semin et al. (2011), to identify and track objects in 2D images captured during a mechanical experiment. With its extended 3D version applied to medical images, a new approach for efficiently building readily usable simulation models with NURBS surface has been developed (Courtecuisse et al., 2020; Jiang et al., 2019, 2015). The advantages remain not only in efficiency, but the method also provides a standardized way to build 3D numerical models for pelvic system and is more suitable for patient-specific application., where the amount of available data continues to increase.

In the next sections, a more detailed review of MRI protocols in the clinical routine will be presented and various geometry representations will be elaborated. Particular attention will then be paid to discuss the advantages and limitations of VIC method in reconstructing the whole 3D geometric numerical model of the pelvic system, an approach designed to provide a streamlined solution to close the gap between medical images and FE simulation.

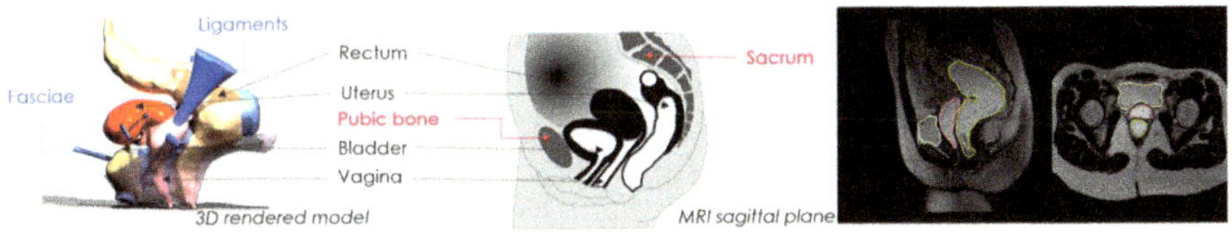

FIGURE 11.1

Pelvic system and MRI. From left to right: 3D rendered model, sagittal view illustration, MRI in sagittal and axial planes. Structures in blue are involved in simulation but not observable in MRI.

11.2 Materials and methods

11.2.1 Magnetic resonance imaging

Magnetic resonance imaging (MRI) has played an important role in the diagnosis of pelvic disorders since the early 1990s (Hamm and Forstner, 2007). The dynamic MRI (or cine-MRI) sequences have become the most essential imaging protocol for examining pelvic mobility. A usual dynamic MRI contains a sequence of 2D temporal images recorded in the sagittal plane. MRI data presented in this chapter will however focus on the static MRI. For modeling the anatomical structures of the pelvic system, 3D spatial information is more relevant and necessary.

Our static MRI acquisition follows a similar protocol that meets the clinical routine requirement (not specifically designed protocols for research) as shown in Fig. 11.1:

- T2-weighted MRI was acquired at resolution of $0.59*0.59$ mm^2 pixel^{-1}, the thickness of slices is 4 mm and with $512*512$ pixels in plane.
- Multi-slice 2D MRIs were scanned separately in three directions (sagittal, axial, and coronal).
- Contrast agent was injected to vaginal and rectal cavities to highlight vagina and rectum, while patient was drinking water to highlight bladder.

There also exists static 3D volumetric MRI: multiple magnetic gradient fields in the three directions are applied simultaneously, while the images are being acquired. At the end of acquisition, a 3D volume with voxels is directly reconstructed, which can be converted to 2D slices as well. This type of protocol is generally used when high-resolution isotropic voxels are required. However, it requires longer acquisition time, thus more suitable for anatomical structures that have less motion and does not meet the clinical routine standard in pelvic disorders diagnosis.

Despite its high contrast and non-invasive nature, MR images have some features not advantageous to patient-specific modeling of the female pelvic system.

- The MR images were acquired using protocols consistent with clinical routine standards. Consequently, the obtained sequences have a limited number of slices, which lead to highly anisotropic voxels when we use multiple slices to gener-

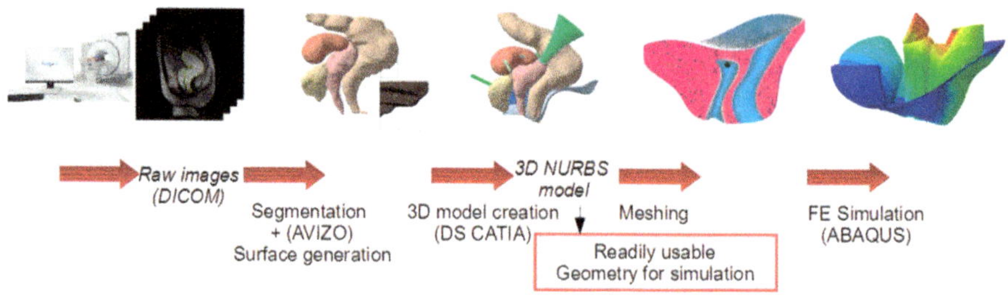

FIGURE 11.2

Pipeline of processing from MRI to 3D models for simulation. Different geometry representations are used.

ate 3D geometric structures. Regardless of geometry representations (parametric surfaces or meshes), the anisotropic property makes it very challenging to create smooth and high-quality surfaces.

- Extra attention needs to be paid to verify that these separately recorded sequences are defined in the same coordinate system setting before downstream processing.
- Suspension structures, such as fascia and ligaments, are important components that affect pelvic mobility and create pathologies, such as genital prolapse (DeLancey, 1992; Tansatit et al., 2013). For the same reason, these structures should be included in the 3D model for FE simulation (Mayeur et al., 2014), but compared to the principal organs and bones, suspension devices are very difficult to be identified in MRI.

In the following sections, we are interested in investigating how to leverage the potential of geometrical modeling methods for the reconstruction of the female pelvic system, applied to the current MRI protocols, to address the listed challenges.

11.2.2 Geometry representation

The objective of modeling the pelvic system is not only to describe the shape or to measure quantities such as volume or length, which is the case of many segmentation tasks. In fact, the idea is to provide a readily usable 3D model for FE simulation. The latter is able to provide objective and precise analysis of pelvic mobility with a mechanical approach. Thus the methods do not focus on how to automatically segment the target structures, but on generating functional 3D models consistent with the organs' anatomy and suitable for simulation.

Before making a decision on the choice of geometry representation and the reconstruction method, it is useful to provide a general description of different geometry representations involved in the following pipeline of processing, from MRI to 3D simulable models (Fig. 11.2):

- Identity target structures: definition of the structures to be extracted from an original image. Considering the image as a set of voxels of intensity values, the

segmented structure is represented by a subset of labeled voxels, *e.g.*, value 1 is assigned to the voxels belonging to the structure, otherwise 0 is assigned.
- Obtain geometric information from volumetric information: the marching cube algorithm and its variety allow to create a discrete iso-surface (defined by surface triangle mesh) from the segmented volume (Lorensen and Cline, 1987; Shekhar et al., 1996). These surfaces usually have terracing effects.
- Generate a smooth surface model: depending on applications, the generated surface model has different quality requirements. One option is to directly repair the triangle mesh using smoothing and fairing algorithms (Botsch et al., 2010). The other option is to create a parametric surface model (*i.e.*, NURBS) that fits the points cloud. The second representation is the standard model definition used in CAD software as well.
- Provide consistent and usable models as the geometric support for numerical simulation: FE simulation normally requires high-quality surface meshes in terms of continuity, curvature, etc. Sometimes it needs volumetric meshes as well (*e.g.*, tetrahedra or hexahedra filling the space enveloped by the surface). Building a parametric surface model has more advantages to FE computation for the following reasons: i) compared to discrete meshes, these continuous models have better mathematical properties, such as continuity and differentiability; ii) it is much more flexible for remeshing and generating volumetric meshes if necessary.

Parametric NURBS surface model is a more powerful geometry representation, and existing CAD software (*e.g.*, CATIA[1]) provide stable, reliable, and easy-to-use tools to create such geometric models. However, constructing complex and complete 3D NURBS models of the pelvic system remains a very time-consuming and operator-dependent procedure. The next sections will present a model-to-image registration framework. The streamlined method enables 3D reconstruction of the pelvic system using NURBS surface directly from volumetric information.

11.2.3 Pelvic organs modeling using virtual image correlation

As aforementioned, the principal pelvic organs (bladder, vagina, rectum) are highlighted by contrast agents (gel and water) during the acquisition. Thus the modeling of these identifiable organs is based on image contrast. A virtual image correlation (VIC) method was developed and customized to conduct this task. This method is based on the following functions and variables:

- an input real image $\mathbf{I_R}$ (3D volume with gray level values or binarized values);
- a well-defined initial geometry (a simple closed B-spline surface that will be deformed later to meet the complexity of organ shape);
- a virtual image $\mathbf{I_V}$ generated from the geometry (a level-set function is used to define intensity values in the neighborhood of the surface);

[1] https://www.3ds.com/products-services/catia/.

- an objective function $\mathbf{F}(\mathbf{I_R}, \mathbf{I_V})$ (a function defining the difference between the virtual image and the real one);
- an optimizer (find the best correlation between the virtual and real images by minimizing **F,** and estimate the parameters for the optimal geometry).

11.2.4 Generic 3D geometry

The generic 3D NURBS surface is defined by a bi-directional $n \times m$ control point net $\{p_{i,j} \mid 0 \le i \le n-1, 0 \le j \le m-1\}$. A single NURBS patch can be computed as a tensor product of control points and B-spline basis functions of degree p and q: $N_{i,p}N_{j,q}$. A surface point at parameters $S(u, v)$ is defined by the following equation:

$$S(u, v) = \sum_{\substack{0 \le j \le m-1 \\ 0 \le i \le n-1}} N_{i,p}(u) N_{j,q}(v) p_{i,j}, \ \forall (u, v) \in [\mathbf{0}, \mathbf{1}]^2 .$$

For more mathematical concepts and methods of NURBS, the reader can refer to Piegl and Tiller (1997).

By setting the value of the degrees of basis functions, *e.g.*, $p = q = 3$, highly continuous surface patch (differentiability everywhere except the boundaries of patches) can be obtained. The shape of the surface can be deformed by modifying the position of the control points $p_{i,j}$. Hence, control points will be considered as degrees of freedom for optimization in the remainder of this section.

The next important issue that needs to be addressed is how to define the control point net to model a pelvic organ. Although in CAD software multi-patch geometries are commonly used, such complex models are difficult to be handled during optimization. Particularly, continuity at patches' boundaries should be maintained. Jiang et al. (2019) proposed a simplified B-spline surface to model the organs, which is a trade-off between complexity and functionality. In the simplified version, a single closed patch is used to define the surface. In one direction, control points at the extremities $\{p_{0...n-1,0}\}$ and $\{p_{0...n-1,m}\}$ are merged to form the two poles closing the top and bottom; in the other direction, $m-2$ pairs of control points are joined together to form the lateral boundary (Fig. 11.3).

To satisfy the continuity C_1 at the boundaries, tangential constraints are applied on these control points during optimization, *e.g.*, lateral points must remain collinear and adjacent points of the poles have to be coplanar. In addition to the definition in Jiang et al. (2019), refinement of control point net by adding control points near the boundaries can make the optimization more stable. In a practical view, more control points near the boundaries can prevent large deformation of the shape when tangential constraints are imposed at each iteration of optimization.

11.2.5 Virtual image

The virtual image links the medical image and the geometry that were presented in the previous section. The virtual image domain is in the neighborhood of the surface

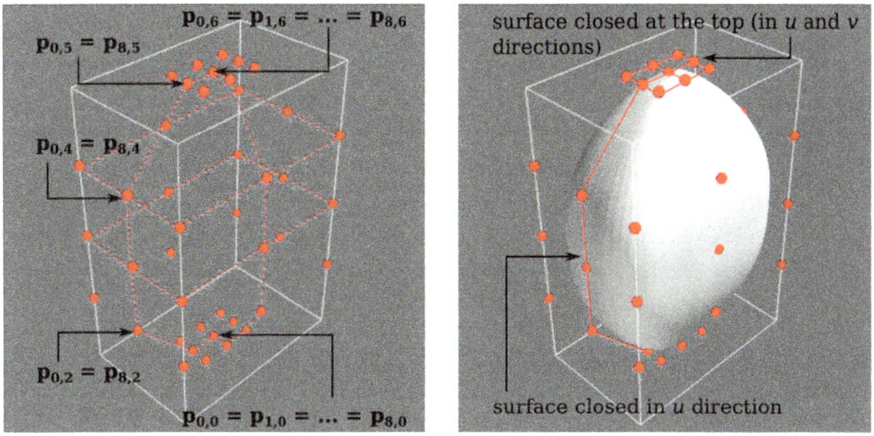

FIGURE 11.3

Example of control point net of the organ model: control points in red and closed surface in white (Jiang et al., 2019).

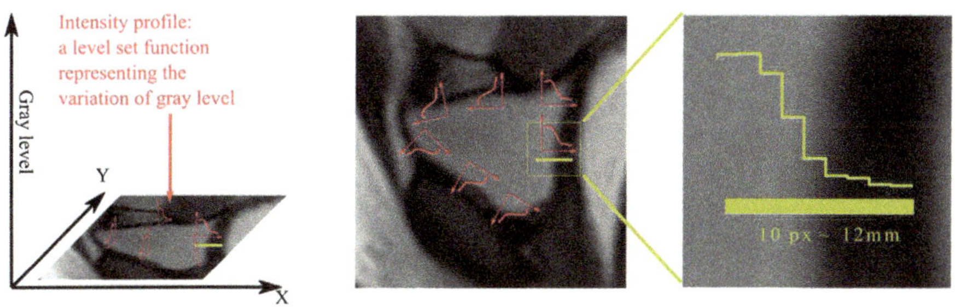

FIGURE 11.4

Illustration of virtual image generation. An intensity profile is defined by a level set function along the contour. The profile simulates the change in gray level (Jiang et al., 2019).

with a given range τ, and the image intensity (gray level) at a position x near the surface is determined by a level set function L:

$$L(d) = \begin{cases} A_{max} \; if \; d < 0, \\ A_{min} + \frac{A_{max} - A_{min}}{2} \times (1 + \cos(\frac{\pi d}{\tau})) \; if \; 0 \leq d \leq \tau, \\ A_{min} \; if \; d > \tau, \end{cases}$$

where d is the distance between the position x and the B-spline surface, A_{max} and A_{min}, are the gray levels inside and outside the organ enveloped by the surface. Thus the virtual image simulates the change in gray levels on the real image, but in a limited zone and in a homogeneous way (Fig. 11.4).

11.2.6 Objective function and optimization

By combining the input image, virtual image, and geometry into an objective function E_{image}, the VIC method can be formulated as an optimization problem. To compute E_{image}, discrete points are sampled in the neighborhood of the surface. A gradient descent method is then applied to find the optimal position of the control points (the degrees of freedom). In a more advanced version, we can directly add regularization terms to the objective function, to control the smoothness and regularity of the surface, for example, by using thin plate energy (Botsch et al., 2010).

$$\begin{aligned} E_{image} &= \sum_{u,v,d} \left[I_R\left(X\left(u,v,d\right)\right) - I_V\left(X\left(u,v,d\right)\right)\right]^2 \\ &= \sum_{u,v,d} \left[I_R\left(X\left(u,v,d\right)\right) - L\left(d\right)\right]^2 \end{aligned}$$

11.2.7 Pelvic floor modeling for simulation

In addition to the pelvic organs that can be modeled by the VIC method, building a numerical model of the pelvic floor is necessary for simulation as well. The pelvic floor is an anatomical structure supporting the pelvic organs and numerically it helps define the boundary conditions and plays an important role in the pelvic mobility simulation (Lecomte-Grosbras et al., 2013). However, it can hardly be identified directly on MRI and has to be modeled using the positions of the pelvic organs.

Since these anatomical structures are not identifiable on MRI, their modeling is principally based on prior knowledge of anatomy and empirical data. Concretely, the generated pelvic floor must satisfy the following conditions:

- cover the lower and lateral parts of the pelvic organs,
- fit the pelvis size (especially the distance ***dsag*** between sacral promontory and inner pubic arch),
- compatible for defining boundary conditions (with same continuity as the organs' model, well fitted to the bottom of the organs, and modeled by symmetric NURBS patches).

Firstly, three measurement points are manually selected by an operator (inner pubic arch, end of the sacrum, and sacral promontory). From the three points, a sagittal symmetric plane of the pelvic floor can be constructed. In the perpendicular direction to the plane, the dimension of the pelvic floor is determined by ***dsag*** (computed from statistic of pelvis). Since the pelvic floor is assumed symmetric, only the left part is modeled using two NURBS patches. Extremities of the organs are used to determine the boundaries of each patch; tangent and curvature of the patches are also controlled. Fig. 11.5 illustrates the generation of the pelvic floor.

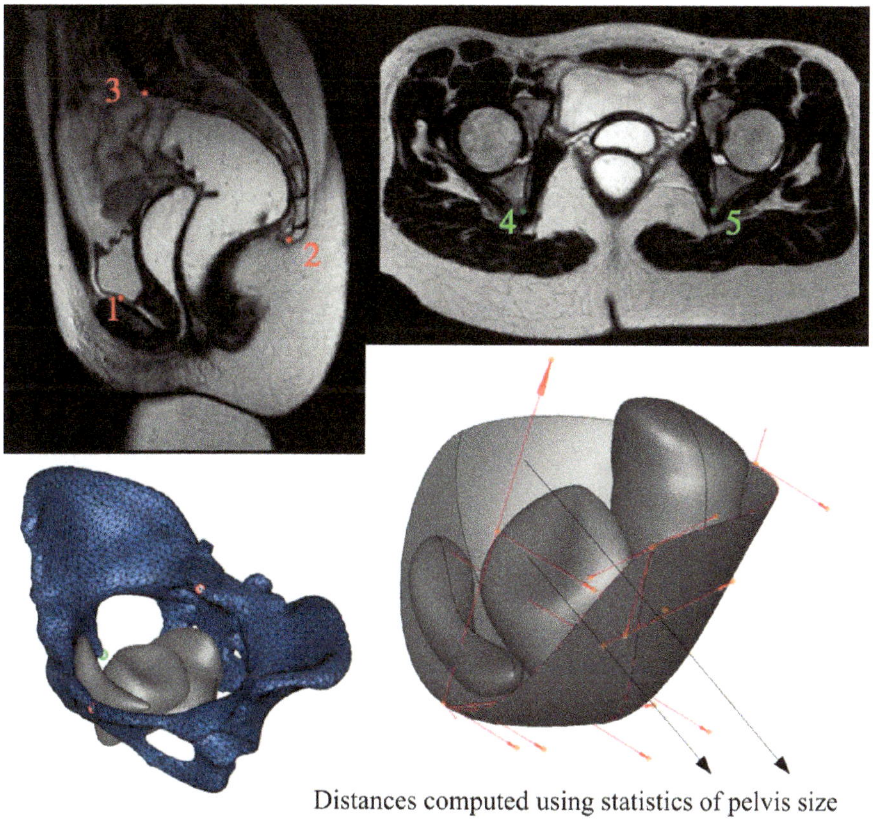

FIGURE 11.5

Generation of pelvic floor and pelvis using five manually selected points. The registered pelvis is shown in blue. The red arrows indicate tangential constraints applied to the patches of the pelvic floor.

11.3 Applications and discussion

Jiang et al. (2019) proposed the VIC method to build the B-spline surface models of the pelvic organs. Firstly, the method was validated using a prototypical image and geometry. In the experiments, a single-patch bell-shaped geometry was used, and an artificial 3D image was created with the presence of a bell-shaped pattern in the middle. The algorithm found the optimal affine registration, which aligned the geometry and image with the VIC method. In this experiment, the degrees of freedom are parameters of the transform matrix (rotation, scaling, etc.). Initially, the geometry was set overlapped with the image pattern, then a random affine transformation made it diverge from its initial position. Finally, the registration is to find the best transformation that can return the geometry to its initial position. Results showed that absolute errors of the transformation parameters ranged from 0.01 to 0.05. Moreover, the method demonstrated a sub-voxel precision and was less influenced by the change in image resolution.

The VIC method was then applied to real MRI datasets to build 3D models of bladder, vagina, and rectum. Due to the large difference between the initial geometry and the shape of the organs, optimization was conducted in two steps: i) a global registration (affine transformation) allowing a better initialization of the geometry; ii) a local registration adjusts the position of control points to obtain the final optimal 3D geometry. The proposed 3D modeling method provided functional geometries readily usable for simulation. To evaluate their overlap with the ground truth (manually segmented volume from MRI), point-wise distances were computed between the generated surface and the points cloud. The average of these distances was used to evaluate the deviation of the surfaces from ground truth, and approximately 10,000 points were projected onto each surface. Results showed that the average deviation is 2.36 mm for bladder, 1.81 mm for vagina, 2.90 mm for rectum, and 2.35 mm for all organs combined. The images have a resolution at 0.59 mm pixel^{-1} or 0.73 mm pixel^{-1}, and the slice thickness ranges from 3 to 4 mm.

Due to intrinsic properties and image acquisition artefacts, anatomical structures in MRI usually have irregular and unwanted details on their surface. However, the objective of the presented registration is to find a trade-off between the precision of the model and its functionality: particular attention has been paid to guarantee the surface quality and its compatibility with FE simulation, moreover, it allows adaptive remeshing with the surface support. A parametric CAD model will be a better support for creating tetrahedral meshes for FE simulation as well. Fig. 11.6 illustrates the generated organs and pelvic floor overlapped with the MRI image; a generic pelvis (red bone structure) was coarsely registered to this image for visual effect and was not taken into account while performing FE simulation.

More validation and evaluation of errors are required to find the best trade-off and to understand whether the reconstruction error will reduce the reliability of simulation, or if it will instead increase the performance of simulation, because it is a mechanically more stable model, even if it is not the closest model to the images.

Another limitation is that with a fixed number of control points and one single patch, the model is not flexible to a certain degree. The generation of the models may have undesirable surfaces in some cases, because it is difficult to deal with extremely degenerated or irregular shapes. For this reason, post-processing with operator's intervention would be necessary. To facilitate this operation, the possibility of a local correction technique with kernel functions (radial basis function) was investigated in Jiang et al. (2020). A limited number of surface points can be moved to modify locally the shape of NURBS surface or mesh.

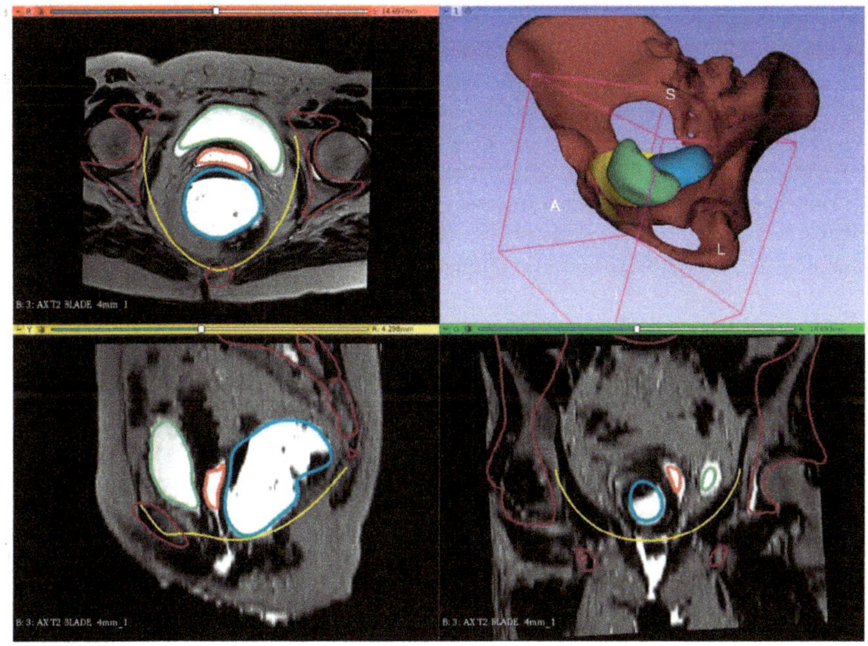

FIGURE 11.6

Reconstructed pelvic organs (green-bladder, pink-vagina, blue-rectum), pelvic floor (yellow) and pelvis (red), visualized in 3D Slicer (https://www.slicer.org).

11.4 Conclusions

To satisfy the increasing needs of 3D reconstruction of female pelvic system for surgical simulation, MRI protocols employed in clinical routine were discussed, and the choice of geometry representation for patient-specific modeling was elaborated. An efficient and simulation-compatible 3D modeling method was presented, which constructs directly from image semi-automatically geometric models of bladder, vagina, and rectum with NURBS representation, flexible and suitable for simulation. Based on the reconstructed principal organs, the pelvic floor is generated to help define the boundary conditions, one of the inputs of the simulation model.

One approach to overcome the limitation of the method is to evaluate the stability and robustness of the optimization with more data from patients, especially in the case of irregular shapes. The idea is to quantify the impact of geometry imprecision on the performance of patient-specific simulation. Eventually, it will be interesting to determine to which extent can geometry be simplified. Since the pelvic system is highly complex, in addition to the pelvic floor, other suspension devices, such as paravaginal ligaments, need to be generated automatically. Moreover, much recent research interest has been taken in developing automatic segmentation tool with available training data. The design of a complete pelvic system reconstruction with all necessary parts and highly automated will be an important improvement.

The problems considered in this chapter are challenging and focus on clinical usage. From application point of view, the development of graphical user interface, including stable APIs, is essential to make the presented reconstruction method more user-friendly and streamlined, which can also assist physicians to express clearly their needs and to collect more efficiently results and data for evaluation. All these perspectives aim to build a bridge between medical images and simulation that will assist in constructing usable and functional models in the clinical context.

Acknowledgments

Financial support for this work was provided by SATT Nord, France (Project CND tissus mous).

References

Botsch, M., Kobbelt, L., Pauly, M., Alliez, P., Levy, B., 2010. Polygon Mesh Processing. CRC Press Taylor & Francis Group.

Bump, R.C., Mattiasson, A., Bø, K., Brubaker, L.P., DeLancey, J.O., Klarskov, P., Shull, B.L., Smith, A.R., 1996. The standardization of terminology of female pelvic organ prolapse and pelvic floor dysfunction. Am. J. Obstet. Gynecol. 175, 10–17.

Courtecuisse, H., Jiang, Z., Mayeur, O., Witz, J.F., Lecomte-Grosbras, P., Cosson, M., Brieu, M., Cotin, S., 2020. Three-dimensional physics-based registration of pelvic system using 2D dynamic magnetic resonance imaging slices. Strain 56. https://doi.org/10.1111/str.12339.

DeLancey, J.O.L., 1992. Anatomie aspects of vaginal eversion after hysterectomy. Am. J. Obstet. Gynecol. 166, 1717–1728. https://doi.org/10.1016/0002-9378(92)91562-O.

Dell'oro, M., Collinet, P., Robin, G., Rubod, C., 2013. Multidisciplinary approach for deep endometriosis: interests and organization. Gynecol. Obstet. Fertil. 41, 58–64.

Hamm, B., Forstner, R. (Eds.), 2007. Diagnostic Imaging - MRI and CT of the Female Pelvis, Diagnostic Imaging. Springer-Verlag Berlin Heidelberg.

He, K., Cao, X., Shi, Y., Nie, D., Gao, Y., Shen, D., 2019. Pelvic organ segmentation using distinctive curve guided fully convolutional networks. IEEE Trans. Med. Imaging 38, 585–595. https://doi.org/10.1109/TMI.2018.2867837.

Heimann, T., Meinzer, H.-P., 2009. Statistical shape models for 3D medical image segmentation: a review. Med. Image Anal. 13, 543–563.

Hoyte, L., Ye, W., Brubaker, L., Fielding, J.R., Lockhart, M.E., Heilbrun, M.E., Brown, M.B., Warfield, S.K., for the Pelvic Floor Disorders Network, 2011. Segmentations of MRI images of the female pelvic floor: a study of inter- and intra-reader reliability. J. Magn. Reson. Imaging 33, 684–691. https://doi.org/10.1002/jmri.22478.

Jiang, Z., Mayeur, O., Patrouix, L., Cirette, D., Witz, J.-F., Dumont, J., Brieu, M., 2020. Patient-specific modeling of pelvic system from MRI for numerical simulation: validation using a physical model. In: Nash, M.P., Nielsen, P.M.F., Wittek, A., Miller, K., Joldes, G.R. (Eds.), Computational Biomechanics for Medicine. Springer International Publishing, Cham, pp. 19–30.

Jiang, Z., Mayeur, O., Witz, J., Lecomte-Grosbras, P., Dequidt, J., Cosson, M., Duriez, C., Brieu, M., 2019. Virtual image correlation of magnetic resonance images for 3D geometric modelling of pelvic organs. Strain 55, e12305. https://doi.org/10.1111/str.12305.

Jiang, Z., Witz, J.-F., Lecomte-Grosbras, P., Dequidt, J., Duriez, C., Cosson, M., Cotin, S., Brieu, M., 2015. B-spline based multi-organ detection in magnetic resonance imaging. Strain 51, 235–247.

Lecomte-Grosbras, P., Nassirou-Diallo, M., Witz, J.-F., Marchal, D., Dequidt, J., Cotin, S., Cosson, M., Duriez, C., Brieu, M., 2013. Towards a better understanding of pelvic system disorders using numerical simulation. In: MICCAI. Nagoya, Japan.

Lecomte-Grosbras, P., Witz, J.-F., Brieu, M., Faye, N., Cosson, M., Rubod, C., 2015. Quantification of pelvic mobility on dynamic magnetic resonance images: using mechanical insight to help diagnose pelvic pathologies. Strain 51, 301–310.

Lei, T., Wang, R., Wan, Y., Zhang, B., Meng, H., Nandi, A.K., 2020. Medical image segmentation using deep learning: a survey. arXiv:2009.13120 [Cs.Eess].

Lorensen, W.E., Cline, H.E., 1987. Marching cubes: a high resolution 3D surface construction algorithm. In: Proceedings of the 14th Annual Conference on Computer Graphics and Interactive Techniques. SIGGRAPH '87. ACM, pp. 163–169.

Mayeur, O., Jeanditgautier, E., Witz, J.-F., Lecomte-Grosbras, P., Cosson, M., Rubod, C., Brieu, M., 2017. Evaluation of strains on levator ani muscle: damage induced during delivery for a prediction of patient risks. In: Computational Biomechanics for Medicine: From Algorithms to Models and Applications. Springer International Publishing, Cham, pp. 135–146.

Mayeur, O., Lamblin, G., Lecomte-Grosbras, P., Brieu, M., Rubod, C., Cosson, M., 2014. FE simulation for the understanding of the median cystocele prolapse occurrence. In: Biomedical Simulation, pp. 220–227.

Meuleman, C., Vandenabeele, B., Fieuws, S., Spiessens, C., Timmerman, D., D'Hooghe, T., 2009. High prevalence of endometriosis in infertile women with normal ovulation and normospermic partners. Fertil. Steril. 92, 68–74.

Pannu, H.K., Kaufman, H.S., Cundiff, G.W., Genadry, R., Bluemke, D.A., Fishman, E.K., 2000. Dynamic MR imaging of pelvic organ prolapse: spectrum of abnormalities 1. Radiographics 20, 1567–1582.

Paragios, N., Chen, Y., Faugeras, O. (Eds.), 2006. Handbook of Mathematical Models in Computer Vision. Springer, Boston, MA.

Piegl, L., Tiller, W., 1997. The NURBS book. In: Monographs in Visual Communication, 2nd edition. Springer-Verlag Berlin Heidelberg.

Rao, G.V., Rubod, C., Brieu, M., Bhatnagar, N., Cosson, M., 2010. Experiments and finite element modelling for the study of prolapse in the pelvic floor system. Comput. Methods Biomech. Biomed. Eng. 13, 349–357.

Réthoré, J., François, M.L.M., 2014. Curve and boundaries measurement using B-splines and virtual images. Opt. Lasers Eng. 52, 145–155.

Samuelsson, E.C., Victor, F.T., Tibblin, G., Svärdsudd, K.F., 1999. Signs of genital prolapse in a Swedish population of women 20 to 59 years of age and possible related factors. Am. J. Obstet. Gynecol. 180, 299–305.

Semin, B., Auradou, H., François, M.L.M., 2011. Accurate measurement of curvilinear shapes by Virtual Image Correlation. Eur. Phys. J. Appl. Phys. 56, 10701.

Shekhar, R., Fayyad, E., Yagel, R., Cornhill, J.F., 1996. Octree-based decimation of marching cubes surfaces. In: Vis. 96 Proc, pp. 335–342.

Swift, S.E., 2000. The distribution of pelvic organ support in a population of female subjects seen for routine gynecologic health care. Am. J. Obstet. Gynecol. 183, 277–285.

Tansatit, T., Apinuntrum, P., Phetudom, T., Phanchart, P., 2013. New insights into the pelvic organ support framework. Eur. J. Obstet. Gynecol. Reprod. Biol. 166, 221–225. https://doi.org/10.1016/j.ejogrb.2012.10.038.

Vallet, A., Witz, J.-F., Rubod, C., Brieu, M., Cosson, M., 2011. Simulation of pelvic mobility: topology optimization of ligamentous system. Comput. Methods Biomech. Biomed. Eng. 14, 159–163.

CHAPTER

12 Quantitative assessment of pelvic mobility in women using MRI image analysis

Pauline Lecomte-Grosbras[a], Jean-François Witz[a], Zhifan Jiang[b], Olivier Mayeur[a], Mathias Brieu[d], Chrystèle Rubod[a,c], and Michel Cosson[a,c]

[a]*Univ. Lille, CNRS, Centrale Lille, UMR 9013-LaMcube – Laboratoire de Mécanique, Multiphysique, Multiéchelle, Lille, France*

[b]*Sheikh Zayed Institute for Pediatric Surgical Innovation, Children's National Hospital, Washington, DC, United States*

[c]*CHU Lille, Service de Chirurgie Gynécologique, Lille, France*

[d]*College of Engineering, Computer Science and Technology, Dept. of Mechanical Engineering, California State University, Los Angeles, CA, United States*

12.1 Introduction

Genital prolapse is a complex multifactorial pathology, whose pathophysiology is not yet fully understood. It is a real public health problem, whose surgical management concerns 12.6% of women during their lifetime in the United States (Wu et al., 2014). Some risk factors are recognized, such as age, parity, or obesity (Vergeldt et al., 2015), but the pathophysiology of their action is not yet fully defined. Pelvic organ prolapse seems to be the result of a weakening or lengthening of the pelvic floor tissue (a complex ligament), fascia, and muscle structure that supports the uterus, bladder, and rectum in the female pelvic cavity. This approach to pelvic organ dynamic support was originally developed by Peter Petros, who developed the "integral theory" in the 1990s to explain certain stress urinary incontinence (Petros and Ulmsten, 1990). The clinical definition of prolapse is currently based on the POP-Q classification (Bump et al., 1996), which has made it possible to standardize the diagnostic criteria and the evaluation of the severity of these disorders. In parallel, MRI analysis methods have been developed since the 1990s (Yang et al., 1991) to study the physiological and pathophysiological behavior of the different organs and suspension elements of the pelvis. The contribution of dynamic imaging is now essential in the management of pelvic static disorders, particularly when the posterior segment is affected (Altman et al., 2005). The current radiological definition of prolapse is mainly based on the position of the pelvic organs in relation to the subpubo-saccro-coccygeal line during

Biomechanics of the Female Reproductive System: Breast and Pelvic Organs. https://doi.org/10.1016/B978-0-12-823403-7.00025-7

an effort to push in symptomatic patients. These examinations then help to determine the stage of prolapse and guide surgical management techniques.

However, new techniques for dynamic MRI analysis have been developed, and it is now possible to define objectively the specific displacements of each organ individually and precisely (Lecomte-Grosbras et al., 2015; van IJsselmuiden et al., 2020) as well as inter-organ relative displacement (Jiang et al., 2017). It seems important to have a better understanding of the physiological mobility in the absence of prolapse to better understand the mechanisms and improve the management of this pathology (Babayi et al., 2019; Chen et al., 2015; Gordon et al., 2019; Jean Dit Gautier et al., 2018; Lamblin et al., 2016; Rubod et al., 2013). Nowadays, there is little quantitative data available on the amplitude and axis of displacement of the pelvic organs during a pushing effort in patients who do not present a pelvic statics' disorder. Moreover, we only partially know the impact of surgeries using medical devices on these mobilities.

Numerous approaches of numerical modeling and simulation, mainly based on finite element (FE) methods, have been developed to better understand the role of suspension structures, such as ligaments, on pelvic mobility and the physiopathology of prolapses, and to predict the motion of an organ (Rubod et al., 2013; Chen et al., 2015; Jean Dit Gautier et al., 2018; Lamblin et al., 2016; Gordon et al., 2019; Babayi et al., 2019). In this context, the use of a FE modeling strategy would help to define the relative importance of each suspension structure more accurately and better understand the mechanisms leading to pathology and its treatment. For the development of numerical simulation methods, the validation stage is crucial and requires experimental data. Developing a full-field measurement protocol based on dynamic MR images taken during the diagnostic procedure in a clinical context would be a useful tool to improve diagnosis and propose more personalized and efficient treatment. In addition, it would provide experimental data to help to validate mechanical model and predictive numerical simulation of the treatment that are currently in development.

12.2 Method for quantifying pelvic mobility

12.2.1 Dynamic MRI protocol

To date, pelvic disease is diagnosed through exhaustive radiological examinations. Dynamic MRI defecography is a classic examination to quantify pathological mobility and establish a grade or level of pathology. Protocols may vary from hospital to hospital but are often very similar. The examination is usually done with a partially filled bladder; contrast gel is often inserted into the vaginal and rectal cavities to improve contrast for a better demarcation of the vaginal walls, cervix, and rectum boundaries. The patient, in the supine position in the MRI, must perform a push effort or Valsalva exercise. A sequence of 2D images in the sagittal plane is then recorded to observe the in-plane movement of the organs.

In general, to observe the entire pelvic system, the field of view is between 300 mm and 250 mm, the slice thickness between 1 mm and 1.2 mm, and the acquisition frequency around 1 Hz.

12.2.2 POP-Q method

The MRI images are analyzed to quantify the position of the organs at maximum effort. This information is used to classify the level of genital prolapse from which the patient suffers.

The standard method to quantify prolapse is to give the position of some characteristic points according to certain artificial lines, usually from the pubic to the sacral bone. The radiologist observes the movement of the pelvic organs and note if and how much they cross the line during the dynamic magnetic resonance (MR) exam (Bump et al., 1996).

The POP-Q Method (Bump et al., 1996; Treszezamsky et al., 2010) is a grading standardized method of level of prolapse. It consists in defining the stage of prolapse by evaluating the position of few points above or proximal to the hymen, as shown in Chapter 2. This diagnostic method can immediately identify whether prolapse is occurring as well as the stage of prolapse. However, this protocol does not really quantify the displacement of pelvic organs, whereas obtaining the whole displacement map may provide more complete and quantitative information to localize pathological areas.

It might be possible to refine the diagnosis using a more complete protocol to analyze MR images with a full-field measurement technique. This will also allow one to have a fine local understanding of physiological mobilities. Thus quantifying the whole displacement field would help to localize possible pathological areas with abnormal displacement when compared to physiological mobilities. It may also help to define more precisely the surgery, this way facilitating the patient being fully informed.

12.2.3 Full field measurement method

With the development and recent improvement of image analysis techniques, such as image registration and digital image correlation, it is possible to propose an objective and reproducible method to quantify pelvic mobility during the clinical MR exam. Full-field measurement techniques based on image registration (Brown, 1992; Zitová and Flusser, 2003) and digital image correlation (Peters and Ranson, 1982; Sutton et al., 1983) are widely used in experimental mechanics. These methods find many applications in biomechanics for the characterization of heterogeneous fields in 2D, 3D, at different scales, in vitro on specimens or in vivo, and under different modalities, such as visible images, ultrasound, or MRI (Naghibi et al., 2019; Ghosh et al., 2012; Rueckert et al., 1999; Mertzanidou et al., 2012; Boyce et al., 2008; Moerman et al., 2009). The main advantages of these full-field measurement techniques are that they are contactless and allow highlighting complex transformations with significant displacements and gradients. This makes it possible to analyze the whole displacement or strain fields to better understand the pelvic kinematics, as well as to localize pathological areas.

These methods are therefore highly relevant for the analysis and quantification of displacement fields in pelvic structures throughout a dynamic MRI sequence. However, to be applicable, these methods need to have a random grayscale distribution

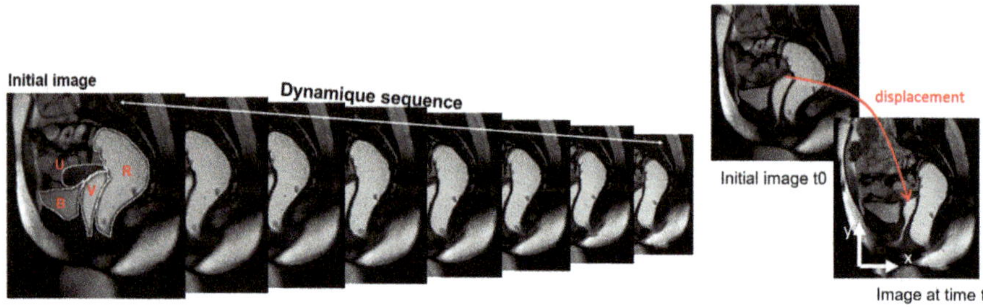

FIGURE 12.1

Image registration on dynamic MRI sequence principle. *No permission required.* a) 2D dynamic sequence; b) displacement of pixels from beginning of the push (time t0) to time t.

on the images to be analyzed, which might not be the case in pelvic observations. Speckle can be artificially created on *ex vivo* laboratory samples by spraying paint, for example, but this is not possible in the context of internal *in vivo* observations (Dong and Pan, 2017). Moreover, occurrence of large displacements, as in the pelvic system, might cause difficulties.

The mobility of pelvic organs are determined by analysis of all images of the dynamic sequence using an image registration method (Lecomte-Grosbras et al., 2015) and Elastix software (Klein et al., 2007). The methodconsists in comparing images of the dynamic sequence during the pushing effort to the initial image before any loading, as shown on Fig. 12.1.

The method is based on the optical flow conservation assumption and consists in comparing the initial image at time t_0 to images at time t. Each image is described by its gray level function f, such as

$$f_{t0}\left(x_0\right) = f_t\left(x_t\right) \text{ with } x_t\left(x_0\right) = x_0 + u,$$

where x_0 and x_t are, respectively, the position of pixels in the initial and current images. The displacement u is calculated by minimizing the gap between the two functions $f_{t0}\left(x_0\right)$ and $f_{t0}\left(x_0 + u\right)$.

The measurement uncertainties have been estimated by applying a rigid known body during MRI observations. The uncertainties are highly linked to image resolution and will change with image resolution and quality. They have been estimated to 0.15 mm for an MRI image of 256 pixel × 256 pixel with 1.2 mm per pixel side and when applying a rigid body displacement of 50 mm (Lecomte-Grosbras et al., 2015).

12.2.4 Displacement analysis

The displacement field is calculated on the whole image, but its values are extracted on the contour of the organs for the analysis on these targeted areas. The three organs (vagina, bladder, and rectum) are segmented on the initial image, and the measured displacement fields are plotted on these contours.

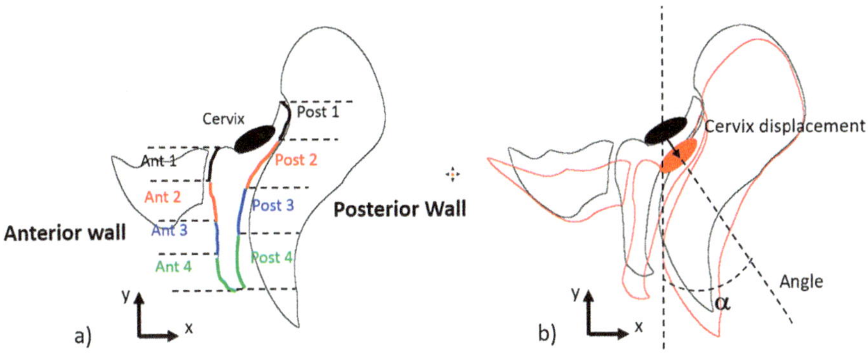

FIGURE 12.2

Analysis areas. *No permission required.* a) Nine areas on anterior and posterior wall where mean displacements are calculated; b) displacement and direction of cervix during the pushing.

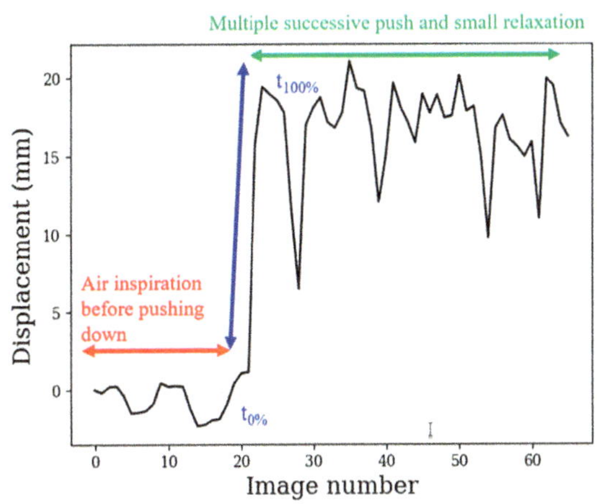

FIGURE 12.3

Pushing chronology. *No permission required.* Displacement measured on cervix during a dynamic image sequence.

To analyze and compare the mobility, displacements are calculated on some areas of interest. Nine areas were defined: 4 segments on the anterior and posterior vagina and cervix Fig. 12.2.

Moreover, the direction of the cervix (angle on Fig. 12.2b) is also calculated.

Despite the guidelines during the MRI to push gradually until maximum and expulsion of the intra-vaginal and intrarectal contrast gel, not all patients push at the same rate. Some push very hard and quickly, whereas others push more gradually (see Fig. 12.3).

Thus to compare results of all patients at the same time, the duration of the push has been converted into percent with the following equation:

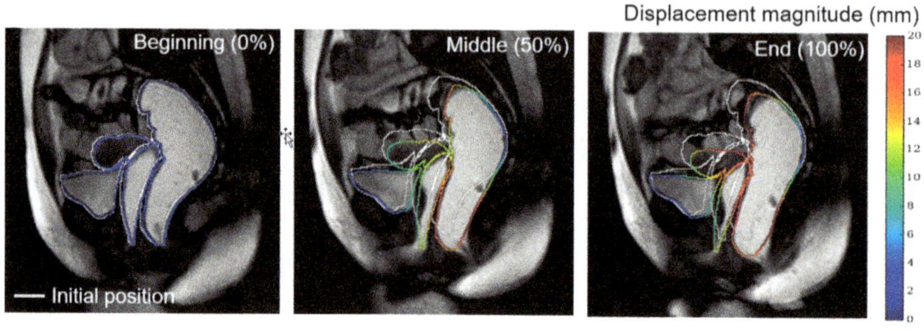

FIGURE 12.4

Displacement map on control patient. *From Lecomte-Grosbras, P., Witz, J.F., Brieu, M., Faye, N., Cosson, M. & Rubod, C. (2015). Quantification of Pelvic Mobility on Dynamic Magnetic Resonance Images: Using Mechanical Insight to Help Diagnose Pelvic Pathologies. Strain, 51(4), 301–310. https://doi.org/10.1111/str.12141.* Displacement measured at different moments of the pushing loading during dynamic MRI examination on control patients without pelvic pathology; in white: initial position of pelvic organs; in color: displacement magnitude on contour of organs on the current dynamic image.

$$push(\%) = \frac{t - t_0}{t_f - t_0}$$

where t is the current time, t_0 and t_f, respectively, the time at the beginning and at the end of the first pushing down.

12.3 Method for quantifying pelvic mobility

12.3.1 Physiological pelvic mobilities

Fig. 12.4 shows the displacement fields measured on the contours of the pelvic organs on a control patient without pelvic pathology (Lecomte-Grosbras et al., 2015).

These results show areas of high and low mobility. In particular, the anterior part of the bladder in contact with the pubis and the posterior part of the rectum in contact with the sacrum appear immobile. These measurements therefore illustrate areas of organ fixity. In addition, variations in mobility can be observed depending on anatomical location. To compare mobility in a more quantitative way, the following graphs show the average displacement over the 9 zones defined on anterior and posterior part of the vagina and on the cervix in Fig. 12.5

Fig. 12.6 shows a comparison of displacement in the y-direction (see Fig. 12.2) at maximum straining on the 9 different areas measured on 8 control patients without any known gynecological pathology. The results do not show significant differences between the anterior and posterior vagina. The mobilities are however more important on the upper parts of the vagina than on the lower parts.

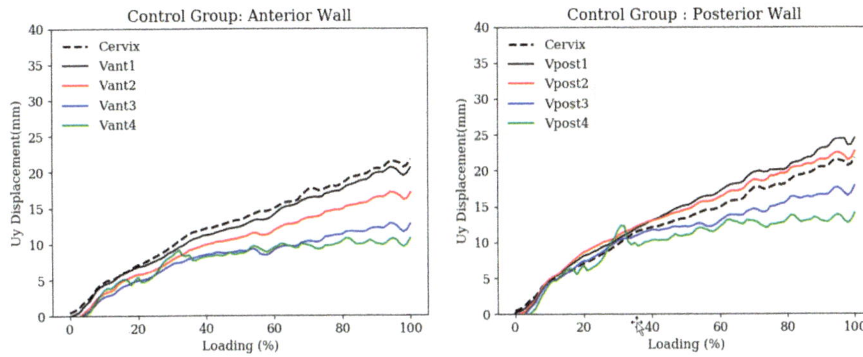

FIGURE 12.5

Displacement graph control patient. *No permission required.* Average displacement (n = 8) during a progressive push measured on control patients without pelvic pathology (n = 8) in different regions 1 and 4 corresponding, respectively, to the upper and lower vagina: a) anterior vagina, b) posterior vagina.

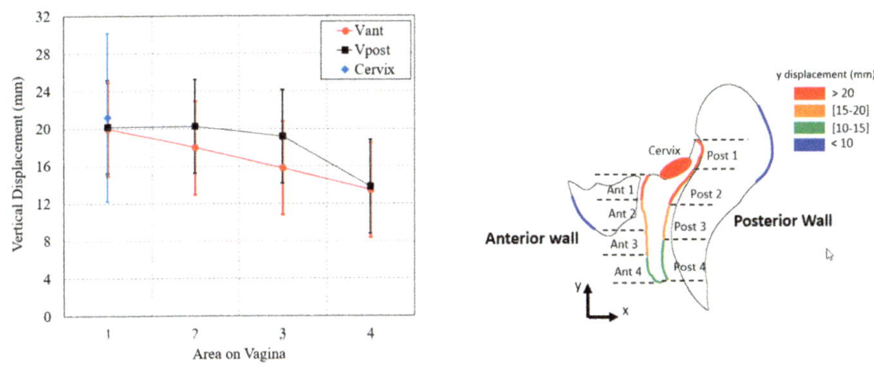

FIGURE 12.6

Displacement per area. *No permission required.* a) Comparison of mobility of different areas on vagina at maximum straining: 1 corresponding to the highest segment and 4 to the lowest segment on the anterior and posterior parts of the vagina (mean value on $n = 8$ young ($23 < \text{age} < 33$) nulliparous control patients without any known pelvic pathology); b) cartography of physiological mobilities.

These measures allow one to compare the relative mobilities of different areas by identifying the most and least mobile areas. It thus makes it possible to establish a cartography of the physiological pelvic mobilities to better understand them as the mechanisms of suspension or support of the pelvic organs. Fig. 12.8 shows the mobilities of the different zones and allows to identify the least mobile zones corresponding to the attachments between the organs and the bony parts (attachment of the bladder to the pubis and the rectum to the sacrum). The most mobile areas are the

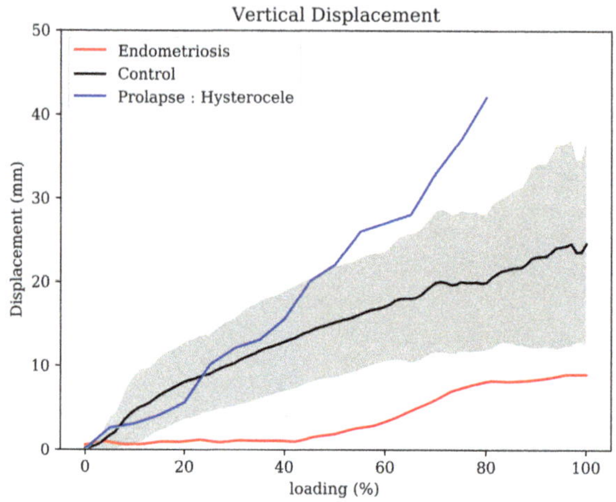

FIGURE 12.7

Comparison mobility. *No permission required.* Comparison of vertical displacement on upper zone of posterior vagina (area Vpost1) in the case of hyper-mobility (hysterocele stage 4) and endometriosis.

upper parts of the vagina and the cervix, which have greater mobility than the lower parts of the vagina.

12.3.2 Pathological mobilities

This mapping of physiological mobilities can be used to compare and identify pathological mobilities. As an example, Fig. 12.7 compares a case of hypo-mobility with a patient suffering from endometriosis and hypermobility (patient with a stage 4 hysterocele).

For cases of hyper-mobility, as is the case of high grade of genital prolapse, the measurement is complex and cannot be carried out to the end of the maximum effort push. Indeed, the mobilities of the organs are very important, making them difficult to evaluate with this method of measurement. In addition, during the pushing effort, the gel is expelled from the organs, the organs are completely emptied; and for high grades of prolapse, the vagina is completely turned inside out and organs might be exteriorized, which is completely outside the framework of the fundamental hypotheses of the measurement (conservation of the optical flow). It is for this reason that in Fig. 12.8 the displacements could not be measured until the end of the push (100%) for the case of genital prolapse corresponding to a stage 4 hysterocele and leading at the end of the push to an exteriorization of the uterus in the vaginal cavity.

One case of endometriosis was also analyzed and compared to the average results obtained on control patients. This patient had severe endometriosis with endometriosis lesion in the sigmoid area. The comparison shows a significant difference in mobilities, with mobilities that are much lower for this patient with endometriosis.

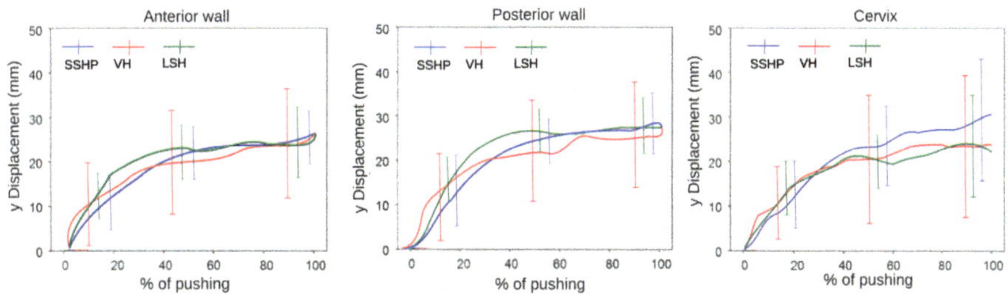

FIGURE 12.8

Post-surgery mobility. *From van IJsselmuiden, M.N., Lecomte-Grosbras, P., Witz, J.F., Brieu, M., Cosson, M. & van Eijndhoven, H.W.F. (2020). Dynamic magnetic resonance imaging to quantify pelvic organ mobility after treatment for uterine descent: differences between surgical procedures. International Urogynecology Journal, 31(10), 2119–2127. https://doi.org/10.1007/s00192-020-04278-5.* Mean Uy (mm) (n = 5) and standard deviation, in anterior vaginal wall (AV), posterior vaginal wall (PV), and cervix/vaginal vault (C/VV) assessed with MRI 6 months after vaginal sacrospinous hysteropexy (SSHP), vaginal hysterectomy (VH), and laparoscopic sacrohysteropexy (LSH).

These results show that this measurement method allows a quantitative local comparison of pelvic mobility. They would thus allow, when applied to a larger number of patients, a better understanding of the physiological mobilities and the mechanisms at the origin of pathologies, such as genital prolapse. They also allow the identification of more or less pathological areas in comparison to physiological mobilities to help in the diagnosis or the choice of the surgical management.

12.3.3 Mobility analysis for evaluation of surgical techniques

This analysis, applied to dynamic MRIs performed after reconstructive surgery, can be used to evaluate surgical techniques aimed at restoring functional mobility. A recent study compared different surgical techniques (van IJsselmuiden et al., 2020). This study was carried out on three surgeries for the treatment of genital prolapse—sacrospinous hysteropexy, vaginal hysterectomy, and laparoscopic sacrohysteropexy—showed (as illustrated in Fig. 12.8) that these three surgeries led to similar results in terms of postoperative mobilities of the pelvic organs.

Fig. 12.8 shows the displacements obtained for the three surgeries on the anterior vagina (average over the entire anterior wall visible on dynamic MRI), the posterior vagina (average over the entire posterior wall), and in the cervix area. Displacements on the vaginal walls measured post-surgery during a dynamic MRI examination during a pushing effort show very similar amplitudes of displacement, even though the standard deviations seem large.

Other analyses carried out on another group of patients also made it possible to compare the postoperative mobilities after low and high approach surgery. The mobilities were also compared to the results obtained on control patients without pelvic pathology.

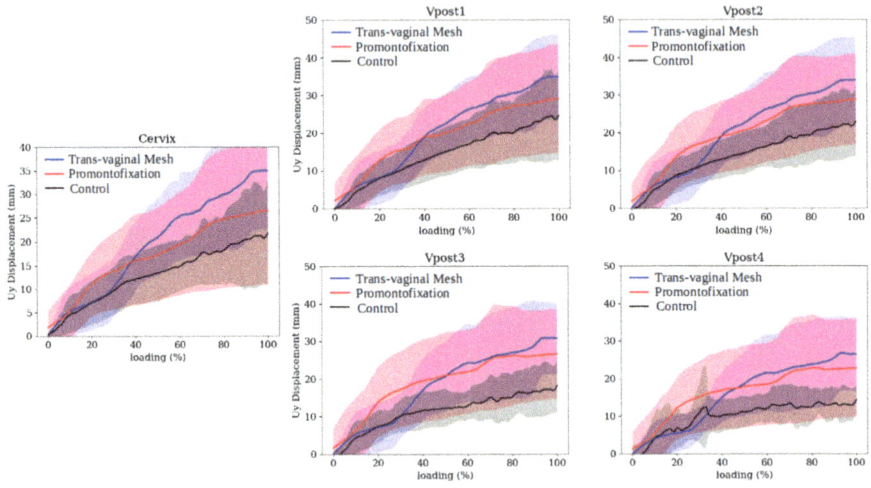

FIGURE 12.9

Comparison zone post-surgery. *No permission required.* Comparison of mobility (displacement in y direction in mm) after lower (transvaginal Mesh (TVM)) and upper (promontofixation) approach for prolapse surgery with physiological mobilities on posterior boundary of the vagina and cervix.

Fig. 12.9 shows the results of the treatment of genital prolapse with physiological mobilities in different areas of the posterior wall of the vagina and in the cervix. The results confirm the results obtained by van IJsselmuiden et al. (2020) and show that the two surgical techniques make it possible to restore similar mobilities, whether after surgery with a prosthetic material placed through the lower part of the body (transvaginal mesh (TVM)) or through the upper part of the body (promontofixation). These mobilities correspond to the observed zone to physiological mobilities. The organs are no longer exteriorized, and there are no longer extreme mobilities of the organs. However, the mobilities remain globally superior to the physiological mobilities, in particular for the lower segments of the vaginal wall. For example, it can be seen in Fig. 12.10 that the displacement at maximum thrust on the lowest part of the posterior vagina is about 10 mm, whereas it is more than twice as great for the postoperative mobilities.

Another interesting observation is the quantification of thrust axes. Indeed, surgical repair with insertion of prosthetic material can modify the conformation of the pelvic system by modifying, for example, the vaginal axes, which can have an impact on the directions of movement of the pelvic organs. Fig. 12.10 shows the directions of movement of the cervix after two types of pelvic surgery by the lower approach (TVM) and the upper approach (promontofixation). The direction of movement of the cervix is quantified by measuring the alpha angle shown in Fig. 12.2b. This angle is calculated from the tangent of the ratio between the displacements in the x-directions and the y-directions. Fig. 12.10a shows the evolution of this angle throughout the

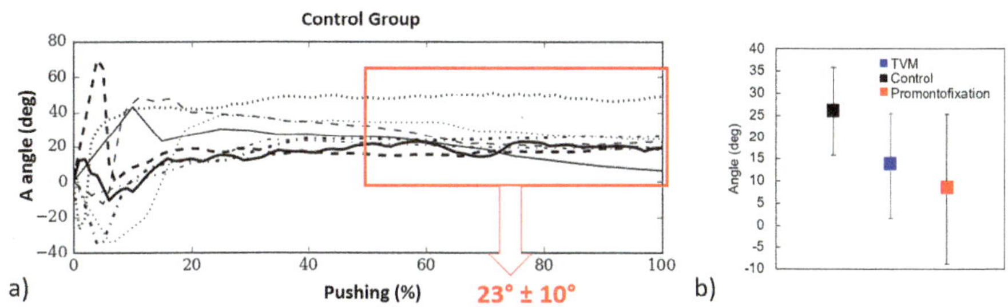

FIGURE 12.10

Angle variation. *No permission required.* Quantification of direction of movement of cervix: a) evolution of the direction during the straining; b) comparison of direction of displacement of cervix after prolapse surgery.

thrust effort. The angle shown in Fig. 12.10b represents the average angle at the end of the push (between 50% and 100% of the push). For the control group without pelvic pathology, this angle is evaluated at $25° \pm 10°$ with respect to the y-direction. However, this angle is significantly different after chirurgy (p value $= 0.05$). This angle becomes smaller, showing the cervix moving more vertically in the y direction of the image. This angle is $14° \pm 12°$ after TVM surgery and $8° \pm 17°$ after promontofixation.

These analyses make it possible to compare quantitatively and locally the mobilities after surgery to the physiological mobilities.

12.4 Inter-organs displacements

Interorgan fasciae are described in various theories as suspension structures having important role in pelvic statics (Petros and Ulmsten, 1990). On one hand, weaknesses of these structures may lead to pathological mobility. On the other hand, these volumes may also be affected by endometriosis lesions. Though endometriotic lesions are difficult to observe on MRI images, they can significantly modify pelvic mobility, as shown in Fig. 12.7. Evaluation of the mobility may provide an objective assessment of these structures, and thus locate pathological areas. However, fasciae structures are very thin, as they are located at the interspace between the organs and are difficult to observe, especially on low-resolution dynamic MRI images. The measurements presented in the previous paragraphs allow to map the displacement fields on the contours of the organs but do not allow quantifying precisely the inter-organ displacements or shear strain and to evaluate the role or the efficiency of the fasciae.

Jiang et al. (2017) proposed an interesting method allowing us to quantify the inter-organ shear strain. The shear strain cannot be directly measured and is to be computed numerically by considering the middle organ, the vagina, as the reference. It provides a model of the behavior of some structures such as ligaments and fascias, as mentioned, hard to observe on images. In the proposed method, virtual image

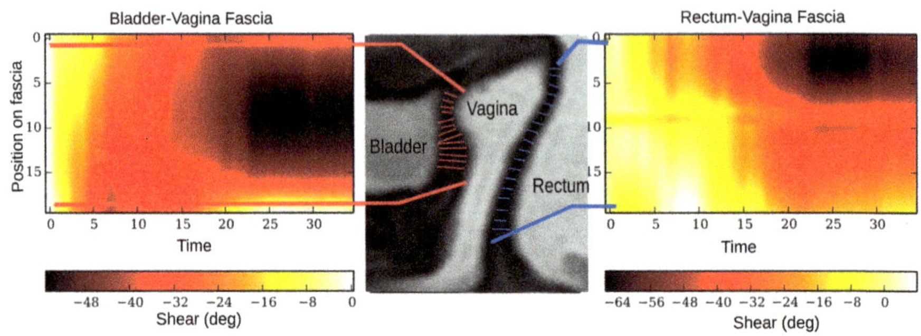

FIGURE 12.11

Interorgan displacements. *From Jiang, Z., Witz, J.-F., Lecomte-Grosbras, P., Dequidt, J., Cotin, S., Rubod, C., Duriez, C. & Brieu, M. (2017). Multiorgan motion tracking in dynamic magnetic resonance imaging for evaluation of pelvic system mobility and shear strain. Strain, 53(2), e12224. https://doi.org/10.1111/str.12224.* Spatial and temporal shear strain map from the beginning to the end of straining, in the context of normal mobility.

correlation (Réthoré et al., 2010; Réthoré and Francois, 2014) is used to identify the contour of the pelvic organs. This method is very efficient to define the contour of the organs but does not allow one to link this contour to material points. To overcome this difficulty, Jiang et al. (2017) use an image registration method based on the hypothesis of conservation of optical flow to identify fixed material points, such as the sacrum and the pubis, and link the segmented contour to material points and quantify the shear between the organs. Displacements are calculated with respect to the vagina (perpendicular and parallel to the contour of the vagina), and shear is calculated from the initial distance between the contours of the bladder or rectum and the vagina. Fig. 12.11 (below) shows the shear between the vagina and bladder (anterior vagina position) and between the vagina and rectum (posterior vagina position) during a dynamic progressive thrust sequence (Jiang et al., 2017) on a control patient without pelvic pathology.

These measurement methods allow the quantification of relative organ mobilities and shear in inter-organ fasciae. To date, they have not yet been tested on a large number of measurements to give average physiological and pathological shear values. However, the feasibility of these measurements has been tested on different pathologies, allowing interesting perspectives on larger scale measurements to evaluate loading and mechanical behavior of inter-organs fasciae to evaluate their role in the suspension or localize pathological areas in case of prolapse or hypo-mobility, such as endometriosis.

12.5 Discussion

The study of the physiological mobility of the pelvic organs is important to improve the overall understanding of pelvic statics. It allows us to better understand the phenomena that can lead to a pathological situation, such as pelvic prolapse, by detailing not only the individual displacements of each organ, but also the relative displacements of the organs with respect to each other and to adjacent structures. All these results show that these measurement methods allow the quantification of relative organ mobilities and shear in inter-organ fasciae. To date, they have not yet been tested on a large number of measurements to give average physiological and pathological shear values. However, the feasibility of these measurements has been tested on different pathologies, allowing interesting perspectives on larger scale measurements to evaluate loading and mechanical behavior of inter-organs fasciae to evaluate their role in the suspension or localize pathological areas in the case of prolapse or hypomobility, for example, endometriosis.

There are relatively fixed areas between the anterior wall of the bladder and the pubis, as well as between the posterior wall of the rectum and the sacrum, corresponding to areas of attachment to bony structures.

There is then a gradient of mobility between the lower part of the vagina, which is less mobile, and the upper part of the vagina associated with the cervix, which is more mobile. This can be explained anatomically by the fact that the lower part of the vagina is maintained by the pelvic floor and in particular by the pubovaginal bundle of the anus elevator muscle, which inserts at the level of the lower part of the vagina, and which is intact and of good quality in the control patients.

Mean mobilities were measured in a cohort of 8 control patients. Average measurements presented should be enriched by a larger number of patients to confirm these results. Because dynamic MRI is a relatively standard and common examination in the evaluation of pelvic pathologies, the evaluation presented in this chapter focuses on the quantification of such types of images. During the dynamic MRI images studied, the patients are requested to generate a pushing force by applying a craniocaudal force, attempting to mimic the usual pelvic floor loading. The effect of gravity on the overall support system probably has an impact that has not been studied in the presented protocol. This bias could be corrected with the use of open MRI, in which standing examinations are possible (Morsinkhof et al., 2021). However, the examinations performed in the context of pelvic statics pathologies, such as prolapse, are performed in the supine position. If one wants to compare the physiological state with the pathological state, it is probably still more relevant to perform all the examinations in the same position.

These MRIs were performed in the supine position and are therefore not exactly representative of mobilities in the standing position, but they nevertheless make it possible, in the same configuration, to compare the mobilities of control patients, with no known pelvic pathology, with other pathological cases of hypomobility or hypermobility, or to check whether a surgical technique makes it possible to restore physiological pelvic kinematics.

These physiological mobilities can be compared to pathological ones, for example, in cases of genital prolapse or endometriosis. A complete mapping enabled by objective measurements helps to better identify, for example, the deficient areas to be strengthened to restore functional mobilities. Endometriosis lesions are also sometimes difficult to observe on medical images and the objective measurement of mobility may allow the identification of the signature on the kinematics of the potential presence of a lesion, enabling to anticipate and duly inform the patient before the surgical treatment.

The analysis of pelvic mobility also allows us to evaluate the surgical techniques and to quantify in an objective way the success of the surgery in terms of restoration of functional mobilities. The surgical management of genital prolapse still poses many difficulties and has high recurrence rates (Swift, 2000; Abed et al., 2011; Maher et al., 2013, 2011). Recently 95% of implants have been withdrawn from the market due to these high complication rates. It is therefore necessary to better understand the physiological mobilities and how to restore them in an optimal and biomimetic way. A better knowledge of functional mobilities and the impact of local surgical treatment on pelvic kinematics should help gynecologists and bioengineers to design new implants with a more biofidelic mechanical role.

Inter-organ fascia are interorgan attachment structures that are described in the literature as having an important role in the suspension system (Petros and Ulmsten, 1990). Failure of these structures can alter the delicate balance of pelvic statics and play a role in the development of pathological hypermobility depending on the areas affected. A better knowledge of their mechanical role and a biomechanical evaluation of these structures may allow a better diagnosis and treatment of these pathologies. Indeed, almost all commercial implants for the treatment of genital prolapse have now been withdrawn from the market. It is therefore necessary to better evaluate the biomechanical roles of the different suspension structures and to propose more optimized and better targeted therapeutic management to treat the defective areas more effectively. A more complete mapping of the physiological referential mobility and the mobility of the patient to be treated could allow a better and more personalized management and open the way to the development of biomimetic and better tolerated treatments, thereby allowing restoration of functional mobility.

To better understand pelvic mobility and the pathophysiology of genital prolapse, numerous numerical studies have been proposed in the literature. These studies make it possible, for example, to identify the most important structures for pelvic mobility, to compare the relative role of these different structures (Rubod et al., 2013; Vallet et al., 2011), to better understand the mechanisms leading to genital prolapse (Chen et al., 2015; Rahim et al., 2011). Thanks to numerical simulation tools, it is possible to better understand the effectiveness of surgical techniques, for example, according to different pathological or topological configurations, fixation methods, etc.

Numerical simulation of pelvic mobility is also a relevant tool for the specific evaluation of surgical techniques for a given patient (Lamblin et al., 2016). These digital tools are thus very interesting and important for gynecological diagnostic and personalized surgical planning and are hence being more widely adopted. Large industrial groups or start-ups are very much engaged in connection to this emerging and expanding subject of personalized surgical planning.

Whatever the objective of understanding a physiopathology or personalized simulation for research or commercial purposes, all digital tools must be validated by objective and quantitative measurements to be usable and effective. Maps of displacements measured on the contour of the pelvic organs allow building and validating relevant numerical models and evaluating their accuracy. This step of validation is thus crucial and the approach of confrontation and quantification of the difference between simulation and experimental measurements seems to be the essential base to any numerical simulation action. Without experimental data for comparison or validation, there can be no relevant numerical simulation.

The coupling between simulations and kinematic field measurements also allows us to identify in vivo mechanical properties by inverse method of minimization of the difference between simulation and measurements (Silva et al., 2018).

To further improve the reliability of numerical models, the coupling between experimental measurements and numerical simulations can be further strengthened and the numerical simulation can be directly guided by the measured kinematics. This is, for example, the case in the work of Courtecuisse et al. (2020), which allows at each simulation step, i.e., at each numerical image of the dynamic sequence, to quantify the deviation from the simulation and to guide the simulation towards the measured position and mobility of the organs, thereby allowing the building of efficient numerical simulation under large displacement and prolapse.

12.6 Conclusion

The quantification of pelvic mobilities allows an objective, quantitative, and local measurement of the mobilities of the pelvic organs on their contours. This enables us to improve our understanding of physiological mobilities from measurements performed on patients without known pelvic pathology. These mobilities are important in the physiological context of enabling the organs to perform their physiological functions, for example, bladder and rectal excretion, or during sexual intercourse. It is already known that bladder mobility is essential for urinary continence during exercise, and that rectal mobility is essential for satisfactory defecation.

This mobility is highly variable from patient to patient and the boundary between normal and pathological mobility should be studied in larger cohorts. Descriptions of pelvic anatomy generally do not take into account this aspect of physiological mobility of the region, which is nevertheless important, as our work indicates.

The modification of mobilities is indirectly the consequence of a pathology. For example, genital prolapse leads to hyper-mobility of the pelvic organs, whereas endometriosis reduces their mobility, or the appearance of a cyst or cancer can locally modify the topology of the system and affect mobility. The measurement of mobility compared to physiological mobility can thus help in the diagnosis of complex pathologies and identify locally defective zones or potential sites of lesion or pathological tissue.

Moreover, they can also allow the evaluation of surgical techniques, for example, for the management of genital prolapse, by assessing whether the treatment has restored physiological mobility. This analysis could also be transposed to the case of endometriosis, allowing one to evaluate the efficiency of a treatment or a surgery by comparing the mobilities before and after treatment, as well as with average physiological mobilities.

The links between mobility and functional symptoms are still to be clarified. Such links will open doors to a better understanding of the physiology of the pelvic region in women, and should also be studied in men.

In addition, improvements in the performance of computing and imaging resources, as well as the development of increasingly efficient numerical methods, will enable the construction of more biofidelic numerical models by coupling simulations and measurements that are validated on real experimental measurements. These comparison and couplings between numerical methods and quantitative measurements allow us to go towards more and more personalized numerical simulations. Such simulation, validation, and coupled approaches can be applied to different pathologies affecting pelvic mobility.

References

Abed, H., Rahn, D.D., Lowenstein, L., Balk, E.M., Clemons, J.L., Rogers, R.G., 2011. Incidence and management of graft erosion, wound granulation, and dyspareunia following vaginal prolapse repair with graft materials: a systematic review. International Urogynecology Journal 22 (7), 789–798. https://doi.org/10.1007/s00192-011-1384-5.

Altman, D., López, A., Kierkegaard, J., Zetterström, J., Falconer, C., Pollack, J., Mellgren, A., 2005. Assessment of posterior vaginal wall prolapse: comparison of physical findings to cystodefecoperitoneography. International Urogynecology Journal 16 (2), 96–103. https://doi.org/10.1007/s00192-004-1220-2.

Babayi, M., Azghani, M.R., Hajebrahimi, S., Berghmans, B., 2019. Three-dimensional finite element analysis of the pelvic organ prolapse: a parametric biomechanical modeling. Neurourology and Urodynamics 38 (2), 591–598. https://doi.org/10.1002/nau.23885.

Boyce, B.L., Grazier, J.M., Jones, R.E., Nguyen, T.D., 2008. Full-field deformation of bovine cornea under constrained inflation conditions. Biomaterials 29 (28), 3896–3904. https://doi.org/10.1016/j.biomaterials.2008.06.011.

Brown, L.G., 1992. A survey of image registration techniques. ACM Computing Surveys 24 (4), 325–376. https://doi.org/10.1145/146370.146374.

Bump, R.C., Mattiasson, A., Bo, K., Brubaker, L.P., DeLancey, J.O.L., Klarskov, P., Shull, B.L., Smith, A.R.B., 1996. The standardization of terminology of female pelvic organ prolapse and pelvic floor dysfunction. American Journal of Obstetrics and Gynecology 175 (1), 10–17. https://doi.org/10.1016/S0002-9378(96)70243-0.

Chen, Z.-W., Joli, P., Feng, Z.-Q., Rahim, M., Pirró, N., Bellemare, M.-E., 2015. Female patient-specific finite element modeling of pelvic organ prolapse (POP). Journal of Biomechanics 48 (2), 238–245. https://doi.org/10.1016/j.jbiomech.2014.11.039.

Courtecuisse, H., Jiang, Z., Mayeur, O., Witz, J.-F., Lecomte-Grosbras, P., Cosson, M., Brieu, M., Cotin, S., 2020. Three-dimensional physics-based registration of pelvic system using 2D dynamic magnetic resonance imaging slices. Strain 56 (3). https://doi.org/10.1111/str.12339.

Dong, Y.L., Pan, B., 2017. A review of speckle pattern fabrication and assessment for digital image correlation. Experimental Mechanics 57, 1161–1181. https://doi.org/10.1007/s11340-017-0283-1.

Ghosh, R., Gupta, S., Dickinson, A., Browne, M., 2012. Experimental validation of finite element models of intact and implanted composite hemipelvises using digital image correlation. Journal of Biomechanical Engineering 134 (8). https://doi.org/10.1115/1.4007173.

Gordon, M., DeLancey, J., Renfroe, A., Battles, A., Chen, L., 2019. Development of anatomically based customizable three-dimensional finite-element model of pelvic floor support system: POP-SIM1.0. Interface Focus.

Jean Dit Gautier, E., Mayeur, O., Lepage, J., Brieu, M., Cosson, M., Rubod, C., 2018. Pregnancy impact on uterosacral ligament and pelvic muscles using a 3D numerical and finite element model: preliminary results. International Urogynecology Journal 29 (3), 425–430. https://doi.org/10.1007/s00192-017-3520-3.

Jiang, Z., Witz, J.-F., Lecomte-Grosbras, P., Dequidt, J., Cotin, S., Rubod, C., Duriez, C., Brieu, M., 2017. Multiorgan motion tracking in dynamic magnetic resonance imaging for evaluation of pelvic system mobility and shear strain. Strain 53 (2), e12224. https://doi.org/10.1111/str.12224.

Klein, S., 2007, Staring, M., Pluim, J.P. Evaluation of optimization methods for nonrigid medical image registration using mutual information and B-splines. IEEE Transactions on Image Processing 16 (12), 2879–2890.

Lamblin, G., Mayeur, O., Giraudet, G., Jean dit Gautier, E., Chene, G., Brieu, M., Rubod, C., Cosson, M., 2016. Pathophysiological aspects of cystocele with a 3D finite elements model. Archives of Gynecology and Obstetrics 294 (5), 983–989. https://doi.org/10.1007/s00404-016-4150-6.

Lecomte-Grosbras, P., Witz, J.F., Brieu, M., Faye, N., Cosson, M., Rubod, C., 2015. Quantification of pelvic mobility on dynamic magnetic resonance images: using mechanical insight to help diagnose pelvic pathologies. Strain 51 (4), 301–310. https://doi.org/10.1111/str.12141.

Maher, C., Feiner, B., Baessler, K., Schmid, C., 2013. Surgical management of pelvic organ prolapse in women. Cochrane Database of Systematic Reviews 2013 (4). https://doi.org/10.1002/14651858.CD004014.pub5.

Maher, C.M., Feiner, B., Baessler, K., Glazener, C.M.A., 2011. Surgical management of pelvic organ prolapse in women: the updated summary version Cochrane review. International Urogynecology Journal 22 (11), 1445–1457. https://doi.org/10.1007/s00192-011-1542-9.

Mertzanidou, T., Hipwell, J., Cardoso, M.J., Zhang, X., Tanner, C., Ourselin, S., Bick, U., Huisman, H., Karssemeijer, N., Hawkes, D., 2012. MRI to X-ray mammography registration using a volume-preserving affine transformation. Medical Image Analysis 16 (5), 966–975. https://doi.org/10.1016/j.media.2012.03.001.

Moerman, K.M., Holt, C.A., Evans, S.L., Simms, C.K., 2009. Digital image correlation and finite element modelling as a method to determine mechanical properties of human soft tissue in vivo. Journal of Biomechanics 42 (8), 1150–1153. https://doi.org/10.1016/j.jbiomech.2009.02.016.

Morsinkhof, L., Witz, J.-F., Mayeur, O., Bellos-Grob, A., Simonis, F., Lecomte-Grosbras, P., 2021. Deformation characterization to quantify pelvic organ mobility in supine and upright position for assessment of pelvic organ prolapse: vol. ISMRM.

Naghibi, H., Mazzoli, V., Gijsbertse, K., Hannink, G., Sprengers, A., Janssen, D., Van den Boogaard, T., Verdonschot, N., 2019. A noninvasive MRI based approach to estimate the mechanical properties of human knee ligaments. Journal of the Mechanical Behavior of Biomedical Materials 93, 43–51. https://doi.org/10.1016/j.jmbbm.2019.01.022.

Peters, W.H., Ranson, W.F., 1982. Digital imaging techniques in experimental stress analysis. Optical Engineering 21 (3), 427–431. https://doi.org/10.1117/12.7972925.

Petros, P.E.P., Ulmsten, U.I., 1990. An integral theory of female urinary incontinence: experimental and clinical considerations. Acta Obstetricia et Gynecologica Scandinavica 69 (153 S), 7–31. https://doi.org/10.1111/j.1600-0412.1990.tb08027.x.

Rahim, M., Bellemare, M.E., Pirro, N., Bulot, R., 2011. A quantitative approach for the assessment of the pelvic dynamics modeling. IRBM 32, 311–316.

Réthoré, J., Elguedj, T., Simon, P., Coret, M., 2010. On the use of NURBS functions for displacement derivatives measurement by digital image correlation. Experimental Mechanics 50 (7).

Réthoré, J., François, M.L.M., 2014. Curve and boundaries measurement using B-splines and virtual images. Optics and Lasers in Engineering 52.

Rubod, C., Lecomte-Grosbras, P., Brieu, M., Giraudet, G., Betrouni, N., Cosson, M., 2013. 3D simulation of pelvic system numerical simulation for a better understanding of the contribution of the uterine ligaments. International Urogynecology Journal and Pelvic Floor Dysfunction 24 (12), 2093–2098. https://doi.org/10.1007/s00192-013-2135-6.

Rueckert, D., Sonoda, L.I., Hayes, C., Hill, D.L.G., Leach, M.O., Hawkes, D.J., 1999. Nonrigid registration using free-form deformations: application to breast MR images. IEEE Transactions on Medical Imaging 18 (8), 712–721. https://doi.org/10.1109/42.796284.

Silva, M.E.T., Parente, M.P.L., Brandão, S., Mascarenhas, T., Natal Jorge, R.M., 2018. Characterization of the passive and active material parameters of the pubovisceralis muscle using an inverse numerical method. Journal of Biomechanics 71, 100–110. https://doi.org/10.1016/j.jbiomech.2018.01.033.

Sutton, M., Wolters, W., Peters, W., Ranson, W., McNeill, S., 1983. Determination of displacements using an improved digital correlation method. Image and Vision Computing 1 (3), 133–139. https://doi.org/10.1016/0262-8856(83)90064-1.

Swift, S., 2000. The distribution of pelvis organ in a population of female subjects seen for routine gynecologic health care. American Journal of Obstetrics and Gynecology 183, 77–285.

Treszezamsky, A.D., Rascoff, L., Shahryarinejad, A., Vardy, M.D., 2010. Use of pelvic organ prolapse staging systems in published articles of selected specialized journals. International Urogynecology Journal 21 (3), 359–363. https://doi.org/10.1007/s00192-009-1044-1.

Vallet, A., Rubod, C., Witz, J.F., Brieu, M., Cosson, M., 2011. Simulation of pelvic mobility: topology optimisation of ligamentous system. Computer Methods in Biomechanics and Biomedical Engineering 14 (1), 161–162. https://doi.org/10.1080/10255842.2011.596351.

van IJsselmuiden, M.N., Lecomte-Grosbras, P., Witz, J.F., Brieu, M., Cosson, M., van Eijndhoven, H.W.F., 2020. Dynamic magnetic resonance imaging to quantify pelvic organ mobility after treatment for uterine descent: differences between surgical procedures. International Urogynecology Journal 31 (10), 2119–2127. https://doi.org/10.1007/s00192-020-04278-5.

Vergeldt, T.F.M., Weemhoff, M., IntHout, J., Kluivers, K.B., 2015. Risk factors for pelvic organ prolapse and its recurrence: a systematic review. International Urogynecology Journal 26 (11), 1559–1573. https://doi.org/10.1007/s00192-015-2695-8.

Wu, J., Matthews, C., Conover, M., Pate, V., Funk, M., 2014. Lifetime risk of stress urinary incontinence or pelvic organ prolapse surgery. Obstetrics and Gynecology 123 (1201), 1206.

Yang, A., Mostwin, J.L., Rosenshein, N.B., Zerhouni, E.A., 1991. Pelvic floor descent in women: dynamic evaluation with fast MR imaging and cinematic display. Radiology 179 (1), 25–33. https://doi.org/10.1148/radiology.179.1.2006286.

Zitová, B., Flusser, J., 2003. Image registration methods: a survey. Image and Vision Computing 21 (11), 977–1000. https://doi.org/10.1016/S0262-8856(03)00137-9.

CHAPTER

13 Patient-specific biomechanical modeling for applications in breast cancer diagnosis and treatment

Marianne Sanchez[a], John Pan[a], Kejia Khoo[a], Matthew French[a], Rhonda-Joy Sweeney[b], Freda Song[b], Anthony Doyle[b], Gonzalo Daniel Maso Talou[a], Poul Michael Fønss Nielsen[a,c], Martyn Peter Nash[a,c], and Thiranja Prasad Babarenda Gamage[a]

[a]*Auckland Bioengineering Institute, University of Auckland, Auckland, New Zealand*
[b]*Department of Anatomy and Medical Imaging, University of Auckland, Auckland, New Zealand*
[c]*Department of Engineering Science, University of Auckland, Auckland, New Zealand*

13.1 Introduction

13.1.1 Breast anatomy

The breast has a tear-drop shape and is composed of many different types of tissues as seen in Fig. 13.1 (McGhee and Steele, 2020). Externally, the breast comprises the nipple, areolar, and skin. Internally, the tissues consist of glandular, adipose, and lactiferous components, in addition to blood vessels, Cooper's ligaments, the retromammary space, and the various fascias with the chest wall at the pectoralis major muscle making up the internal boundary (Hogg et al., 2015). The reader is directed to McGhee and Steele (2020) for a comprehensive review of breast anatomy.

13.1.2 Breast cancer

Globally, breast cancer is the most common cancer that impacts women, and is predominantly seen in females, whereas malignancy is less frequently observed in men (Momenimovahed and Salehiniya, 2019). It is a multifactorial disease with risks correlated with lifestyle factors, family history, age, as well as hormonal, environmental, and breast-related factors, such as breast density or length of lactation (Momenimovahed and Salehiniya, 2019). Globally, there is an overall 5-year survival rate of 73% from the time of diagnosis, which reduces to 61% at 10 years, with survival rates varying significantly depending on country and disease stage at diagnosis (Maajani et al., 2019). Breast cancers can be characterized in many ways, including, grade,

Biomechanics of the Female Reproductive System: Breast and Pelvic Organs. https://doi.org/10.1016/B978-0-12-823403-7.00026-9

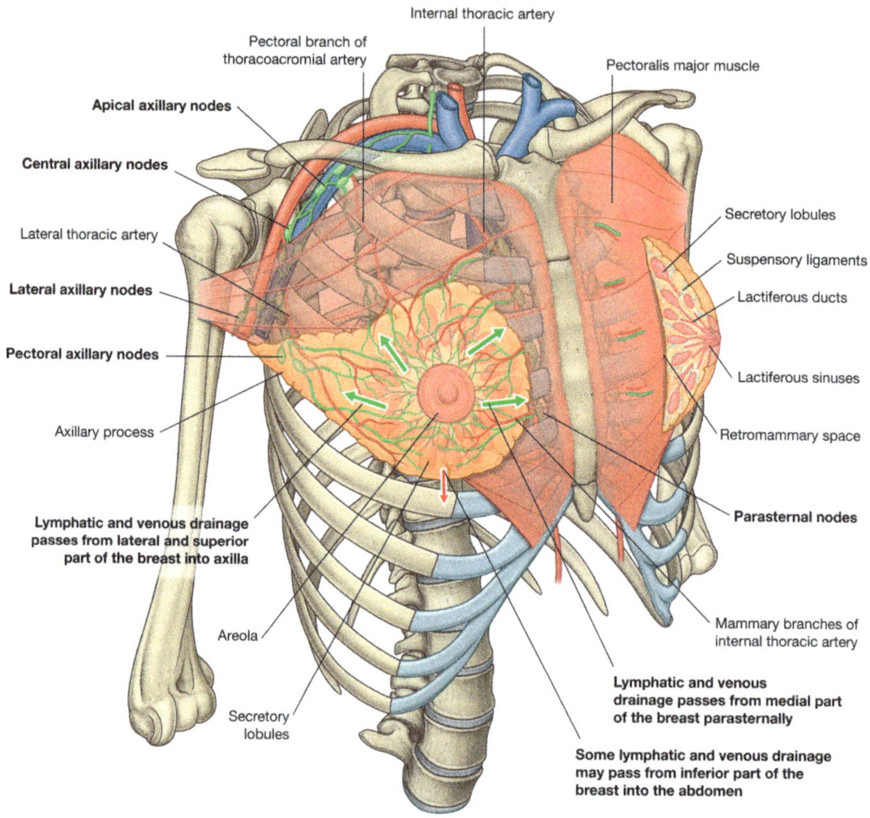

FIGURE 13.1

The anatomy of the female breast. Reprinted from Gray's Anatomy for Students 4th Edition, Richard Drake, A. Wayne Vogl, Adam Mitchell, Thorax, Page 133, Copyright 2019, with permission from Elsevier.

histology, physical size, and nodal involvement (Zhang et al., 2020). Diagnosis is typically based on histology, with invasive ductal carcinoma accounting for 75% of breast cancers at diagnosis (Zhang et al., 2020).

Categorization of the different stages, describing the extent of breast cancer, is one of the most important factors in determining prognosis and treatment options. Two stages, T1 and T2, are defined for early breast cancer. The T1 stage describes tumor sizes up to 20 mm, whereas the T2 stage describes tumors sizes between 20 mm and 50 mm. A review by the American Society for Radiation Oncology, investigating surgical margins for BCT with whole breast irradiation (WBI), reported a median tumor size of 16 mm, with an interquartile range of 15 mm to 21 mm, for the combination of T1 and T2 stage breast cancers (Moran et al., 2014). Breast tumors are most prevalent in the upper lateral region of the breast (Rummel et al., 2015).

Further information on awareness and knowledge of breast cancer can be found in Akram et al. (2017).

13.1.3 Breast cancer diagnosis and treatment

X-Ray mammography is currently the main screening tool for breast cancer (De Paredes, 2007). This imaging procedure involves generating a 2D projection through the patient's breast, while the breast is compressed between two plates. The two most common views for X-Ray mammography are craniocaudal (CC) and mediolateral oblique (MLO) images. The CC view gives a cranial view of the breast, whereas the MLO view is angled 40° to 60° from vertical to include the axillary tail and axilla tissues. Identifying the same feature of interest between views can help with strengthening disease diagnosis. However, X-Ray mammography has limitations in breast screening with lower specificity for women with dense breasts, for whom supplementary imaging is often necessary (Lebron-Zapata and Jochelson, 2018).

Ultrasound (US) imaging of the breasts involves examination with a handheld high-frequency transducer to sequentially sweep the surface of the breast and record lesions, typically used for follow up after a mammogram (Brem et al., 2015). Breast US is widely accessible, relatively inexpensive, does not involve exposure to ionizing radiation, and is most commonly used as an adjunct to X-ray mammography or MRI to localize lesions. When combined with X-ray mammography screening, the addition of US has been shown to increase sensitivity to breast cancer from 74.7% to 97.3%, with an increased sensitivity of 47.6% to 76.1% for women with dense breasts (Brem et al., 2015). The clinical benefit of US screening has been evaluated through several studies, reporting that cancers detected with US are often not visible on X-ray mammography, even retrospectively (Lebron-Zapata and Jochelson, 2018; Bae et al., 2014).

Magnetic resonance imaging (MRI) can provide high-resolution morphological imaging and discriminate breast cancer from other conditions (Mann et al., 2019). Though this modality has high sensitivity, it lacks specificity, leading to a greater number of biopsies (Morrow et al., 2011; Lebron-Zapata and Jochelson, 2018). MRI sensitivity for women with dense breasts ranges from 71% to 100% compared to 35% to 50% with X-ray mammography (Lebron-Zapata and Jochelson, 2018). As a result, the current standard for supplementary screening by the American Cancer Society recommends MRI for the screening of high-risk individuals (e.g., those with a family history of breast cancer) or individuals with dense breasts (Saslow et al., 2007). Clinically, breast MRI is acquired in the prone position (with the patient face down), enabling gravitational forces to separate out the tissues in the breast (Mann et al., 2019). This position also reduces imaging artefacts due to respiratory motion. A contrast agent is injected into the patient's blood stream prior to imaging to highlight areas of high blood flow that may help identify tumors. A dynamic contrast enhanced MRI sequence (DCE-MRI) is used to monitor the passage of the contrast agent through the tissue. T1 and T2 weighted MRI sequences can be acquired during breast imaging. In T1 weighted imaging, anatomical characteristics are outlined relatively well with the signal ranging from dark/hypointense for fluid structures, tendons, and ligaments, to bright/hyperintense for fat, whereas muscle shows as gray mid-intensities (McMahon et al., 2011). For T2 imaging, structures with high water content (such as adipose) appear hyperintense, whereas less hydrated, solid structures, such as tendons and cartilage, appear hypointense. T2 is useful for showing pathology as it is usually associated with edema or the presence of abnormal fluid, e.g., fluid-filled cysts (McMahon et al., 2011). MRI can also be acquired in the supine position. However, obtaining high quality imaging in this position is challenging as it requires specific setups to position the coils above the patient (Siegler et al., 2011). As a result, supine MRI is typically not acquired as part of standard clinical imaging of the breast.

Collocating suspicious tissue features between mammography, ultrasound, and MR images may help reduce the false positive rate associated with MR imaging. This also helps diagnose cancer with a significantly higher sensitivity, and at a more favorable stage in high risk women, compared with screening with only mammography and ultrasound imaging (Malur et al., 2001; Berg et al., 2004; Kuhl et al., 2005). However, the significantly different loading conditions that the breast experiences during these different imaging procedures makes collocation of regions of interest across the images a difficult task. Addressing this issue has the potential to improve diagnosis of breast cancer.

Following diagnosis, surgery is the typical treatment for women with early-stage nonmetastatic breast cancer. This involves either breast conserving surgery (lumpectomy) plus radiotherapy or mastectomy (Waks and Winer, 2019). With breast conserving surgery, it is necessary to excise the tumor with an adequate clear margin of tissue surrounding the tumor. This is made difficult due to differences in patient positioning during surgery compared with diagnostic imaging. Breast conserving surgeries have a 20% to 30% re-operation rate globally van Leeuwen et al., 2018. This is shown to impact the patient negatively because of longer treatment durations, higher cost of care, psychological distress, increased morbidity, and increasing the need for ongoing cosmetic surgery (Pleijhuis et al., 2009). Advances in navigational guidance for treatment procedures, e.g., using biomechanics for predicting breast tissue motion may assist clinicians with localizing tumors to help reduce recurrence and any long-term visible, functional, and psychological sequelae of lumpectomies (Waks and Winer, 2019; Wang et al., 2020).

13.2 Anatomical modeling

The first step towards performing patient-specific breast biomechanical simulations is the generation of personalized anatomical models of the key tissues that contribute to breast biomechanics during clinical procedures. These include the skin surface of the breast, the fibroglandular tissues, the pectoral muscles, and the chest wall. This section describes approaches used to segment these tissues from medical images, and how this information has been used to construct personalized anatomical models for biomechanics simulations. Particular emphasis is given to MRI, which has been the focus of recent developments, as it has been the primary imaging modality used for constructing patient-specific anatomical models of the breast and surrounding tissues.

13.2.1 Segmenting medical images of the breast

Many image processing techniques have been used to segment breast tissue from medical images. Over the past decade, automated techniques have been developed to replace existing semi-automatic or manual approaches. Table 13.1 highlights some of these techniques along with examples of the modalities and tissues they have been applied to.

One of the earliest methods proposed was simply thresholding the image intensities. A challenge in applying these techniques has been the selection of the threshold values, which typically requires a histogram analysis of the image intensities to be performed. This technique has been successfully applied to images with prominent edges and structures with large contrast with respect to the background, e.g., in mammograms (Wei et al., 2016) or CT images (Drukker et al., 2002). However, this approach does not consider spatial information or prior anatomical knowledge regarding the shape of the tissues. As a consequence, lower accuracy is achieved for lower quality images with low contrast or low signal-to-noise ratio, or when the structures of interest present complex texture patterns containing similar intensities as other tissues (Chen and Zwiggelaar, 2012).

Probabilistic atlas-based methods incorporate prior anatomical knowledge of tissue shape (Gubern-Mérida et al., 2015) and can yield accurate segmentations, even for images with low contrast between the tissues. These methods provide an approach to map tissue structures across individuals to support anatomical modeling, for example, from MR images. However, such methods are less resilient to anatomical or topological differences, such as implants, or in cases where the patient has undergone mastectomy.

Unsupervised K-means segmentation, which partitions images into a specified number of groups via distance metrics, has also been used to segment breast tissues, such as fibroglandular tissue and skin boundary (Niukkanen et al., 2018). Fuzzy C-means clustering (FCM) is another technique to cluster pixels into groups with common feature characteristics. To do this, it computes the distance between the position of a cluster's center and any given voxel position in a feature space. As FCM relies exclusively on image intensity, its performance decreases as image noise levels increase (Miao et al., 2020). Panigrahi et al. (2019) developed a multi-scale FCM algorithm, more robust to noise than conventional FCM, for use with ultrasound breast images.

The development of machine learning methods in recent years has provided many new methods for automated segmentation of medical images. Convolutional neural networks (CNNs) are the most popular machine learning methods for segmenting medical images of the brain (Moeskops et al., 2016) or prostate (Zhu et al., 2017). Neural networks for image recognition and processing use architectures inspired by

Table 13.1 Overview of breast segmentation studies. Fibroglandular tissue: FGT; Non-contrast enhanced T1-weighted MRI: T1; Non-contrast enhanced T2-weighted MRI: T2; Non-fat-suppressed:NFS; Fat-suppressed: FS; Accuracy: Acc = (TN + TP)/(TN + TP + FN + FP); Dice score: DSC; True-positive volume fraction: TPFV; Number of cases in the training dataset: Ntr; Number of cases in the testing dataset: Nte; Number of cases in the validation dataset: Nval.

Author	Methods	Types of tissue	Images	Case numbers and results
Gubern-Mérida et al. (2015)	Automated / Atlas-based	Breast	28 T1 NFS	Ntr = 1 Nte = 27 DSC = 0.94
Saltanat et al. (2010)	Semi-automated / Pixel value mapping	Pectoral	322 mammograms	Accuracy for each expert 0.84 and 0.94
Kwok et al. (2004)	Semi-automated / Line detection	Pectoral	322 mammograms	Acc = 0.839
Wu et al. (2013)	Automated / Edge detection	Breast	60 sagittal NFS T1	N = 60 DSC = 0.95
Lin et al. (2013)	Semi-automated / K-means clustering	Breast	31 NFS T1	Ntr = 1 Nte = 30 DP = 3.05
Piantadosi et al. (2020)	Semi-automated / Deep CNN	Breast	109 DCE-T1	DSC = 0.96
Dalmış et al. (2017)	Semi-automated /Deep learning	Breast, FGT	66 NFS T1	Ntr = 39 Nte = 22 Nval = 5 DSC (N = 66) = 0.944 DSC FGT(N = 66) = 0.811
Wu et al. (2015)	Semi-automated /Artificial neural network	Tumor	74 NFS T1	Ntr = 60 Nval = 60 DSC (Nval = 60) = 0.612
Ertas et al. (2017)	Semi-automated / Fuzzy C-means clustering	Breast	82 T1	Ntr = 50 Nte = 32 TPFV = 0.97
Ha et al. (2019a)	Semi-automated / Modified 3D CNN 2D U-Net	FGT, Parenchymal Enhancement	137 sagittal FST2	DSC FGT = 0.813 DSC BPE = 0.829
Moeskops et al. (2016)	Automated / CNN	Pectoral, brain	34 T1	DSC brain (Ntr = 14) = 0.91 DSC pectoral (Ntr = 14) = 0.77
Ha et al. (2019b)	Semi-automated / U-Net	Tumor	141 sagittal FS T1	Acc (Ntr = 37) = 0.877 Acc (Ntr = 46) = 0.877 Acc (Ntr = 30) = 0.878
Ivanovska et al. (2019)	Semi-automated / 2D U-Net	FGT	40 T1	Nte = 8 DSC = 0.925
Zhang et al. (2019)	Automated / U-Net	Breast, FGT	28 Non-contrast NFS T1	Ntr = 286 Nte = 224 DSC Breast(Nval = 224) = 0.86 DSC FGT (Nval = 224) = 0.83

the human visual cortex connectivity. They have been specifically designed to process structured data as images, sound signals, and videos. CNNs consist of multiple convolutional layers usually combined with activation, batch-normalization, down-sampling, and up-sampling layers. Many CNN architectures have demonstrated state-of-the-art accuracy for different tasks, outperforming the previously described traditional image-processing segmentation techniques (Ha et al., 2019a; Moeskops et al., 2016). One of the most successful CNN architectures for medical image processing is the U-Net (Ronneberger et al., 2015), which has been widely adopted by the medical imaging community (Ivanovska et al., 2019; Fashandi et al., 2019). The original U-Net implementation was proposed for application to 2D images (Ronneberger et al., 2015), followed by their 3D counterpart (Çiçek et al., 2016), which explores larger contextual data, but with greater demand on computational resources. The U-Net network is composed of an encoder, which reduces the image resolution while extracting global information from a wide field of view, and a decoder with skip connections, which repositions the global information at the original image resolution.

U-Net CNNs have been applied to segment the skin boundary, fibroglandular tissues, pectoral muscles, and tumors in the breast from MRI. Though there is sufficient contrast to distinguish between adipose and muscular tissues from MRI, it is usually difficult to determine the boundaries between muscular and fibroglandular tissues (e.g., segmentation of the pectoral muscle (Marrone et al., 2016; Rosado-Toro et al., 2015)). Low signal-to-noise ratio can also result in loss of accuracy, particularly meaningful for the segmentation of muscles surrounding the breast. Despite CNNs having shown promising results in segmenting tissues from breast MRI, several challenges remain. Included amongst these is segmentation of anatomical features that are necessary for constructing patient-specific anatomical models, such as the outer rib surface, for which training data in clinical MRI is difficult to obtain due to challenges in identifying this boundary. Furthermore, as the architectures of CNNs become increasingly complex, the number of model variables that need to be identified increases, requiring large datasets to obtain efficient and accurate performance.

13.2.2 Constructing anatomical models from segmentations

Once the breast tissues of interest have been segmented from medical images, patient-specific anatomical models can be constructed to provide a parametric description of the geometry of the breast tissues (see Fig. 13.2). Such models have predominantly been created from MR images as they provide the most detailed 3D information about the soft tissues in the breast and torso. Other imaging modalities, such as computed tomography (CT), or surface imaging techniques, such as structured light or laser scanners, have also been used to construct anatomical models. A summary of the anatomical models generated for recent biomechanical modeling applications is presented in Table 13.2.

Several approaches have been used for constructing parametric meshes to describe the geometries of the breast boundaries and internal tissues. These include marching cube algorithms (del Palomar et al., 2008; Mertzanidou et al., 2012; Han et al., 2014;

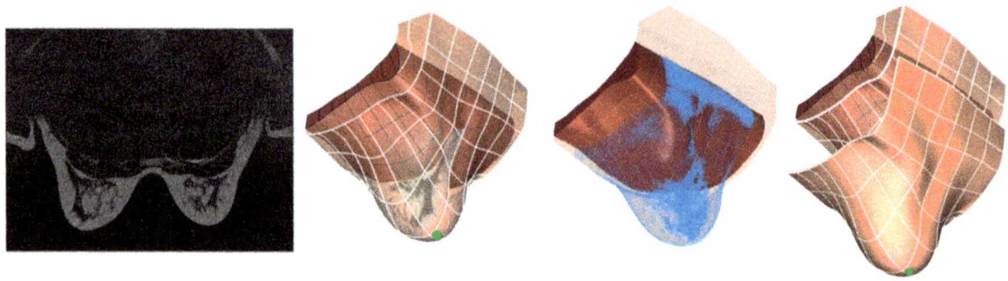

FIGURE 13.2

Example of patient-specific anatomical model of the breast generated from MRI using nonlinear geometric fitting and cubic Hermite hexahedral mesh elements (Babarenda Gamage et al., 2017b) (a, b). The fibroglandular tissue, segmented using a standard thresholding approach, is shown in (c). A skin mesh of constant thickness, extracted from the skin surface boundary of the fitted breast FE mesh is shown in (d) (the skin mesh has been offset from the breast surface for visualization purposes). Reprinted from Babarenda Gamage et al. (2017b) with permission from Elsevier.

Eiben et al., 2013; Mertzanidou et al., 2014; García et al., 2019; Na et al., 2019), Delaunay triangulation (Said et al., 2021), and nonlinear geometric fitting (Babarenda Gamage et al., 2019). Tetrahedral mesh elements are commonly employed for constructing anatomical models from the segmentation due the availability of existing tools to support their construction (Han et al., 2012; Hopp et al., 2013; Mertzanidou et al., 2014; Eiben et al., 2016a; García et al., 2017; Mîra et al., 2018; García et al., 2019; Said et al., 2021). A number of commercial software tools facilitate automatic generation of meshes from segmentations, for example, TetGen has been used to generate tetrahedral meshes in a number of studies (Han et al., 2012; Mertzanidou et al., 2012; Eiben et al., 2013; Mertzanidou et al., 2014; García et al., 2019) (see Table 13.2 for a list of commonly used software). Such algorithms typically generate many thousands of elements, to which constitutive parameters can be specified per element directly from segmented tissue labels, for subsequent biomechanical modeling applications. Other element types include higher-order cubic Lagrange (Babarenda Gamage et al., 2019) or cubic Hermite (Babarenda Gamage et al., 2017a) elements. Various error metrics have been used to assess the quality of the surface mesh generation procedure, such as the root-mean-squared-error (RMSE) or Hausdorff distances between the meshed tissues boundaries and the surfaces of the generated meshes.

13.3 Applications

Patient-specific anatomical models of the breast have been used in a number of applications for assisting clinicians with diagnosing and treating breast cancer. This typically involves solving the equilibrium equations for large deformation mechanics using the finite element method (FEM). A summary of recent studies and the ap-

Table 13.2 Applications of patient-specific anatomical models of the breast for breast cancer diagnosis and treatment.

Author	Application	Tissues considered	Mesh generation software	Simulation software
Assisting image registration for pre-operative treatment planning				
Eiben et al. (2013)	Prone / supine	Adipose, fibroglandular, pectoral	Meshlab and TetGen	NiftySim
Han et al. (2014)	Prone / supine	Adipose, fibroglandular, skin, pectoral	TetGen	NiftySim
Babarenda Gamage et al. (2017a)	Prone / supine	Homogeneous breast tissue, pectoral muscle	In-house	OpenCMISS
Multimodal image registration for supporting diagnostic imaging				
Han et al. (2012)	MR-Mammo	Adipose, fibroglandular, pectoral, tumor	TetGen	NiftySim and ABAQUS
Hopp et al. (2013)	MRI / X ray	Adipose, fibroglandular	Iso2mesh, TetGen	ABAQUS
Lee et al. (2013)	MRI / X ray	Homogeneous	In-house	CMISS
Solves-Llorens et al. (2014)	MRI / X ray	Adipose, fibroglandular, skin	Simpleware	ANSYS
Mertzanidou et al. (2014)	MRI / X ray	Homogeneous	VTK, TetGen	In-house
García et al. (2017)	MR-Mammo	Adipose, fibroglandular	TetGen	NiftySim
Martínez-Martínez et al. (2017)	MRI / X ray	Adipose, fibroglandular skin	Simpleware	ANSYS
García et al. (2019)	MRI / X ray	Homogeneous	TetGen	NiftySim
Said et al. (2021)	MRI / X ray	Adipose, fibroglandular, pectoral	In-house	ABAQUS
Predicting tumor positions during treatment procedures				
Mîra et al. (2018)	Prone / supine / supine titled	breast, pectoral, skin, suspensory ligaments, pectoral fascias	ANSYS Space-Claim	ANSYS
Babarenda Gamage et al. (2019)	Prone / supine	Homogeneous	In-house	OpenCMISS
Predicting outcomes following breast cancer treatment				
del Palomar et al. (2008); del Palomar et al. (2012)	Cosmetic outcome prediction	Adipose, fibroglandular	Harpoon	ABAQUS
Eiben et al. (2016a)	Cosmetic outcome prediction	Adipose, fibroglandular	VTK, Gmsh and Netgen	FEB3
Vavourakis et al. (2016)	Cosmetic outcome prediction	Adipose, fibroglandular	Gmsh	FEB3
Na et al. (2019)	Pre-mastectomy breast shape (prone to upright)	Adipose, fibroglandular	Gmsh	NiftySim

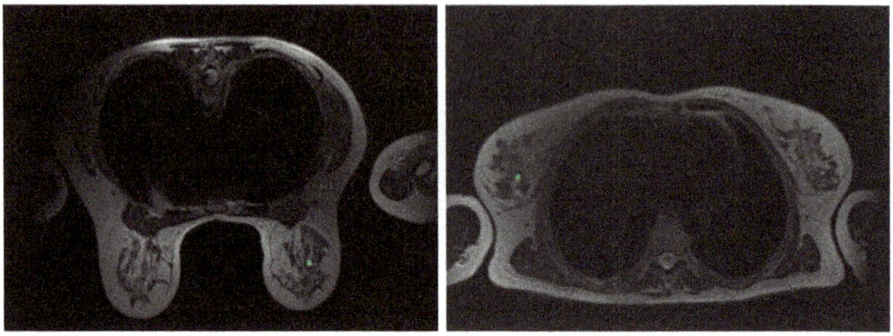

FIGURE 13.3

Example of breast MRI acquired in the prone (left) and supine (right) positions from volunteers. These images are generally of higher resolution than typical clinical images.

plications considered is presented in Table 13.2. These are described further in the sections that follow.

13.3.1 Assisting image registration for pre-operative treatment planning

One of the most active areas of research for the application of biomechanical modeling of the breast has been to assist with registration of breast MRI. As described in Section 13.2.2, diagnostic DCE-MR images are typically acquired while the patient is in a prone position. However, treatment procedures, such as lumpectomies or radiotherapy, are typically performed with patients oriented in the supine position. As the breast is a highly deformable and heterogeneous organ, there is substantial displacement of tissues as a patient is repositioned from prone to supine. As dynamic image guidance is generally not available during treatment procedures, it is difficult for clinicians to determine where tumors identified in the diagnostic images would move to during treatment procedures. Complete excision of tumors and their margins depends on the ability of a surgeon to palpate the lesion in its entirety. However, intra-operative localization of these tumors can be challenging, depending on the composition of the breast and the characteristics of the cancer. Guide wires may be embedded within the breast to help localize tumors before interventional procedures. However, these approaches cannot outline tumors in their entirety. To provide additional guidance, studies have investigated if acquiring pre-operative MRI in the supine position may help with localizing tumors during treatment procedures (Siegler et al., 2011; Gombos et al., 2016). However, the large deformations the tissues undergo between these two positions can make it difficult to map corresponding features between the images (see Fig. 13.3).

Non-rigid, free-form deformation (FFD) image-intensity-based registration approaches have traditionally been used to map features between pairs of images in different organs of the body. Though this approach has seen success in organs such as the brain and prostate, the application of these approaches to register prone and supine breast MRI has had limited success. This has been primarily due to the large deformation of the soft tissue observed between these two positions. This typically results in standard intensity-based registration methods, which are based on nonlinear optimization approaches, failing to correctly map the tissues between the two MRI.

Patient-specific biomechanical models have been applied to provide physics-based constraints to aid the registration of diagnostic prone breast MRI and pre-operatively acquired supine MRI, as illustrated in Fig. 13.4b (Babarenda Gamage et al., 2017b). This has been achieved by using physics-based models to predict the changes in breast shape that occurs as a patient is repositioned from prone to supine. The diagnostic MRI can be warped using the deformation field from the simulation to predict the configuration of the tissues in the supine position that has increased intensity overlap with the pre-operatively acquired supine MRI. This can provide a robust initial estimate of the tissue motion to enable standard intensity-based non-rigid registration algorithms to complete the mapping.

Novel approaches that further exploit the benefits of biomechanical modeling have also been used to assist with registering diagnostic prone MRI and pre-operatively acquired supine MRI. For example, Eiben et al. (2013) used a patient-specific biomechanical model to estimate the gravity-free unloaded configuration of the breast. Unloading was performed using anatomical models generated from both the prone and supine MRI. Image intensity-based non-rigid registration was then applied to map features in this virtual unloaded configuration to reduce deformations that need to be applied to the images compared with directly simulating prone to supine repositioning. Eiben et al. (2016b) proposed a surface driven registration method that combines FE-based biomechanical models with image warping techniques to match MR image volumes acquired from prone to supine positions. In an iterative process, the unknown material parameters are estimated to register the biomechanical model with the target, followed by an iterative closest point algorithm for surface registration. Their strategy also allowed discrepancies in simulation predictions caused by effects other than gravity, such as breast compression due to skin contact with the MR coil, to be reduced for improving the performance of the non-rigid registration procedure.

13.3.2 Multimodal image registration for supporting diagnostic imaging

Co-locating suspicious tissue features between imaging modalities is important for diagnosing breast cancer. X-ray mammograms only provide 2D projections of the tissues between the two plates. MRI on the other hand provides 3D information while a patient lies prone with their breasts pendant. Therefore co-locating suspicious lesions between these modalities is not straightforward.

Registration of MRI and X-ray mammograms for supporting breast cancer diagnosis has been an active area of research in the breast biomechanical modeling community (Hipwell et al., 2016). Previous research has aimed to construct anatom-

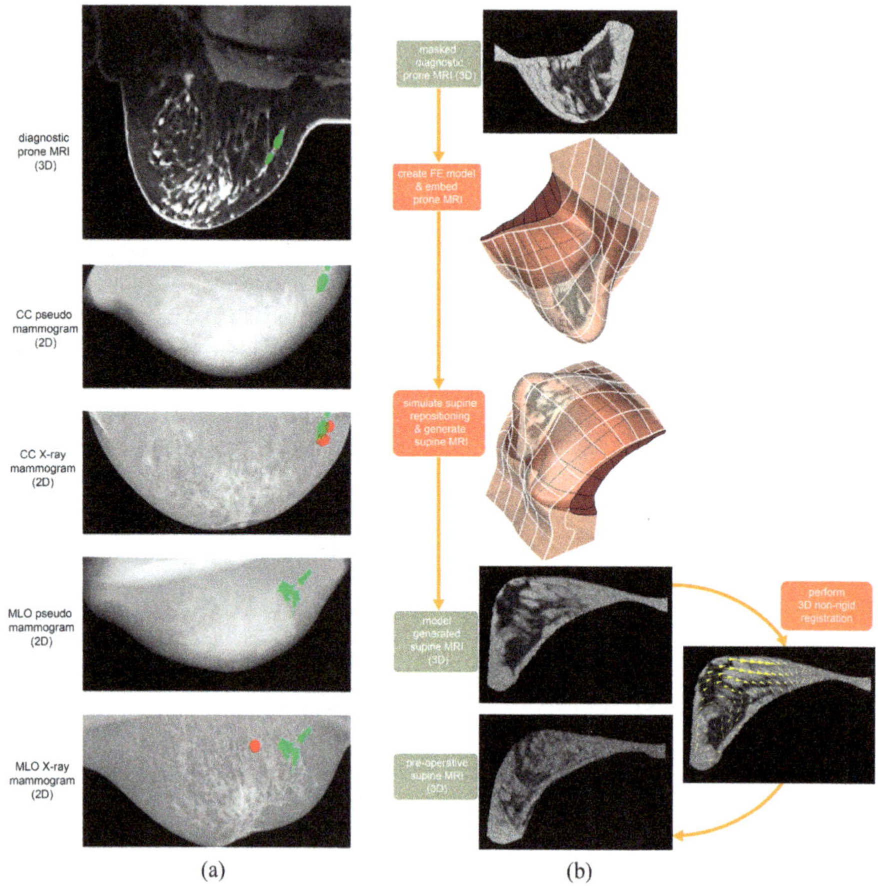

FIGURE 13.4

(a) A comparison between pseudo mammograms generated from diagnostic prone MR images of a patient using the modeling framework described by Lee et al. (2013), and X-ray mammograms. Green regions represent expert segmentation of lesions in prone diagnostic breast MR images (top) that were tracked to the pseudo-mammograms using biomechanical models. Red regions represent expert manual segmentations of the lesions in X-ray mammograms. (b) An example of how a biomechanical model can be used to simulate repositioning of the breast from the prone to the supine positions, to help register regions of interest between diagnostic prone MRI and supine MRI that could be used for pre-operative planning. This involves creating a breast model from the prone diagnostic MRI, which is then embedded into the model. The supine configuration can then be predicted using physics-based simulation. The deformation field from this prone-to-supine repositioning simulation can be used to warp the prone MRI to create a model-generated supine MRI, which can then be more easily registered with the pre-operative acquired supine MRI using standard intensity-based non-rigid registration techniques. Reprinted from Babarenda Gamage et al. (2017b) and Lee et al. (2013) with permission from Elsevier.

ical models of the breast from MR images and simulate tissue compression during an X-ray mammogram procedure. MR images can be transformed to generate a synthetic X-ray mammogram image that can be compared with a real mammogram to assist with co-locating suspicious features of interest as shown in Fig. 13.4a.

Hopp et al. (2013) proposed an automatic FEM-driven intensity-based 2D/3D registration via image similarity measures. Lee et al. (2013) proposed a patient-specific biomechanical FE model-based registration technique that estimates unknown parameters, such as compression plate orientation, which results in the best match between a pseudo and real X-ray image (Fig. 13.4a). Similarly, Mertzanidou et al. (2014) presented an intensity-based registration framework, based on a model for mapping information between the two modalities that had seven degrees of freedom, including parameters for the rigid pose of the breast and constitutive parameters for the biomechanical model. Whereas most approaches have focused on projecting information from an MRI volume into the 2D mammographic space, García et al. (2017) introduced an efficient method capable of back-projection, locating suspicious lesions in MRI space from mammograms. García et al. (2019) presented two gradient-based approaches for registering breast MRI and X-ray mammograms without needing to segment the internal tissues of the breast.

Though FEM is one of the most popular numerical methods used for simulating deformation of the breast, due to its high accuracy, the lengthy compute times have limited the clinical applications of these models. Indeed, the complexity is related to the number of elements included in the model, leading to a trade-off between accuracy and speed. To address this issue, machine learning surrogate models have been developed in recent years to accelerate simulation speeds. For example, Rupérez et al. (2018) used finite element models to simulate off-line behavior of the breast under compression, and trained three machine learning surrogate models from the simulation results to predict the deformation of the breast tissues in real-time. A similar approach was used in the work of Martínez-Martínez et al. (2017), where the authors trained machine learning models based on clinical data from 10 patients. The accuracy of the machine learning models typically depends on how well the FE model simulates breast compression.

13.3.3 Predicting tumor positions during treatment procedures

Physics-based biomechanical models have also been developed to simulate breast deformation from prone to supine to assist clinicians in localizing breast tumors for treatment planning (see Table 13.3). For example, Mîra et al. (2018) developed and evaluated an advanced biomechanical FE model for the breast containing a range of tissues, including the skin, pectoral muscle fasciae, and suspensory ligaments of the breast. The model used a neo-Hookean constitutive relation to represent the mechanical behavior of the tissues and was used to simulate breast deformation in prone, supine, and tilted supine positions.

As mentioned in previous sections, diagnostic procedures and treatment procedures are usually performed in different positions. Traditionally, the simulated breast

Table 13.3 Result of inter-modality registration algorithms using patient-specific biomechanical models in recent years (SMAD: symmetric mean absolute surface distance; TRE: target registration error).

Author	Modalities	Dataset	Error metrics	Error metric values
Lee et al. (2013)	MRI/X-ray	6 patients	SMAD	10 of 12 mammograms achieved < 14 mm error
Hopp et al. (2013)	MRI/X-ray	78 patients	TRE	13.2 mm
Solves-Llorens et al. (2014)	MRI/X-ray	14 patients	Euclidean distance	MRI to CC: 4.2 ± 1.3 mm MRI to MLO: 4.8 ± 1.3 mm CC and MLO to MRI: 4.1 ± 1.3 mm
Mertzanidou et al. (2014)	MRI / X-ray	10 patients	Euclidean distance	CC view: 11.6 ± 3.8 mm MLO view: 11 ± 5. 4 mm
Eiben et al. (2016c)	MRI/CT	3 patients	TRE	10.9 mm (internal structure)
García et al. (2019)	MRI/X-ray	10 patient	TRE	CC-mammograms: 5.65 ± 2.76 mm MLO-mammograms: 7.83 ± 3.04 3D MRI 7.33 ± 3.62 mm
Said et al. (2021)	MRI/ X-ray	1 patients	TRE	7.3 mm

tissues and tumor locations are visualized pre-operatively for surgery planning with the results typically presented on a computer monitor. However, with the development and wider availability of augmented reality (AR) and mixed-reality (MR) technologies, there has been increasing attention towards visualizing results from biomechanical models directly on patients, particularly for applications in breast surgery (Rancati et al., 2018; Gouveia et al., 2021).

Automated frameworks have been developed to construct and visualize patient-specific models. Georgii et al. (2014) developed a prototype that simulates breast tissue deformation when changing position from prone to supine. The software provides intuitive 3D visualization of the breast and lesion tracking for treatment planning. Babarenda Gamage et al. (2019) proposed a biomechanical modeling framework to simulate breast tissue deformation from prone to supine positions. This fully automated workflow integrated state-of-the-art image processing and machine learning techniques for segmentation. A population-based statistical shape model was combined with nonlinear geometric fitting to construct patient-specific anatomical models of the breast. Prone to supine simulation results were visualized using a purpose-built interface to assist clinicians by indicating positions of tumors relative to surrounding landmarks (see Fig. 13.5).

Esslinger et al. (2019) proposed a novel navigation system that integrates pre-operatively captured information through medical imaging, and simulates the deformation of the breast under different gravity orientations using a FE model. The workflow also includes an indoor positioning system to track medical devices during surgical procedures. Gouveia et al. (2021) described the first successful case of a breast conservative surgery guided with AR. A patient-specific 3D digital breast

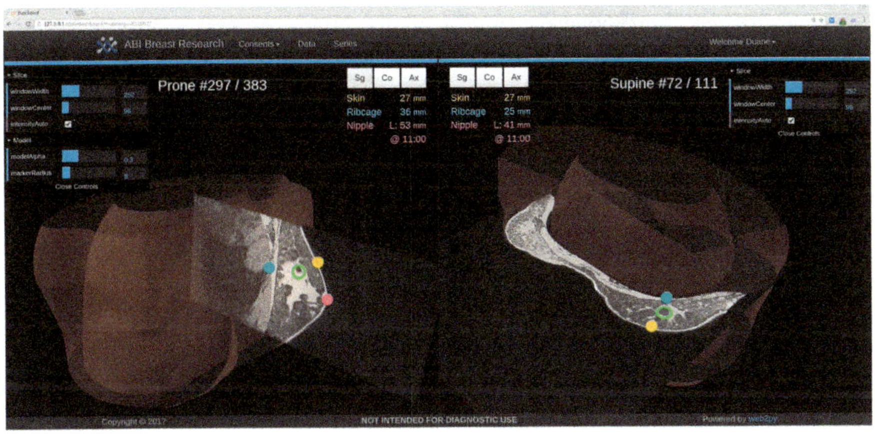

FIGURE 13.5

Interface for visualizing results from an automated workflow proposed by Babarenda Gamage et al. (2019), showing a patient-specific FE model generated from diagnostic prone MRI (left), and a simulated supine MRI created using a biomechanical simulation (right). The simulated MRI was generated by warping the prone MRI with the displacement field from the biomechanics model. A clinician can select a feature of interest, such as the location of a tumor, and the distances from the nipple, and closest points on the skin and rib cage (pink, yellow, and blue markers, respectively) are automatically calculated and displayed. Used with permission of The Royal Society (U.K.), from Babarenda Gamage et al. (2019); permission conveyed through Copyright Clearance Center, Inc.

model was processed and projected through a Microsoft Hololens mixed-reality AR headset, providing guidance on pre-operative positioning of a tumor.

13.3.4 Predicting cosmetic outcomes following breast cancer treatment

Patient-specific biomechanical models can be also used to predict the cosmetic outcome of breast treatment procedures. The predicted outcome is not only able to provide information to assist in making surgical decisions, but it also provides the ability to predict outcomes from such procedures and thereby facilitate communication with patients. A randomized controlled-trial study of 117 women showed that those participants who were provided with 3D post-operative simulation of the breast have a greater level of confidence going into surgery, when compared to those who were only provided with verbal descriptions or 2D photographic information (Godden et al., 2021).

Eiben et al. (2016a) present a workflow that integrates MRI and 3D optical surface scans to predict surgical results. A patient-specific biomechanical model generated from prone MRI was used to simulate breast tissue deformations in the upright configuration. The upright simulation was aligned with surface scans to tune the estimates of the constitutive parameters of the model, thereby improving the simulation accuracy. Vavourakis et al. (2016) proposed a numerical framework based on MRI scans to predict the outcome of breast-conserving therapy. The framework integrated a 3D multiscale FE model to simulate breast tissue deformation associated with wound healing and wound contraction after surgery. In the work of Na et al. (2019), a biomechanical model was used to describe the breast shape while standing, to facilitate accurately representing the natural, pre-mastectomy state of the breast.

Moreover, there is a growing demand for non-clinical breast augmentation procedures, which rely on patient-specific models to predict the post-treatment cosmetic outcomes. For example, Georgii et al. (2014) developed a software that uses a FE model to simulate the insertion of a breast implant using a biomechanical model built from skin surface measurements.

13.4 Challenges and opportunities

Over the past decade there have been many advances that demonstrate the value of patient-specific biomechanical modeling for assisting clinicians with breast cancer diagnosis and treatment. These range from image registration for improving diagnosis and treatment planning, to predicting outcomes following treatment procedures. There are many opportunities that remain to address the remaining challenges to enable patient-specific modeling of the breast to be routinely applied in clinical settings.

In some applications, such as image registration, patient-specific models are only used to provide initial estimates of tissue motion to support mapping of tissue features between different medical images. In such applications, highly accurate models may not be essential for achieving acceptable registration performance. However, in applications where these models are used in a predictive capacity, to predict tumor locations during surgery for example, requires greater performance. The performance of biomechanical simulations depends on how well the geometry of the breast tissues have been modeled, the accuracy of the boundary and loading conditions applied during the simulations, and the identification of patient-specific model parameters that describe the behavior of the different breast tissues. The order of importance of these aspects of the modeling procedure is expected to be application-specific and dependent on the magnitude of deformation observed.

While MRI has been predominantly used for constructing biomechanical models, this imaging modality is generally expensive and resource-intensive to acquire, and is typically not acquired clinically, except in high-risk patients. However, most research studies are based on using MR images for constructing 3D models. Opportunities exist for developing approaches for building 3D models without requiring MRI, for example, using information from more accessible imaging modalities such as 3D ultrasound or from skin surface measurements. The application of boundary conditions also presents challenges. For example, the changes in skeletal pose, particularly of the shoulder and arms, play a significant role in determining the deformation of the breast, because the shape of the underlying pectoral muscle changes with the movement of the arms. The positions of the arms vary depending on the procedure being performed, from arms raised above the head or against the side of the body during MRI imaging, to being directed away from the body during radiation therapy procedures. Patient-specific biomechanical models that accurately predict breast deformation across these different ranges of motion have not yet been developed and present an opportunity for further research.

The accuracy of patient-specific biomechanical model predictions depends on how well the model parameters represent the mechanical properties of the breast tissues. The current state-of-the-art for calibrating these model parameters requires the acquisition of MRI in multiple positions, such as the prone and supine positions. For example, the mechanical properties of the breast can be identified by tuning the stiffness parameters of the biomechanical model to best match the deformed breast shape seen in the supine MRI. Non-rigid free-form deformation registration algorithms are typically used to quantitatively map motion between diagnostic prone images and supine MRI to obtain the necessary information for constraining calibration, especially for tissues within the breast. Furthermore, the application of nonlinear constitutive relationships has not been extensively investigated despite most soft tissues such as the skin exhibiting significant nonlinear behavior. This presents avenues for further research to compare the performance of such models with the neo-Hookean constitutive relation, which is extensively used across all applications in which breast biomechanical models are considered.

The validation of biomechanical model predictions is essential for establishing confidence required for clinical application. This is often challenging due to the limited availability of data for testing model predictions. For example, when developing models to predict breast deformation in the supine position, prone MR images can be acquired to construct patient-specific models. Supine MRI can be acquired to calibrate the model's constitutive parameters using non rigid image registration. This approach enables the accuracy of the models to be evaluated by comparing the simulation results for predicting the supine position from the calibrated model against observations of tissue displacements between the prone and supine MRI. However, this does not constitute a validation of the model predictions, as both parameter estimation and model evaluation were performed using the same dataset. Additional MRI datasets in different positions (for example, with the torso resting on its side or a tilted body position (Mîra et al., 2018)) are required to independently validate the predictive accuracy of the models. Acquiring such data for demonstrating performance over a large population represents a significant challenge that needs to be addressed before these models can be routinely applied in clinical practice.

This review described the current state-of-the-art in patient-specific biomechanical modeling from medical images. Clinical adoption of these patient-specific models of the breast in clinical practice has significant potential to increase efficiency and accuracy of tumor localization, reduce variation in care, and improve outcomes following treatment procedures.

Acknowledgments

The authors are grateful for financial support from the New Zealand Government Ministry for Business, Innovation and Employment (UOAX1004), the University of Auckland Foundation (F-IBE-BIP), and the New Zealand Breast Cancer Foundation (R1704).

References

Akram, Muhammad, Iqbal, Mehwish, Daniyal, Muhammad, Khan, Asmat Ullah, 2017. Awareness and current knowledge of breast cancer. Biological Research 50 (1), 33. https://doi.org/10.1186/s40659-017-0140-9.

Babarenda Gamage, Thiranja P., Baluwala, Habib Y., Nash, Martyn P., Nielsen, Poul M.F., 2017a. Registration of prone and supine breast MRI for breast cancer treatment planning. In: Computational Biomechanics for Medicine. Springer, Cham, pp. 123–134.

Babarenda Gamage, Thiranja P., Malcolm, Duane T.K., Talou, Gonzalo Maso, Mîra, Anna, Doyle, Anthony, Nielsen, Poul M.F., Nash, Martyn P., 2019. An automated computational biomechanics workflow for improving breast cancer diagnosis and treatment. Interface Focus 9 (4), 20190034. https://doi.org/10.1098/rsfs.2019.0034.

Babarenda Gamage, Thiranja P., Nielsen, Poul M.F., Nash, Martyn P., 2017b. Chapter 10 - clinical applications of breast biomechanics. In: Payan, Yohan, Ohayon, Jacques (Eds.), Biomechanics of Living Organs. Academic Press, Oxford, pp. 215–242.

Bae, Min Sun, Moon, Woo Kyung, Chang, Jung Min, Koo, Hye Ryoung, Kim, Won Hwa, Cho, Nariya, Yi, Ann, et al., 2014. Breast cancer detected with screening US: reasons for nondetection at mammography. Radiology 270 (2), 369–377. https://doi.org/10.1148/radiol.13130724.

Berg, Wendie A., Gutierrez, Lorena, NessAiver, Moriel S., Carter, W. Bradford, Bhargavan, Mythreyi, Lewis, Rebecca S., Ioffe, Olga B., 2004. Diagnostic accuracy of mammography, clinical examination, US, and MR imaging in preoperative assessment of breast cancer. Radiology 233 (3), 830–849. https://doi.org/10.1148/radiol.2333031484.

Brem, Rachel F., Lenihan, Megan J., Lieberman, Jennifer, Torrente, Jessica, 2015. Screening breast ultrasound: past, present, and future. American Journal of Roentgenology 204 (2), 234–240. https://doi.org/10.2214/AJR.13.12072.

Chen, Zhili, Zwiggelaar, Reyer, 2012. A combined method for automatic identification of the breast boundary in mammograms. In: 2012 5th International Conference on BioMedical Engineering and Informatics, pp. 121–125.

Çiçek, Özgün, Abdulkadir, Ahmed, Lienkamp, Soeren S., Brox, Thomas, Ronneberger, Olaf, 2016. 3D U-Net: learning dense volumetric segmentation from sparse annotation. In: Ourselin, Sebastien, Joskowicz, Leo, Sabuncu, Mert R., Unal, Gozde, Wells, William (Eds.), Medical Image Computing and Computer-Assisted Intervention – MICCAI 2016. In: Lecture Notes in Computer Science. Springer International Publishing, Cham, pp. 424–432.

Dalmış, Mehmet Ufuk, Litjens, Geert, Holland, Katharina, Setio, Arnaud, Mann, Ritse, Karssemeijer, Nico, Gubern-Mérida, Albert, 2017. Using deep learning to segment breast and fibroglandular tissue in MRI volumes. Medical Physics 44 (2), 533–546. https://doi.org/10.1002/mp.12079.

De Paredes, Ellen Shaw De, 2007. Atlas of Mammography. Lippincott Williams & Wilkins.

del Palomar, A.P., Calvo, B., Herrero, J., López, J., Doblaré, M., 2008. A finite element model to accurately predict real deformations of the breast. Medical Engineering & Physics 30 (9),

1089–1097. https://doi.org/10.1016/j.medengphy.2008.01.005. Epub 2008 Mar 10. PMID: 18329940.

del Palomar, A.P., Calvo, B., Lapuebla-Ferri, A., 2012. Numerical modelling of human breast deformation. In: Preedy, V. (Ed.), Handbook of Anthropometry. Springer, New York, NY.

Drukker, Karen, Giger, Maryellen L., Horsch, Karla, Kupinski, Matthew A., Vyborny, Carl J., Mendelson, Ellen B., 2002. Computerized lesion detection on breast ultrasound. Medical Physics 29 (7), 1438–1446. https://doi.org/10.1118/1.1485995.

Eiben, Björn, Han, L., Hipwell, J., Mertzanidou, T., Kabus, S., Buelow, T., Lorenz, C., et al., 2013. Biomechanically guided prone-to-supine image registration of breast MRI using an estimated reference state. In: 2013 IEEE 10th International Symposium on Biomedical Imaging (ISBI), pp. 214–217.

Eiben, Björn, Lacher, Rene, Vavourakis, Vasileios, Hipwell, John H., Stoyanov, Danail, Williams, Norman R., Sabczynski, Jörg, et al., 2016a. Breast conserving surgery outcome prediction: a patient-specific, integrated multi-modal imaging and mechano-biological modelling framework. In: Tingberg, Anders, Lång, Kristina, Timberg, Pontus (Eds.), Breast Imaging. In: Lecture Notes in Computer Science. Springer International Publishing, Cham, pp. 274–281.

Eiben, Björn, Vavourakis, Vasileios, Hipwell, John H., Kabus, Sven, Buelow, Thomas, Lorenz, Cristian, Mertzanidou, Thomy, et al., 2016b. Symmetric biomechanically guided prone-to-supine breast image registration. Annals of Biomedical Engineering 44, 154–173. https://doi.org/10.1007/s10439-015-1496-z.

Eiben, Björn, Vavourakis, Vasileios, Hipwell, John H., Kabus, Sven, Lorenz, Cristian, Buelow, Thomas, Williams, Norman R., Keshtgar, M., Hawkes, David J., 2016c. Surface driven biomechanical breast image registration. In: Medical Imaging 2016: Image-Guided Procedures, Robotic Interventions, and Modeling, vol. 9786. SPIE, pp. 282–291.

Ertas, Gokhan, Doran, Simon J., Leach, Martin O., 2017. A computerized volumetric segmentation method applicable to multi-centre MRI data to support computer-aided breast tissue analysis, density assessment and lesion localization. Medical & Biological Engineering & Computing 55 (1), 57–68. https://doi.org/10.1007/s11517-016-1484-y.

Esslinger, Dominik, Rapp, Philipp, Knödler, Luzia, Preibsch, Heike, Tarín, Cristina, Sawodny, Oliver, Brucker, Sara Y., Hahn, Markus, 2019. A novel finite element model–based navigation system–supported workflow for breast tumor excision. Medical & Biological Engineering & Computing 57 (7), 1537–1552. https://doi.org/10.1007/s11517-019-01977-0.

Fashandi, Homa, Kuling, Gregory, Lu, Yingli, Wu, Hongbo, Martel, Anne L., 2019. An investigation of the effect of fat suppression and dimensionality on the accuracy of breast MRI segmentation using U-Nets. Medical Physics 46 (3), 1230–1244. https://doi.org/10.1002/mp.13375.

García, E., Oliver, A., Diaz, O., Diez, Y., Gubern-Mérida, A., Martí, R., Martí, J., 2017. Mapping 3D breast lesions from full-field digital mammograms using subject-specific finite element models. In: Medical Imaging 2017: Image-Guided Procedures, Robotic Interventions, and Modeling, vol. 10135. SPIE, pp. 26–33.

García, Eloy, Diez, Yago, Diaz, Oliver, Lladó, Xavier, Gubern-Mérida, Albert, Martí, Robert, Martí, Joan, Oliver, Arnau, 2019. Breast MRI and X-ray mammography registration using gradient values. Medical Image Analysis 54 (May), 76–87. https://doi.org/10.1016/j.media.2019.02.013.

Georgii, Joachim, Eder, Maximilian, Bürger, Kai, Klotz, Sebastian, Ferstl, Florian, Kovacs, Laszlo, Westermann, Rüdiger, 2014. A computational tool for preoperative breast augmentation planning in aesthetic plastic surgery. IEEE Journal of Biomedical and Health Informatics 18 (3), 907–919. https://doi.org/10.1109/JBHI.2013.2285308.

Godden, A.R., Micha, A., Wolf, L.M., Pitches, C., Barry, P.A., Khan, A.A., Krupa, K.D.C., Kirby, A.M., Rusby, J.E., 2021. Three-dimensional simulation of aesthetic outcome from breast-conserving surgery compared with viewing photographs or standard care: randomized clinical trial. British Journal of Surgery 108 (10), 1181–1188. https://doi.org/10.1093/bjs/znab217.

Gombos, Eva C., Jayender, Jagadeesan, Richman, Danielle M., Caragacianu, Diana L., Mallory, Melissa A., Jolesz, Ferenc A., Golshan, Mehra, 2016. Intraoperative supine breast MR imaging to quantify tumor deformation and detection of residual breast cancer: preliminary results. Radiology 281 (3), 720–729. https://doi.org/10.1148/radiol.2016151472.

Gouveia, Pedro F., Costa, Joana, Morgado, Pedro, Kates, Ronald, Pinto, David, Mavioso, Carlos, Anacleto, João, et al., 2021. Breast cancer surgery with augmented reality. Breast 56 (April), 14–17. https://doi.org/10.1016/j.breast.2021.01.004.

Gubern-Mérida, Albert, Kallenberg, Michiel, Mann, Ritse M., Martí, Robert, Karssemeijer, Nico, 2015. Breast segmentation and density estimation in breast MRI: a fully automatic framework. IEEE Journal of Biomedical and Health Informatics 19 (1), 349–357. https://doi.org/10.1109/JBHI.2014.2311163.

Ha, Richard, Chang, Peter, Mema, Eralda, Mutasa, Simukayi, Karcich, Jenika, Wynn, Ralph T., Liu, Michael Z., Jambawalikar, Sachin, 2019a. Fully automated convolutional neural network method for quantification of breast MRI fibroglandular tissue and background parenchymal enhancement. Journal of Digital Imaging 32 (1), 141–147. https://doi.org/10.1007/s10278-018-0114-7.

Ha, Richard, Chin, Christine, Karcich, Jenika, Liu, Michael Z., Chang, Peter, Mutasa, Simukayi, Van Sant, Eduardo Pascual, Wynn, Ralph T., Connolly, Eileen, Jambawalikar, Sachin, 2019b. Prior to initiation of chemotherapy, can we predict breast tumor response? Deep learning convolutional neural networks approach using a breast MRI tumor dataset. Journal of Digital Imaging 32 (5), 693–701. https://doi.org/10.1007/s10278-018-0144-1.

Han, Lianghao, Hipwell, J.H., Eiben, B., Barratt, D., Modat, M., Ourselin, S., Hawkes, D.J., 2014. A nonlinear biomechanical model based registration method for aligning prone and supine MR breast images. IEEE Transactions on Medical Imaging 33 (3), 682–694. https://doi.org/10.1109/TMI.2013.2294539.

Han, Lianghao, Hipwell, John H., Tanner, Christine, Taylor, Zeike, Mertzanidou, Thomy, Cardoso, Jorge, Ourselin, Sebastien, Hawkes, David J., 2012. Development of patient-specific biomechanical models for predicting large breast deformation. Physics in Medicine and Biology 57 (2), 455. https://doi.org/10.1088/0031-9155/57/2/455.

Hipwell, John H., Vavourakis, Vasileios, Han, Lianghao, Mertzanidou, Thomy, Eiben, Björn, Hawkes, David J., 2016. A review of biomechanically informed breast image registration. Physics in Medicine and Biology 61 (2), R1. https://doi.org/10.1088/0031-9155/61/2/R1.

Hogg, Peter, Kelly, Judith, Mercer, Claire, 2015. Digital Mammography: A Holistic Approach. Springer.

Hopp, T., Dietzel, M., Baltzer, P.A., Kreisel, P., Kaiser, W.A., Gemmeke, H., Ruiter, N.V., 2013. Automatic multimodal 2D/3D breast image registration using biomechanical FEM models and intensity-based optimization. Medical Image Analysis 17 (2), 209–218. https://doi.org/10.1016/j.media.2012.10.003.

Ivanovska, Tatyana, Jentschke, Thomas G., Daboul, Amro, Hegenscheid, Katrin, Völzke, Henry, Wörgötter, Florentin, 2019. A deep learning framework for efficient analysis of breast volume and fibroglandular tissue using MR data with strong artifacts. International Journal of Computer Assisted Radiology and Surgery 14 (10), 1627–1633. https://doi.org/10.1007/s11548-019-01928-y.

Kuhl, Christiane K., Schrading, Simone, Leutner, Claudia C., Morakkabati-Spitz, Nuschin, Wardelmann, Eva, Fimmers, Rolf, Kuhn, Walther, Schild, Hans H., 2005. Mammography, breast ultrasound, and magnetic resonance imaging for surveillance of women at high familial risk for breast cancer. Journal of Clinical Oncology 23 (33), 8469–8476. https://doi.org/10.1200/JCO.2004.00.4960.

Kwok, Sze Man, Chandrasekhar, R., Attikiouzel, Y., Rickard, M.T., 2004. Automatic pectoral muscle segmentation on mediolateral oblique view mammograms. IEEE Transactions on Medical Imaging 23 (9), 1129–1140. https://doi.org/10.1109/TMI.2004.830529.

Lebron-Zapata, Lizza, Jochelson, Maxine S., 2018. Overview of breast cancer screening and diagnosis. In: Breast Cancer: Uses and Opportunities for Molecular Imaging. PET Clinics 13 (3), 301–323. https://doi.org/10.1016/j.cpet.2018.02.001.

Lee, Angela W.C., Rajagopal, Vijayaraghavan, Babarenda Gamage, Thiranja P., Doyle, Anthony J., Nielsen, Poul M.F., Nash, Martyn P., 2013. Breast lesion co-localisation between X-ray and MR images using finite element modelling. Medical Image Analysis 17 (8), 1256–1264. https://doi.org/10.1016/j.media.2013.05.011.

Lin, Muqing, Chen, Jeon-Hor, Wang, Xiaoyong, Chan, Siwa, Chen, Siping, Su, Min-Ying, 2013. Template-based automatic breast segmentation on MRI by excluding the chest region. Medical Physics 40 (12), 122301. https://doi.org/10.1118/1.4828837.

Maajani, Khadije, Jalali, Arash, Alipour, Sadaf, Khodadost, Mahmoud, Tohidinik, Hamid Reza, Yazdani, Kamran, 2019. The global and regional survival rate of women with breast cancer: a systematic review and meta-analysis. Clinical Breast Cancer 19 (3), 165–177. https://doi.org/10.1016/j.clbc.2019.01.006.

Malur, Sabine, Wurdinger, Susanne, Moritz, Andreas, Michels, Wolfgang, Schneider, Achim, 2001. Comparison of written reports of mammography, sonography and magnetic resonance mammography for preoperative evaluation of breast lesions, with special emphasis on magnetic resonance mammography. Breast Cancer Research 3 (1), 55–60.

Mann, Ritse M., Cho, Nariya, Moy, Linda, 2019. Breast MRI: state of the art. Radiology 292 (3), 520–536. https://doi.org/10.1148/radiol.2019182947.

Marrone, Stefano, Piantadosi, Gabriele, Fusco, Roberta, Petrillo, Antonella, Sansone, Mario, Sansone, Carlo, 2016. Breast segmentation using fuzzy C-means and anatomical priors in DCE-MRI. In: 2016 23rd International Conference on Pattern Recognition (ICPR), pp. 1472–1477.

Martínez-Martínez, F., Rupérez-Moreno, M.J., Martínez-Sober, M., Solves-Llorens, J.A., Lorente, D., Serrano-López, A.J., Martínez-Sanchis, S., Monserrat, C., Martín-Guerrero, J.D., 2017. A finite element-based machine learning approach for modeling the mechanical behavior of the breast tissues under compression in real-time. Computers in Biology and Medicine 90 (November), 116–124. https://doi.org/10.1016/j.compbiomed.2017.09.019.

McGhee, Deirdre E., Steele, Julie R., 2020. Breast biomechanics: what do we really know? Physiology 35 (2), 144–156. https://doi.org/10.1152/physiol.00024.2019.

McMahon, Katie L., Cowin, Gary, Galloway, Graham, 2011. Magnetic resonance imaging: the underlying principles. The Journal of Orthopaedic and Sports Physical Therapy 41 (11), 806–819. https://doi.org/10.2519/jospt.2011.3576.

Mertzanidou, Thomy, Hipwell, John, Johnsen, Stian, Han, Lianghao, Eiben, Bjoern, Taylor, Zeike, Ourselin, Sebastien, et al., 2014. MRI to X-ray mammography intensity-based registration with simultaneous optimisation of pose and biomechanical transformation parameters. Medical Image Analysis 18 (4), 674–683. https://doi.org/10.1016/j.media.2014.03.003.

Mertzanidou, Thomy, Hipwell, John H., Han, Lianghao, Taylor, Zeike, Huisman, Henkjan, Bick, Ulrich, Karssemeijer, Nico, Hawkes, David J., 2012. Intensity-based MRI to X-ray mammography registration with an integrated fast biomechanical transformation. In: Maidment, Andrew D.A., Bakic, Predrag R., Gavenonis, Sara (Eds.), Breast Imaging. In: Lecture Notes in Computer Science. Springer, Berlin, Heidelberg, pp. 48–55.

Miao, Jiaqing, Zhou, Xiaobing, Huang, Ting-Zhu, 2020. Local segmentation of images using an improved fuzzy C-means clustering algorithm based on self-adaptive dictionary learning. Applied Soft Computing 91 (June), 106200. https://doi.org/10.1016/j.asoc.2020.106200.

Mîra, Anna, Carton, Ann-Katherine, Muller, Serge, Payan, Yohan, 2018. A biomechanical breast model evaluated with respect to MRI data collected in three different positions. Clinical Biomechanics 60 (December), 191–199. https://doi.org/10.1016/j.clinbiomech.2018.10.020.

Moeskops, Pim, Viergever, Max A., Mendrik, Adriënne M., de Vries, Linda S., Benders, Manon J.N.L., Išgum, Ivana, 2016. Automatic segmentation of MR brain images with a convolutional neural network. IEEE Transactions on Medical Imaging 35 (5), 1252–1261. https://doi.org/10.1109/TMI.2016.2548501.

Momenimovahed, Zohre, Salehiniya, Hamid, 2019. Epidemiological characteristics of and risk factors for breast cancer in the world. Breast Cancer: Targets and Therapy 11 (April), 151–164. https://doi.org/10.2147/BCTT.S176070.

Moran, Meena S., Schnitt, Stuart J., Giuliano, Armando E., Harris, Jay R., Khan, Seema A., Horton, Janet, Klimberg, Suzanne, et al., 2014. Society of surgical oncology–American Society for radiation oncology consensus guideline on margins for breast-conserving surgery with whole-breast irradiation in stages I and II invasive breast cancer. International Journal of Radiation Oncology, Biology, Physics 88 (3), 553–564. https://doi.org/10.1016/j.ijrobp.2013.11.012.

Morrow, Monica, Waters, Janet, Morris, Elizabeth, 2011. MRI for breast cancer screening, diagnosis, and treatment. The Lancet 378 (9805), 1804–1811. https://doi.org/10.1016/S0140-6736(11)61350-0.

Na, Gun-Yeol, Yang, Jeongsam, Cho, Sungwoo, 2019. Development of a 3D breast shape generation and deformation system for breast implant fabrication. Journal of Mechanical Science and Technology 33 (3), 1293–1303. https://doi.org/10.1007/s12206-019-0230-4.

Niukkanen, Anton, Arponen, Otso, Nykänen, Aki, Masarwah, Amro, Sutela, Anna, Liimatainen, Timo, Vanninen, Ritva, Sudah, Mazen, 2018. Quantitative volumetric K-means cluster segmentation of fibroglandular tissue and skin in breast MRI. Journal of Digital Imaging 31 (4), 425–434. https://doi.org/10.1007/s10278-017-0031-1.

Panigrahi, Lipismita, Verma, Kesari, Singh, Bikesh Kumar, 2019. Ultrasound image segmentation using a novel multi-scale Gaussian kernel fuzzy clustering and multi-scale vector field convolution. Expert Systems with Applications 115 (January), 486–498. https://doi.org/10.1016/j.eswa.2018.08.013.

Piantadosi, Gabriele, Sansone, Mario, Fusco, Roberta, Sansone, Carlo, 2020. Multi-planar 3D breast segmentation in MRI via deep convolutional neural networks. Artificial Intelligence in Medicine 103 (March), 101781. https://doi.org/10.1016/j.artmed.2019.101781.

Pleijhuis, Rick G., Graafland, Maurits, de Vries, Jakob, Bart, Joost, de Jong, Johannes S., van Dam, Gooitzen M., 2009. Obtaining adequate surgical margins in breast-conserving therapy for patients with early-stage breast cancer: current modalities and future directions. Annals of Surgical Oncology 16 (10), 2717–2730. https://doi.org/10.1245/s10434-009-0609-z.

Rancati, Alberto, Angrigiani, Claudio, Nava, Maurizio B., Catanuto, Giuseppe, Rocco, Nicola, Ventrice, Fernando, Dorr, Julio, 2018. Augmented reality for breast imaging. Minerva Chirurgica 73 (3), 341–344. https://doi.org/10.23736/s0026-4733.18.07659-9.

Ronneberger, Olaf, Fischer, Philipp, Brox, Thomas, 2015. U-Net: convolutional networks for biomedical image segmentation. In: Navab, Nassir, Hornegger, Joachim, Wells, William M.,

Frangi, Alejandro F. (Eds.), Medical Image Computing and Computer-Assisted Intervention – MICCAI 2015. In: Lecture Notes in Computer Science. Springer International Publishing, Cham, pp. 234–241.

Rosado-Toro, José A., Barr, Tomoe, Galons, Jean-Philippe, Marron, Marilyn T., Stopeck, Alison, Thomson, Cynthia, Thompson, Patricia, et al., 2015. Automated breast segmentation of fat and water MR images using dynamic programming. Academic Radiology 22 (2), 139–148. https://doi.org/10.1016/j.acra.2014.09.015.

Rummel, Seth, Hueman, Matthew T., Costantino, Nick, Shriver, Craig D., Ellsworth, Rachel E., 2015. Tumour location within the breast: does tumour site have prognostic ability? Ecancermedicalscience 9, 552. https://doi.org/10.3332/ecancer.2015.552.

Rupérez, M.J., Martínez-Martínez, F., Martínez-Sober, M., Lago, M.A., Lorente, D., Bakic, P.R., Serrano-López, A.J., Martínez-Sanchis, S., Monserrat, C., Martín-Guerrero, J.D., 2018. Modeling the mechanical behavior of the breast tissues under compression in real time. In: Tavares, João Manuel R.S., Jorge, R.M. Natal (Eds.), VipIMAGE 2017. In: Lecture Notes in Computational Vision and Biomechanics. Springer International Publishing, Cham, pp. 583–592.

Said, S., Clauser, P., Ruiter, N.V., Baltzer, P.a.T., Hopp, T., 2021. Image registration between MRI and spot mammograms for X-ray guided stereotactic breast biopsy: preliminary results. In: Medical Imaging 2021: Image-Guided Procedures, Robotic Interventions, and Modeling, vol. 11598. SPIE, pp. 354–361.

Saltanat, N., Hossain, M.A., Alam, M.S., 2010. An efficient pixel value based mapping scheme to delineate pectoral muscle from mammograms. In: 2010 IEEE Fifth International Conference on Bio-Inspired Computing: Theories and Applications (BIC-TA), pp. 1510–1517.

Saslow, Debbie, Boetes, Carla, Burke, Wylie, Harms, Steven, Leach, Martin O., Lehman, Constance D., Morris, Elizabeth, et al., 2007. American cancer society guidelines for breast screening with MRI as an adjunct to mammography. CA: A Cancer Journal for Clinicians 57 (2), 75–89. https://doi.org/10.3322/canjclin.57.2.75.

Siegler, Peter, Holloway, Claire M.B., Causer, Petrina, Thevathasan, Gilbert, Plewes, Donald B., 2011. Supine breast MRI. Journal of Magnetic Resonance Imaging 34 (5), 1212–1217. https://doi.org/10.1002/jmri.22605

Solves-Llorens, J.A., Rupérez, M.J., Monserrat, C., Feliu, E., García, M., Lloret, M., 2014. A complete software application for automatic registration of X-ray mammography and magnetic resonance images. Medical Physics 41 (8), 081903. https://doi.org/10.1118/1.4885957.

van Leeuwen, Marina T., Falster, Michael O., Vajdic, Claire M., Crowe, Philip J., Lujic, Sanja, Klaes, Elizabeth, Jorm, Louisa, Sedrakyan, Art, 2018. Reoperation after breast-conserving surgery for cancer in Australia: statewide cohort study of linked hospital data. BMJ Open 8 (4), e020858. https://doi.org/10.1136/bmjopen-2017-020858.

Vavourakis, Vasileios, Eiben, Bjoern, Hipwell, John H., Williams, Norman R., Keshtgar, Mo, Hawkes, David J., 2016. Multiscale mechano-biological finite element modelling of oncoplastic breast surgery—numerical study towards surgical planning and cosmetic outcome prediction. PLoS ONE 11 (7), e0159766. https://doi.org/10.1371/journal.pone.0159766.

Waks, Adrienne G., Winer, Eric P., 2019. Breast cancer treatment: a review. JAMA 321 (3), 288–300. https://doi.org/10.1001/jama.2018.19323.

Wang, Chuan-Bing, Lee, Sangwook, Kim, Taehun, Hong, Dayeong, Kim, Guk Bae, Yoon, Ga Young, Kim, Hak Hee, Kim, Namkug, Ko, BeomSeok, 2020. Breast tumor movements analysis using MRI scans in prone and supine positions. Scientific Reports 10 (1), 4858. https://doi.org/10.1038/s41598-020-61802-9.

Wei, Chia-Hung, Gwo, Chih-Ying, Huang, Pai Jung, 2016. Identification and segmentation of obscure pectoral muscle in mediolateral oblique mammograms. The British Journal of Radiology 89 (1062), 20150802. https://doi.org/10.1259/bjr.20150802.

Wu, H, Gallego-Ortiz, C., Martel, A., 2015. Deep artificial neural network approach to automated lesion segmentation in breast DCE-MRI. In: Proceedings of the 3rd MICCAI Workshop on Breast Image Analysis, pp. 73–80.

Wu, Shandong, Weinstein, Susan P., Conant, Emily F., Schnall, Mitchell D., Kontos, Despina, 2013. Automated chest wall line detection for whole-breast segmentation in sagittal breast MR images. Medical Physics 40 (4), 042301. https://doi.org/10.1118/1.4793255.

Zhang, Hanwen, Zhang, Ning, Moran, Meena S., Li, Yaming, Liang, Yiran, Su, Peng, Haffty, Bruce G., Yang, Qifeng, 2020. Special subtypes with favorable prognosis in breast cancer: a registry-based cohort study and network meta-analysis. Cancer Treatment Reviews 91 (December), 102108. https://doi.org/10.1016/j.ctrv.2020.102108.

Zhang, Yang, Chen, Jeon-Hor, Chang, Kai-Ting, Park, Vivian Youngjean, Kim, Min Jung, Chan, Siwa, Chang, Peter, et al., 2019. Automatic breast and fibroglandular tissue segmentation in breast MRI using deep learning by a fully-convolutional residual neural network U-Net. Academic Radiology 26 (11), 1526–1535. https://doi.org/10.1016/j.acra.2019.01.012.

Zhu, Qikui, Du, Bo, Turkbey, Baris, Choyke, Peter L., Yan, Pingkun, 2017. Deeply-supervised CNN for prostate segmentation. In: 2017 International Joint Conference on Neural Networks (IJCNN), pp. 178–184.

CHAPTER

Ultrasound elastography: in vivo assessment of tissue stiffness

14

O. Bou Matar and P. Pernod

Centrale Lille, Institute of Electronics, Microelectronics and Nanotechnology, IEMN UMR 8520, Lille, France

14.1 Introduction

Manual palpation, which consists of measuring the stiffness (i.e., elasticity) of soft tissues, is a diagnostic technique dating back to early medicine and is still commonly used today in physical examination to assess pathological tissue (such as tumors) with altered elastic properties. Ultrasound is one of the most used medical imaging due to its noninvasive nature, its real-time capability, its portability, its ease of use, and its low cost. Unfortunately, conventional ultrasound B-mode imaging measures the acoustic properties of tissues and is not able to differentiate tissues with different mechanical properties. Elastography, whose development began in the early 90s (Yamakoshi et al., 1990; Ophir et al., 1991; Fatemi and Greenleaf, 1998; Sarvazyan et al., 1998; Nightingale et al., 2001; Bercoff et al., 2004), aims at noninvasively imaging tissue stiffness. To measure the elasticity of the tissue, all elastography techniques use an external force applied to the studied tissue and follow the resulting movements. Moreover, this technique can measure the elasticity of tissues deep within the body, unlike manual palpation, which requires a direct contact and can only be applied to superficial organs, thus opening up new clinical applications.

This chapter presents first the principles of elastography, and then reviews its well-established application to tumor detection in breast and its potential applications to the cervix for preterm birth prediction or successful induction of labor, and for intraepithelial neoplasia and cancer diagnostics.

14.2 Principles of elastography

14.2.1 Introduction

There are different methods of ultrasound elastography in the literature, which can be grouped into two main families: static elastography techniques (or strain elastography) and dynamic elastography techniques (or shear wave elastography). Current ultrasound scanners have options for performing one or the other (or even both) of

Biomechanics of the Female Reproductive System: Breast and Pelvic Organs. https://doi.org/10.1016/B978-0-12-823403-7.00027-0

these techniques and to visualize not only acoustic impedance contrast images, but also images of tissue elasticity, whose changes can be correlated with soft tissue pathologies. The principles underlying each of these techniques will now be presented, highlighting their respective advantages and disadvantages.

14.2.2 Static elastography / strain elastography (SE)

Static elastography consists of measuring the deformations undergone by soft tissues when they are subjected to static (or quasi-static) compression (by hand or using physiological movements of the organs such as cardiovascular pulsations or respiratory motion), hence the name strain elastography (SE), though the more logical name strain imaging is sometimes used. The three steps for performing an elastogram with these techniques are as follows:

- Measurement of tissue displacements following compression: 1D (axial), 2D (axial and lateral) or 3D;
- Calculation of the deformations undergone by the tissues;
- Determination of the elasticity of tissues.

The measurement of tissue displacements between two images before and after deformation is obtained by speckle-tracking techniques. Tissue deformation is usually performed directly with the ultrasound probe. There are three main families of speckle-tracking techniques: windowing or block methods, regularized optimization methods, and more recently deep learning-based methods.

Windowing methods (Hall et al., 2003; Luo and Konofagou, 2010; Wang et al., 2018; Mirzaei et al., 2020) slice the (usually RF) images into blocks and look for the positions of maximum correspondence between the windows of the pre- and post-deformation images to determine the tissue displacements. The maximum correspondence between blocks can be determined by normalized cross-correlation (Zahiri-Azar and Salcudean, 2006) or zero-phase crossing techniques (Chen et al., 2004; Ara et al., 2013). Windowing techniques, although sensitive to acquisition noise and signal decorrelation, remain the most frequently used techniques due to their easy implementation.

Regularized optimization- or energy-based speckle tracking methods (Jiang and Hall, 2009; Pan et al., 2014; Ashikuzzaman et al., 2021) estimate tissue displacements by optimizing a nonlinear cost function that imposes data fidelity and spatial continuity constraints. These methods, though computationally heavy, have an interesting noise reduction capability. The computational load can be reduced by using the dynamic programming technique (Rivaz et al., 2008; Jiang and Hall, 2009).

More recently, promising techniques using deep learning algorithms, such as deep convolutional neural networks, have appeared in the literature (Gao et al., 2019; Peng et al., 2020; Tehrani and Rivaz, 2020). These methods, currently in the process of development, require clinical datasets that are not yet available for the learning phase of the algorithms.

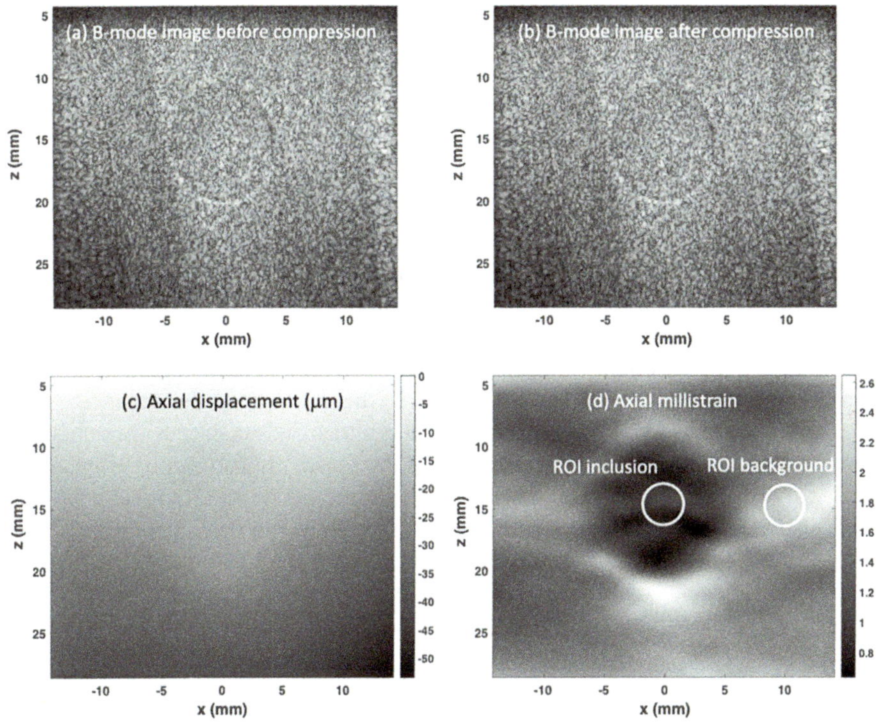

FIGURE 14.1

Stain elastography of a 10 mm diameter hard inclusion of a CIRS elasticity QA phantom model 049. (a) and (b), respectively: B-mode images obtained with a Verasonics vantage 128 research ultrasound scanner before and after compression with the L11-4v probe. (c) Axial displacement (in μm) obtained by the SOUL regularized optimization-based speckle tracking method proposed by Ashikuzzaman et al. (2021). (d) Axial millistrain obtained by a least squares method from the axial displacement.

Once the axial and lateral displacements (or 3D displacements in the best case) of the tissues have been measured, the deformation fields can be obtained by deriving with respect to the spatial dimensions. Unfortunately the direct derivation operation introduces a lot of noise, so it is usually performed by least squares methods (Kallel and Ophir, 1997) or derivation techniques using Golay filters (Oliphant et al., 2001).

An example of stain elastography of a 10 mm diameter hard inclusion of a CIRS elasticity QA phantom model 049, obtained by the SOUL regularized optimization-based speckle tracking method proposed by Ashikuzzaman et al. (2021) from two B-mode images measured with a Verasonics vantage 128 research ultrasound scanner before and after compression by hand with the L11-4v probe, is shown in Fig. 14.1.

The most used parameter in strain elastography is the strain ratio (SR), which is the ratio of the mean strain measured in the region of interest (ROI) to the one measured in a reference ROI (ROI background). In the example of Fig. 14.1, the mean strain in the ROI placed in the hard inclusion is 0.94×10^{-3}, and the one in the background ROI is 2.2×10^{-3}. The SR obtained here is then 2.34.

The step of passing from strain measurements to a measurement of the elasticity of the tissues is often very delicate, because it is difficult to measure (or even to know) the deformation undergone at any point by these same tissues when they are compressed by pressing the ultrasound probe on them. This is one of the reasons why commercial systems often propose simple strain imaging (sometimes called strain elastography), which allows to highlight the hardest areas on the images without giving quantitative values of the elasticity. In addition, it is often difficult to access deep organs because of the surface application of stress with the ultrasound probe. This problem can be overcome in some cases by using the physiological movements of the organs. Although it is difficult to quantify, this method has the advantage that it can be implemented simply on a conventional ultrasound scanner without having to modify the image acquisition principle. As the strain can be calculated between two successive ultrasound images, it is possible to follow the strain in real time and to calculate the strain rate. Thus strain elastography and strain rate imaging (SRI) are implemented on most ultrasound machines, on which the strain images are generally superimposed in color on the gray level ultrasound images as it has been done for a long time for the Doppler.

14.2.3 Dynamic elastography / shear wave elastography (SWE)

Although there are many different names or acronyms in the literature, the methods of ultrasonic dynamic elastography are all based on the measurement of the propagation velocity of shear waves in biological soft tissues. These shear waves propagate at a relatively slow speed (between 1 and 10 m/s in most biological tissues). It is therefore possible to follow these waves, thanks to ultra-fast ultrasound imaging techniques, which allow to obtain rates up to ten kilo images per second, instead of a hundred images per second for classical ultrasound scanners (Tanter and Fink, 2014). The link between the shear wave propagation speed c_s and the Young's modulus E is given by

$$c_s = \sqrt{\frac{E}{3\rho}}, \tag{14.1}$$

for soft tissue, for which we have $E \simeq 3\mu$, where μ is the shear modulus. ρ is the density of the tissue. It is often considered constant and equal to $1000\ \mathrm{kg/m^3}$ in soft tissue. For this reason, ultrasound scanners that have a shear wave elastography (SWE) option usually offer the ability to display either shear wave velocity images (in m/s) or Young's modulus images (in kPa).

The steps required to image soft tissue elasticity using shear wave elastography are as follows:

- Excitation of shear waves inside the tissue;

- Measurement of displacements induced by shear wave propagation using ultrafast imaging techniques;
- Estimation of shear wave velocity;
- Determination of tissue elasticity.

In practice, shear waves can be produced in soft tissue in three different ways: by radiation pressure induced inside the tissue by an (usually focused) ultrasound beam of long duration (several hundred periods at the transducer center frequency), by a vibration induced at the surface of the tissues by an external vibrator (vibrating pot for example), and which propagates in the tissues, or by the physiological movements of the organs.

Most SWE techniques implemented on commercial ultrasound scanners use shear waves generated by the radiation pressure created by ultrasound beams emitted by the same probe used to perform B-mode ultrasound imaging. The advantage of this technique is that we control the characteristics (frequency, location, ...) of the shear waves used to produce SWE images. The ultrasound beams used range from a single focused beam inducing an acoustic radiation force impulse (ARFI - Siemens, Philips, Hitachi-Aloka, Esaote) to multiple beams with focus swept over depth faster than shear-wave speed to create a Mach cone (SSI: SuperSonic Imagine) or focused successively at different depth (Siemens, Toshiba, Philips, Mindray Zonare), by way of simultaneous focused beams as in Comb-Push Ultrasound Shear Elastography (CUSE - GE).

When the vibration is induced at the surface of the tissues it can be difficult to access deep organs even though it is the technique used by the transient elastography (TE) developed by Echosens, which allows to give an average value on a line of measurement of the liver elasticity.

The use of physiological organ movements allows the generation of high-amplitude shear waves in deep or hard-to-reach organs of the body surface, but requires the synchronization of the induced displacement measurement system and the physiological movements. On the other hand, the frequency content of the generated shear waves is not controllable by the user. This can limit the resolution of the obtained elasticity images, which decreases with the frequency of the waves as in any imaging system.

The displacements induced by shear wave propagation, of the order of a few micrometers, are much smaller than those induced by the compression used for strain imaging. The techniques for measuring displacements or particle velocities are often different from those used in static elastography and are often derived from techniques used in ultrasound Doppler (Pinton et al., 2006): RF echo correlation tracking, such as those used in static elastography, Doppler processing (Kasai et al., 1985; Loupas et al., 1995), and Clutter Filter Wave Imaging (Salles et al., 2019). The measurement of displacements induced by shear wave propagation can be improved by techniques used in ultrasonic echography, such as angular compounding (Tanter et al., 2002), nonlinear harmonic imaging by pulse inversion (Song et al., 2013), or coded pulses (Song et al., 2015). An example of the axial particle velocity (in mm/s), induced by two radiation force pushes (focused on the edges of the image) in a CIRS elasticity QA phantom model 049 with a 10 mm diameter hard inclusion and measured

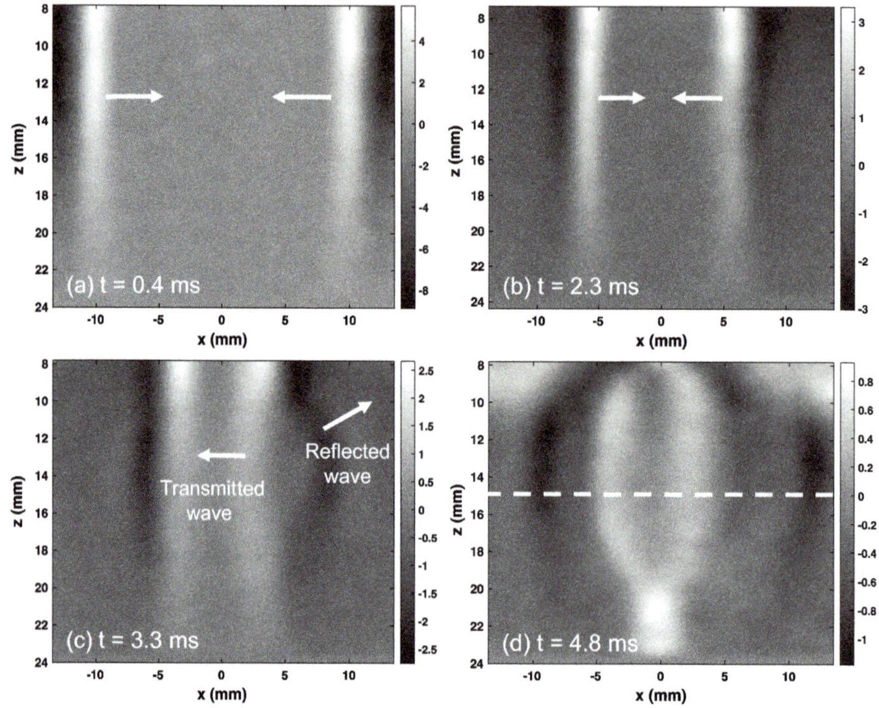

FIGURE 14.2

Axial particle velocity (in mm/s), induced by two radiation force pushes (focused on the edges of the image), versus depth and lateral position at four different times: (a) t = 0.4 ms, (b) t = 2.3 ms, (c) t = 3.3 ms, and (d) t = 4.8 ms in a CIRS elasticity QA phantom model 049 with a 10 mm diameter hard inclusion. Arrows indicate shear waves directions of propagation and show waves propagating in both directions (left and right) and waves reflected by and transmitted across the hard inclusion.

with ultra-fast ultrasound imaging using Loupas algorithm (Loupas et al., 1995), is displayed versus depth and lateral position at four different times on Fig. 14.2. The arrows on the figure indicate shear waves directions of propagation and show waves propagating in both directions (left and right) and waves reflected by and transmitted across the hard inclusion. Fig. 14.3 shows the same axial particle velocity (in mm/s) versus time and lateral position at a constant depth (a) where the phantom is homogeneous and (b) where the phantom is heterogeneous (indicated in Fig. 14.2d by a dotted line).

Many methods have been proposed for shear wave velocity estimation from these measured displacement, such as algebraic inversion method (Oliphant et al., 2001), including a faster 1D version using directional filters (Engel and Bashford, 2015), correlation-based methods (McLaughlin and Renzi, 2006; Tanter et al., 2008; Song et al., 2012), the time to peak method (Palmeri et al., 2008), the Radon transform

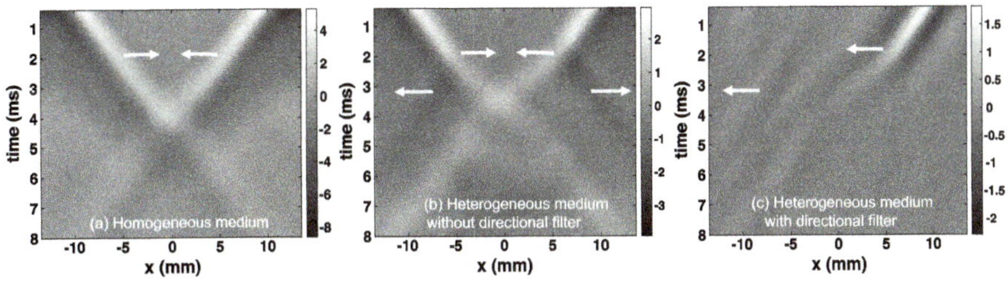

FIGURE 14.3

Axial particle velocity (in mm/s), induced by the same two radiation force pushes as in Fig. 14.2, versus time and lateral position at a constant depth: (a) where the phantom is homogeneous and (b) where the phantom is heterogeneous (indicated in Fig. 14.2d by a dotted line). (c) The same axial particle velocity as in (b) but after applying the directional filter keeping only the waves propagating towards the left. Arrows indicate shear waves directions of propagation.

method (Rouze et al., 2010), or more recently the plane wave elastography method (Khodayi-mehr et al., 2021). Most of these techniques are based on the assumption that shear waves are plane and propagate in a single direction. To approach these conditions it is often necessary to use directional filters (Manduca et al., 2003; Deffieux et al., 2011), as illustrated in Fig. 14.3(c). Once the speed of propagation of the shear waves is known, it is possible to determine the Young's modulus using Eq. (14.1). The reconstruction of the shear wave propagation velocity or the Young's modulus is performed only on a part of the ultrasound image called field of view (FOV). Commercial ultrasound equipment generally offers the option of displaying either the shear wave velocity (in m/s) or the Young's modulus (in kPa) of the tissue. As in the case of SE, SWE images are superimposed in color on the gray level ultrasound images. If the FOV remains localized just around the excitation point of the shear wave, the realized elasticity measurement is punctual, we speak then of point shear wave elastography (pSWE), also known as ARFI quantification. We can then make an image by moving the point of excitation. An example of SWE images is shown in Fig. 14.4: (a) an image of the propagation velocity (in m/s) of the shear wave and (b) the Young's modulus (in kPa) obtained from the measurements shown in Figs. 14.2 and 14.3. As for strain elastography, we can place ROIs on the images and measure different parameters. Popular quantitative SWE parameters in the differential diagnosis of lesions visible on ultrasonography are the mean (E_{mean}), the min (E_{min},), the max (E_{max}), and standard deviation (E_{SD}) of the measured Young modulus in the ROIs, as well as the elasticity ratio (E_{ratio}), which is the ratio of the mean value of Young modulus of the hard inclusion (a breast lesion for example) to the reference surrounding tissue. In the ROIs of Fig. 14.4 we can measure $E_{mean} = 100$ kPa, $E_{min} = 94$ kPa, $E_{max} = 106$ kPa, $E_{SD} = 9$ kPa, and E_{mean} (background) $= 43$ kPa, giving $E_{ratio} = 2.32$, same value as the SR obtained for the same phantom in the previous part.

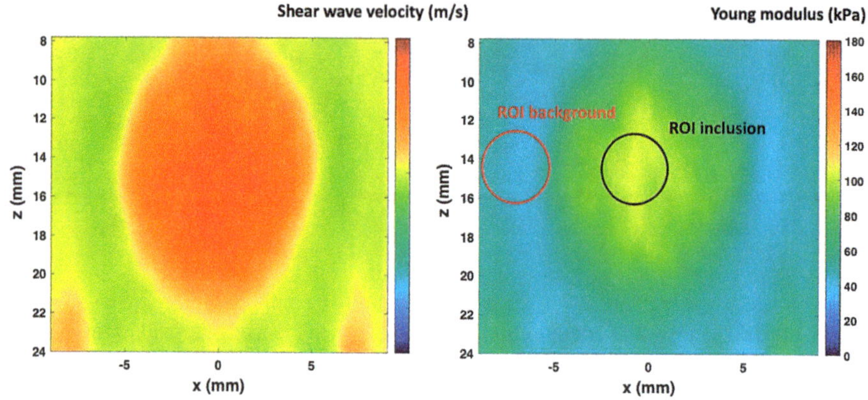

FIGURE 14.4

(a) Shear wave velocity (in m/s) and (b) Young modulus (in kPa) images obtained in the CIRS elasticity QA phantom model 049 showing the ROI used in the 10 mm diameter hard inclusion and the one used as a reference (ROI background).

It may be difficult to image in SWE very hard inclusions relative to the background, because the wave may in this case not penetrate the inclusion, making it impossible to determine the speed of propagation, and therefore their elasticity.

It is important to remember that the relationship given by Eq. (14.1) between the shear wave velocity and Young's modulus is only valid for homogeneous, isotropic, and linear elastic soft tissues, and therefore has many limitations, because biological tissues are often anisotropic, viscoelastic, and heterogeneous. Due to the anisotropy of some soft tissues (such as muscles due to the orientation of their fibers), the shear wave velocities vary, depending on the direction, and it is therefore important to perform the measurements properly by positioning the probe always along the same orientation (Chen et al., 2009; Gennisson et al., 2010; Amador et al., 2012). The viscoelasticity of tissues induces a variation of the propagation speed of shear waves with frequency, which is called dispersion (Amador et al., 2012; Yamakoshi et al., 1990; Deffieux et al., 2009; Gennisson et al., 2010; Kanai, 2005). Thus the Young's moduli determined using Eq. (14.1) can vary as a function of the frequency of the shear wave. By modifying the relationship used to determine the Young's modulus from the shear wave propagation velocity to account for these dispersion effects, and by performing the measurements at multiple frequencies, it is possible to obtain a measure of shear wave attenuation that provides additional information about the tissue. Finally, in stratified tissues, shear waves can propagate as guided modes, whose velocities are no longer related to Young's modulus by Eq. (14.1) but by more complex relationships, depending also on the geometry of the tissues (Bernal et al., 2011; Kanai, 2005).

Finally, we can notice that it has been shown very recently that it is possible to use SE and SWE techniques simultaneously to measure nonlinear stress-strain curves in vivo without contact (Wang et al., 2022).

14.3 Clinical applications for breast and cervix

The different variants of strain elastography and shear wave elastography were all studied for clinical applications during the last two decades. In relation with the purpose of the present book, they were considered in particular for the diagnostic of breast, the analysis of cervical elasticity during pregnancy for the prediction of preterm delivery or successful induction of labor, and for the diagnosis of cervical cancer and cervical intraepithelial neoplasia. The next paragraphs summarize the main clinical advances and present state in these different fields.

14.3.1 Breast elastography for tumor detection

According to the Breast Cancer Research Foundation, breast cancer is the most common cancer worldwide, overtaking lung cancer for the first time in 2020 with more than 2.3 million women diagnosed with the disease. This has stimulated a lot of research efforts on the subject in particular in the field of diagnostics. Mammography and B-mode ultrasound echography are the two classical imaging methods for breast cancer diagnostic but suffer from limits due to false negative results for mammography, and poor specificity for echography (Sigrist et al., 2017). On the other hand, it is well known that palpation is used to detect the presence of breast masses with high stiffness as a first sign of breast cancer detection. For this reason, many studies have been made with successful results, in particular this last decade, on application of ultrasound elastography (both strain elastography and shear wave elastography) as a method of detection of the presence of high stiffness masses for the early detection and characterization of breast cancer. The following paragraphs provide a summary on advances on the topic. The reader who wishes additional detailed information on the subject may refer to the following review papers, which were used in part for this section: Barr et al. (2015); Sigrist et al. (2017); Youk et al. (2017); Barr (2020).

14.3.1.1 Strain elastography (SE) of breast

As described in the first part of this chapter, strain elastography (SE) provides a qualitative image, where colors represent the relative displacement of the tissues when a compression is applied, externally by the operator or the imaging equipment, or internally by natural pulsating sources from the patient himself. Since the applied compression is unknown, the absolute value of the stiffness cannot be measured. In addition, color scales in the images are also dependent on the equipment without standardization at the date. The SE elastograms using only the color of the chosen ROI in the image does not provide adequate information for high sensitivity and specificity of characterization of breast lesions (Barr, 2020). For this reason, three main strategies were developed and are presently usually used for the analysis of SE images of the breast.

A first method of analysis of SE images for breast diagnostics is the elasticity score (also called Tsukuba score or 5-point color score) (Itoh et al., 2006). It is based on a stiffness color map of the tissue inside and around the lesion with integer values ranging from 1 to 5, increasing with higher probability of malignancy: score 1: the stiffness lesion is lower or equal to the stiffness of the surrounding tissues; score 2: the lesion includes areas of mixed (higher, lower, or equal) stiffnesses relatively to the surrounding tissues; score 3: the lesion has a higher stiffness than the surrounding tissues and a smaller size on the elastographic image than in the B-mode one; score 4: the lesion is stiff and has the same size on the elastographic image than in the B-mode one; score 5: the lesion has a higher stiffness than the surrounding tissues and a larger size on the elastographic image than in the B-mode one. Scores in the range 1–3 correspond to a probably benign lesion, whereas for scores in the range 4–5 a biopsy is recommended. Typical cutt-off values between 3 and 4 were used to discriminate between benign and malignant lesions and typical sensitivities of 86.5%–93% and specificities of 83%–90% were obtained by this method (Sigrist et al., 2017; Barr, 2020).

Another method of analysis of SE images for breast cancer, usually named EI/B-mode ratio, also provides a semi-quantitative measure and consists in considering the ratio defined as the size of the lesion measured on the elastographic image divided by the size measured on the B-mode image (Barr, 2012). This ratio was shown to be related to the aggressiveness of the malignancy, because it exploits the specific feature of malignant lesions to appear larger on elastographic images than on B-mode ones, contrary to benign lesions, which appear smaller (Hall et al., 2003). It was shown in several single center studies (123 biopsy-proven lesions (Barr, 2010), 230 lesions (Destounis et al., 2013)) and large multicenter study (635 biopsy-proven lesions (Barr et al., 2012)) that a cutt-off value of 1 provides a good sensitivity (98%–100%) and specificity (85%–91.5%) in differentiating between benign (ratio < 1) to malignant (ratio ≥ 1) lesions (Barr et al., 2015).

The third method also provides a semi-quantitative measure of stiffness by considering the strain ratio (SR), which compares the stiffness of the lesion (with ROI totally included in the lesion) to a reference tissue (fat without fibroglandular breast tissue at a similar depth to the lesion) (Ueno et al., 2007). The SR has been used in a number of clinical studies and has demonstrated good ability to discriminate between benign and malignant lesions (Ueno et al., 2007; Thomas et al., 2010; Farrokh et al., 2011). The reported cutoff value of SR varies in a typical range 2.3–4.8, depending on studies with typical sensitivities of 90%–94.4% and specificities of 89%–87.3%. Very recently, Mutala et al. (2022) published a systematic review and meta-analysis seeking to determine the optimal strain ratio cut off value for differentiating benign from malignant breast lesions. This work shows that breast SR measurement is still laden with multiple heterogeneities, and that the general optimal SR cut-off value could not be resolved yet. The authors stress that at this stage more robust evidence should be obtained before being able to define a machine predetermined cut-off value, and they propose more research on the manufacturer model calibration aspects.

14.3.1.2 Shear wave elastography (SWE) of breast

As described in the first part of this chapter, shear wave elastography (SWE) provides a quantitative measure of the stiffness (i.e., Young modulus, such as E_{max}, E_{min}, E_{mean}, E_{SD}, or E_{ratio}) using shear wave velocity. The measured values are usually displayed in color maps in the chosen field of view (FOV) or by providing the desired value measured in a small region of interest (ROI). During the last decade, a number of clinical studies have analyzed the benefits of SWE relative to B-mode ultrasound to better characterize breast lesions. The paragraphs that follow provide a summary of the results from the last decades.

In a BE1 multinational study involving 958 women, Berg et al. show that adding any of the SWE features (such as homogeneity and Eratio) to B-mode breast imaging reporting and data system (BI-RADS) feature analysis improved specificity from 61.1% to 78.5% ($P < 0.001$) of breast ultrasound mass assessment without loss of sensitivity (Berg et al., 2012). This allowed to selectively upgrade BI-RADS category 3 masses by identifying malignancies among oval, circumscribed masses without other suspicious features on B-mode echography, and downgrade category 4a masses, which could reduce unnecessary biopsies. With a similar approach, Schafer et al. analyzed 303 breast lesions from a multicenter BE1 prospective study population and have shown that SWE could help to downgrade benign BI-RADS category 3 lesions to category 2, in particular for values of Emax lower than 20 kPa or using "black to dark blue" colors in the stiffness image, thus allowing to reduce the number of unnecessary initial short-term follow-up visits (Schäfer et al., 2013).

Feldmann et al. made an analysis based on 83 breast lesions (38 malignant and 45 benign) (Feldmann et al., 2015). They assessed both qualitative and quantitative criteria to differentiate between benign and malignant lesions. The qualitative used criterium was the homogeneous or heterogeneous appearance in the SWE color map, and the quantitative ones were size of the lesion, maximum stiffness of the lesion E_{lesion}, maximum stiffness in the close peri-lesion area Eperilesion, and their respective ratios with the stiffness of the softest healthy fatty tissue Efat. They observed a higher heterogeneity with higher stiffness and ratio values for malignant lesions than for benign lesions and a better evaluation of size of malignant lesions using SWE than ultrasound B-mode. The optimal cutt-off value obtained for the perilesion ratio was 4.32 with 89% sensitivity and 73% specificity. The sensitivity was increased to 94% when combining the latter quantitative criterium with the qualitative criterium of heterogeneity appearance, this combination providing the best discrimination. As for the benefit of adding SWE to B-mode ultrasound, the best performance was obtained when downgrading BI-RADS 4a and 4b lesions with benign SWE features to BI-RADS 3.

If the maximum stiffness value E_{max} is often recommended to be used to characterize a lesion, a recent study made by Bulum et al. analyzed the influence of the ROI and lesion size on the performance of shear wave elastography in differentiating malignant breast lesions and discussed the related choice of stiffness parameters (Bulum et al., 2021). Their study involved 150 women with histopathologically confirmed malignant breast lesions. ROI diameters considered were 2 mm, 4 mm, and 6 mm. They have found in this study that E_{max} was independent on the ROI size. On contrary, it was shown that the smallest ROI size should be used when measuring E_{min}, E_{mean}, and E_{ratio} because of their highest values and lowest variability for malignant breast lesions. Higher E_{mean} and E_{max} values were observed for larger lesions. In that latter case, the authors thus suggested to use the E_{ratio} for uniform results regardless the lesion size.

Very recently, Pillai et al. made a new systematic review and meta-analysis encompassing 17,810 women and a total of 19,043 lesions on diagnostic accuracy of SWE for breast lesion characterization (Pillai et al., 2022). This meta-analysis confirmed the conclusions from previous works on the power of SWE in the diagnostic of breast cancer and its promise as an adjunct to mammographic and B-mode ultrasound. The obtained sensitivity and specificity were 86% and 87%, respectively, for the mean shear wave speed (SWS), without significant differences found when using the maximum of SWS, or standard deviation of SWS. The literature shows a range of thresholds for the mean SWS with a mean value found to be 3.89 ± 1.03 m/s. Nevertheless, the optimal value is still undetermined up to now. Comparing the results from this meta-analysis with their own results, the authors suggest that SE alone, or SE combined with SWE, may provide better diagnostic performance than SWE alone.

In a single center study, with a sample size of 81 women and 101 masses, and using a single model of machine (Siemens Acuson S2000), Tay et al. also suggested that strain elastography (SE) based on the elastogram to B-mode (E/B) size ratio provides a better diagnostic performance than shear-wave elastography (SWE), incorporating assessment of boundary tissue stiffness (sensitivity of 94.7% for SE and 73.7% for SWE; specificity of 81% for SE and 82.5% for SWE; AUC 0.878 for SE higher than 0.697 for SWE) (Tay et al., 2022), in line with studies made by Cantisani et al. (2021). The accuracy of SR was statistically significantly higher than SWE (sensitivity, specificity, and AUC were 89.2%, 76.6%, and 0.83 for SR and 72.3%, 66.0% and 0.69 for SWE, respectively, $p = 0.003$)

14.3.2 Cervical elastography

14.3.2.1 Cervical elastography for the prediction of preterm delivery or successful induction of labor

It is well known that remodeling and shortening of the cervix before delivery are associated with its stiffness change. Temporal abnormalities of cervical stiffness changes can be assessed by ultrasound elastography and were studied by several groups for the prediction of preterm delivery. Main results of these studies are presented in the paragraphs that follow.

The feasibility of strain elastography (SE) for this purpose was considered in 2007 by Yamaguchi et al. using a transvaginal probe, which was also used to apply the compression (Yamaguchi et al., 2007). Some improvements were made in 2012 by Fruscalzo et al., who proposed to use the full thickness of the anterior cervical lip as a standard region of interest (ROI) to reduce incongruences due to tissue inhomogeneity and dispersion of the stress related to the distance from the vaginal probe, and who also proposed recommendations to improve the control of the applied stress by pushing the probe perpendicularly to the longitudinal axis of the organ and increasing the applied force until saturation of the observed deformation on the anterior cervical lip (Fruscalzo et al., 2012; Fruscalzo and Schmitz, 2012). In parallel, in 2011, Swiatkowska-Freund & Price used the patient's breathing and pulsatile arterial movements as a natural internal source of compression of pregnant cervices for the prediction of labor induction success (Swiatkowska-Freund and Preis, 2011). The method has the advantage of being relatively independent from the operator. Using the same approach, Fuchs et al. showed in 2013 a good correlation between the elasticity of the cervical lip and its length (Fuchs et al., 2013).

Concerning application of shear wave elastography (SWE) for the purpose, Hernandez-Andrade et al. observed in 2014 a correlation between softening and gestation progress and spontaneous preterm delivery through a cross-sectional study of pregnant women at 11–36 weeks of gestation (Hernandez-Andrade et al., 2014). Softening of the cervix has been also observed by Sabiani et al. in 2015 in pregnancies complicated by preterm delivery (Sabiani et al., 2015). The same year, Carlson et al. observed a statistically significant difference in shear wave velocity speed before and after cervical ripening prior to induction of labor in pregnant women (Carlson et al., 2015).

More recently, in 2019, Bo Wang et al. made a systematic review of papers published in English before 2019 and provided their meta-analysis on 7 selected studies (in the frame 2014 to 2018) involving 1488 pregnant women to evaluate the diagnostic accuracy of cervical elastography in predicting preterm delivery (Wang et al., 2019). The authors concluded that cervical elastography is a promising and reliable method to predict the preterm delivery with a better diagnostic performance of prediction than the use of cervical length. They stressed that in this study subgroup analysis showed no significant difference between strain elastography and shear wave elastography to predict preterm delivery ($P = 0.94$). They also found no significant difference between transvaginal and transabdominal elastography of the cervix ($P = 0.12$). They finally underlined that given the relatively small sample size in their analysis, the conclusions should be considered with caution and be confirmed by future multicenter studies with a large sample size.

14.3.2.2 Cervical elastography for intraepithelial neoplasia (CIN) and cancer diagnostics

The diagnosis of the cervix in gynecology is another subject of interest for the use of ultrasound elastography. Marina Dudea-Simon et al. made in 2021 a systematic review of the bibliography from January 2000 until May 2020 (23 papers) (Dudea-Simon et al., 2021), which was used in part for this section.

In 2007, Thomas et al. used the color analysis elastography to compare cancer and normal cervices on 102 women (Thomas et al., 2007). They observed a significantly higher proportion of hard tissue for patients with cervix cancer ($34 \pm 15\%$) than for the normal group ($26 \pm 13\%$). In another study made in 2015 on 87 women with elastic color scores, Bakay et al. also observed a higher stiffness of malignant cervices than in nonmalignant ones (Bakay and Golovko, 2015). From 2012 to 2020, several groups applied the SE using the strain ratio (SR) to provide a semi-quantitative assessment of the cervix (Sun et al., 2012; Shady et al., 2015; Dudea-Simon et al.,

2021). They all found that the SR of the tumors were significantly higher than the SR of normal tissues and could define some cutoff values of the SR for the differentiation. Nevertheless, Dudea-Simon et al. emphasize that, at the date of their paper (2021), it was not possible to define a universal cu-off for diagnosing malignant lesions of the cervix since it depends on several parameters, such as the choice of the study group, its size, the choice of the reference for comparison for SR evaluation or the type of ultrasound equipment and settings chosen by the operator (Dudea-Simon et al., 2021).

Regarding SWE, Liu et al. also demonstrated in 2019 the ability to differentiate between normal and abnormal cervical tissues as well as benign and malignant lesions using transvaginal real-time shear wave elastography in a study involving 246 women: the maximum shear wave velocity was 5.24 ± 1.11 m/s for malignant cervix (138 cases), 3.93 ± 0.39 m/s for benign cervix (40 cases), and 3.27 ± 0.31 m/s for normal cervix (68 cases); the area under the curve (AUC) for both differential diagnosis "normal versus benign cervical tumors" and "benign versus malignant" was 0.909 ($P < 0.001$) (Liu et al., 2019).

Very recently, in 2022, Guo et al. also demonstrated the applicability of real-time shear wave elastography to cervix cancer diagnosis in a study involving 285 female patients, divided in three groups: 94 with cervix cancer (CC), 91 with cervical intraepithelial neoplasia (CIN), and 100 for the normal control group (Guo et al., 2022). The results show no significant differences in cervical hardness between the patients with CIN and the normal population. On the contrary, the stiffness differences were significant for patients with CC. The maximum Young's modulus was found to provide the highest accuracy for CC diagnosis (89.7%) with a cutoff value of 43.48 kPa.

Some recent works interestingly considered elastography as a way to detect and predict tumor response to therapy in cervical cancer with encouraging results. The aim is to provide a cheaper and more simple method than magnetic resonance imaging, diffusion-weighted imaging and 18F-fluorodeoxyglucose positron emission tomography (Mabuchi et al., 2015). In 2015, Mabuchi et al. found through a study of 14 participants a SR about three times higher for cervical tumors (mean SR value of 3.8) than for normal cervical tissues (mean SR value of 1.2) and a reduction of SR to the same level as the normal cervix for patients having a complete response to definitive radiotherapy (Mabuchi et al., 2015), which was not the case among patients who developed residual disease after the treatment. They also found that the tumor SR reduction preceded the tumor volume reduction.

14.4 Conclusion

Studies and meta-analysis of the last three decades have demonstrated the benefits of elastography for differentiating benign breast and cervix lesions from malignant lesions with high sensitivity and high specificity, as well as for predicting preterm delivery with better diagnostic performance than cervix length measurements, or predicting successful induction of labor. For breast lesion characterization, the EF-

SUMB guidelines and recommendations for the clinical practice of elastography in nonhepatic applications indicate that elastography methods have been evaluated for improving the generally high sensitivity and specificity of the breast imaging reporting and data system (BIRADS) and are recommended to be used as add-ons to the regular B-mode examination (Săftoiu et al., 2019). One of the main benefits of elastography relies on the possibility to downgrade some BI-RADS category 4a masses, thus allowing to reduce unnecessary biopsies, or upgrade BI-RADS category 3 masses for histological evaluation. Elastography for the characterization of the cervix was less studied up to now and is still not mentioned as a possible application in the EFSUMB guidelines and recommendations for the clinical practice of elastography in nonhepatic applications. Nevertheless, interesting results have been obtained recently and have shown its relevance and usefulness in gynecology.

Strain elastography can only provide relative or semi-quantitative information on the tissue stiffness. However, recent works proposed to solve this problem by adding a force-measuring device on the ultrasound probe to overcome this limitation (Thomsen et al., 2022). Shear wave elastography, on the other hand, provides the absolute value of stiffness and is less dependent upon the operator. As an example, Castro et al. obtained an intra- and interobserver correlation of 0.996–0.999 of stiffness in their studies evaluating both normal and pathological cervix ($p < 0.005$) (Castro et al., 2021). Also, some lesions prevent or alter the propagation of the shear wave, inducing in these cases absence or false measurements of the lesion stiffness. The meta-analyses have shown for both methods a large variation of cutt-off values among the studies due to a lack of standardization.

Research on elastography is still continuing, both on the contribution of innovations to diagnostic systems and associated analysis methods, and on their clinical applications, for which more large-scale studies are still needed. It also seems important for the future to converge towards a standardization of the methods to avoid the present variability between systems and techniques.

References

Amador, C., Urban, M.W., Chen, S., Greenleaf, J.F., 2012. Loss tangent and complex modulus estimated by acoustic radiation force creep and shear wave dispersion. Phys. Med. Biol. 57, 1263–1282.

Ara, S.R., Mohsin, F., Alam, F., Rupa, S.A., Lee, S.Y., Hasan, Md.K., Awwal, R., 2013. Phase-based direct average strain estimation for elastography. IEEE Trans. Ultrason. Ferroelectr. Freq. Control 60, 2266–2283.

Ashikuzzaman, M., Sadeghi-Naini, A., Samani, A., Rivaz, H., 2021. Combining first- and second-order continuity constraints in ultrasound elastography. IEEE Trans. Ultrason. Ferroelectr. Freq. Control 68 (7), 2407–2418.

Bakay, O.A., Golovko, T.S., 2015. Use of elastography for cervical cancer diagnostics. Exp. Oncol. 37, 139–145.

Barr, R.G., 2010. Real-time ultrasound elasticity of the breast: initial clinical results. Ultrasound Q. 26 (2), 61–66.

Barr, R.G., 2012. Sonographic breast elastography: a primer. J. Ultrasound Med. 31 (5), 773–783.

Barr, R.G., et al., 2012. Evaluation of breast lesions using sonographic elasticity imaging: a multicenter trial. J. Ultrasound Med. 31 (2), 281–287.

Barr, Richard G., 2020. Chapter 2 - breast elastography. In: Tissue Elasticity Imaging, Volume 2: Clinical Applications, pp. 21–46.

Barr, Richard G., Nakashima, Kazutaka, Amy, Dominique, Gosgrove, David, Farrokh, Andre, Schafer, Fritz, Bamber, Jeffrey C., Castera, Laurent, Choi, Byung Ihn, Chou, Yi-Hong, Dietrich, Christoph F., Ding, Hong, Ferraioli, Giovanna, Filice, Carlo, Friedrich-Rust, Mireen, Hall, Timothy J., Nightingale, Kathryn R., Palmeri, Mark L., Shiina, Tsuyoshi, Suzuki, Shinichi, Sporea, Ioan, Wilson, Stephanie, Kudo, Masatoshi, 2015. WFUMB guidelines and recommendations for clinical use of ultrasound elastography: Part 2: breast. Ultrasound Med. Biol. 41 (5), 1148–1160.

Bercoff, J., Tanter, M., Fink, M., 2004. Supersonic shear imaging: a new technique for soft tissue elasticity mapping. IEEE Trans. Ultrason. Ferroelectr. Freq. Control 51, 396–409.

Berg, W.A., Cosgrove, D.O., Dore, C.J., Schafer, F.K., Svensson, W.E., Hooley, R.J., Ohlinger, R., Mendelson, E.B., Balu-Maestro, C., Locatelli, M., Tourasse, C., Cavanaugh, B.C., Juhan, V., Stavros, A.T., Tardivon, A., Gay, J., Henry, J.P., Cohen-Bacrie, C., 2012. Shear-wave elastography improves the specificity of breast US: the BE1 multinational study of 939 masses. Radiology 262, 435–449.

Bernal, M., Nenadic, I., Urban, M.W., Greenleaf, J.F., 2011. Material property estimation for tubes and arteries using ultrasound radiation force and analysis of propagating modes. J. Acoust. Soc. Am. 129 (3), 1344–1354.

Bulum, A., Ivanac, G., Divjak, E., Biondić Špoljar, I., Džoić Dominković, M., Bojanić, K., Lucijanić, M., Boris Brkljačić, B., 2021. ElasticModulus and elasticity ratio of malignant breast lesions with ShearWave ultrasound elastography: variations with different region of interest and lesion size. Diagnostics 11, 1015. https://doi.org/10.3390/diagnostics11061015.

Cantisani, V., David, E., Barr, R.G., et al., 2021. US-elastography for breast lesion characterization: prospective comparison of US BIRADS, strain elastography and shear wave elastography. Ultraschall Med. 42, 533–540.

Carlson, L.C., Romero, S.T., Palmeri, M.L., Munoz Del Rio, A., Esplin, S.M., Rotemberg, V.M., et al., 2015. Changes in shear wave speed pre- and post-induction of labor: a feasibility study. Ultrasound Obstet. Gynecol. 46, 93–98.

Castro, Laura, García-Mejido, Jose Antonio, Holgado, Ainhoa, García, Rocio, Fernández-Palacín, Ana, Sainz, Jose Antonio, 2021. Reliability of shear-wave elastography (SWE) for investigating cervix elastic properties in normal and benign pathological situations. Clin. Exp. Obstet. Gynecol. 48 (3), 583–589. https://doi.org/10.31083/j.ceog.2021.03.2420.

Chen, S., Urban, M.W., Pislaru, C., Kinnick, R., Zheng, Y., Yao, A., Greenleaf, J.F., 2009. Shearwave dispersion ultrasound vibrometry (SDUV) for measuring tissue elasticity and viscosity. IEEE Trans. Ultrason. Ferroelectr. Freq. Control 56 (1), 55–62.

Chen, X., Zohdy, M.J., Emelianov, S.Y., O'Donnell, M., 2004. Lateral speckle tracking using synthetic lateral phase. IEEE Trans. Ultrason. Ferroelectr. Freq. Control 51 (5), 540–550.

Deffieux, T., Gennisson, J.L., Bercoff, J., Tanter, M., 2011. On the effects of reflected waves in transient shear wave elastography. IEEE Trans. Ultrason. Ferroelectr. Freq. Control 58 (10), 2032–2035.

Deffieux, T., Montaldo, G., Tanter, M., Fink, M., 2009. Shear wave spectroscopy for in vivo quantification of human soft tissues visco-elasticity. IEEE Trans. Med. Imaging 28 (3), 313–322.

Destounis, S., et al., 2013. Clinical expérience with elasticity imaging in a community-based breast center. J. Ultrasound Med. 32 (2), 297–302.

Dudea-Simon, Marina, Dudea, Sorin, Ciortea, Razvan, Malutan, Andrei, Mihu, Dan, 2021. Elastography of the uterine cervix in gynecology: normal appearance, cervical intraepithelial neoplasia and cancer. A systematic review. Med. Ultrason. 23 (1), 74–82.

Engel, A.J., Bashford, G.R., 2015. A new method for shear wave speed estimation in shear wave elastography. IEEE Trans. Ultrason. Ferroelectr. Freq. Control 62 (12), 2106–2114. https://doi.org/10.1109/TUFFC.2015.007282.

Farrokh, A., Wojcinski, S., Degenhardt, F., 2011. Diagnostic value of strain ratio measurement in the differentiation of malignant and benign lesions. Ultraschall Med. 32 (4), 400–405.

Fatemi, M., Greenleaf, J.F., 1998. Ultrasound-stimulated vibro-acoustic spectrography. Science 280, 82–85.

Feldmann, Alice, Langlois, Carole, Dewailly, Marion, Martinez, Elise Freart, Boulanger, Loic, Kerdraon, Olivier, Faye, Nathalie, 2015. Shear wave elastography (SWE): a, analysis of breast lesion characterization in 83 breast lesions. Ultrasound Med. Biol. 41 (10), 2594–2604.

Fruscalzo, A., Schmitz, R., 2012. Quantitative cervical elastography in pregnancy. Ultrasound Obstet. Gynecol. 40, 612.

Fruscalzo, A., Schmitz, R., Klockenbusch, W., Steinhard, J., 2012. Reliability of cervix elastography in late first and second trimester of pregnancy. Ultraschall Med. 33, 1–7.

Fuchs, T., Woyton, R., Pomorski, M., Wiatrowski, A., Slejman, N., Tomialowicz, M., et al., 2013. Sonoelastography of the uterine cervix as a new diagnostic tool of cervical assessment in pregnant women— preliminary report. Ginekol. Pol. 84, 12–16.

Gao, Z., Wu, S., Liu, Z., Luo, J., Zhang, H., Gong, M., Li, S., 2019. Learning the implicit strain reconstruction in ultrasound elastography using privileged information. Med. Image Anal. 58, 101534.

Gennisson, J.-L., Deffieux, T., Macé, E., Montaldo, G., Fink, M., Tanter, M., 2010. Viscoelastic and anisotropic mechanical properties of in vivo muscle tissue assessed by supersonic shear imaging. Ultrasound Med. Biol. 36 (5), 789–801.

Guo MM, Yan-Jing, Yan MM, Xiao-Jing, Zhao MM, Hai-Xia, Wen MM, Xiao-Duo, Wang MM, Jing, Yuan-Yuan MM, Du, Wang MM, Nan, Yang MM, Yi, 2022. Quantitative evaluation of the normal cervix, cervical cancer, and cervical precancerous changes via real-time shear wave elastography. J. Ultrasound Med. (ISSN 0278-4297) 9999, 1–10.

Hall, T.J., Zhu, Y., Spalding, C.S., 2003. In vivo real-time freehand palpation imaging. Ultrasound Med. Biol. 29 (3), 427–435. https://doi.org/10.1016/s0301-5629(02)00733-0.

Hernandez-Andrade, E., Aurioles-Garibay, A., Garcia, M., Korzeniewski, S.J., Schwartz, A.G., Ahn, H., et al., 2014. Effect of depth on shearwave elastography estimated in the internal and external cervical os during pregnancy. J. Perinat. Med. 42, 549–557.

Itoh, Ako, Ueno, Ei, Tohno, Eriko, Kamma, Hiroshi, Takahashi, Hideto, Shiina, Tsuyoshi, Yamakawa, Makoto, Matsumura, Takeshi, 2006. Breast disease: clinical application of US elastography for diagnosis. Radiology 239 (2), 341–350. https://doi.org/10.1148/radiol.2391041676.

Jiang, J., Hall, T.J., 2009. A generalized speckle tracking algorithm for ultrasonic strain imaging using dynamic programming. Ultrasound Med. Biol. 35 (11), 1863–1879.

Kallel, F., Ophir, J., 1997. A least-squares strain estimator for elastography. Ultrason. Imag. 19, 195–208.

Kanai, H., 2005. Propagation of spontaneously actuated pulsive vibration in human heart wall and in vivo viscoelasticity estimation. IEEE Trans. Ultrason. Ferroelectr. Freq. Control 52 (11), 1931–1942.

Kasai, C., Namekawa, K., Koyano, A., Omoto, R., 1985. Real-time two-dimensional blood flow imaging using an autocorrelation technique. IEEE Trans. Sonics Ultrason. 32 (3), 458–464. https://doi.org/10.1109/T-SU.1985.31615.

Khodayi-mehr, R., Urban, M.W., Zavlanos, M.M., Aquino, W., 2021. Plane wave elastography: a frequency-domain ultrasound shear wave elastography approach. Phys. Med. Biol. 66, 125017.

Liu, C., Li, T.T., Hu, Z., et al., 2019. Transvaginal real-time shear wave elastography in the diagnosis of cervical disease. J. Ultrasound Med. 38, 3173–3181.

Loupas, T., Powers, J.T., Gill, R.W., 1995. An axial velocity estimator for ultrasound blood flow imaging, based on a full evaluation of the Doppler equation by means of a two-dimensional autocorrelation approach. IEEE Trans. Ultrason. Ferroelectr. Freq. Control 42 (4), 672–688. https://doi.org/10.1109/58.393110.

Luo, J., Konofagou, E.E., 2010. A fast normalized cross-correlation calculation method for motion estimation. IEEE Trans. Ultrason. Ferroelectr. Freq. Control 57 (6), 1347–1357.

Mabuchi, S., Sasano, T., Kuroda, H., Takahashi, R., Nakagawa, S., Kimura, T., 2015. Real-time tissue sonoelastography for earl response monitoring in cervical cancer patients treated with definitive chemoradiotherapy: preliminary results. J. Med. Ultrason. (2001) 42, 379–385.

Manduca, A., Lake, D.S., Kruse, S.A., Ehman, R.L., 2003. Spatio-temporal directional filtering for improved inversion of MR elastography images. Med. Image Anal. 7, 465–473.

McLaughlin, J., Renzi, D., 2006. Shear wave speed recovery in transient elastography and supersonic imaging using propagating fronts. Inverse Probl. 22, 681.

Mirzaei, M., Asif, A., Fortin, M., Rivaz, H., 2020. 3D normalized cross-correlation for estimation of the displacement field in ultrasound elastography. Ultrasonics 102, 106053.

Mutala, T.M., Mwango, G.N., Aywak, A., Cioni, D., Neri, E., 2022. Determining the elastography strain ratio cut off value for differentiating benign from malignant breast lesions: systematic review and meta-analysis. Cancer Imaging 22, 12. https://doi.org/10.1186/s40644-022-00447-5.

Nightingale, K.R., Palmeri, M.L., Nightingale, R.W., Trahey, G.E., 2001. On the feasibility of remote palpation using acoustic radiation force. J. Acoust. Soc. Am. 110, 625–634.

Oliphant, T.E., Manduca, A., Ehman, R.L., Greenleaf, J.F., 2001. Complex-valued stiffness reconstruction for magnetic resonance elastography by algebraic inversion of the differential equation. Magn. Reson. Med. 45 (2), 299–310. https://doi.org/10.1002/1522-2594(200102)45:2<299::AID-MRM1039>3.0.CO;2-O.

Ophir, J., Cespedes, I., Ponnekanti, H., Yazdi, Y., Li, X., 1991. Elastography: a quantitative method for imaging the elasticity of biological tissues. Ultrason. Imag. 13, 111–134. https://doi.org/10.1016/0161-7346(91)90079-W.

Palmeri, M.L., Wang, M.H., Dahl, J.J., Frinkley, K.D., Nightingale, K.R., 2008. Quantifying hepatic shear modulus in vivo using acoustic radiation force. Ultrasound Med. Biol. 34 (4), 546–558. https://doi.org/10.1016/j.ultrasmedbio.2007.10.009.

Pan, X., Gao, J., Tao, S., Liu, K., Bai, J., Luo, J., 2014. A two-step optical flow method for strain estimation in elastography: simulation and phantom study. Ultrasonics 54 (4), 990–996.

Peng, B., Xian, Y., Zhang, Q., Jiang, J., 2020. Neural network-based motion tracking for breast ultrasound strain elastography: an initial assessment of performance and feasibility. Ultrason. Imag. 42 (2), 74–91.

Pillai, Aishwarya, Voruganti, Teja, Barr, Richard, Langdon, Jonathan, 2022. Diagnostic accuracy of shear-wave elastography for breast lesion characterization in women: a systematic review and meta-analysis. J. Am. Coll. Radiol. 19, 625–634.

Pinton, G.F., Dahl, J.J., Trahey, G.E., 2006. Rapid tracking of small displacements with ultrasound. IEEE Trans. Ultrason. Ferroelectr. Freq. Control 53 (6), 1103–1117. https://doi.org/10.1109/TUFFC.2006.1642509.

Rivaz, H., Boctor, E., Foroughi, P., Zellars, R., Fichtinger, G., Hager, G., 2008. Ultrasound elastography: a dynamic programming approach. IEEE Trans. Med. Imaging 27 (10), 1373–1377.

Rouze, N.C., Wang, M.H., Palmeri, M.L., Nightingale, K.R., 2010. Robust estimation of time-of-flight shear wave speed using a radon sum transformation. IEEE Trans. Ultrason. Ferroelectr. Freq. Control 57 (12), 2662–2670. https://doi.org/10.1109/TUFFC.2010.1740.

Sabiani, L., Haumonte, J.B., Loundou, A., Caro, A.S., Brunet, J., Cocallemen, J.F., et al., 2015. Cervical HI-RTE elastography and pregnancy outcome: a prospective study. Eur. J. Obstet. Gynecol. Reprod. Biol. 186C, 80–84.

Săftoiu, Adrian, Gilja, Odd Helge, Sidhu, Paul S., Dietrich, Christoph F., Cantisani, Vito, Amy, Dominique, Bachmann-Nielsen, Michael, Bob, Flaviu, Bojunga, Jörg, Brock, Marko, Calliada, Fabrizio, Clevert, Dirk André, Correas, Jean-Michel, D'Onofrio, Mirko, Ewertsen, Caroline, Farrokh, André, Fodor, Daniela, Fusaroli, Pietro, Havre, Roald Flesland, Hocke, Michael, Ignee, André, Jenssen, Christian, Klauser, Andrea Sabine, Kollmann, Christian, Radzina, Maija, Ramnarine, Kumar V., confienza, Luca Maria, Solomon, Carolina, Sporea, Ioan, Ştefănescu, Horia, Tanter, Mickael, Vilmann, Peter, 2019. The EFSUMB guidelines and recommendations for the clinical practice of elastography in non-hepatic applications: update 2018. Ultraschall Med. 40, 425–453.

Salles, S., Løvstakken, L., Aase, S.A., Bjåstad, T.G., Torp, H., 2019. Clutter filter wave imaging. IEEE Trans. Ultrason. Ferroelectr. Freq. Control 66 (9), 1444–1452. https://doi.org/10.1109/TUFFC.2019.2923710.

Sarvazyan, A.P., Rudenko, O.V., Swanson, S.D., Fowlkes, J.B., Emelianov, S.Y., 1998. Shear wave elasticity imaging: a new ultrasonic technology of medical diagnostics. Ultrasound Med. Biol. 24, 1419–1435.

Schäfer, F.K.W., Hooley, R.J., Ohlinger, R., Hahne, U., Madjar, H., Svensson, W.E., Balu-Maestro, C., Juhan, V., Athanasiou, A., Mundinger, A., Order, B., Locatelli, M., Cosgrove, D., Wolf, O.J., Henry, J.P., Moutfi, M., Gay, J.M., Cohen-Bacrie, C., 2013. ShearWave elastography BE1 multinational breast study: additional SWE features support potential to downgrade BI-RADS 3 lesions. Ultraschall Med. 34 (3), 254–259. https://doi.org/10.1055/s-0033-1335523.

Shady, M., Latif, M.A., Nabil, H., El Sadda, W, 2015. Could transvaginal sono-elastography help benign-malignant differentiation of cervical masses? Egypt. J. Radiol. Nucl. Med. 46, 1291–1299.

Sigrist, Rosa M.S., Liau, Joy, El Kaffas, Ahmed, Chammas, Maria Cristina, Willmann, Juergen K., 2017. Ultrasound elastography: review of techniques and clinical applications. Theranostics 7 (5), 1303–1329.

Song, P., Urban, M.W., Manduca, A., Greenleaf, J.F., Chen, S., 2015. Coded excitation plane wave imaging for shear wave motion detection. IEEE Trans. Ultrason. Ferroelectr. Freq. Control 62 (7), 1356–1372.

Song, P., Zhao, H., Manduca, A., Urban, M.W., Greenleaf, J.F., Chen, S., 2012. Comb-push ultrasound shear elastography (CUSE): a novel method for two-dimensional shear elasticity imaging of soft tissues. IEEE Trans. Med. Imaging 31 (9), 1821–1832. https://doi.org/10.1109/TMI.2012.2205586.

Song, P., Zhao, H., Urban, M.W., Manduca, A., Pislaru, S.V., Kinnick, R.R., Pislaru, C., Greenleaf, J.F., Chen, S., 2013. Improved shear wave motion detection using pulse-inversion harmonic imaging with a phased array transducer. IEEE Trans. Med. Imaging 32 (12), 2299–2310.

Sun, L.T., Ning, C.P., Liu, Y.J., et al., 2012. Is transvaginal elastography useful in pre-operative diagnosis of cervical cancer? Eur. J. Radiol. 81, e888–e892.

Swiatkowska-Freund, M., Preis, K., 2011. Elastography of the uterine cervix: implications for success of induction of labor. Ultrasound Obstet. Gynecol. 38, 52–56.

Tanter, M., Bercoff, J., Athanasiou, A., Deffieux, T., Gennisson, J.-L., Montaldo, G., Muller, M., Tardivon, A., Fink, M., 2008. Quantitative assessment of breast lesion viscoelasticity: initial clinical results using supersonic shear imaging. Ultrasound Med. Biol. 34 (9), 1373–1386. https://doi.org/10.1016/j.ultrasmedbio.2008.02.002.

Tanter, M., Bercoff, J., Sandrin, L., Fink, M., 2002. Ultrafast compound imaging for 2-D motion vector estimation: application to transient elastography. IEEE Trans. Ultrason. Ferroelectr. Freq. Control 49 (10), 1363–1374.

Tanter, M., Fink, M., 2014. Ultrafast imaging in biomedical ultrasound. IEEE Trans. Ultrason. Ferroelectr. Freq. Control 61 (1), 102–119.

Tay, Ian Wei Ming, Sim, Llewellyn Shao-jen, Moey, Tammy Hui Lin, Tan, Karen Pei Pei, Lai, Lily Mei San, Leong, Lester Chee Hao, 2022. Shear wave versus strain elastography of breast lesions—the value of incorporating boundary tissue assessment. Clin. Imaging 82, 228–233. https://doi.org/10.1016/j.clinimag.2021.11.030.

Tehrani, A.K.Z., Rivaz, H., 2020. Displacement estimation in ultrasound elastography using pyramidal convolutional neural network. IEEE Trans. Ultrason. Ferroelectr. Freq. Control 67 (12), 2629–2639.

Thomas, A., et al., 2010. Significant differentiation of focal breast lesions: calculation of strain ratio in breast sonoelastography. Acad. Radiol. 17 (5), 558–563.

Thomas, A., Kümmel, S., Gemeinhardt, O., Fischer, T., 2007. Real-time sonoelastography of the cervix: tissue elasticity of the normal and abnormal cervix. Acad. Radiol. 14, 193–200.

Thomsen, Christine Rohr, Jensen, Maria Strandbo Schmidt, Leonhard, Anne Katrine, Mortensen, Tanja Østerlund, Bor, Pinar, Sandager, Puk, Hinge, Mogens, Uldbjerg, Niels, 2022. A force-measuring device combined with ultrasound-based elastography for assessment of the uterine cervix. Acta Obstet. Gynecol. Scand. 101, 241–247.

Ueno, E., Umemoto, T., Bando, H., Tohno, E., Waki, K., Matsumura, T., 2007. New quantitative method in breast elastography: fat lesion ratio (FLR). In: Radiological Society of North America 2007 Scientific Assembly and Annual Meeting. Chicago, IL.

Wang, Bo, Zhang, Yong, Chen, Shuangshuang, Xiang, Xiaowei, Wen, Juan, Yi, Mei, He, Baiyun, Hu, Bing, 2019. Diagnostic accuracy of cervical elastography in predicting preterm delivery: a systematic review and meta-analysis. Medicine 98, e16449.

Wang, Y., Jacobson, D.S., Urban, M.W., 2022. A Non-Invasive method to estimate the stress-strain curve of soft tissue using ultrasound elastography. Ultrasound Med. Biol. 48 (5), 786807.

Wang, Y., Jiang, J., Hall, T.J., 2018. A 3-D region-growing motion-tracking method for ultrasound elasticity imaging. Ultrasound Med. Biol. 44, 1638–1653.

Yamaguchi, S.I., Kamei, Y., Kozuma, S., Taketani, Y., 2007. Tissue elastography imaging of the uterine cervix during pregnancy. J. Med. Ultrason. 34, 209–210.

Yamakoshi, Y., Sato, J., Sato, T., 1990. Ultrasonic imaging of internal vibration of soft tissue under forced vibration. IEEE Trans. Ultrason. Ferroelectr. Freq. Control 37 (1), 45–53.

Youk, Ji Hyun, Gweon, Hye Mi, Son, Eun Ju, 2017. Shear-wave elastography in breast ultrasonography: the state of the art. Ultrasonography 36 (4).

Zahiri-Azar, R., Salcudean, S., 2006. Motion estimation in ultrasound images using time domain cross correlation with prior estimates. IEEE Trans. Biomed. Eng. 53 (10), 1990–2000.

PART

4

From biomechanical models to medical devices, and patients treatments

CHAPTER

Numerical simulation of vaginal delivery

15

Florence Zara[a,b]

[a]*Univ Lyon, UCBL, CNRS, INSA Lyon, LIRIS, UMR5205, Villeurbanne, France*

[b]*Domaine scientifique de la Doua, Bâtiment Nautibus, Villeurbanne, France*

15.1 Introduction

We cannot make a book about the female reproductive system without talking about childbirth and more particularly about the interest of knowing how to numerically simulate a vaginal delivery. Indeed, there are several interests in knowing how to simulate in 3D this complex phenomena.

One objective can be pedagogical: it is then a question of improving the learning of the obstetrical gestures required during a childbirth by allowing the midwives and obstetricians to see what occurs during the vaginal delivery. The visualization of the organs involved in the delivery enables a better understanding of the gestures and their implications, as well as the role of each of the interacting organs. For this purpose, the technological aim is to develop a medical simulator, based on new virtual reality technologies, to learn childbirth obstetrical gestures for instrumental extraction (using forceps or vacuum cup for example). The scientific challenge is then to simulate in 3D the complex phenomenon of childbirth, in a realistic way and in interactive time.

Another objective may be to have a better knowledge of the mechanism of childbirth and its subsequent implications into the physiology of the parturient. In this second context, it is a question of simulating as finely as possible all the anatomical structures of the reproductive system to simulate as realistically as possible the different stages of childbirth and to evaluate the damages caused by this childbirth on the tissues of the parturient. In this context, numerous studies have been conducted to determine, as accurately as possible, the deformation of the fetal head or injury of the pelvic floor muscle during the second stage of labor (Li et al., 2010b). These models do not necessarily account for the entire process of childbirth, but they provide valuable information on the functional and mechanical aspects of the target organs and aid in the understanding of the mechanisms related to childbirth.

Regarding what follows: Section 15.2 addresses the three phases of a vaginal delivery. Section 15.3 presents the interest to perform a simulator of childbirth to learn obstetrical gestures required using forceps for the extraction of the fetus. This section highlights the importance to propose a medical training simulator, including a numerical simulation of vaginal delivery. In this context, Section 15.4 presents an approach to obtain a real-time childbirth simulation. This section both details the

Biomechanics of the Female Reproductive System: Breast and Pelvic Organs. https://doi.org/10.1016/B978-0-12-823403-7.00029-4

mechanical modeling of the fetus and one of each organ involved in the simulation, i.e., the bony pelvis, the pelvic floor and the uterus. It ends by presenting the complete numerical simulation that has been coupled to an haptic device to enable the trainee to interact with the simulation. Section 15.5 presents the interest to perform accurate simulations of vaginal delivery and focuses on relative work in this second context. The aim is to have a better understanding of implications bearing on the parturient for a vaginal delivery to prevent clinical risks. At last, Section 15.6 concludes this chapter on delivery simulation.

15.2 The three phases of a vaginal birth

Let us first recall the three important phases of childbirth, which usually occurs between 38^e and 41^e weeks of pregnancy. The first phase of labor, which is called dilation, starts with the appearance of the first regular uterine contractions. These contractions are involuntary and permit the uterine walls to retract, decreasing the volume of the uterus and generating a pushing force allowing the descent of the fetus into the pelvic canal (also called the birth canal). During this phase, the neck of the uterus is going to collapse and is going to dilate progressively until about 10 cm to be enough open to allow the passage of the head of the fetus.

The second phase corresponds to the expulsion of the newborn. It begins when the fetus engages in the large basin and ends when it is expelled after having crossed the soft basin. This process lasts between thirty minutes and two hours. The uterine contractions (UC), higher during this phase, will allow the descent of the fetus. These contractions are still involuntary. They occur about 3 or 4 times every ten minutes and the average duration of a contraction during a period is 90 seconds. The amplitude of the contraction will vary between the "baseline tone" (pressure in the uterus caused by its strong deformation) and the "intensity of the UC." The true intensity is the difference between these two amplitudes: it corresponds to the actual pushing efforts of the UC during delivery. This is illustrated in Fig. 15.1. But this push is insufficient to allow the effacement of the pelvic muscles and the exit of the fetus. This is why the UC must be added to the voluntary maternal thrusts so as to generate a sufficient force to permit the progression of the fetus through the pelvic floor. The role of the midwife during this phase is to guide the parturient to perform a series of strong abdominal thrusts at the same time as the UC. Indeed, though these forces (called expulsion forces) are about 4 times more important, they must absolutely be combined with the UC to allow the pelvic floor resistance threshold to be crossed. This is illustrated by Fig. 15.2. Moreover, these forces of expulsion are caused by contraction of the abdominal and the diaphragm. The abdominal muscles are those muscles located in the anterior part of the abdomen, but they are enhanced by the presence of the fetus. The stretched muscles encompass the uterine surface and exert even pressure on the upper part of the uterus. The diaphragm also pushes the fetus into the vaginal canal. Ultimately, the descent of the fetus is caused by a combination of the forces of UC and the forces of expulsion (abdominal and diaphragm forces) applied to the uterus,

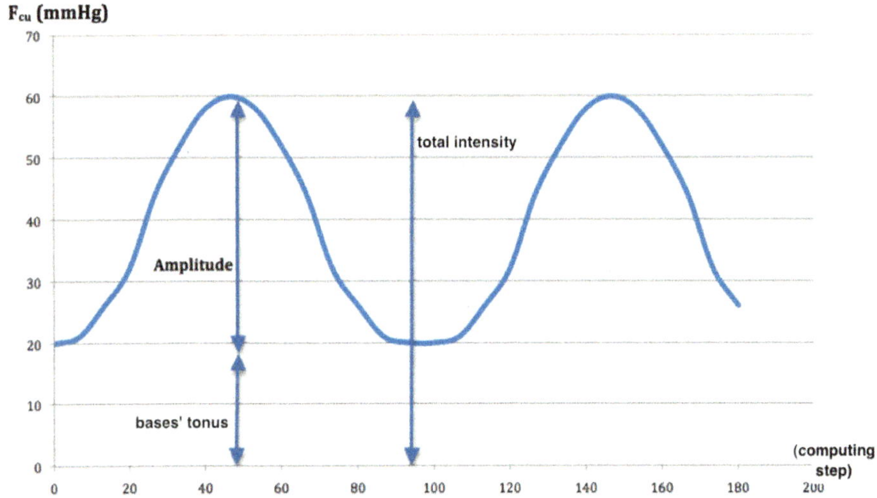

FIGURE 15.1

Uterine Contraction (mmHg) according to time (Buttin et al., 2013).

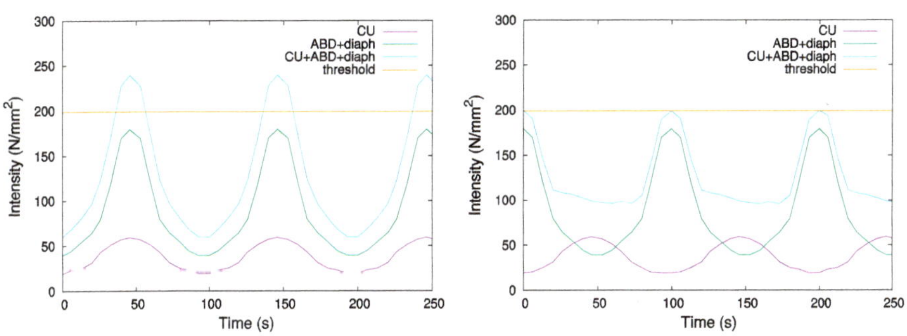

FIGURE 15.2

Evolution of the different uterine forces: synchronized forces (left) and non-synchronized forces (right). The line corresponds to the threshold allowing the expulsion of the fetus.

constricting the uterine walls and causing a force to expel the fetus into the vaginal canal.

The third phase of the birth, called delivery, corresponds to the expulsion of the placenta. This third phase begins after the expulsion of the newborn and ends in a few minutes. The placenta detaches from the uterine wall and is expelled usually in one uterine contraction.

For more details on these three phases, readers may refer to Chapter 4 of this book or to Chapter 4 of Parente (2008). The phase which interests us in this chapter is the second phase of the vaginal delivery. Thus we consider that the first phase is completed and the neck is open by about 10 cm. No aspect of third phase of labor is simulated in this chapter.

15.3 Interest to perform a childbirth training simulator

During the second phase, the fetus passes through the pelvic floor to exit the maternal abdomen. This phase should not exceed 30 to 40 minutes so that the fetus is not in pain. In this context, it is sometimes necessary to assist in the extraction of the newborn with forceps or vacuum instruments. It requires the adequate gesture to not damage neither the tissues of the parturient nor those of the fetus. Moreover, learning this gesture directly from the parturient is delicate as the future mother needs to be aware of the same and as the second parent is generally also present in the delivery room. However, it is important that obstetricians master this procedure so as not to resort to a cesarean delivery just to avoid doing this gesture. Indeed, a Cesarean section is a non-trivial surgical procedure and it can involve some harmful consequences on the parturient. In this context, a certain number of simulators have been developed to train obstetrical gestures for vaginal delivery.

Existing delivery simulators

In 2007, Gardner (2007) proposed a state of the art on the use of simulators in the field of obstetrics gynecology. It can be completed by the PhD thesis of Moreau (2007) centered on childbirth simulators and by Lepage et al. (2016). The simulators most used (as they are the least expensive) are the *anatomical simulators,* which reproduce the anatomical landmarks used in obstetrics gynecology. For example, the simulator proposed by Graves and Savannah (1951) was composed of a woman's mannequin on a scale of one, with three levels of gestation, a placenta and a fetal mannequin, which can be positioned in different ways in the abdomen. Then the *instrumented anatomical simulators* allow to offer a haptic return in addition to the characteristics of the anatomical simulators. Simulators of this type are Knapp and Eads (1974); Eggert et al. (2003); Sielhorst et al. (2004); Riener and Burgkart (2003); Allen et al. (1995). Among these simulators, the Victoria simulator of the Gaumard Scientific (2022) company represents a very complete product in the philosophy "learning dummy." It proposes several simulations in various typical situations, which allow the learning of the reflexes to get. Finally, the last class of simulators concerns the *simulators based on virtual reality* with a visualization of the descent of the fetus in the pelvic canal as, for example, Boissonnat and Geiger (1993); Geiger (1993); Kheddar et al. (2004); Forster et al. (2001). In this sense, Letterie (2002) highlighted the beneficial contributions of the virtual reality for the training in obstetric gynecology.

It is important to note that anatomical simulators, instrumented or not, are based on plastic mannequins and do not allow the simulation of tissue interactions and therefore do not offer a realistic sensory rendering. However, the precision of the gesture depends on the reaction of tissues and anatomical structures. This is why it is necessary to design simulators that permit to restore realistic sensations, while taking into account various parameters (such as uterine contractions, abdominal thrust efforts, or the parturient's morphology as well as that of the fetus). In relation to these objectives, four simulators hold our attention.

- The Geburtensimulator patented by Riener and Burgkart (2003); Obst et al. (2004); Sielhorst et al. (2004) is a device of augmented reality. It couples a haptic interface to a software part. The software part simulates physiological functions (blood pressure, heart rate, pain, lack of oxygen) and biomechanical functions (position of the fetal head calculated from the contraction forces but with an imposed direction) of the parturient and the fetal body. The physical part is composed of a mannequin of the pelvis, a model of fetal head, using actuators and a programmable control unit. In sum, this simulator provides sensations similar to those felt by the obstetrician during the use of forceps. But the use of a 6-axis robot for the positioning of the fetal head makes it bulky, heavy, non-transportable, and very expensive. Moreover, the trajectory followed by the fetal head in the vaginal canal is imposed. Thus it does not take into account the morphology of the fetus and of the parturient, nor the interaction of the user with the system.
- The simulator developed by Lapeer and Prager (2001); Lapeer et al. (2004); Lapeer (2005) is an augmented reality device for training in forceps insertion. The use of forceps equipped with sensors allows their visualization in 3D during their positioning around the virtual head of the fetus also visualized in 3D. The objective was to highlight the contacts that take place between the head and forceps, but the deformation of the skull of the fetus at the points of contact was not yet modeled in this version. However, it is a fundamental aspect of the training to learn how to control the pressure exerted by the instruments on the skull. In Lapeer et al. (2014), they improved their initial version by starting to analyze the mechanical contacts between head and forceps in several situations using a finite element analysis.
- The simulator designed by Boissonnat and Geiger (1993); Boissonnat and Yvinec (1995) was only based on a software part. It allows the realization of a precise prognosis of the childbirth by analyzing various parameters (size, shape of the maternal abdomen, shape of the fetal head, uterine contractions, and position of the fetal head). This simulator simulates the rotation movement of the baby inside the maternal abdomen, but using imposed theoretical trajectories. This simulator thus relates more to an upstream phase of the childbirth.
- The simulator developed by Kheddar et al. (2004) is composed of two parts: a software part, allowing the 3D visualization of the fetus and the bony and soft pelvis of the parturient; and a haptic part composed of a three-axis force feedback device and a virtual hand representing the practitioner's hand. The fetus advances in the pelvic canal under the effect of a force representing the uterine contractions, the abdominal effort, and the friction between fetus and soft pelvis. This simulator allows the realization of obstetrical gestures in a concrete way, but the use of an arm with return of effort is far from the initial gesture of the childbirth, and the use of imposed trajectories of the fetus does not make it possible to take into account the interactions and organs morphology, nor the gesture of the practitioner.

To sump up, none of these simulators offer yet the possibility to simulate different clinical cases, nor to propose several scenarios. But this aspect is essential to understand different types of situations for training. In this sense, we will present in next section, the advantages of using a numerical simulations compared to a simulator only based on a physical device, including a single type of anthropomorphic mannequin.

Interest to provide a simulator including a numerical childbirth simulation

The use of a numerical simulation inside a training simulator of childbirth gestures offers several possibilities.

- First of all, a numerical simulation permits to easily manage the different variables of the learning scenarios and to multiply and target the situations. We can reproduce both the usual and rare situations met during a vaginal delivery. This permits learning the different maneuvers made by the midwife during a childbirth. For example, the numerical simulation enables to take into account several morphology of bony pelvic and fetal skull (thanks to the use of a set of 3D meshes to represent the anatomy). It also enables to carry out a various levels of engagement of the fetus in the pelvic canal. It also allows to manage a rocking of the pelvis and a rotation of the head of the fetus during the childbirth.
- Another benefit of the simulation comes from the possibility of analyzing forces involved by maternal pushes, as well as the amplitude of the uterine contractions necessary to release the fetus. Moreover, for an instrumental childbirth training, we can measure for which intensity the use of the forceps becomes necessary: this can be as soon as the parturient is tired (the trainee must then understand that the maternal pushes will be insufficient and that an external help is necessary); or for a certain morphology of pelvis and fetal head (for which an external force is necessary in any case).
- Another benefit is that the simulation allows the trainee to see the descent of the fetus during the childbirth. This visualization helps the apprentice to acquire the gestures by watching what is normally not visible, with a direct understanding of the implications of each of the gestures performed (for example, the view of consequences related to a bad positioning of the forceps; or consequences of an extremely important extraction force applied to the fetus head).
- Finally, the main use of the numerical simulation, when it is coupled with a haptic device, is to send information to the haptic device. These data are either related to the movements performed by the trainee or they are relative to the positions and forces computed during the fetal descent, according to the scenario parameters and gestures made by the trainee. It allows us to not use predefined descent trajectories within the physical device, but to keenly take into account the learning scenarios and apprentices movements.

To offer such training simulators, coupling a numerical simulation to a haptic device, we have to succeed in carrying out an adequately realistic simulation, allowing to bring the adequate information to the physical device (position of the fetal head, forces applied on different strategic points) at a speed fast enough to ensure the interactivity of the device. In this sense, we have (i) to simplify the modeling, while keeping a realistic result; (ii) to couple the numerical simulation to the haptic interface to reflect the user's actions on the simulation. Moreover, the challenge is to perform the simulation without imposing a trajectory to the fetus, i.e., to compute the descent of the fetus by the single application of the contractions forces and maternal pushes into the uterus. In this context, Chen and Grimm (2021) proposed a state of art about childbirth computational models. They summed up the studies by highlighting several aspects: (i) the creation of the geometrical models of pelvic floor; (ii) the in vivo measurements of pelvic tissues; (iii) the fetus anatomical models; (iv) the associated applications proposed, such as forceps delivery simulation. Moreover, the authors propose a table that summaries the main characteristics of maternal and fetal models presented in the cited papers. This gives an overview of the work proposed in this field. We can note that few research teams have proposed a simulation of the fetus descent without imposed trajectory, which is the main purpose of the numerical simulation presented in next section.

15.4 A real-time childbirth simulation

In this section, we will present the approach proposed by Buttin (2010); Buttin et al. (2013) to obtain a real-time childbirth simulation. The trajectory of the fetus was computed according to the morphology of the parturient and fetus, considering uterine contractions and maternal pushes. The anatomical structures that interests us are located in the pelvic cavity of the parturient, which contains part of the digestive system, the urinary system, the reproductive system (with the vagina, the gravid uterus, and the uterine tubes and ovaries), and the fetus. In the following, we firstly present the process to obtain the geometrical model of organs involved in the vaginal delivery. Secondly, we detail their biomechanical models to reproduce their behavior during the childbirth. For this purpose, for each organ involved in the childbirth simulation, we will: (i) briefly recall the anatomical notions required for their modeling; (ii) present the existing models proposed in the literature; (iii) present our approach adopted in the context of a real-time simulation and our results. For more details about the anatomy of the female pelvis, readers may refer to Chapter 1 of this book or to the Chapter 2 of Parente (2008), which presents a lot of illustrations.

15.4.1 Geometrical representations of the organs

The first step to elaborate the childbirth simulation consists in having a geometrical representation of each organ involved in the vaginal delivery. The meshes of the organs were elaborated from medical imaging data (scans and MRI (magnetic resonance imaging)) of women at the end of pregnancy. The precision of these medical data is about 5 mm (space between two slices), with a precision of 2 mm for the bony

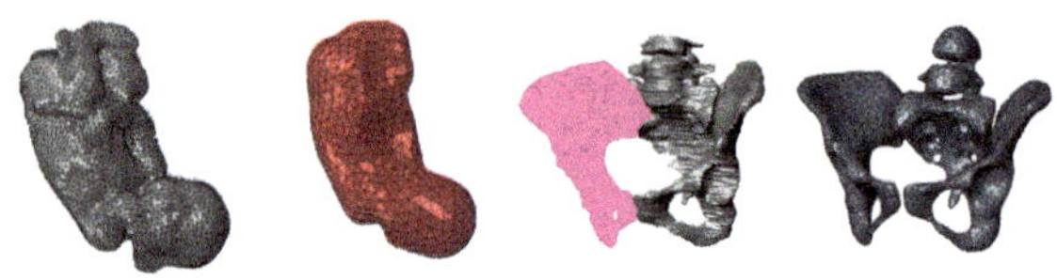

(a) (b)

	Number of nodes	
	Before	After
Fetus	21,500	2,800
Pelvis	1,750,000	3,500
Uterus	42,811	2,348
Abdomen	38,863	3,268

(c)

FIGURE 15.3

(a) Surface meshes of the fetus. (b) Bony pelvis (Image from Buttin et al., 2013). (c) Number of nodes present in the meshes before and after their simplification.

pelvis. It thus constitutes the limit of precision of the numerical simulation. But this limit is largely sufficient in the context of the realization of an interactive simulator.

After the segmentation of medical images, the surface meshes of soft tissues and pelvis can be generated from segmented images using adequate software (such as the MRIcro software). Generally, the initial resulting meshes have a large number of elements. Thus they have to be simplified to limit the execution time of the numerical simulation. For this step, the ReMESH software developed by Attene and Falcidieno (2006) could be used. It allows us to simplify the areas with a small angular variation between mesh elements. To illustrate this simplification, Fig. 15.3 shows the mesh of (a) the fetus, (b) the bony pelvis, and (c) the number of nodes present in the surface meshes before and after their simplification. For this simplification, the main characteristics of the organs have to be kept by setting up a precision criterion. The MESH software proposed by Aspert et al. (2002) allows us to calculate the average distance d_m between the initial mesh and the simplified mesh; this is done by varying the rate of deletion of the elements. For each of the organs, a characteristic quantity d_o was considered to determine the accuracy criterion α defined by $\alpha = 100 \times (d_o - d_m)/d_o$. As characteristic magnitudes, the thickness of the iliac wing was chosen for the bony pelvis, the thickness of the uterine membrane was chosen for the uterus, and the diameter of the fetal head was chosen for the fetus. Fig. 15.4 presents the evolution of the accuracy of the meshes as a function of the percentage of simplification. By simplifying the meshes to 90%, an accuracy higher than 85% can be kept. This simplification allowed us to considerably reduce the number of nodes of the surface meshes.

The volumetric meshes could then be created from the surface meshes using the Abaqus simulation software (developed by Dassault Systems). The method used to create the tetrahedrons from surface meshes was based on the Delaunay algorithm. Using this software, we obtained the volume meshes of considered organs, i.e., the bony pelvis, the maternal abdomen, the uterus, and the fetus.

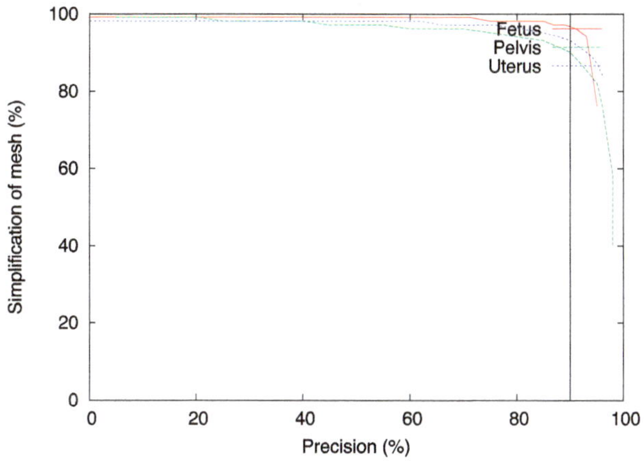

FIGURE 15.4

Comparison of the evolution of mesh accuracy as a function of the percentage of mesh simplification for the pelvis (green), fetus (red) and uterus (blue) (Buttin et al., 2013).

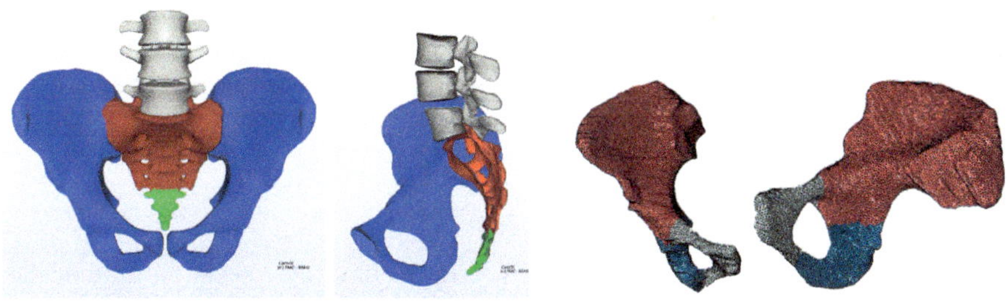

(a) (b)

FIGURE 15.5

(a) Bony pelvis consisting of the coxal-iliac bones (blue), sacrum (red), and coccyx (green) (Images from Bailet, 2014). (b) The two coxal-iliac bones are made up of three parts (Images from Buttin, 2010).

15.4.2 Modeling of bony pelvis and pelvic floor

If we want to do a simple simulation of childbirth, a first test consists in trying to make the head of the fetus pass in the basin. The pelvis is made up of two parts: the *bony pelvis* and the soft pelvis, which is also called *pelvic floor*. The bony pelvis is made up of several bones visible in Fig. 15.5: (a) bones which are connected by a number of ligaments: the sacrum (in red), which is located in the lower part of the spine; the coccyx (in green), which is an extension of the sacrum; the two coxal-iliac bones (in blue), which form the lateral and anterior parts of the pelvis. These latter two bones consist of three parts detailed in Fig. 15.5 (b): the ischium (blue), the ilium (red), and the pubis (white). The iliac wings of these bones consist of a smooth, hollow area called the iliac fossa. This area accommodates the head of the fetus as it descends. The soft pelvis (or pelvic floor) is made up of two main muscles: the levator ani muscle and the coccygeal muscle. These muscles are inserted in the lateral walls of the pelvis, forming a muscular diaphragm on which the pelvic organs rest, preventing the descent of the fetus during pregnancy. Due to its density, it is difficult to identify the different muscles of the muscular network.

During the second stage of labor, the fetus enters the pelvic inlet (called the upper strait) completely, and then passes through the pelvic inlet to exit the pelvic inlet (called the lower strait). To cross these various obstacles, the fetus will carry out a series of movements to adapt itself to the shape of the pelvic cavity. Thus at the same time as the head engages in the higher strait, it is flexed on the chest, then it will turn slightly towards the right or left side to enter the pelvis (to present its smaller size). Once this first strait is crossed, the head then descends into the pelvis to make a second rotation. To accompany these movements, the pelvis also carries out a particular movement called "nutation," consequence of the push of the fetal head thereon.

Thus during the descent of the fetus, the upper part of the iliac wings performs a movement of adduction accompanied by an abduction of the lower part of the ischium. During these rotations, the head will press on the coccyx, which then performs a flexion movement. The coccyx will then perform a counter-nutation movement when the head is released from the pelvis. It is thus important to model in a relevant way both the movements of the bony pelvis and those of the fetal head to obtain a realistic trajectory of the fetal descent.

15.4.2.1 Bony pelvis

First, we consider the model of the bony pelvis. To limit the execution time of the simulation, we have to considerably simplify the surface mesh of the bony pelvis by removing the sharp edges, while keeping its essential features, such as the ischiatic spines, the tip of the coccyx, and the pubis area. To do this, one solution consists in making a very coarse mesh based on the bounding boxes of the different related parts of the pelvis. Fig. 15.6 (a) shows the resulting mesh having only 1750 nodes. For the boundary conditions (shown in Fig. 15.6 (b)): the iliac wings were considered as fixed and undeformable organs; the rachis was fixed; the rocker at the lower level was mobile. For the mechanical behavior, Fung (1993) recalls the experiments performed by Yamada (1970) suggesting that Hooke's law is applicable to bones in the framework of limited stresses. Thus the Hooke law is used for modeling the bone pelvis allowing for small deformations. For the Young's modulus of the bone, Dufour and Pillu (2007) present an average value of 15 to 19 MPa, including the trabecular and cortical parts of the bone. Since cortical bone is much denser than trabecular bone (the spongy part of the bone), a Young's modulus $E = 23$ MPa was chosen to focus on cortical bone. Since the Poisson's ratio for bone is between 0.2 and 0.3 according to Dalstra et al. (1993), a value $\nu = 0.3$ was chosen for the bony basin. In addition, the value of density was estimated at 1000 kg/m3, closest to that of water. Note that the muscular part of the pelvis was directly incorporated in the modeling of the par-

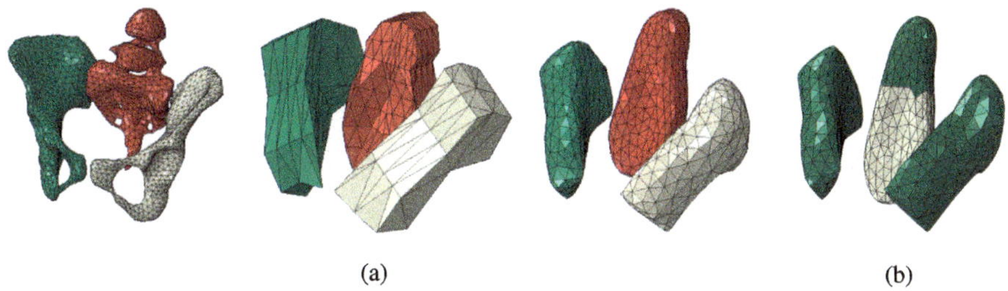

FIGURE 15.6

(a) Simplification of the bone basin mesh (from left to right): smoothed mesh (3500 nodes), bounding boxes, final mesh (1750 nodes). (b) Boundary conditions of the bone basin with fixed parts in green. Images from Buttin (2010).

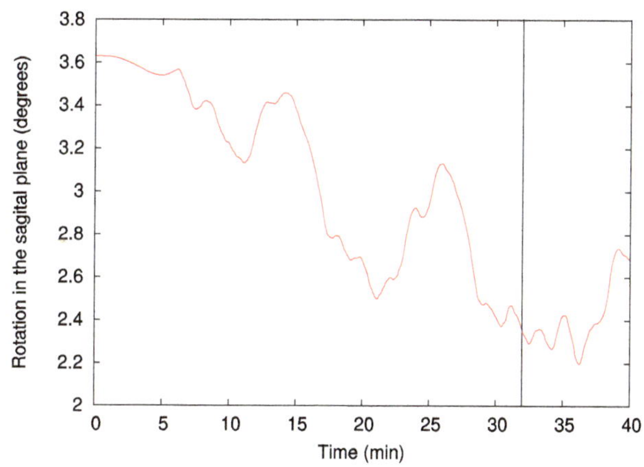

FIGURE 15.7

Angular evolution of the tip of the sacrum in the sagittal plane (Buttin et al., 2013).

turient abdomen: it was therefore not modeled as an independent organ. This strategy enables to simplify the biomechanical modeling of the childbirth.

Then, to analyze the behavior of the pelvis during the simulation, the tilt of the sacrum was observed. Fig. 15.7 shows the angular evolution of the tip of the sacrum in the sagittal plane. Two peaks are visible on this curve. The first one corresponds to the first contact of the head with the sacrum, which is pushed backwards by the skull bones of the fetus. Then, as the head begins to enter the vaginal canal, the second peak is caused by the passage of the rest of the fetal body. Moreover, we can note that at the end of the labor (32 minutes later), the pelvis did not return to its initial position. This simulation thus provided results in agreement with reality.

However, even if we have taken care to keep the tip of the sacrum (constituting one of the most important zones of the pelvis), Fig. 15.8 shows that the simplification of the bony pelvis mesh has reduced the promontory line running along the ilio-

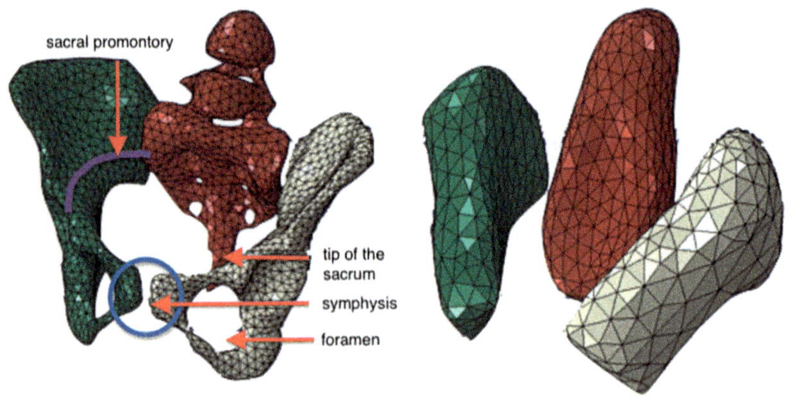

FIGURE 15.8

Promontory line too shallow (purple line), pubic symphysis much too large (blue circle). Images from Buttin (2010).

pubic interior at the level of the superior strait. Thus the head of the fetus is placed in this zone during the childbirth, conditioning the initial positioning and the arrival in the genital tract of the fetus. The trajectory of the fetal descent is consequently inevitably disturbed by this simplification. Moreover, the simplification of the bony pelvis has also significantly enlarged the pubic symphysis. However, its role during delivery is important, causing rotation and repositioning around the sagittal/frontal axis. An improvement of the geometrical model of the bony pelvis would thus allow the correction of the phenomenon of head deflection at the exit of the sacrum tilt, a phenomenon not currently visible in the simulation.

15.4.2.2 Pelvic floor

Let us now consider the pelvic floor. Several studies have focused on modeling the anus elevator muscles (AEM) of the pelvic floor. The idea is to study their stretching during the passage of the fetal head to better understand their damage during childbirth, and thus prevent the phenomenon of prolapse. In this sense, Martins et al. (2007); d'Aulignac et al. (2005) have proposed a finite element model based on the Neo-Hooke law of behavior for the AEMs. These muscles were considered quasi-incompressible and isotropic with a single direction for the muscle fiber, and the material properties of the pelvic floor muscles were approximated using data obtained on cardiac tissue. A realistic model of the fetus represented by tetrahedral elements was also used by considering it as a quasi-undeformable material with very high stiffness. Finally, the maximum muscle stretch ratio was calculated by varying the geometry of the fetal pelvis and head, as well as parameters such as head presentation.

More recently, Li et al. (2011, 2010a) have investigated the effect of mechanical anisotropy on the biomechanical response of AEMs during delivery. They simulated the passage of a fetal skull through two pelvic floor models, incorporating the levator ani muscles at different ratios of anisotropy. Interactions between the AEMs and the fetal head were modeled by considering elastic deformation and frictionless mechanical contact. The results showed a substantial decrease in the force required for delivery as fiber anisotropy increased.

Regarding the measurement of pelvic organ characteristics, Janda et al. (2003) published a database containing geometrical parameters of the pelvic floor and mechanical parameters of the muscle fibers from a 72-year-old female cadaver. Rubod et al. (2007, 2008); Brieu et al. (2010) developed an experimental protocol for the study of the mechanical properties of the vaginal tissues highlighting the elastic, non-linear, and highly deformable character of the organic tissues involved in pelvic statics. Then, Rubod et al. (2013); Lepage et al. (2014) continued their work by proposing a model of the pelvic system based on a linear elastic finite element model in large displacements. Their aim was to better understand the role of the uterine ligaments in the prolapse phenomenon. Their method consists in analyzing the constraints of pelvic components during a childbirth. In this context, they proposed a pelvic model of a pregnant woman, including muscles, ligaments, and pelvic organs. The finite element method was used for the simulation with an imposed trajectory of the fetal head. First results showed the constraints and deformations of pelvic structures during vaginal delivery. Moreover, the uterosacral ligaments are submitted to an important traction during the passage of the fetal head, and its intensity naturally depends on the fetal head size. Gatellier et al. (2020) has completed the geometrical model of the parturient, integrating the pelvic floor and its ligamentous and muscular components. The difficulty remains in tracking anatomical structures from MRI medical images (Jiang et al., 2017). Their model was very complete, but it requires several hours of computation to perform a simulation based on an imposed trajectory for the fetal head.

A strategy to obtain a real-time childbirth simulation

During pregnancy the volume of the abdomen increases significantly and the organs that are present in it (i.e., bladder, rectum, spine, ribs, liver) change position to make room for the fetus. Then, during delivery, the volume of the abdomen will decrease and the internal organs will return to their initial position, thanks to the presence of an internal pressure ensuring the cohesion of the organs. For obvious reasons of computation time, it is not possible to simulate all the organs individually. Thus the abdomen was considered as a single organ, whose contour was defined from the MRI images. Moreover, its interior was meshed using tetrahedrons, by removing the other organs modeled (i.e., pelvis, uterus and fetus), as we can see on Fig. 15.9 (a). Then, as the pelvic floor was integrated into the model of the abdomen, it took on a mechanical behavior quite close to the properties of the muscular tissues of the soft pelvis. So it was considered elastic and compressible. These properties allow the repositioning of the abdomen elements around the uterus during the descent of the fetus.

In the end, the abdomen was modeled as a hyper-elastic material using Neo-Hooke's law with a density of 2500 kg/m^3 and C10 $=$ 5 kPa. For the boundary conditions, the back of the abdomen was fixed to take into account the position of

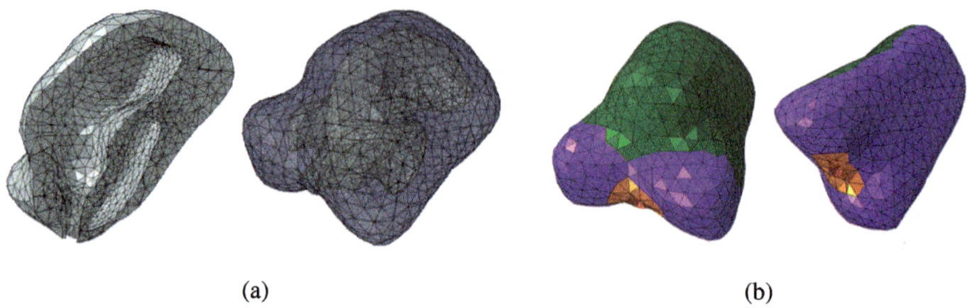

FIGURE 15.9

(a) Mesh of the parturient's abdomen (right) with visualization of its sagittal section (left). (b) Boundary conditions of the abdomen: (purple) fixed parts, (orange) possible stretching, (red) vertical displacements forbidden, (green) no constraint. Images from Buttin et al. (2013).

the parturient (i.e., sitting with her back fixed) during delivery. But we cannot impose a zero displacement to the whole contour of the maternal abdomen. Indeed, by fixing the lower part of the maternal abdomen, the exit of the fetus would not be allowed. This last condition was solved by allowing only lateral displacements on the lower part of the maternal abdomen around the vaginal area. These boundary conditions are illustrated in Fig. 15.9 (b). Furthermore, to ensure cohesion between the modeled organs inside the abdomen (uterus, pelvis, and fetus), a slight pressure was added. It represents the residual muscle tone. For contact management, we did not have to deal with collisions as when the bony pelvis is being included in the abdomen with several nodes in common.

A real modeling of the pelvic floor would allow the simulation of its tears caused during childbirth and a better understanding of the efforts that are applied to it. On the other hand, this simulation is costly in terms of calculation time and should be optimized. But it would undeniably be a plus in the learning of childbirth gestures to better understand and preserve the pelvic floor tissues. Moreover, the external deformation of the perineum were not taken into account. In reality, at the end of the fetal exit, the vulva inflates and causes a temporary horizontal exit of the organs. However, in the simulation, the movements allowed for the external region of the vulva were only in the transverse plane to allow the dilation of the vagina and the descent of the fetus, because the absence of modeling of the skin associated with the strong simplification of the abdomen forced us to block the movements of the vulva along the cranio-caudal axis. To remedy these shortcomings, we can refer to the work proposed in the literature on pelvic floor modeling.

15.4.3 Modeling of uterus

The uterus is a hollow muscular sac located in the pelvic cavity. During pregnancy, it is brought towards the front of the parturient's abdomen and its size will increase from approximately 8 to 35 cm long at the end of pregnancy. This important change

FIGURE 15.10

Geometric model of the uterus: (left) mesh obtained from medical data, (right) mesh after smoothing and reconstruction of the cervix.

modifies its mechanical characteristics. Moreover, these parameters are very variable; they depend on the number of previous pregnancies and from one woman to another, hence it is very difficult to identify them.

The cervix is separated from the uterine body by a constriction at the base of the uterus called the isthmus. During pregnancy, the cervix remains closed with a length of 30 to 40 mm. Then, at the end of pregnancy, it will soften and shorten to about 13 mm.

Mizrahi and Karni (1975) presented a mechanical model of the uterus using a kinematic approach. As boundary conditions, they considered that the displacements of the cervix are zero during a contraction and remain constant during the second phase of labor, i.e., from the moment when the cervix is fully dilated. In addition, they assumed that during deformation, the volume delimited by the organ remained constant, due to the incompressibility of the intrauterine fluid. Regarding mechanical properties, Mazza et al. (2006) present a study performed with a suction device to characterize the mechanical properties of the cervix *in vivo*. The mean values obtained for stiffness vary from 0.095 to 0.24 bar/mm for tests carried out on eight patients aged 47 to 69 years and having 1 to 4 children. However, it is difficult to use these values obtained in non-pregnant women because of the significant changes that occur in mechanical properties during pregnancy. In this sense, Bauer et al. (2009) present another study carried out with the same device, but on pregnant women (between 21 and 36 weeks of gestation), thus highlighting the physiological differences between the tissues of a non-pregnant and pregnant woman. The stiffness values obtained vary between 0.013 and 0.068 bar/mm confirming that the tissues of non-pregnant women are significantly stiffer (in traction and compression) than those of pregnant women. Further work followed to improve the modeling of the uterus. Readers can refer to the state of the art proposed by Zara and Dupuis (2017) on uterus modeling for more details.

A strategy to obtain a real-time childbirth simulation

For the geometric model of the uterus, an approach consist in modeling as a single object, the uterus, the cervix and the vaginal canal. Fig. 15.10 presents this geometrical model: on the left, the initial mesh from medical data with 42,811 nodes; on

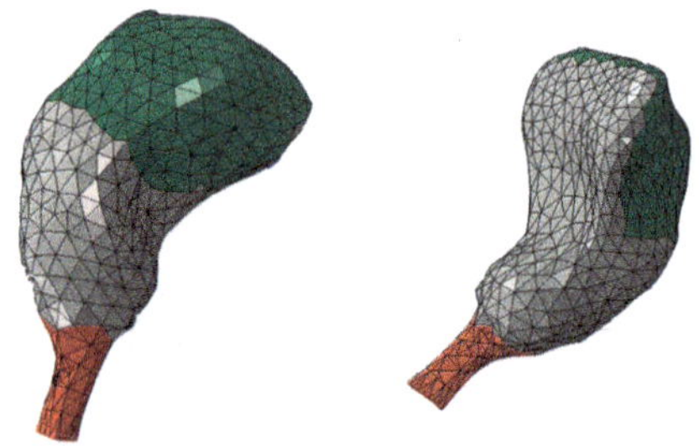

FIGURE 15.11

Fields of forces applied on the uterus: in gray the part on which the UC is applied; in green the part on which the UC, the abdominal thrust forces and the diaphragm are applied; in red the area fixed on the pelvis. Images from Buttin et al. (2013).

the right, the mesh obtained after smoothing and manual reconstruction of the cervix from anatomy (Schaal et al., 2007; Kamina, 2006) considering a complete dilation. For the biomechanical modeling of the uterus, the work of Mizrahi et al. (1980) showed that the behavior of the uterine muscles changes during labor, with isotropic behavior in the early stages of labor and anisotropic behavior during the progression of labor.

To simplify the model, an anisotropic behavior was considered for the uterine membrane. Thus the uterus was modeled as a hyperelastic material using the Neo-Hooke law of behavior, with a density of 950 kg/m3 and $C10 = 30$ kPa. As boundary conditions, the displacements of the vaginal canal were limited in the transverse plane to allow the opening and closing of the vaginal canal, avoiding the descent of the organs. For the treatment of collisions, the contacts between the uterus and the fetus were considered as frictionless. This assumption is due to the fact that when the labor phase begins the uterine sac empties of amniotic fluid, and the internal walls remain very wet during labor, generating an extremely lubricated behavior. The contacts between uterus and abdomen were also considered without friction, to take into account the viscous phenomenon existing between all the organs of the abdomen. However, instead of modeling the behavior of the muscle, that is, the uterus, we modeled its consequences, namely uterine contractions (UC). Thus in accordance with reality, UC and expulsion forces were modeled as two periodic pressure fields, uniform on the internal and external surface of the uterus, with 12 periods for a labor duration of 30 to 40 minutes. Fig. 15.11 shows these two force fields.

As a validation parameter, we know that uterine contractions exerted on the uterus cause its volume to decrease during fetal descent. To verify this behavior, two points of the uterus chosen in the transverse plane have been considered. Fig. 15.12 shows the movement of these two points during the simulation in the frontal/transversal axis. We can see that the displacements of these two points follow opposite directions, indicating that the behavior of the uterus well corresponds to the behavior of the UC. In addition, the size of the uterus was decreased by about 2/3 at the end of a delivery. This phenomenon is verified by following the evolution of the sagittal-

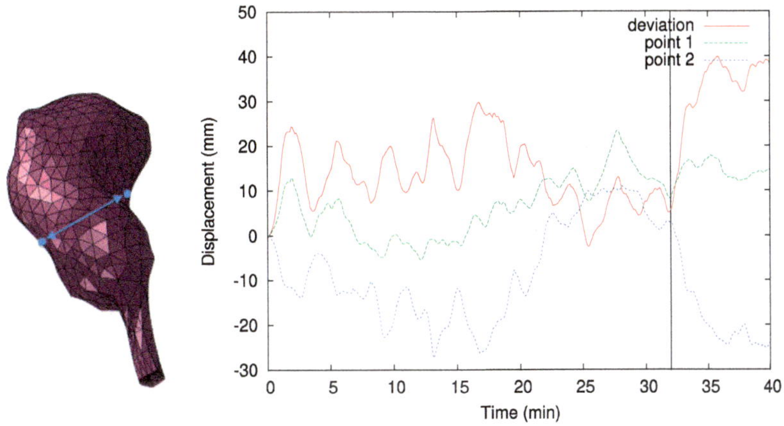

FIGURE 15.12

Points of the uterus followed during the simulation (left). Displacement of 2 points of the uterus during the simulation (right) (Buttin et al., 2013).

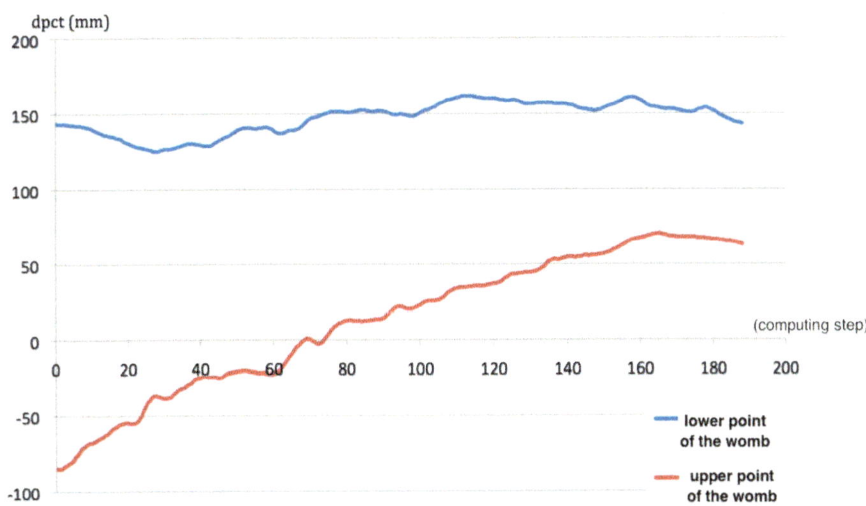

FIGURE 15.13

Evolution of the sagito-frontal trajectory of a point in the uterus (Buttin et al., 2013).

frontal trajectory of a point at the top of the uterus and comparing it to a point at the bottom of the uterus. In Fig. 15.13, we can observe that during the simulation this difference is 230 mm at the beginning of the labor phase and 80 mm at the exit of the fetus. It gives us a decrease of its size of the order of 2/3, in agreement with reality.

The push and UC efforts were modeled as periodic functions from contraction force patterns (maximum value, periodicity, cycles (from the literature)). It would be interesting to analyze the sensitivity of the fetal trajectory as a function of muscle forces. The modeling of these forces can also be modified to take into account the fact that these forces originate from the bottom of the uterus, and then propagate throughout the uterus.

15.4.4 Modeling of fetus

For the simulation of the delivery, we consider only the case where the delivery occurs at term, i.e., from the 37^e week of pregnancy. The fetus then weighs about 3.5 kg with a height of about 50 cm. The fetus has a flexion position made possible by a great flexibility of its joints. Moreover, the proportion of the fetus organs is not the same one as that of the adults or a child: its skull has a more important size in proportion. The fetal head is thus the most important part of the fetus. Thus during a delivery, we consider that if the head passes, the rest of the body also passes. On the other hand, the rotations of the head are essential during the delivery, to cross the cavities of the basin. But it is also necessary that the skull can deform. This is made possible by the mobility of the bones of the cranial vault in relation to each other. As sutures of the cranial vault, these are membranous bands that connect the different bones of the skull and meet at the fontanel allowing the "shaping of the fetal head" as it passes through. Of course, the intracranial volume is almost constant due to the movement of cerebrospinal fluid in the vertebral canal, ensuring the preservation of the brain during delivery. For more details, readers can refer to Chapter 4 of Parente (2008) explaining the several positions of the fetus and implications for the delivery, as well as the determination of these positions, thanks to a palpation of the fetal skull.

A strategy to obtain a real-time childbirth simulation

From a complexity point of view, it is not possible to model all the organs of the fetus. One approach consists in creating three zones within the fetus (as illustrated on Fig. 15.14): the skull, the body, and the skin tissue. Moreover, the skull and the body were included in the skin tissue (with several nodes in common) to avoid having to deal with collisions between these three parts. From a mechanical point of view, the skull was considered as a deformable object since it undergoes significant deformation during delivery. The body was considered as a slightly deformable object to allow the back of the fetus to move freely and to simulate the different joints. The skin tissue was considered more elastic than the body and skull, i.e., with a lower modulus of elasticity. Furthermore, the skin tissue was the only compressible organ to decrease the repulsion forces induced by the contact between fetus and uterus.

The three parts were modeled with the Neo-Hooke's behavior law with C10 = 130 kPa for the skin tissue, C10 = 75 kPa for the skull, and C10 = 70 kPa for the body. Moreover, considering that a fetus has a lower muscle density than an adult, the average density has to be slightly lower than 1000 kg/m3, with a density of 400 kg/m3 for the skin tissue, and a similar density of 950 kg/m3 for the skull and body. This modeling allows us to simplify the fetus considerably, while keeping an articulation at the head level induced by the deformation of the skin tissue (as shown on

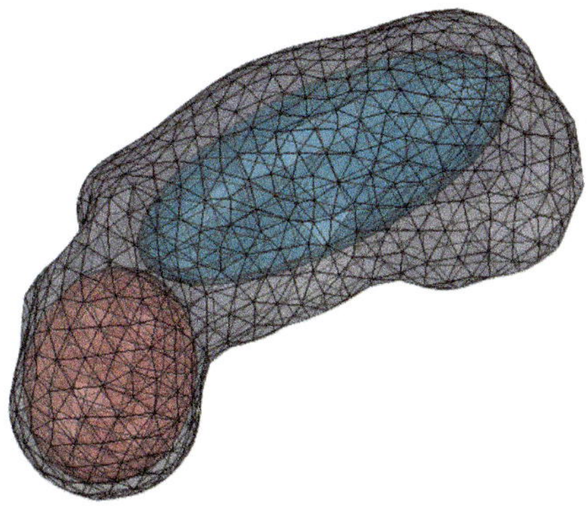

FIGURE 15.14

Model of the fetus in 3 parts: skull (red), body (blue) and skin tissue (white). Image from Buttin et al. (2013).

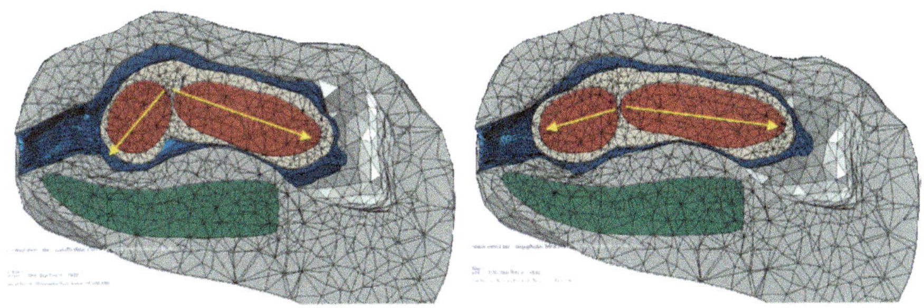

FIGURE 15.15

Articulation of the fetal head induced by skin tissue deformation. Images from Buttin et al. (2013).

Fig. 15.15). Fig. 15.16 shows the deformations of the head of the fetus during the simulation caused by pelvic muscle compression. Thus even if we did not model a plastic behavior for the fetal head, the hyper-elastic model employed enables this type of deformation during the descent of the fetus.

However, a more detailed modeling has to be performed to take into account the different plates of the fetal head, allowing its deformation during the descent into the pelvic canal. In this context, the initial mesh of the fetal head is presented in Fig. 15.17 (a): it has 4466 nodes and 8928 triangles. Then, this mesh has been separated into two zones to be able to model the different parts of the skull with different mechanical properties. We can see in blue the bones of the skull, and in green the fontanels and sutures. Fig. 15.17 (b) presents the measurements used to evaluate the deformations undergone by the fetal head. These measurements were put forward in the work of Sorbe and Dahlgren (1983) in Table 15.2. They correspond to the diameters that most reflect the deformations resulting from the intrauterine

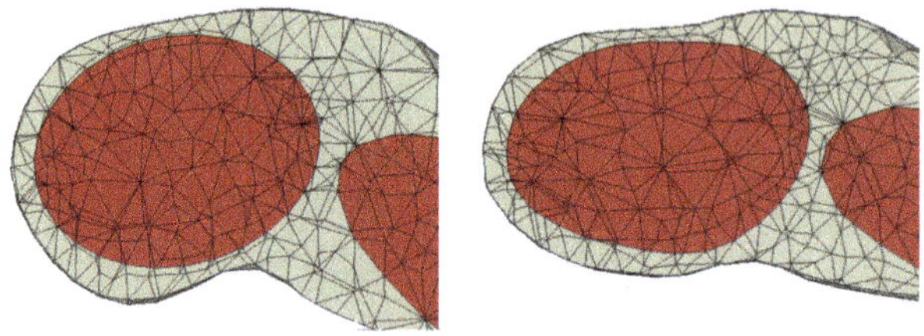

FIGURE 15.16

Crushing of the fetal head during the simulation. Images from Buttin et al. (2013).

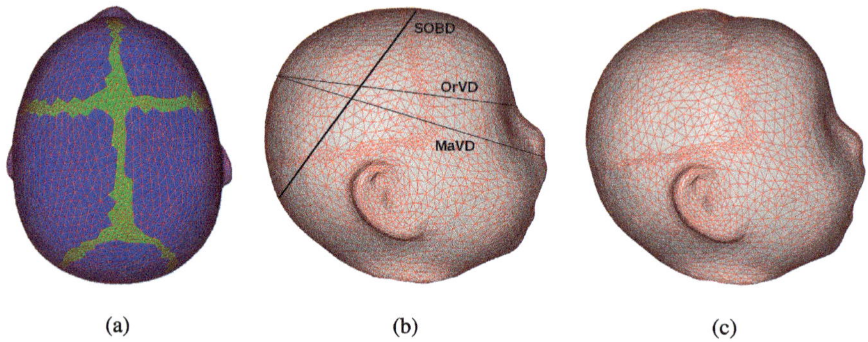

FIGURE 15.17

(a) Fetal head's mesh incorporating the skull bones (blue); fontanels and sutures (green). (b) Head diameter measurements. (c) Deformation of the fetal head undergoing intrauterine pressures and contact with the cervix. Images from Bailet et al. (2014).

Table 15.1 Mechanical properties of the fetal head parts.

	Young modulus	Poisson ratio	Density	Thickness
Skull bones	E = 2 GPa	$\nu = 0.22$	$\rho = 1.8$	0.8 mm
Fontanels and sutures	E = 31.5 MPa	$\nu = 0.45$	$\rho = 1$	0.6 mm

pressure applied to the head during delivery: the SOB (suboccipito-bregmatic), the OrVD, and the MaVD. The mechanical modeling of the head is based on the CST-DKT shell model proposed by Bailet (2014), which also incorporates a constraint for volume preservation. The mechanical properties employed were derived from the work of McPherson and Kriewall (1980a) and McElhaney et al. (1970). They were summarized in Table 15.1. In terms of boundary conditions, the nodes at the base of the neck were fixed to avoid rigid displacements.

Fig. 15.18 shows the pressures applied to the fetal head to simulate those experienced during childbirth. They correspond to intrauterine pressures and cervical pressures. The fontanels and sutures were not subjected to any pressure. The area around the SOB plane corresponds to the area in contact with the cervix. Three dif-

FIGURE 15.18

Intra-uterine and cervical pressures applied to the fetal skull. Image from Bailet et al. (2014).

Table 15.2 Measurement of the fetal head diameter and its variations (Bailet et al., 2014).

	Sorbe 1983		Lapeer 2001		Bailet et al. (2013)	
Measures	**ø**	**variation**	**ø**	**variation**	**ø**	**variation**
MaVD	140.5	+1.90	129.3	+1.43	124.6	+1.29
OrVD	126.9	+2.20	119.3	+1.24	110.9	+1.13
SOBD	117.1	−1.70	88.7	−2.52	107.4	-3.57

ferent pressure fields were applied: 45 kPa for the highest part of the skull (yellow), 40 kPa for the middle part (brown), and 30 kPa for the lowest field (red). The rest of the skull was subjected to a pressure of 7 kPa corresponding to the amniotic pressure. We have relied on the measurements of Rempen and Kraus (1991) to establish these pressure fields. Fig. 15.17 (c) shows the deformations of the fetal head obtained by simulation, as well as the measurements of these deformations compared with those made by Sorbe and Dahlgren (1983) and those of Lapeer and Prager (2001) obtained by simulation with an anisotropic model for the cranial bones and hyperelastic for the fontanels and sutures. We obtained similar results to those of Lapeer, with a larger variation for the SOB. We also observed an overlap of the parietal bones (bones on the sides of the skull), which is a classic phenomenon during childbirth (Schaal et al., 2007). We will detail the approach of Lapeer in the next section.

15.4.5 The complete delivery simulation coupled to a haptic device

Fig. 15.19 illustrates the entire 3D model with the boundary conditions of the different organs. Table 15.3 sums up the mechanical properties and behavior laws of the different organs involved. Fig. 15.20 shows images from the complete simulation of the delivery. We succeeded in simulating the descent of the fetus caused by uterine

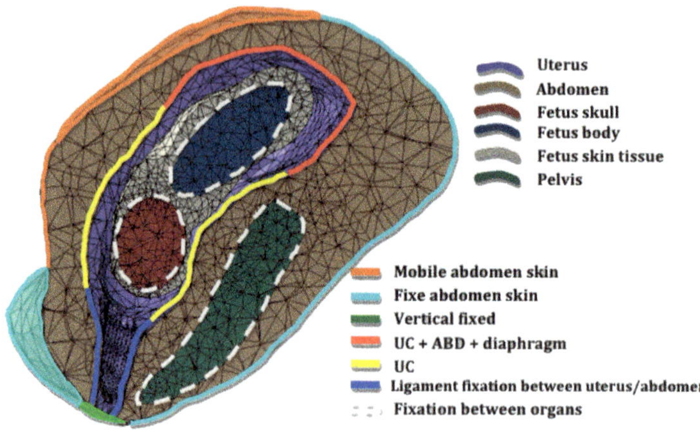

FIGURE 15.19

Global diagram of the delivery model (ABD: abdominal thrusts). Image from Buttin et al. (2013).

Table 15.3 Mechanical properties and behavior laws of the modeled organs.

Organs		Constitutive law	Mechanical parameters
Pelvis		Elastic (Hooke)	$E = 23$ MPa $\nu = 0.3$ $\rho = 1000$ kg/m3
Fetus	skin	Hyper-elastic (Neo-Hooke)	$C10 = 130$ kPa $\rho = 400$ kg/m3
	skull		$C10 = 75$ kPa $\rho = 950$ kg/m3
	body		$C10 = 70$ kPa $\rho = 950$ kg/m3
Abdominal		Hyper-elastic (Neo-Hooke)	$C10 = 5$ kPa $\rho = 2500$ kg/m3
Uterus		Hyper-elastic (Neo-Hooke)	$C10 = 30$ kPa $\rho = 950$ kg/m3

contractions and maternal pushing forces. Previous work on childbirth simulation approximates the forces on the uterus and abdomen by imposing kinematic boundary conditions, i.e., by imposing the trajectory of the fetal head. In our work, the kinematic behavior of the fetal skull was computed as a function of the forces exerted. This approach is required to propose a training simulator, including several scenarios, one aim being to see the consequences of the user's gestures.

Finally, to propose a complete simulator (to help midwifes and obstetricians to make decisions and learn the gestures for instrumental childbirth), the numerical

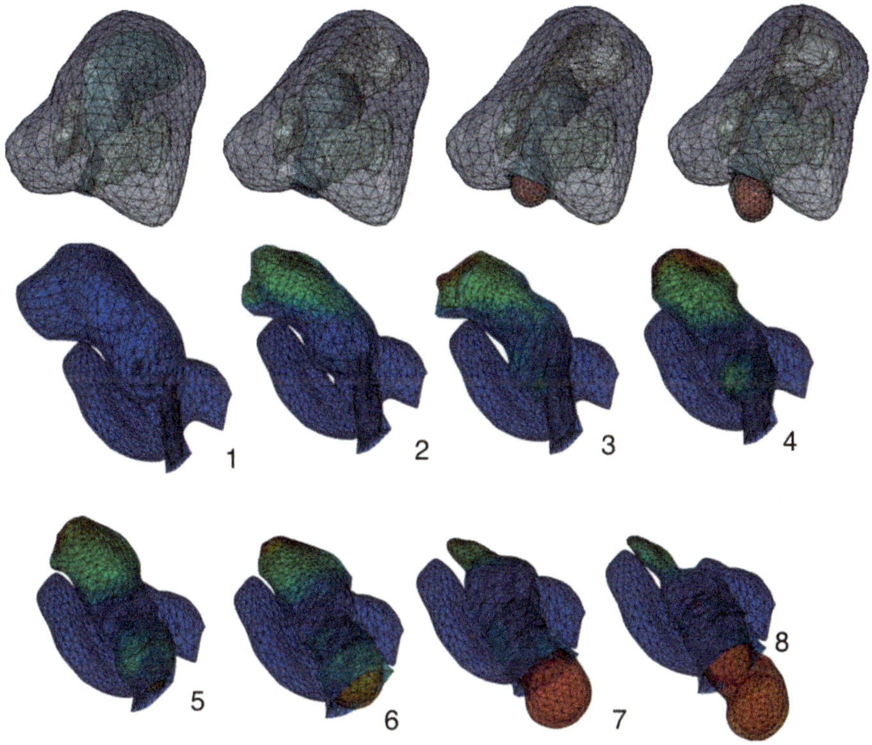

FIGURE 15.20

Different steps of childbirth simulation. Images from Buttin et al. (2013).

simulation was coupled to a physical device designed using anthropomorphic mannequin of the parturient's pelvis and the head of the fetus. This coupling allows (i) the user's actions to be taken into account in the numerical part; (ii) the integration of the forces calculated by the numerical model into the physical device. The haptic part was based on the BirthSIM simulator (shown in Fig. 15.21) developed by Moreau (2007). This simulator consists of anthropomorphic mannequins (usually used by obstetricians and midwives for their training). The use of a pneumatic actuator allows the descent of the fetal head through the pelvis. Moreover, six degrees of freedom position sensors (located on the forceps and the fetal head) allow the visualization of the movements. This prototype was then improved to increase its number of degrees of freedom, with the consideration of possible rotations of the fetal head during its descent. To continue on a kinematic analysis of the simulation, Fig. 15.22 shows the evolution of the fetal head according to the 3 anatomical axes. This phase lasts 32 minutes in our simulation, with an average speed of descent of the fetus of 0.09 mm/s. It is in conformity with reality with an average duration of 30 to 45 minutes. The main displacement was along the sagittal/frontal axis. The head displacement amplitude was indeed 51.5 mm along the frontal/transversal axis; 63.5 mm along the

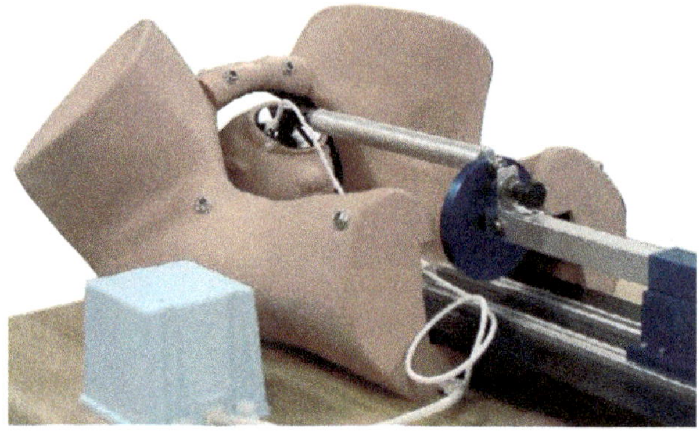

FIGURE 15.21

The BirthSIM physical simulator developed by the Ampère laboratory. Images from Buttin et al. (2013).

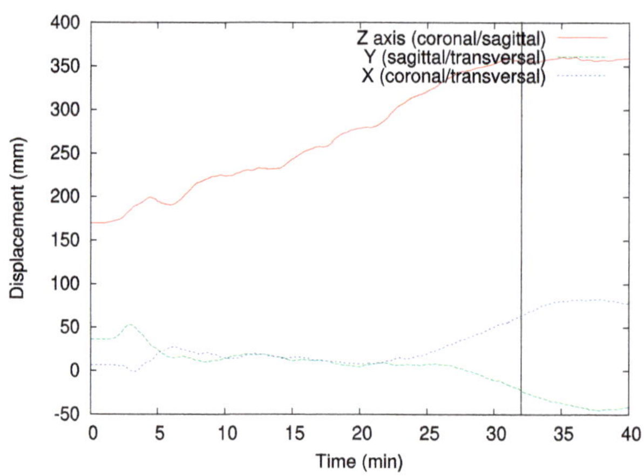

FIGURE 15.22

Displacement of the fetal head along the anatomical axes (Buttin et al., 2013).

sagittal/transversal axis; 184 mm for the sagittal/frontal axis. If we focus on the behavior according to the sagittal/frontal axis, we can observe that the speed of the fetal head was not linear. Indeed, the parturient does not push continuously, thus implying increases or decreases of the fetal head speed during the delivery. This behavior is more visible in Fig. 15.23, which presents the evolution of the fetal head speed in the sagittal/frontal axis. The negative parts of this curve thus corresponds to the inverse movement of the head.

To finish, Fig. 15.24 shows a comparison between our simulation-calculated trajectory and the one usually used in BirthSIM (before the coupling with the simulation). The trajectory initially used in BirthSIM (in red) did not take into account the

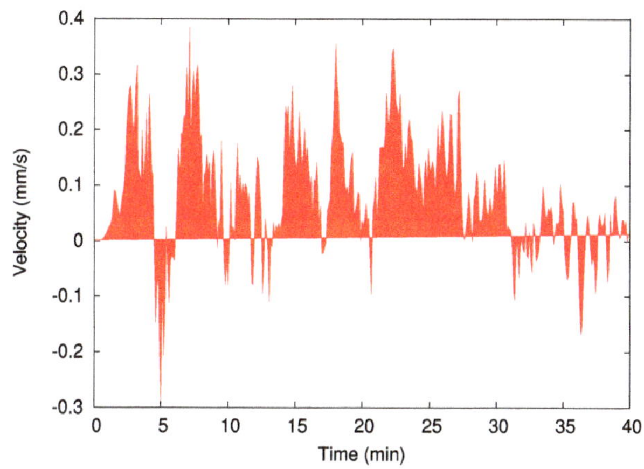

FIGURE 15.23

Speed of the fetal head along the sagittal/frontal axis (right) (Buttin et al., 2013).

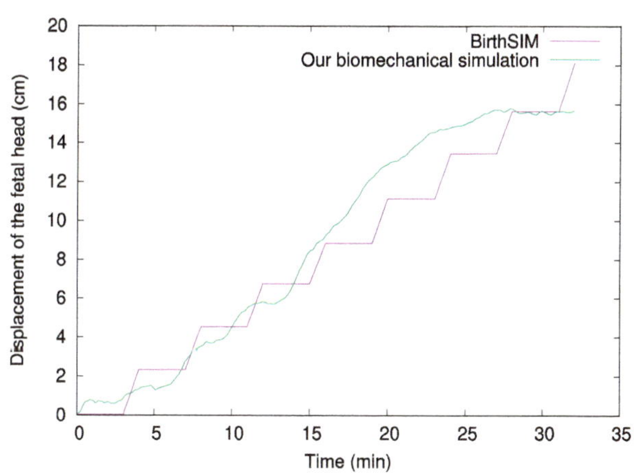

FIGURE 15.24

Comparison of the fetal head trajectory used in BirthSIM with our biomechanical simulation (Buttin et al., 2013).

morphology of the parturient and the fetus. Its displacement is thus assumed to be linear and cyclic, whereas on the green curve from the simulation, a change corresponding to the passage of the head into the pelvic floor, is visible at 15 minutes.

15.4.6 Conclusion

We have just presented how it was possible to simulate in real time, the descent of the fetus during a childbirth without imposed trajectory. This interactive time was nec-

Table 15.4 Summary of modeled parameters and areas for improvement.

Delivery time	V	Average time between 30 and 40 min. Simulation lasts 32 min.
Trajectory of the fetus	V	Realistic trajectory with phases of slight upward movement of the fetus that should be amplified (average upward movement of the order of one centimeter; our model is at 3-4 mm).
Fetal expulsion speed	V	The average speed of deliveries is about 0, 1 mm/s. Our model is at 0, 09 mm/s.
Choice of organs modeled	V	Organs not considered (placenta, bladder, rectum) had little enough impact to be hidden
Frictionless contacts	V	Lubrication justified by the presence of amniotic fluid and the addition of soap at the beginning of the delivery by the nurses.
Periodicity and cycle of contractions	V	Consistent with the average.
Perineal deformity	X	Consider the scalability of the skin.
Sacrum tilt	V	Realistic behavior of the sacrum with angular movements of a few degrees.
Behavior of the sacrum at the end of the delivery	V	Deviation of 1.5 degrees from the initial position. Conforms to reality.
Uterus behavior	V	Volume of the uterine fundus decreased by 66% at the end of the delivery. Conforms to reality.
Behavior of the fetal head	X	Current elastic modeling. Addition of plastic law and modeling of bone plates and fontanels.
Geometric simplification	~	Good for all organs except the pelvis (review pelvic symphysis and promontory).

essary to develop a learning simulator for instrumental obstetric gestures of vaginal childbirth. This real time is obtained, thanks to various simplifications, from the geometrical representation of the organs implied in the simulation to their biomechanical modeling and interaction between them. Moreover, in the proposed childbirth simulation, the following organs present in the pelvic cavity were not modeled: the placenta, the bladder, or the rectum. Indeed, the placenta is a relatively thin organ, which is located inside the uterus. At the mechanical level, this organ brings only a partial increase in the thickness of the uterine wall, at the price of a strong increase in the treatment of the contacts. The bladder is a rather large organ; it can contain about 350 mL of liquid. However, before the beginning of the delivery, it is emptied, reducing significantly its size, and thus limiting its implication on the movements of the others organs. Consequently, it seems unnecessary to add it in the childbirth simulation. In the same way, rectum does not have a great role during the childbirth. Table 15.4 summarizes all the parameters of the organs and gives some improvement perspectives.

15.5 Accurate childbirth simulations

After presenting a real-time childbirth simulation, we will now present researches that aim to better understand the implications of a vaginal delivery on the parturient's health and the newborn. Indeed, an instrumental delivery using forceps or vacuum cup could imply some damages on the head of the fetus, or it could imply episiotomy or some rips for the mother. Moreover, some injuries inflicted on the pelvic floor of the parturient may involved some urinary incontinence or prolapses in her future life. The challenge is then to obtain a better understanding of all the implications of a vaginal delivery, thanks to an accurate modeling of the pelvic anatomy and a numerical simulation of the childbirth. This simulation enables to visualize and measure the stress applied on tissues during the descent and delivery of the fetus. Generally, the simulation focuses on a specific organ due to its execution time and accuracy. In what follow, we will present studies that focus on the pelvic floor, or on the deformation suffered by fetal head.

15.5.1 Stresses applied on pelvic floor muscles

In this section, we will principally refer to the work carried out at the University of Porto by Natal Jorge and his team with the supervision of the PhD thesis of Parente (2008) and Oliveira (2016). The objective of their work is to better understand the changes that take place during childbirth, to obtain a better knowledge of the risk factors on pelvic floor dysfunctions related to a vaginal delivery. These dysfunctions could provide urinary incontinence or pelvic organ prolapse in the future life of parturients. Moreover, without a complete understanding of their causes, it seems difficult to provide adequate therapy. For more details on pelvic floor dysfunctions, readers can refer to the Chapter 3 of Parente (2008), which gives some important explanations.

Parente (Parente et al., 2009, 2010; Parente, 2008) proposed a numerical simulation of delivery, including the pelvic bone, the pelvic floor, and the fetus. The simulation has been realized using the Abaqus finite elements software. Its aim is to analyze the effects of the passage of the fetal head on the muscles of the pelvic floor. The geometrical model for the pelvic floor muscles were constructed using information obtained by Janda et al. (2003) from experimental measurements performed on old female cadavers. As constitutive equation used for the muscles of the pelvic floor, they used a modified form of the incompressible transversely isotropic hyperelastic model (proposed by Humphrey and Yin, 1987). The parameters of the material model were obtained from the experimental data provided by Janda (2006). Then, the pelvic floor was connected to a model of the pelvic skeletal structure. At the end, two positions of the fetus were studied in the simulation: occipito-anterior and occipito-posterior. The movements of the fetus were performed by imposing a vertical displacement of its head of 60 to 90 mm. The stresses and deformations along the pelvic floor were then been analyzed. As conclusion, authors observed that the complete flexion of the fetus head seems the most favorable position during birth as the stretches and stresses were smaller in this situation. Moreover, the simulation will permit a non-invasive procedure used in the future to estimate the damage that a vaginal delivery can induce on a specific pelvic floor.

Oliveira (Oliveira et al., 2016; Oliveira, 2016; Silva et al., 2015) continue the previous work to model damage process in pelvic tissues during vaginal delivery. Some of his contributions concern the influence of the fetal head molding on the biomechanical behavior of the pelvic floor muscles. The finite element model used for the fetal head included the skin and soft tissues, the skull with sutures and fontanels, and the brain, which constitutes a deformable tissue. Thus the reaction forces, stretch, stress, and strain fields were evaluated on the pelvic floor muscles. The results obtained highlight the fact that the molding of the fetal head reduces the amount of stretching undergone by the pelvic floor muscles during delivery. Moreover, several simulations of incisions on the pelvic floor muscles were conducted during this PhD thesis to obtain real effects of the episiotomy and its consequences during a vaginal delivery. First results showed that mediolateral episiotomies performed during normal vaginal delivery might facilitate birth, protecting both the pelvic floor as the fetus itself, as the episiotomy can be associated with a reduction in the forces that oppose the fetus descent. The authors recalled all the limitations associated to these first results. Indeed their simulations were based on a finite element model which was simple compared to the real pelvic female structure. Moreover, it was still difficult to evaluate the long term implications of episiotomies that damage the pelvic floor muscle strength. Finally, several structures of the pelvic cavity have been added to complete the finite element model. The model was created based on MRI data of nulliparous 24-year female without pelvic dysfunction complaints, including the levator ani and coccygeus muscles, organs of the pelvic cavity, and supportive structures. The results highlighted the importance of the ligaments improving the research previously presented about this aspect.

To conclude, we can hope that within a few years the implementation of these simulations (Oliveira et al., 2019; Margarida et al., 2020) will allow a better understanding of the forces and constraints undergone by the pelvic floor muscles during childbirth. The authors hope to be able to study different ways of performing episiotomies, to better understand their implications and perhaps improve this gesture (which is often badly perceived by the parturient). But the female pelvic system remains complex, as well as childbirth itself, involving a large number of anatomical structures, positions of the fetus, and forces that depend on uterine contractions and maternal thrusts specific to the parturient. Furthermore, the mechanical parameters of the different tissues differ from one patient to another and from one patient to another (depending on the number of deliveries she has already experienced), making the simulations and their interpretations complex.

15.5.2 Uterine pressures applied on fetal head

In this section, we will principally refer to the work carried out at the University of East Anglia (UK) by Lapeer and his team. The first objective of their work is to simulate the fetal head deformation during a vaginal delivery. This deformation is due to uterine pressures and use of forceps in the case of an instrumented extraction. Their goal is to propose a medical simulator that enables the learning of obstetrical use of forceps. In this context, Lapeer and Prager (2001) proposed in 2001 an extension of the work of McPherson and Kriewall (1980a,b) to model the deformation of the fetal skull due to cervical pressures during the early phase of delivery. A static nonlinear finite element model was used to model the deformation of the fetal skull with the use of a theoretical model to model the intrauterine and cervical pressures. The simulations were static and the large number of elements did not allow a real-time simulation. This work was then continued by proposing an augmented reality platform based on a simulation of instrumented delivery with forceps (Lapeer et al., 2004). The goal was to place a real forceps, tracked by passive optical markers, around a virtual fetal head. The contacts between the virtual skull and the forceps were then detected on the simulator. Then, the feasibility of a real-time mechanical contact model was tested in Lapeer (2005) to describe the interaction between the instrument and the fetal head. The conclusion was: "it was probably possible to achieve real-time performance by employing an explicit dynamic model to calculate the deformation of the major bones of the fetal skull, or by employing a quasi-static model to calculate the deformation of the entire fetal head." In 2014, Lapeer et al. (2014) proposed a finite element analysis performed to better study contacts between forceps and skull in several situations. They observed that asymmetric placement of forceps involves deformation with a bigger magnitude than for symmetric placement. This observation confirmed the obstetricians' presentiment when they practice this gesture.

The PhD thesis of Sadulaev (2019) then focused on the development of fetal neck models. The proposed model consists in a ball and socket joint for intervertebral discs, spring-and-damper systems for ligaments, and six-degrees-of-freedom bushing element to simulate a coupled behavior of the discs, ligaments and neck muscles. The idea was to complete previous work to be able to experiment a higher variety of biomechanical parameters (such as the neck's length, strength, and a full range of fetal biparietal diameters, mechanical properties of the fetal spine). The simulation was coupled to the use of two haptic devices (Sadulaev et al., 2017), which enable the manipulation of a virtual fetal head. This coupling permits to validate the resistance of the fetal head, during its flexion, extension, and rotation. It was clinically tested by midwives and obstetricians at the hospital to check its possibility to replicate the biomechanical properties of a newborn's head motion.

For the biomechanical model, the total Lagrangian explicit dynamics (TLED) contact method was used (Lapeer et al., 2019). A set of experiments were conducted for resolving a direct occipito-posterior position (OP) of the fetal head, which is considered to be a difficult birth scenario. In OP midwives advise their patients to tilt their pelvises anteriorly to help with labor. This method was experimentally tested in the childbirth simulator, and the results showed that tilting a pelvis anteriorly could potentially ease the dilation during the first stage of labor. However, no significant difference was observed as compared to the non-tilted pelvis during the second stage of labor.

Moreover, experiments were run to observe shoulder dystocia, thanks to a maternal pelvis with a mobile sacrum attached to the rest of the pelvis using the six degrees-of-freedom bushing element. It permits to study the effects of the sacrum mobility during a childbirth. The results showed that a mobile sacrum contributes toward the full internal rotation of the fetal head during a childbirth, whereas the absence of the mobility leads to the arrest of the head in the anteroposterior diameter of the pelvis.

Some other work starts to model the use of vacuum cup for extraction. In this context, Goordyal et al. (2020) presented a review of the literature of vacuum-assisted delivery devices from an engineering perspective. This review assesses the interest of this device as an option to a vaginal delivery requiring the assistance of the obstetrician (while avoiding the use of a Caesarean section). It also highlights the traumatic character of this medical device. The authors believe that a better understanding of this medical device would help improving its use, particularly by increasing the understanding of traction forces at play on the fetal scalp. In this regard, Chen et al. (2021) proposed a finite element analytical model to study the influence of vacuum extractors designed with different materials on the fetal head under various extractive pressures. They measured von Mises stress of vacuum extractor and one applied on the skull of fetal head. These results could permit a better knowledge for obstetricians regarding pressure applied during an operative delivery with vacuum extractor.

15.5.3 Conclusion

We reviewed modeling and simulating work, as precisely as possible, based on some anatomical parts of the female pelvic system or fetus during a vaginal delivery. These numerical simulations enable to compute forces and stresses involved during a childbirth. Their aims can be to analyze the stress involved by the fetus on pelvic floor muscles to facilitate the study of different strategies of episiotomies by computing its consequences on the pelvic floor. An other goal can be to measure the stresses involved on fetal head using vacuum extractor or forceps to improve these gestures with a better knowledge. The clinical challenge is then to predict if the childbirth could involve injuries on the parturient or the newborn. These damages can be due to several sources: the anatomical morphology of the parturient/fetus (e.g., a big fetal head with a narrow bony pelvis); the biomechanical parameters of pelvic floor tissues, whose elastic properties vary according to age, number of deliveries; the number of previous vaginal delivery carried out by the parturient; the medical extractor device used during an instrumented delivery; etc. The clinical aims are many: (i) to avoid some future pelvic floor dysfunctions of the mother (as they could involve incontinence or prolapse); (ii) to improve the therapeutic management of these dysfunctions, with a better knowledge of the episiotomy consequences; (iii) to improve the obstetrician gestures for instrumented deliveries.

15.6 Conclusion & perspective

The simulation of a vaginal delivery represents a complex scientific challenge. In this chapter, we have presented two approaches. The first one has the ambitious challenge to obtain a real-time childbirth simulation that is adequately realistic to be integrated inside a learning simulator for obstetrical gestures of extraction using forceps. The second one has the difficult challenge to increase the clinical knowledge of the effects of a vaginal delivery on the parturient or newborn. These two strategies are complementary.

If we want to go further by combining these two approaches, the idea would be to start from an ideal biomechanical model and to simplify it progressively until obtaining a model adapted to the interactive time. This would allow to determine, for each organ involved, its adequate law of behavior in agreement with the precision and desired execution time. It would also permit to evaluate the impact of these choices on the precision of the fetal trajectory during the delivery.

In a few years, we can then hope to be able to make a diagnosis before a delivery to facilitate determining the risk for the parturient to make a vaginal delivery. In addition to the recovery of medical images (allowing the geometrical modeling of the parturient's organs), this implies being able to easily measure the biomechanical parameters of each organ involved in the simulation.

Acknowledgments

The realization of the childbirth simulation has been performed during the PhD thesis of R. Buttin under the supervision of F. Zara, prof. B. Shariat (LIRIS, Université Lyon 1), and prof. T. Redarce (laboratoire Ampère, INSA de Lyon). The simulation of deformation of the fetal head has been performed during the PhD thesis of M. Bailet under the supervision of E. Promayon (TIMC-IMAG laboratory, Université Grenoble Alpes) and F. Zara. This work was financed by grants from ARC 6, GMCAO and SIMED projects of the ISLE cluster of the French Rhône-Alpes region, and supported by French state funds managed by the ANR within the SAGA project (ANR-12-MONU-0006). The medical images used have been provided by prof. C. Adamsbaum of the St. Vincent de Paul Hospital (AP-HP), Paris. The segmentation of these medical data has been performed by Jérémie Anquez (TELECOM ParisTech, CNRS, UMR-5141, LTCI). This work has been performed in collaboration with prof. Olivier Dupuis (obstetrician gynecologist, Université de Lyon, Maternité CHU Lyon Sud) and prof. Gilles Grangé from the Port Royal Maternity Hospital (AP-HP).

References

Allen, R.H., Bankoski, B.R., Nagey, D.A., 1995. Simulating birth to investigate clinician-applied loads on newborns. Medical Engineering & Physics (ISSN 1350-4533) 17, 380–384.

Aspert, N., Santa-Cruz, D., Ebrahimi, T., 2002. MESH: measuring errors between surfaces using the Hausdorff distance. In: IEEE International Conference on Multimedia and Expo, pp. 705–708.

Attene, M., Falcidieno, B., 2006. ReMESH: an interactive environment to edit and repair triangle meshes. In: Shape Modeling and Applications (SMI), pp. 271–276.

Bailet, M., 2014. Modèles biomécaniques pour la simulation interactive de l'accouchement. Thèse de doctorat en informatique. Université Joseph Fourier. https://hal.archives-ouvertes.fr/tel-01158380.

Bailet, M., Zara, F., Promayon, E., 2013. Shell finite element model for interactive fetal head deformation during childbirth. Computer Methods in Biomechanics and Biomedical Engineering 16, 312–314.

Bailet, M., Zara, F., Promayon, E., 2014. Biomechanical model of the fetal head for interactive childbirth simulation. In: SURGETICA 2014 Computer-Assisted Medical Interventions: Scientific Problems, Tools and Clinical Applications, pp. 116–119.

Bauer, M., Mazza, E., Jabareen, M., Sultan, L., Bajka, M., Lang, U., Zimmermann, R., Holzapfel, G., 2009. Assessment of the in vivo biomechanical properties of the human uterine cervix in pregnancy using the aspiration test: a feasibility study. European Journal of Obstetrics & Gynecology and Reproductive Biology 144, S77–S81.

Boissonnat, J.D., Geiger, B., 1993. 3D simulation of delivery. In: Nielson, G.M., Bergeron, D. (Eds.), Visualization 93. IEEE Computer Society Press, San Jose CA, pp. 416–419.

Boissonnat, J.D., Yvinec, M., 1995. Géométrie algorithmique, Ediscience ed.

Brieu, M., Bhatnagar, N., Boukerrou, M., Cosson, M., Rubod, C., 2010. Experiments and finite element modelling for the study of prolapse in the pelvic floor system. Computer Methods in Biomechanics and Biomedical Engineering 13, 349–357.

Buttin, R., 2010. Modélisation biomécanique du système reproductif féminin et du foetus pour la réalisation d'un simulateur virtuel d'accouchement. Thèse de doctorat en informatique. Université Claude Bernard Lyon 1.

Buttin, R., Zara, F., Shariat, B., Redarce, T., Grangé, G., 2013. Biomechanical simulation of the fetal descent without imposed theoretical trajectory. Computer Methods and Programs in Biomedicine 111, 389–401. https://doi.org/10.1016/j.cmpb.2013.04.005. https://www.sciencedirect.com/science/article/pii/S0169260713001132.

Chen, S., Grimm, M.J., 2021. Childbirth computational models: characteristics and applications. Journal of Biomechanical Engineering 143, 050801. https://doi.org/10.1115/1.4049226.

Chen, Y.H., Su, K.M., Tsai, M.T., Lin, C.K., Chang, C.C., Su, K.C., 2021. Investigating the vacuum extractors of biomedical devices of different materials and pressures on the fetal head during delivery. Applied Sciences 11. https://doi.org/10.3390/app11178237. https://www.mdpi.com/2076-3417/11/17/8237.

Dalstra, M., Huiskes, R., Odgaard, A., Van Erning, L., 1993. Mechanical and textural properties of pelvic trabecular bone. Journal of Biomechanics 26, 523–535.

d'Aulignac, D., Boubaker, B., Pires, E., Mascarenhas, T., Natal Jorge, R., 2005. A shell finite element model of the pelvic floor muscles. Computer Methods in Biomechanics and Biomedical Engineering 8, 339–347. https://doi.org/10.1080/10255840500405378.

Dufour, M., Pillu, M., 2007. Biomécanique fonctionnelle: Membres-Tête-Tronc. Masson.

Eggert, J.S., Eggert, M.S., Vallejo, P., 2003. Interactive education system for teaching patient care. Patent No. US2003/0081968A1.

Forster, C., Bechtold, I., Melchert, F., 2001. Clinical use of the birth simulation program "anapelvis 2.0" for prediction of feto-maternal disproportion. Zentralblatt fur gynecakologie 123, 465–468.

Fung, Y., 1993. Biomechanics. Mechanical Properties of Living Tissues, second ed. Springer.

Gardner, R., 2007. Simulation and simulator technology in obstetrics: past, present and future. Expert Review in Obstetrics & Gynecology 2, 775–790.

Gatellier, M.A., dit Gautier, E.J., Mayeur, O., Brieu, M., Cosson, M., Rubod, C., 2020. Complete 3 dimensional reconstruction of parturient pelvic floor. Journal of Gynecology Obstetrics and Human Reproduction 49. https://doi.org/10.1016/j.jogoh.2019.101635.

Gaumard Scientific, 2022. Victoria® s2200. http://www.gaumard.com.

Geiger, B., 1993. Three-dimensional modeling of human organs and its application to diagnosis and surgical planning. Ph.D. thesis. Ecole des Mines de Paris.

Goordyal, D., Anderson, J.L., Alazmani, A., Culmer, P.R., 2020. An engineering perspective of vacuum assisted delivery devices in obstetrics: a review. Proceedings of the Institution of Mechanical Engineers. Part H, Journal of Engineering in Medicine 235, 3–16.

Graves, J., Savannah, G., 1951. Educational apparatus for teaching obstetrics and midwifery.

Humphrey, J.D., Yin, F.C., 1987. On constitutive relations and finite deformations of passive cardiac tissue: a pseudostrain- energy function. Journal of Biomechanical Engineering 109, 298–304.

Janda, S., 2006. Biomechanics of the pelvic floor musculature. Ph.D. thesis. Mechanical Maritime and Materials Engineering. http://resolver.tudelft.nl/uuid:68ded298-7765-4490-9c7f-1d079a9a9c7e.

Janda, S., van der Helm, F., de Blok, S., 2003. Measuring morphological parameters of the pelvic floor for finite element modelling purposes. Journal of Biomechanics 36, 749–757.

Jiang, Z., Witz, J., Lecomte-Grosbras, P., Dequidt, J., Cotin, S., Rubod, C., Duriez, C., Brieu, M., 2017. Multi-organ motion tracking in dynamic magnetic resonance imaging for evaluation of pelvic system mobility and shear strain. Strain 53, e12224. https://doi.org/10.1111/str.12224.

Kamina, P., 2006. Anatomie clinique: anatomie générale, membre, third ed. Maloine.

Kheddar, A., Devine, C., Brunel, M., Duriez, C., Sidony, O., 2004. Preliminary design of a childbirth simulator haptic feedback. In: IEEE/RSJ, International Conference on Inteligent Robots and Systems, pp. 3270–3275.

Knapp, C.F., Eads, G.S., 1974. Dynamic childbirth simulator for teaching maternity patient care. Patent n: US 3,824,709.

Lapeer, R., Audinis, V., Gerikhanov, Z., Dupuis, O., 2014. A computer-based simulation of obstetric forceps placement. In: Golland, P., Hata, N., Barillot, C., Hornegger, J., Howe, R. (Eds.), Medical Image Computing and Computer-Assisted Intervention – MICCAI 2014. MICCAI 2014, 14-09-2014 – 18-09-2014. Springer, Germany, pp. 57–64.

Lapeer, R., Gerikhanov, Z., Sadulae, S., Audinis, V., Rowland, R., Crozier, K., Morris, E., 2019. A computer-based simulation of childbirth using the partial Dirichlet–Neumann contact method with total Lagrangian explicit dynamics on the gpu. Biomechanics and Modeling in Mechanobiology 18, 681–700. https://doi.org/10.1007/s10237-018-01109-x.

Lapeer, R.J., 2005. A mechanical contact model for the simulation of obstetric forceps delivery in a virtual/augmented environment. Studies in Health Technology and Informatics 111, 284–289.

Lapeer, R.J., Chen, M., Villagrana, J., 2004. An augmented reality based simulation of obstetric forceps delivery. In: Third IEEE and ACM International Symposium on Mixed and Augmented Reality (ISMAR 2004), pp. 274–275.

Lapeer, R.J., Prager, R.W., 2001. Fetal head moulding: finite element analysis of a fetal skull subjected to uterine pressures during the first stage of labour. Journal of Biomechanics 34, 1125 1133.

Lepage, J., Cosson, M., Mayeur, O., Brieu, M., Rubod, C., 2016. Pedagogical childbirth simulators: utility in obstetrics. European Journal of Obstetrics & Gynecology and Reproductive Biology 197, 41–47. https://doi.org/10.1016/j.ejogrb.2015.11.011. https://hal.archives-ouvertes.fr/hal-01293128.

Lepage, J., Jayyosi, C., Lecomte-Grosbras, P., Brieu, M., Duriez, C., Cosson, M., Rubod, C., 2014. Biomechanical pregnant pelvic system model and numerical simulation of childbirth: impact of delivery on the uterosacral ligaments, preliminary results. International Urogynecology Journal 26, 497–504.

Letterie, G., 2002. How virtual reality may enhance training in obstetrics and gynecology. American Journal of Obstetrics and Gynecology 187, S37–S40.

Li, X., Kruger, J., Nash, M., Nielsen, P., 2011. Anisotropic effects of the levator ani muscle during childbirth. Biomechanics and Modeling in Mechanobiology 10 (4), 485–494.

Li, X., Kruger, J.A., Nash, M.P., Nielsen, P.M., 2010a. Effects of fetal head motion on pelvic floor mechanics. In: Miller, K., Nielsen, P.M. (Eds.), Computational Biomechanics for Medicine. Springer New York, New York, NY, pp. 129–137.

Li, X., Kruger, J.A., Nash, P., Nielsen, M., 2010b. Modeling childbirth: elucidating the mechanisms of labor. Wiley Interdisciplinary Reviews: Systems Biology and Medicine 2, 460–470.

Margarida, B., Rita, M., Dulce, O., M, P., Teresa, M., Natal Jorge, R., 2020. Effect of the birthing position on its evolution from a biomechanical point of view. Computer Methods and Programs in Biomedicine 200, 105921. https://doi.org/10.1016/j.cmpb.2020.105921.

Martins, J.A.C., Pato, M.P.M., Pires, E.B., Jorge, R.M.N., Parente, M., Mascarenhas, T., 2007. Finite element studies of the deformation of the pelvic floor. Annals of the New York Academy of Sciences, 316–334.

Mazza, E., Nava, A., Bauer, M., Winter, R., Bajka, M., Holzapfel, G., 2006. Mechanical properties of the human uterine cervix: an in vivo study. Medical Image Analysis 10, 125–136.

McElhaney, J.H., Fogle, J.L., Melvin, J.W., Haynes, R.R., Roberts, V.L., Alem, N.M., 1970. Mechanical properties of cranial bone. Journal of Biomechanics 3, 495–511.

McPherson, G., Kriewall, T.J., 1980a. The elastic modulus of fetal cranial bone: a first step towards an understanding of the biomechanics of fetal head molding. Journal of Biomechanics 13, 9–16.

McPherson, G., Kriewall, T.J., 1980b. Fetal head molding: an investigation utilizing a finite element model of the fetal parietal bone. Journal of Biomechanics 13, 17–26.

Mizrahi, J., Karni, Z., 1975. A mechanical model for uterine muscle activity durng labor and delivery. Israel Journal of Technology 13, 185–191.

Mizrahi, J., Karni, Z., Polishuk, W., 1980. Isotropy and anisotropy of uterine muscle during labor contraction. Journal of Biomechanics 13, 211–218.

Moreau, R., 2007. Le simulateur BirthSIM : un outil complet pour la formation sans risque en obstétrique. Ph.D. thesis. Institut National des Sciences Appliquées, Lyon, France.

Obst, T., Burgkart, R., Ruckhäberle, E., Riener, R., 2004. The delivery simulator: a new application of medical VR. In: Annual Medicine Meets Virtual Reality Conference (MMVR'04), pp. 281–287.

Oliveira, D., Parente, M., Alvo, B., Mascarenhas, T., Natal Jorge, R., 2016. Numerical simulation of the damage evolution in the pelvic floor muscles during childbirth. Journal of Biomechanics 49, 594–601. https://doi.org/10.1016/j.jbiomech.2016.01.014.

Oliveira, D., Pouca, M., Ferreira, J., Mascarenhas, T., 2019. Episiotomy: the biomechanical impact of multiple small incisions during a normal vaginal delivery. Interface Focus 9. https://doi.org/10.1098/rsfs.2019.0027.

Oliveira, D.A., 2016. Modeling damage process in pelvic tissues during vaginal delivery. Ph.D. thesis. Universidade do Porto.

Parente, M., Natal Jorge, R., Mascarenhas, T., Fernandes, A., Martins, J., 2009. The influence of the material properties on the biomechanical behavior of the pelvic floor muscles during vaginal delivery. Journal of Biomechanics 42, 1301–1306. https://doi.org/10.1016/j.jbiomech.2009.03.011. https://www.sciencedirect.com/science/article/pii/S002192900900133X.

Parente, M., Natal Jorge, R., Mascarenhas, T., Fernandes, A., Martins, J., 2010. The influence of pelvic muscle activation during vaginal delivery. Obstetrics and Gynecology 115, 804–808. https://doi.org/10.1097/AOG.0b013e3181d534cd.

Parente, M.P.L., 2008. Biomechanics of the pelvic floor during vaginal delivery. Ph.D. thesis. Universidade do Porto.

Rempen, A., Kraus, M., 1991. Pressures on the fetal head during normal labor. Journal of Perinatal Medicine 19, 199–206.

Riener, R., Burgkart, R., 2003. Birth simulator (geburtensimulator). Patent No: WO 03/001482 A1.

Rubod, C., Boukerrou, M., Brieu, M., Clay, J.C., Dubois, P., Cosson, M., 2008. Biomechanical properties of vaginal tissue: preliminary results. International Urogynecology Journal 19. https://doi.org/10.1007/s00192-007-0533-3.

Rubod, C., Boukerrou, M., Brieu, M., Dubois, P., Cosson, M., 2007. Biomechanical properties of vaginal tissue. Part 1: new experimental protocol. The Journal of Urology 178, 320–325.

Rubod, C., Lecomte-Grosbras, P., Brieu, M., Giraudet, G., Betrouni, N., Cosson, M., 2013. 3d simulation of pelvic system numerical simulation for a better understanding of the contribution of the uterine ligaments. International Urogynecology Journal 24, 2093–2098.

Sadulaev, S.M., 2019. Towards increased realism of a computer simulation of human childbirth. Ph.D. thesis. University of East Anglia.

Sadulaev, S.M., Lapeer, R., Gerikhanov, Z., Morris, E., 2017. A haptic user interface to assess the mobility of the newborn's neck. In: Proceedings of the 2017 21st International Conference Information Visualisation (IV). 21st International Conference on Information Visualisation iV2017, 11-07-2017 – 14-07-2017. IEEE Press, pp. 218–223.

Schaal, J.P., Riethmuller, D., Maillet, R., Uzan, M., 2007. Mécanique et Technique Obstétricales, third ed. Sauramps medical.

Sielhorst, T., Obst, T., Burgkart, R., Riener, R., Navab, N., 2004. An augmented reality delivery simulator for medical training. In: International Workshop on Augmented Environments for Medical Imaging - MICCAI Satellite Workshop, vol. 141, pp. 11–20.

Silva, E., Oliveira, D., Da Roza, T., Brandao, S., Parente, M., Mascarenhas, T., Natal Jorge, R., 2015. Study on the influence of the fetus head molding on the biomechanical behavior of the pelvic floor muscles, during vaginal delivery. Journal of Biomechanics 48. https://doi.org/10.1016/j.jbiomech.2015.02.032.

Sorbe, B., Dahlgren, S., 1983. Some important factors in the molding of the fetal head during vaginal delivery - a photographic study. International Journal of Gynaecology and Obstetrics 21, 205–212. https://doi.org/10.1016/0020-7292(83)90081-4.

Yamada, H., 1970. Strength of Biological Materials. Williams & Wilkins in Baltimore.

Zara, F., Dupuis, O., 2017. Chapter 15: Uterus. In: Biomechanics of Living Organs: Hyperelastic Constitutive Laws for Finite Element Modeling. Elsevier/Joe Hayton, pp. 325–346.

CHAPTER

Numerical models for breast surgery and reconstruction

16

David Hawkes[a], Vasileios Vavourakis[b,c], Bjoern Eiben[d,b], and John Hipwell[e,b]

[a]*Medical Physics and Biomedical Engineering, Centre for Medical Image Computing (CMIC) and Wellcome/EPSRC Centre for Surgical and Interventional Science (WEISS), University College London, London, United Kingdom*

[b]*Centre for Medical Image Computing, Department of Medical Physics and Biomedical Engineering, University College London, London, United Kingdom*

[c]*Department of Mechanical and Manufacturing Engineering, University of Cyprus, Nicosia, Cyprus*

[d]*Joint Department of Physics, The Institute of Cancer Research and The Royal Marsden NHS Foundation Trust, London, United Kingdom*

[e]*Toshiba Medical Visualization Systems, Edinburgh, United Kingdom*

16.1 Introduction

In 2018 there were about 56,000 new cases of breast cancer in the UK and of these approximately 11,500 died from the disease, making breast cancer the 2nd highest cause of cancer mortality in the UK, after lung cancer (https://www.cancerresearchuk.org). Nevertheless, diagnosis and successful treatment of breast cancer has improved dramatically in the last few decades, with 76% of those diagnosed now surviving more than 10 years. National screening programmes and better patient awareness have led to earlier diagnosis and more successful treatments using a range of interventional, surgical, radiation therapy, and systemic therapies. However, it is clear that breast cancer remains a significant health problem and successful treatment remains a stressful experience and can leave the patient with long recovery times and long-term issues of pain, scarring, and disfigurement.

The physical nature of the breast means that it is accessible to a wide range of imaging and sensing modalities, and treatments, including surgery, can be focused on excising or treating just the lesion with minimal collateral damage. Such minimally invasive procedures do, however, risk leaving residual cancer, which may require re-excision or additional treatment, with the associated stress and risks of side effects.

Biomechanics of the Female Reproductive System: Breast and Pelvic Organs. https://doi.org/10.1016/B978-0-12-823403-7.00030-0

16.2 Previous work on breast biomechanical modeling to aid registration and surgical planning

This chapter focusses on surgical planning and navigation for the surgical treatment of breast cancer, specifically lumpectomy, which makes use of at least some knowledge of breast biomechanics. Ramiao et al. (2016) provided a review of our understanding of the biomechanical properties of the breast, with McGhee and Steele (2020) providing a more recent update. A review of biomechanically informed breast image registration is provided by Hipwell et al. (2016). Janzekovic et al. (2022) review the biomechanics of breast augmentation. The latter addresses one of the other main applications of breast biomechanics in surgery.

It is clear from these recent reviews that much remains to be understood on the biomechanics of the breast. The female breast is a highly deformable structure, with a very wide range of size, shape, and composition across the population, over an individual's menstrual cycle pre-menopause, during pregnancy, post-menopause (with or without hormone replacement therapy), and as a result of patient weight gain or loss. Comprehensive biomechanical breast models, validated to be accurate across the full range of size, shape, and composition of the breast, remain elusive. However a pragmatic approach can lead to useful technologies that provide computationally tractable solutions, able to impact patient management.

Perez del Palomar et al. (2008) reported an early system for surgical planning, while Carter et al. (2006) outlined a framework for image-guided breast surgery with initial clinical experience given in Carter et al. (2008). Babarenda Gamage et al. (2019) proposed one of the first end-to-end, automated workflows incorporating biomechanical models and population-based statistical analysis to assist clinicians during breast cancer detection and treatment. The system is implemented and clinical studies should follow.

A major problem that needed to be addressed is that 3D imaging of the breast for surgical planning and navigation is usually undertaken with the patient standing and breast compressed (digital tomosynthesis) or prone with the breast pendulus (MRI). There are protocols for supine MRI with the breast position in a similar position to that of surgery, for example, Siegler et al. (2011). Alderliesten et al. (2010) and Conley et al. (2014) explored the feasibility of MRI-guided navigation using supine MRI. Unfortunately, supine MRI on its own is unsuitable for diagnosis and surgical planning as the lesion is often poorly visualized. Key to predicting how the breast deforms in different positions is the derivation of the unloaded, or gravity-free, state (Rajagopal et al., 2008a; Gavaghan et al., 2008). Lee et al. (2010) provided a method for image registration combining finite elements with free-form deformations. Babarenda Gamage et al. (2011) provided an overview of methods available at that time, and it was clear that improvement was required before effective surgical planning and guidance could be provided.

Our groups at UCL and previously at KCL, London have worked on improving breast imaging and related technologies for two decades from 1996 to 2016. Our aim was to use imaging and numerical models to better detect, characterize and locate cancerous tissue in the breast and hence guide interventions to completely remove cancerous lesions, while minimizing side-effects, scarring, and disfigurement.

Our introduction to breast imaging was via Professor Martin Leach and colleagues at the Institute of Cancer Research in London. At that time he was developing protocols for dynamic contrast enhanced (DCE) MRI for screening high-risk women (Griebsch et al., 2006). He set us the challenge to align the dynamic MRI sequences to compensate for patient motion during the acquisition of MR volumes of the breast taken at approximately one volume per minute. Patients lie prone in an MRI scanner with their breasts within a specially designed breast RF Coil. Visualization and the quality of the dynamic analysis of the scans can be severely affected by patient movement and respiratory motion. We devised a three-dimensional (3D) B-spline based non-rigid registration algorithm based on maximizing mutual information (Rueckert et al., 1999). This algorithm subsequently proved to be robust in a wide range of diverse medical imaging applications, such as cardiac imaging, lung imaging, and building statistical atlases of the brain from MR images of multiple individuals. The algorithm was subsequently extended to incorporate volume preservation (Tanner et al., 2000). The quality of the DCE MR images of the detected lesions appeared excellent, with a significant reduction in motion artefacts in images created by subtraction of the unenhanced images from the maximally enhanced images. Despite the visual appearance of high-quality subtraction images, we lacked an effective independent and quantitative validation of the registration accuracy. We had the idea that if we could artificially deform MR images of the breast to simulate motion and deformation, we could then create a set of realistically deformed images to assess how accurately our algorithm could retrieve the imposed deformations, allowing us to derive a registration quality metric. This motivated our initial interest in numerical models to provide realistic deformations of the breast.

Our original non-rigid registration algorithm did of course have a model of breast deformation implicitly embedded within it in the form of the 3D cubic B-spline interpolant used to construct the trial deformed image from the displacement of the B-spline node points. Though this produced a smooth deformation, it had no basis in tissue mechanics. We therefore devised a simple deformation system based on a linear elastic finite element model to provide a set of deformed breast images for validation purposes (Schnabel et al., 2003). Our motivation was to devise a set of displacements that approximately encompass those found in-vivo. We were not trying to predict accurately a displacement vector field in the deforming breast of a particular patient. With this technology, we were then able to provide a metric of non-rigid registration algorithm performance and, most importantly, conduct experiments to fine tune our registration algorithm. We showed that including a volume conserving constraint dramatically improved the performance of the original registration algorithm. This made sense as the breast is highly unlikely to undergo volume changes during the dynamic contrast MRI examination. We were confident that we could compensate for tissue deformation and motion with a mean target registration accuracy (TRE) of $\sim$1 mm, which was well within the tolerance of the application and was comparable to the acquisition resolution of the dynamic MRI data. We also showed that registration improves the quality of pharmacokinetic parameter estimation from a dynamic MRI sequence, using an extension to the algorithm incorporating progressive principle component registration (Melbourne et al., 2007).

Our next challenge, posed by the breast radiologists we were collaborating with, was to better align the data contained within a standard compression X-ray mammogram and the corresponding MRI examination. The clinical motivation was two-fold: to improve geometric alignment to better characterize a potential lesion on both MRI and X-ray, and secondly to better locate lesions in one modality given a visible lesion in the other, for example, to improve the accuracy of placement of a biopsy needle.

Our specification was to provide alignment accurate to within 10 mm when projecting a feature from a 3D MR volume, with the patient prone, to a 2D compressed X-ray mammogram. The consensus of our collaborating radiologists was that 10 mm would be sufficient to provide a useful clinical tool. We explored simple transformations to estimate the deformation of the breast in the compressed X-ray mammogram, including the affine transformation (Mertzanidou et al., 2012). We then constructed a numerical model of the breast and devised a registration strategy that simultaneously optimized the parameters of the registration and of the physical model. As the computational cost of running commercial finite element packages would have been impractical in such an optimization scheme, we used a fast explicit finite element solver that runs on a graphics card, enabling the FEM-based transformation model to be fully integrated into the optimization scheme (Taylor et al., 2009). We assessed the system on 10 clinical cases and achieved a mean registration error of 11.6 mm $\pm$ 3:8 mm for the cranio-caudal x-ray mammographic view and 11 mm $\pm$ 5:4 mm for the medial lateral oblique view, indicating that this could be a useful clinical tool (Mertzanidou et al., 2014).

Other avenues of breast cancer research pursued in our laboratory included the investigation of stromal patterns associated with cancer risk. There may be patterns that might predict cancer risk in the tumor and peritumoral microstructure. MRI provides information on these patterns of tissue microstructure via diffusion imaging. We have extended our MRI work on microstructural models for diffusion MRI originally developed for prostate cancer to breast cancer. We examined microstructural models for diffusion MRI in breast cancer and showed correlations with stromal patterns in histology (Bailey et al., 2016). We also developed multiscale models of tumor growth based on the effect of collagen biomechanics (Wijeratne et al., 2016). We hypothesized that the resulting patterns might be associated with breast cancer stroma maturity, and hence cancer risk. We then devised a method for automated classification of breast cancer stroma maturity, that might be related to these collagen patterns, from histological images (Reis et al., 2017). To integrate this work, we developed methods for relating 3D MRI images to mastectomy histology (Mertzanidou et al., 2017). Work continues on modeling breast tissue and cancer microstructure to explore relationships with cancer risk and to devise imaging signals that may be detectable in diagnostic imaging.

16.3 Towards surgical guidance

As described above, MRI can be acquired with the patient supine to provide an estimate of the breast anatomy and lesion location to guide surgery. Unfortunately, the breast, especially in the older, post-menopausal patient, is very deformable and the supine image lacks anatomical detail on lesion location. One solution is to acquire the prone image, and then map the lesion location and outline to the supine image.

Over several years, we developed increasingly accurate and more robust methods of aligning prone and supine MRI of the female breast as input to surgical planning. An initial implementation of the concept of aligning prone MRI with the surgical scene was presented by Carter et al. (2008). A hybrid registration technique was proposed, which uses a biomechanical model to initialize an intensity-based non-rigid registration. Although a registration accuracy of 5 mm was reported on the 3 patient datasets examined, the method lacked sufficient rigor and robustness, but feasibility of the concept was demonstrated.

We were encouraged by discussions with Professor Mo Keshtgar, breast surgeon at the Royal Free Hospital, to explore technologies that might improve on the current rate of approximately 20% of lumpectomy cases that require re-excision as the tissue sample obtained had positive margins. After a lumpectomy, the sample is examined to ensure that no cancer cells are detected within a certain distance, e.g., 2 mm for ductal carcinoma in-situ (DCIS) (The American Society of Breast Surgeons, 2018), of the outer edge of the specimen. If cancer cells are discovered in this region, this is designated a positive margin, usually resulting in further surgery, with the associated pain and distress for the patient. One proposal is that location of the cancer in the surgical, supine position, projected from imaging collected in the prone position, might improve surgical planning, and hence outcome. We were also encouraged to develop technology to model the wound healing process after lumpectomy with the aim of using this information to improve planning and cosmetic outcome.

To this end, we first developed a more robust biomechanical approach to align prone and supine breast MRI (Han et al., 2014) to improve surgical planning. This paper proposed a simultaneous optimization procedure for both the material parameters of breast tissues within a finite element framework and the reversing of the direction of the gravitational force. Overall, the proposed algorithm achieved a registration performance on fiducial markers (target registration error, 8.44 mm $\pm$ 5.5 mm for 45 fiducial markers) and higher overlap rates on segmentation propagation of fibroglandular tissues (Dice similarity score value $>$ 82%). Despite these promising results, further evaluation showed that this method lacked robustness across a wide range of breast shapes and sizes and relied on a physically unrealistic model of tissue deformation that failed to allow for the pre-stressing of the breast tissue, primarily due to the force of gravity.

This chapter describes in more detail the extension of this work by Eiben et al. (2016), in which we developed a more physically realistic system using the concept of the unloaded state of the breast and applied gravity in each direction from this unloaded state to simulate the position of the breast, prone and supine. This method is described in more detail below.

As lumpectomy has become more effective at completely removing the cancerous tissue, achieving long term cure rates, there is now greater emphasis on optimizing the cosmetic effects of surgery. The shape of the female breast after surgery depends both on the insult of the surgical excision itself and the remodeling of the tissue that occurs during the healing phase. To provide a means of estimating the cosmetic effect of surgery we proposed an initial attempt at a modeling system that would predict the appearance after tissue removal and healing. This should allow improved communication between surgeon and patient, including explaining the pay-off between more radical surgery and more optimal cosmetic outcome. In this chapter, we describe this preliminary work and demonstrate its feasibility, as presented in Vavourakis et al. (2016).

16.4 Symmetric biomechanically guided prone-to-supine breast image registration for surgical and radiotherapy planning

This section is derived from the paper by Eiben et al. (2016). More details of the precise formulation and experiments are provided in the paper, here we outline the main aims, methods used, and results achieved. This work aimed to provide a prone-supine registration approach that (i) simultaneously considers pre-stresses in the initial patient position, (ii) allows constrained motion of the breast tissue along the chest wall, (iii) uses a biomechanically constrained deformation model, (iv) optimizes the material parameters, and (v) incorporates image information to correct residual misalignment.

A schematic of the method is provided in Fig. 16.1. First the prone and supine images are segmented into skin surface, chest wall (including the pectoral muscle), and the adipose and glandular components of the breast to enable building biomechanical models, for which we use a neo-Hookean material description for the different tissue classes. This uses only two material coefficients, the Lamé parameters: λ and μ (Bonet and Wood, 2008). Eder et al. (2014) studied breast deformation using finite element simulation from prone MRI to the patient standing and showed that material relations reported by Tanner et al. (2006) and Rajagopal et al. (2008b) using a simple neo-Hookean material constitutive relationship produced the most accurate simulations. Motion along the chest wall is considered by constraining the retro-mammary region to lie on the chest wall. During the simulation, internal forces act as a regularizer for displacements parallel to the surface, while small displacements are applied perpendicular to the surface to enable a sliding-type motion. These 3D finite difference models are then used to simulate the gravity free states from both the prone and the supine positions. The iterative unloading procedure uses a forward biomechanical simulation, during which the unloaded reference configuration is updated by utilizing the same vector field as the simulation but in the sampling (pull) direction. As the literature provides a very wide range of possible material properties, our simulation starts with a too-stiff estimate and lowers the material stiffness repeatedly. This

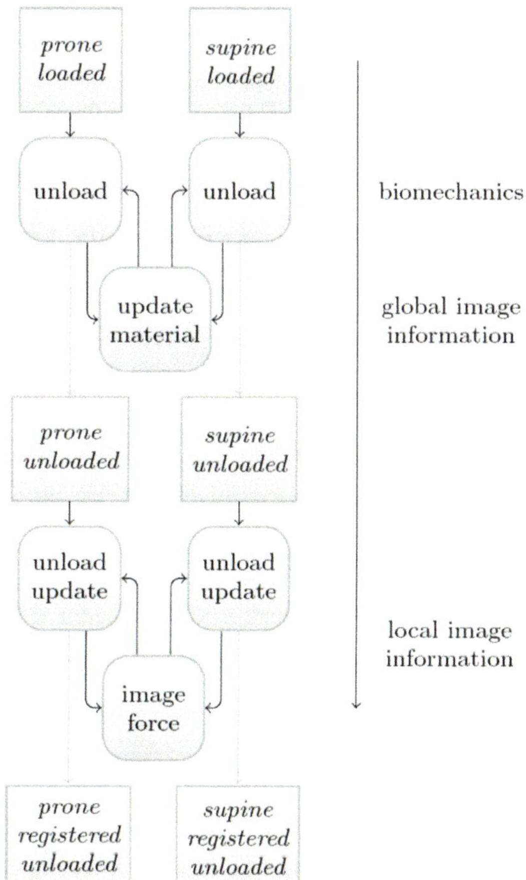

FIGURE 16.1

Schematic of method for symmetric biomechanically guided prone-to-supine breast image registration for surgical and radiotherapy planning. In a first step, the effect of gravity is removed from the prone and supine breast image assuming generic material parameters. In a second step, the material parameters are repeatedly updated until the image similarity no longer improves. The final step involves activation of image-derived forces, which aim to correct modeling inaccuracies and generate the final aligned images in the unloaded configuration. The figure is reproduced from Eiben et al. (2016).

process continues until the sum of square of differences (SSD) is minimized between the resulting image intensities of the estimated unloaded configurations. Finally, local forces are added from the SSD measurement to account for residual misalignment between the unloaded states, which resembles an iterative registration of the resulting resampled images, where the underlying hyperplastic material law acts as an implicit regularizer.

The method was tested on a digital phantom generated from simple geometric models with a cylinder representing the chest wall and a Gaussian function used to generate the outline of the skin surface of the breast. The breast tissue was represented by texture mapping a patient's MRI data into the breast volume. Tests with the digital phantom showed that the mean target registration error (TRE) reduced from 19.3 mm in the unregistered simulated data to 0.9 mm after gravity unloading and image registration, as described above.

The method was also evaluated retrospectively on 10 patient datasets. For 9 of these patient datasets, MR images were acquired for diagnostic purposes with the patient in the prone position, and CT images were acquired post lumpectomy with the patient supine to plan post-surgery radiotherapy. For a subset of 4 of these patients, an additional supine MR volume was collected. The 10th patient had both prone and supine CT, post-surgery. For the first 9 patients, all analyses were done on the contralateral breast to that on which the lumpectomy had been performed. In both modalities, image contrast within the breast was provided by the difference between adipose and non-adipose tissue, and so a simple intensity inversion and rescaling was sufficient, prior to computation of the sum of squares of differences, in the image registration component of the algorithm. Registration was assessed by manually identifying between 8 and 14 landmarks on the internal structures of the breast for each case. To assess the quality of the landmark identification a second observer was presented with the prone landmarks only and asked to identify the same landmarks in the supine image. The overall mean inter-observer distance between these two sets landmarks was 7.1 mm before and 3.8 mm after the exclusion of the outliers. Outliers were defined as an inter-observer distance greater than 10 mm. More details of acquisition, pre-processing, and the inter-observer assessment of landmark identification are provided in Eiben et al. (2016). Landmark identification is a difficult, painstaking, and time-consuming task, not recommended for routine clinical image analysis, but valuable to evaluate an algorithm's performance against a human observer, which is often regarded as the gold-standard. Using these landmarks, the initial mean TRE across all datasets ranged from 38.8 mm to 133.1 mm, reducing to an overall mean TRE of 5.9 mm with a range between patients from 5.3 mm to 9.6 mm. For the 4 prone to supine MR datasets, the mean TRE was reduced to 3.6 mm. Fig. 16.2 provides views of orthogonal slices through the initial images and the registration results for the patients with the lowest and highest target registration error.

16.5 Multiscale mechano-biological finite element modeling of oncoplastic breast surgery: numerical study towards surgical planning and cosmetic outcome prediction

This section is based on the paper by Vavourakis et al. (2016). In this contribution, we proposed a three-dimensional finite element procedure that simulates both breast tissue deformations due to surgery, i.e., lumpectomy, and physiological wound healing.

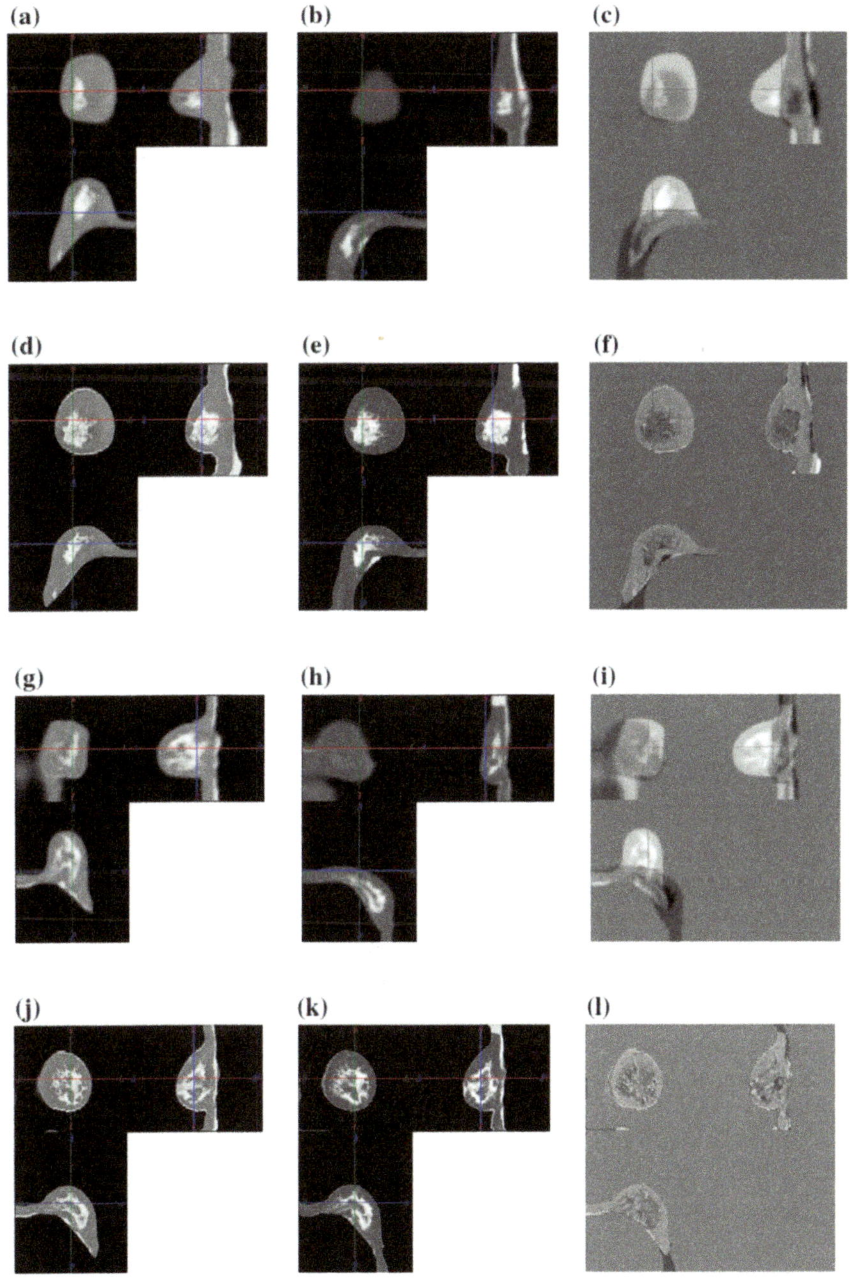

FIGURE 16.2

Provides views of orthogonal slices through the initial images and the registration results for the two (of ten) patients with the lowest and highest target registration error using the symmetric biomechanically guided prone-to-supine breast image registration method. The first column shows the prone, the second the supine, and the third the corresponding difference images. The first two rows show a patient for which a landmark distance of 4.4 mm was measured, whereas the lower two rows show results for a patient where a mean landmark distance of 9.6 mm was measured after registration. The figure is reproduced from Eiben et al. (2016).

Our proposed numerical methodology comprises a multiscale mathematical model for tissue recovery and angiogenesis, together with a model for breast tissue deformation.

Though there have been attempts to model outcomes of breast surgery, these have been restricted to planning and evaluation of breast implants (Georgii et al., 2014), two-dimensional approximations (Garbey et al., 2013), or 3D models focusing on breast organ deformation (Kuhlmann et al., 2013; Eder et al., 2014). The work described here is, to the best of our knowledge, the first to provide a detailed modeling of the surgical resection and wound healing, all within a highly deformable structure, the female breast. The model follows Sherratt & Murray and Javierre et al. (2009); it encompasses key processes involved in physiological wound healing, i.e., tissue remodeling and angiogenesis. Coupled reaction-diffusion equations model the balance between breast tissue cell density, the microvascular network density, the concentration of biochemical cues that control cell mitosis, and angiogenesis, as well as the oxygen saturation and nutrients level. Governing equations are discretized using finite elements, while time-integration is accomplished using an explicit numerical scheme.

Using quantities related to a reference configuration of the breast geometry, equilibrium of the breast organ/tissue, subjected to finite deformations, is described through the linear momentum equation. The internal breast tissue is modeled as a Mooney–Rivlin, nearly incompressible medium, whereas the breast skin is approximated by an incompressible, hyperelastic constitutive law that accounts for large deformations. Stresses within the breast are calculated as the superposition of the corresponding passive stresses in the tissues owing to mechanical deformations and the active stresses effectively appearing in wound contraction during tissue remodeling and recovery.

Fig. 16.3 shows a flow-diagram of the 3D FE (breast organ deformation/transformation, angiogenesis, wound healing and contraction) procedure that integrates with a surgical planning computer-based tool to annotate "digitally" the excision of the lesion during lumpectomy. The prone MRI is segmented into pertinent tissue types and a labeled volumetric FE mesh is created, which forms the basis of the biomechanical model. From the prone configuration of the "digital patient" model, the unloaded state is computed, from which re-application of gravity simulates the breast shape for the patient supine and standing upright prior to surgery. Surgical planning, i.e., specifying the resected lesion and margins, is performed on the computer in the supine pose and results are then transformed into the upright pose for presentation to the end-user for cosmetic outcome assessment. The pre- and post-surgical simulations of the "digital patient" can then be compared to reveal the predicted deformation of the breast due to surgery, i.e., lumpectomy.

To communicate the predictions to the patient, her breasts were scanned with the patient in the standing position prior to surgery, using a 3D optical scanner. The pre-surgery, standing-upright configuration is rigidly registered to the post-op optical scan and residual discrepancies are removed by warping the optically derived surface. The simulation of the wound recovery is assessed by rigidly registering the resulting

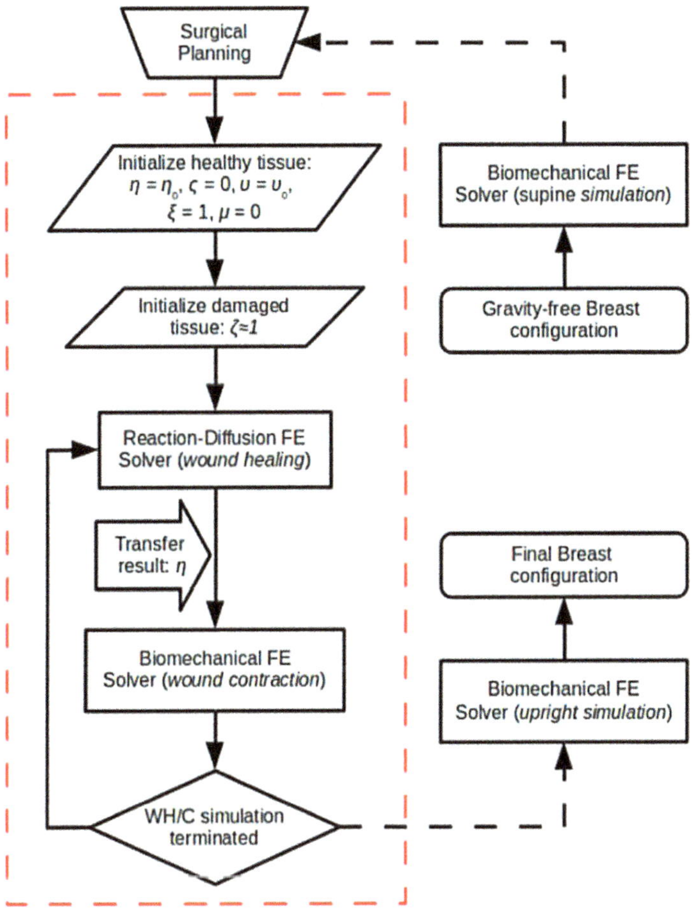

FIGURE 16.3

Flowchart of the multiscale finite element solver to predict surgical outcome. The red-dash line frames the mechano-biological wound healing, angiogenesis, and wound contraction numerical methodology (where WH/C is wound healing/contraction). The figure is reproduced from Vavourakis et al. (2016).

"digital" post-surgery breast surface (in standing position) with that of the patient surface scan 3 months after surgery. The surgical simulation system accuracy was tested against retrospective data on 4 patients, with the residual differences in surface position shown in Fig. 16.4. The surgical excision was estimated by the collaborating surgeon (MK), whereas the outcome of wound healing and tissue recovery after lumpectomy is predicted from the simulator. The results were also compared with the optical scans taken between 6 and 12 months. A photorealistic rendering uses the image texture from the optical scan superimposed on the predicted mesh. A color

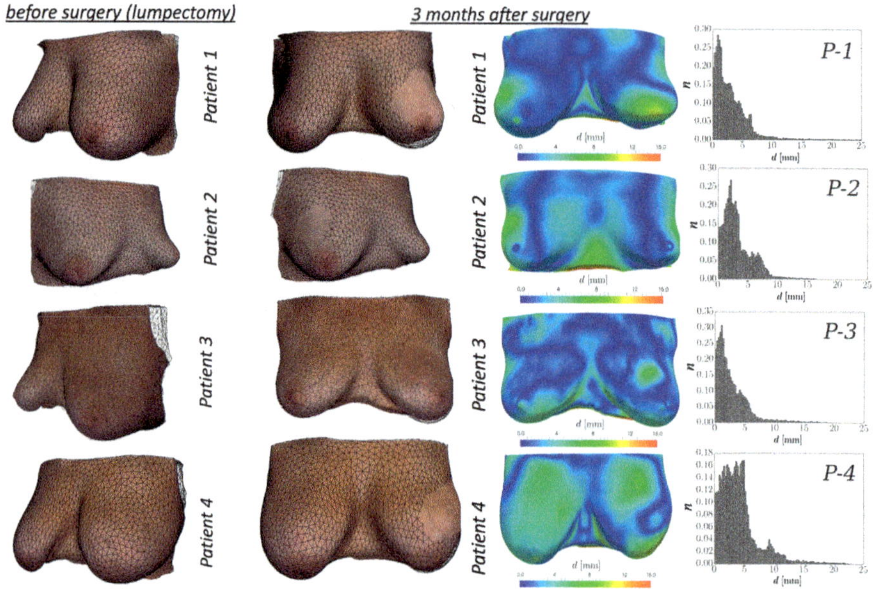

FIGURE 16.4

Comparison of simulated breast shape (with texture) before and 3 months after surgery. Baseline biomechanical model (wireframe) with fitted upright surface scan that visualizes the skin and oreola texture before surgery (far left column). In view of the recorded surgical plan, the surgical simulation tool predicted the breast shape following lumpectomy (middle left column). The breast deformations resulted from the in-silico analysis are then projected onto the original surface scan to visualize the textured post-surgical breast shape. In the middle right column, the breast shape 3 months after lumpectomy—predicted by the surgical simulation tool—is shown. The color map represents the measured distance between the rigidly aligned follow-up scan and the simulated surgical outcome. On the far right column, the histogram to the normalized frequency of occurrence against measured distance is presented. (Vavourakis et al., 2016.)

map of the discrepancies and histograms of the distances between predicted and actual breast surface is illustrated on the right of Fig. 16.4. The measured mean surface distance is reported at approximately 4 mm max, which includes errors due to the surgical simulation, the gross prone to unloaded to upright simulation, population-based material parameter selection, and external factors (such as weight gain or loss). Some discrepancies are noted, e.g., in one patient the actual breast deformity is more pronounced than that of the simulation, and the change of the orientation of the left nipple from anterior to inferior was not accurately predicted by the simulation. In another patient the simulator predicted a slightly larger influence of surgery in the upper inner quadrant of the left breast than actually occurred.

16.6 Conclusion

This chapter summarizes our work over the last 2 decades to develop image registration and tissue modeling technologies to better characterize breast cancer, better select appropriate treatment and, if surgery is recommended, to improve the accuracy of the surgery, while optimizing both cancer control and cosmetic outcome.

In principle, a similar system could also be used for radiotherapy planning, where a mapping of the presurgical anatomy of the breast after surgery is required to estimate the tumor bed location, as radiotherapy is usually applied after lumpectomy.

The method for symmetric biomechanically guided prone-to-supine breast image registration was shown to reduce the mean TRE across 9 prone MR and supine CT datasets and one prone CT to supine CT dataset from an overall mean TRE of 69.7 mm to 5.6 mm, with a range between patients from 5.3 mm to 9.6 mm. For the 4 prone to supine MR datasets the mean TRE was 3.6 mm. The largest discrepancies occur at the boundary of the lateral edge of the breast, an area likely to be of less interest clinically. Though we decided to use simple neo-Hookean material constitutive relations, this is clearly an approximation, and Mira et al. (2018), for example, demonstrated its limitations on one volunteer imaged in the prone, supine, and supine tilted through 45 degrees. Nevertheless, the figures for registration accuracy are within the tolerance of 10 mm specified by our surgeon as being likely to be of clinical benefit in planning surgery and subsequent radiotherapy.

The method for finite element modeling of oncoplastic breast surgery showed that our method for modeling wound healing coupled with a model of soft tissue deformation in the breast could predict the overall shape of the breast after wound healing following lumpectomy. A retrospective study was undertaken on data from 4 patients. The skin surface was computed for the patient standing with breasts unsupported and compared with the actual surface reconstructed from a 3D optical scanner 6 months to 12 months after surgery. The mean difference between the simulated surface predicted by the modeling and the optically scanned surface ranged between 2.8 mm and 4.1 mm. Although this is well within tolerances required to inform the patient of the likely cosmetic outcome, this is only a preliminary study on a small number of patients, so must be interpreted with caution. The study did show some discrepancy in nipple position, implying some "sliding" of the skin in this region. The effect of surgery was underestimated in one case and overestimated in another, although the overall simulation was judged as acceptable.

Our original aim was to build a complete surgical planning system for lumpectomy to assist the surgeon's confidence in achieving cancer-free margins at lumpectomy, and thus reducing the likelihood of repeat surgery. Coupling this with a realistic system for predicting the cosmetic outcome of lumpectomy would enable the surgeon, in consultation with their patient, to establish an acceptable balance between risk of needing additional surgery and cosmetic appearance after wound healing.

All the components of such a system have been shown to perform to the original clinical specification, but they have yet to be put together into a complete clinical system. The complete system would also need to comply with appropriate regulatory requirements and would require careful health technology assessment to provide economic justification for the investment required to complete development and roll out a clinical system. This would then need to undergo appropriate validation and clinical trial to prove efficacy and patient benefit. As a step in this direction, we developed a low-cost vision-based system to assess the shape and visual appearance of the breast in the out-patient environment (Lacher et al., 2019).

Elsewhere Conley et al. (2015) have presented their biomechanical model-assisted image guidance system using supine MR. They use a 3D linear elastic model in their registration after rigid alignment to fiducial markers and ultrasound images of the chest wall. They reported accuracy in locating the tumor centroid on the two patients examined as being 5.5 mm and 5.3 mm, respectively, which is sufficient to be of clinical relevance. Babarenda Gamage et al. (2019) have reported development of a complete, automated workflow for diagnosis and image guidance, including compliance with data access requirements for clinical data. They model breast deformations using an incompressible, hyperelastic, neo-Hookean constitutive relation implemented within the OpenCMISS open-source computational modeling software package (Bradley et al., 2011). They propose a robotic tissue indenter to measure breast stiffness and an optical system to measure skin strains. The system also incorporates a U-Net architecture for learning segmentations.

Meanwhile Bessa et al. (2020) have devised a pragmatic approach based on attaching a number of MR compatible, optically visible markers to the breast surface. The system is based on their surface registration algorithm incorporating free-form deformation. This was tested on 16 patients and in their experiments performed as well as more complex systems based on biomechanical modeling. The same group have presented a Hololens-based display that might be used with such a system to provide the surgeon with augmentation of the surgical view showing the lesion location (Gouveia et al., 2021).

As stated at the beginning of the chapter, it is clear that a complete, arbitrarily accurate biomechanical model, applicable to all sizes, shapes, and composition of the female breast, has yet to be devised and validated. There is still much we cannot predict as to how the breast deforms and how internal structures move relative to each other, change shape, and change composition. However, for specific clinical applications, there may well be pragmatic solutions that can provide sufficient accuracy for surgical planning and navigation, so as to dramatically improve the current, relatively poor rate of achieving a complete excision, while minimizing the likelihood of poor cosmetic outcomes.

As seen above, there are now a significant number of breast models all tested on different synthetic and clinical data, making comparison difficult. A strong recommendation would be to generate a number of standard breast datasets from clinical MRI that have been accurately segmented and use this in a computational challenge to evaluate different approaches and implementations. Such challenges have had a significant impact in other application areas, such as image registration and lung segmentation.

Dedication

This chapter is dedicated to the memory of Professor Mo Keshtgar (1962–2017), Breast Cancer Surgeon at the Royal Free Hospital in London. Mo inspired the work described in this chapter. He performed the first keyhole surgery for breast cancer in the UK and strove hard to improve his patients' experience. He clearly articulated the problems to be solved and enthusiastically supported our efforts to bring advanced imaging and numerical modeling to address these problems. He is sorely missed.

Acknowledgments

The work presented in this chapter was supported by funding from the European Community's Virtual Physiological Human Seventh Framework Programme under grant agreement "PICTURE" (FP7-ICT-2011-9, 600948) and "VPH-PRISM" (FP7-ICT-2011-9, 601040); the EPSRC Programme Grant in Intelligent Imaging, EP/H0404610; EPSRC Grant "MIMIC": EP/K020439/1; EPSRC grants EP/N013220/1, EP/N022750/1, EP/N027078/1 and NS/A000027/1; Philips Research Laboratories Hamburg; the Wellcome/EPSRC Centre for Interventional and Surgical Sciences (WEISS) at UCL (203145Z/16/Z); the Department of Health via the National Institute for Health Research (NIHR) Comprehensive Biomedical Research Centre (168) award to University College Hospital NHS Foundation Trust in partnership with University College London; and Cancer Research UK and EPSRC via the Comprehensive Cancer Imaging Centre Grant C1519/A10331 awarded to University College London and King's College London; VV was funded by a Marie-Curie Fellowship (FP7-PEOPLE-2013-IEF, 627025).

References

Alderliesten, T., Loo, C., Paape, A., Muller, S., Rutgers, E., Peeters, M.-J.V., Gilhuijs, K., 2010. On the feasibility of MRI-guided navigation to demarcate breast cancer for breast-conserving surgery. Medical Physics 37 (6), 2617–2626. https://doi.org/10.1118/1.3429048.

Babarenda Gamage, T.P., Malcolm, D.T.K., Maso Talou, G., Mira, A., Doyle, A., Nielsen, P.M.F., Nash, M.P., 2019. An automated computational biomechanics workflow for improving breast cancer diagnosis and treatment. Interface Focus 9, 20190034. https://doi.org/10.1098/rsfs.2019.0034.

Babarenda Gamage, T.P., Rajagopal, V., Nielsen, P.M.F., Nash, M.P., 2011. Patient-specific modeling of breast biomechanics with applications to breast cancer detection and treatment. In: Gefen, A. (Ed.), Patient-Specific Modeling in Tomorrow's Medicine. In: Studies in Mechanobiology, Tissue Engineering and Biomaterials, vol. 09. Springer, Berlin, Heidelberg.

Bailey, C., Siow, B., Panagiotaki, E., Hipwell, J.H., Mertzanidou, T., Owen, J., Gazinska, P., Pinder, S.E., Alexander, D.C., Hawkes, D.J., 2016. Microstructural models for diffusion MRI in breast cancer and surrounding stroma: an ex vivo study. NMR in Biomedicine 30 (2), e3679. https://doi.org/10.1002/nbm.3679.

Bessa, S., Gouveia, P.F., Carvalho, P.H., Rodrigues, C., Silva, N.L., Cardoso, J.S., Oliveira, H.P., Cardoso, M.J., 2020. 3D digital breast cancer models with multimodal fusion algorithms. Breast 49, 281–290.

Bonet, J., Wood, R.D., 2008. Nonlinear Continuum Mechanics for Finite Element Analysis, 2nd ed. Cambridge University Press, Cambridge.

Bradley, C., et al., 2011. OpenCMISS: a multi-physics and multi-scale computational infrastructure for the VPH/Physiome project. Progress in Biophysics and Molecular Biology 107, 32–47. https://doi.org/10.1016/j.pbiomolbio.2011.06.015.

Carter, T., Tanner, C., Beechey-Newman, N., Barratt, D., Hawkes, D., 2008. MR navigated breast surgery: method and initial clinical experience. In: Metaxas, D., Axel, L., Fichtinger, G., Székely, G. (Eds.), Medical Image Computing and Computer-Assisted Intervention – MICCAI 2008. In: Lecture Notes in Computer Science, vol. 5242. Springer, Berlin, Heidelberg.

Carter, T.J., Tanner, C., Crum, W.R., Beechey-Newman, N., Hawkes, D.J., 2006. A framework for image-guided breast surgery. In: Yang, G.Z., Jiang, T., Shen, D., Gu, L., Yang, J. (Eds.), Medical Imaging and Augmented Reality. MIAR 2006. In: Lecture Notes in Computer Science, vol. 4091. Springer, Berlin, Heidelberg.

Conley, R.H., Meszoely, I.M., Pheiffer, T.S., Weis, J.A., Yankeelov, T.E., Miga, M.I., 2014. Image to physical space registration of supine breast MRI for image guided breast surgery. In: Proceedings of SPIE 9036, Medical Imaging 2014: Image-Guided Procedures, Robotic Interventions, and Modeling, pp. 90362N-1–90362N-8.

Conley, R.H., Meszoely, I.M., Weis, J.A., Pheiffer, T.S., Arlinghaus, L.R., Yankeelov, T.E., Miga, M.I., 2015. Realization of a biomechanical model-assisted image guidance system for breast cancer surgery using supine MRI. International Journal of Computer Assisted Radiology and Surgery 10, 1985–1996. https://doi.org/10.1007/s11548-015-1235-9.

Eder, M., Raith, S., Jalali, J., Volf, A., Settles, M., Machens, H-G., Kovacs, L., 2014. Comparison of different material models to simulate 3-D breast deformations using finite element analysis. Annals of Biomedical Engineering 42, 843–857. https://doi.org/10.1007/s10439-013-0962-8.

Eiben, B., Vavourakis, V., Hipwell, J.H., Kabus, S., Buelow, T., Lorenz, C., Mertzanidou, T., Reis, S., Williams, N.R., Keshtgar, M., Hawkes, D.J., 2016. Symmetric biomechanically guided prone-to-supine breast image registration. Annals of Biomedical Engineering 44, 154–173. https://doi.org/10.1007/s10439-015-1496-z.

Garbey, M., Salmon, R., Thanoon, D., Bass, B.L., 2013. Multiscale modeling and distributed computing to predict cosmesis outcome after a lumpectomy. Journal of Computational Physics 244, 321–335. https://doi.org/10.1016/j.jcp.2012.08.002.

Gavaghan, D.J., Whiteley, J.P., Chapman, S.J., Brady, J.M., Pathmanathan, P., 2008. Predicting tumor location by modeling the deformation of the breast. IEEE Transactions on Biomedical Engineering 55 (10), 2471–2480. https://doi.org/10.1109/TBME.2008.925714.

Georgii, J., Eder, M., Bürger, K., Klotz, S., Ferstl, F., Kovacs, L., Westermann, R., 2014. A computational tool for preoperative breast augmentation planning in aesthetic plastic surgery. IEEE Journal of Biomedical and Health Informatics 18 (3), 907–919. https://doi.org/10.1109/JBHI.2013.2285308.

Gouveia, P.F., Costa, J., Morgado, P., Kates, R., Pinto, D., Mavioso, C., Anacleto, J., Martinho, M., Simões Lopes, D., Ferreira, A.R., Vavourakis, V., Hadjicharalambous, M., Silva, M.A., Papanikolaou, N., Alves, C., Cardoso, F., Cardoso, M.J., 2021. Breast cancer surgery with augmented reality. Breast 56, 14–17. https://doi.org/10.1016/j.breast.2021.01.004.

Griebsch, I., Brown, J., Boggis, C., Dixon, A., Dixon, M., Easton, D., Eeles, R., Evans, D.G., Gilbert, F.J., Hawnaur, J., Kessar, P., Lakhani, S.R., Moss, S.M., Nerurkar, A., Padhani,

A.R., Pointon, L.J., Potterton, J., Thompson, D., Turnbull, L.W., Walker, L.G., Warren, R., Leach, M.O., 2006. Cost-effectiveness of screening with contrast enhanced magnetic resonance imaging vs X-ray mammography of women at a high familial risk of breast cancer. British Journal of Cancer 95, 801–810. https://doi.org/10.1038/sj.bjc.6603356.

Han, L., Hipwell, J.H., Eiben, B., Barratt, D., Modat, M., Ourselin, S., Hawkes, D.J., 2014. A nonlinear biomechanical model based registration method for aligning prone and supine MR breast images. IEEE Transactions on Medical Imaging 33 (3), 682–694. https://doi.org/10.1109/TMI.2013.2294539.

Hipwell, J.H., Vavourakis, V., Han, L., Mertzanidou, T., Eiben, B., Hawkes, D.J., 2016. A review of biomechanically informed breast image registration. Physics in Medicine and Biology 61, R1–R31. https://doi.org/10.1088/0031-9155/61/2/R1.

Janzekovic, J., Hunt, J., Peltz, T., Wagels, M., Brown, T., Hutmacher, D.W., 2022. Biomechanical principles of breast implants and current state of research in soft tissue engineering for cosmetic breast augmentation. Aesthetic Plastic Surgery 46, 1–10. https://doi.org/10.1007/s00266-021-02559-y.

Javierre, E., Moreo, P., Doblaré, M., García-Aznar, J.M., 2009. Numerical modeling of a mechano-chemical theory for wound contraction analysis. International Journal of Solids and Structures 46 (20), 3597–3606. https://doi.org/10.1016/j.ijsolstr.2009.06.010.

Kuhlmann, M., Fear, E.C., Ramirez-Serrano, A., Federico, S., 2013. Mechanical model of the breast for the prediction of deformation during imaging. Medical Engineering & Physics 35 (4), 470–478. https://doi.org/10.1016/j.medengphy.2012.06.012.

Lacher, R.M., Vasconcelos, F., Williams, N.R., Rindermann, G., Hipwell, J., Hawkes, D.J., Stoyanov, D., 2019. Nonrigid reconstruction of 3D breast surfaces with a low-cost RGBD camera for surgical planning and aesthetic evaluation. Medical Image Analysis 53, 11–25. https://doi.org/10.1016/j.media.2019.01.003.

Lee, A.W.C., Schnabel, J.A., Rajagopal, V., Nielsen, P.M.F., Nash, M.P., 2010. Breast image registration by combining finite elements and free-form deformations. In: Martí, J., Oliver, A., Freixenet, J., Martí, R. (Eds.), Digital Mammography. IWDM 2010. In: Lecture Notes in Computer Science, vol. 6136. Springer, Berlin, Heidelberg.

McGhee, D.D., Steele, J.R., 2020. Breast biomechanics: what do we really know? Physiology 35, 144–156. https://doi.org/10.1152/physiol.00024.2019.

Melbourne, A., Atkinson, D.D., White, M.J., Collins, D., Leach, M.O., Hawkes, D.J., 2007. Registration of dynamic contrast-enhanced MRI using a progressive principal component registration (PPCR). Physics in Medicine and Biology 52, 5147–5156.

Mertzanidou, T., Hipwell, J., Cardoso, M.J., Zhang, X., Tanner, C., Ourselin, S., Bick, U., Huisman, H., Karssemeijer, N., Hawkes, D.J., 2012. MRI to X-ray mammography registration using a volume-preserving affine transformation. Medical Image Analysis 16 (5), 966–975. https://doi.org/10.1016/j.media.2012.03.001.

Mertzanidou, T., Hipwell, J., Johnsen, S., Han, L., Eiben, B., Taylor, Z., Ourselin, S., Huisman, H., Mann, R., Bick, U., Karssemeijer, N., Hawkes, D.J., 2014. MRI to X-ray mammography intensity-based registration with simultaneous optimisation of pose and biomechanical transformation parameters. Medical Image Analysis 18 (4), 674–683. https://doi.org/10.1016/j.media.2014.03.003.

Mertzanidou, T., Hipwell, J.H., Reis, S., Hawkes, D.J., Bejnordi, B.E., Dalmis, M., Vreemann, S., Platel, B., van der Laak, J., Karssemeijer, N., Hermsen, M., Bult, P., Mann, R., 2017. 3D volume reconstruction from serial breast specimen radiographs for mapping between histology and 3D whole specimen imaging. Medical Physics 44 (3), 935–948. https://doi.org/10.1002/mp.12077.

Mira, A., Carton, A., Muller, S., 2018. A biomechanical breast model evaluated with respect to MRI data collected in three different positions. Clinical Biomechanics 60, 191–199. https://doi.org/10.1016/j.clinbiomech.2018.10.020.

Perez del Palomar, A., Calvo, B., Herrero, J., Lopez, J., Doblare, M., 2008. A finite element model to accurately predict real deformations of the breast. Medical Engineering & Physics 30 (9), 1089–1097. https://doi.org/10.1016/j.medengphy.2008.01.005.

Rajagopal, V., Lee, A., Chung, J.-H., Warren, R., Highnam, R.P., Nash, M.P., Nielsen, P.M.F., 2008b. Creating individual specific biomechanical models of the breast for medical image analysis. Academic Radiology 15 (11), 1425–1436. https://doi.org/10.1016/j.acra.2008.07.017.

Rajagopal, V., Nash, M.P., Highnam, R.P., Nielsen, P.M.F., 2008a. The breast biomechanics reference state for multimodal image analysis. In: Krupinski, E. (Ed.), Digital Mammography. In: Lecture Notes in Computer Science, vol. 5116. Springer, Berlin, pp. 385–392.

Ramiao, N.G., Martins, P.S., Rynkevic, R., Fernandes, A.A., Barroso, M., Santos, D.C., 2016. Biomechanical properties of breast tissue, a state-of-the-art review. Biomechanics and Modeling in Mechanobiology 15, 1307–1323. https://doi.org/10.1007/s10237-016-0763-8.

Reis, S., Gazinska, P., Hipwell, J.H., Mertzanidou, T., Naidoo, K., Williams, N., Pinder, S., Hawkes, D.J., 2017. Automated classification of breast cancer stroma maturity from histological images. IEEE Transactions on Biomedical Engineering 64 (10), 2344–2352. https://doi.org/10.1109/TBME.2017.2665602.

Rueckert, D., Sonoda, L.I., Hayes, C.C., Hill, D.L.G., Leach, M.O., Hawkes, D.J., 1999. Nonrigid registration using free-form deformations: application to breast MR images. IEEE Transactions on Medical Imaging 18 (8), 712–721. https://doi.org/10.1109/42.796284.

Schnabel, J.A., Tanner, C., Castellano-Smith, A.D., Degenhard, A., Leach, M.O., Hose, D.R., Hill, D.L.G., Hawkes, D.J., 2003. Validation of nonrigid image registration using finite-element methods: application to breast MR images. IEEE Transactions on Medical Imaging 22 (2), 238–247. https://doi.org/10.1109/TMI.2002.808367.

Siegler, P., Holloways, C.M.B., Causer, P., Thevathasan, G., Plewes, D.B., 2011. Supine breast MRI. Journal of Magnetic Resonance Imaging 34 (5), 1212–1217. https://doi.org/10.1002/jmri.22605.

Tanner, C., Schnabel, J.A., Chung, D., Clarkson, M.J., Rueckert, D., Hill, D.L.G., Hawkes, D.J., 2000. Volume and shape preservation of enhancing lesions when applying non-rigid registration to a time series of contrast enhancing MR breast images. In: Delp, S.L., DiGoia, A.M., Jaramaz, B. (Eds.), Medical Image Computing and Computer-Assisted Intervention – MICCAI 2000. In: Lecture Notes in Computer Science, vol. 1935. Springer, Berlin, Heidelberg, pp. 327–337.

Tanner, C., Schnabel, J.A., Hill, D.L.G., Hawkes, D.J., Leach, M.O., Hose, D.R., 2006. Factors influencing the accuracy of biomechanical breast models. Medical Physics 33, 1758. https://doi.org/10.1118/1.2198315.

Taylor, Z.A., Comas, O., Cheng, M., Passenger, J., Hawkes, D.J., Atkinson, D., Ourselin, S., 2009. On modelling of anisotropic viscoelasticity for soft tissue simulation: numerical solution and GPU execution. Medical Image Analysis 13 (2), 234–244. https://doi.org/10.1016/j.media.2008.10.001.

The American Society of Breast Surgeons, 2018. Consensus Guideline on Breast Cancer Lumpectomy Margins. https://www.breastsurgeons.org/docs/statements/Consensus-Guideline-on-Breast-Cancer-Lumpectomy-Margins.pdf.

Vavourakis, V., Eiben, B., Hipwell, J.H., Williams, N.R., Keshtgar, M., Hawkes, D.J., 2016. Multiscale mechano-biological finite element modelling of oncoplastic breast surgery—

numerical study towards surgical planning and cosmetic outcome prediction. PLoS ONE 11 (7), e0159766. https://doi.org/10.1371/journal.pone.0159766.

Wijeratne, P.A., Vavourakis, V., Hipwell, J.H., et al., 2016. Multiscale modelling of solid tumour growth: the effect of collagen micromechanics. Biomechanics and Modeling in Mechanobiology 15, 1079–1090. https://doi.org/10.1007/s10237-015-0745-2.

CHAPTER

A numerical model for prolapse surgery

17

Olivier Mayeur[a], Mathias Brieu[c], Pauline Lecomte-Grosbras[a], Jean-François Witz[a], and Michel Cosson[b]

[a] *Univ. Lille, CNRS, Centrale Lille, UMR 9013-LaMcube – Laboratoire de Mécanique, Multiphysique, Multiéchelle, Lille, France*

[b] *CHU Lille, Service de Chirurgie Gynécologique, Lille, France*

[c] *College of Engineering, Computer Science and Technology, Dept. of Mechanical Engineering, California State University, Los Angeles, CA, United States*

Abbreviations

CAD computer assisted design
FEM finite element method
FEA finite element analysis
POP pelvic organ prolapse

17.1 Introduction

Pelvic organ prolapse (POP) is a frequent pathology. We can estimate from the literature that this disease affects 60% of the over 60 years old women's population (DeLancey, 1992; Samuelsson et al., 1999; Olsen et al., 1997; Petros and Woodman, 2008; Weintraub et al., 2020). This pathology is characterized by hypermobility of the pelvic organs, causing discomfort, pain, and incontinence (Jelovsek et al., 2018); for more details, the reader can refer to Chapter 2. To treat and cure this pelvic static disorder, there are several techniques: some only with pelvic floor muscle training and rehabilitation (Radzimińska et al., 2018), for early stages; some in more critical POP stages require more invasive intervention with the use of surgical procedures, with or without prosthesis (Mourtialon et al., 2010); for more information, read Chapter 3. Unfortunately, in this last case, we observe a significant recurrence rate three years after the surgery (Qatawneh et al., 2013; Long et al., 2012). The factors of failure are varied. The quantification assessment of prolapse is well documented in the literature with consensus methods, such as POP-Q score (Bump et al., 1996; Bland et al., 1999) or more recent techniques (Gordon et al., 2019; Lamblin et al., 2016). Treatment is always based on a classic clinical examination. It is usually adapted to the patient's anatomy but rarely, if not never, to the material properties of involved soft tissues (DeLancey, 1992). Research on the development of new and more biofidelic prostheses for the cure of prolapse goes in this direction (Guenin et al., 2011; Chanda et al., 2018; Morch et al., 2021). However, few tools help evaluate surgical techniques and

Biomechanics of the Female Reproductive System: Breast and Pelvic Organs. https://doi.org/10.1016/B978-0-12-823403-7.00031-2

patient-specific benefits in this POP context (Jeanditgautier et al., 2016). These tools, linked to numerical simulation, would provide a significant response for improving POP's patient-specific treatment.

To perform personalized numerical simulation of the surgical treatment and, consequently, improve the quality of the treatment, it is first required to determine each patient's anatomy, taking into account their organs' geometrical specificities and supporting structures. Methods of generation of patient-specific geometric models based on medical imaging analysis are now robust when applied to the pelvic system (Venugopala Rao et al., 2010; Chen et al., 2013; Mayeur et al., 2014); see Chapter 11 for more information. In a second step, patient-specific simulation of treatment, well known in the scientific community, is completed using the finite elements method (FEM) (Chen et al., 2009; Venugopala Rao et al., 2010; Vallet et al., 2011; Jeanditgautier et al., 2016) and hyperelastic mechanical properties of pelvic soft tissue (Goh, 2002; Rubod et al., 2008; Sanches et al., 2009; Peña et al., 2010; Jean-Charles et al., 2010; Rubod et al., 2012; Chantereau et al., 2014; Elahi et al., 2019), see Chapters 6, 8, and 10 for added information. This method, combining image reconstruction and FEM simulations, tends more and more to be used in a medical context and adapt to the patient specificities (Wu et al., 2020). The development of tools for personalized diagnosis or treatment of pelvic static disorders requires the specific knowledge of geometry (Silva et al., 2017), mechanical properties of constitutive tissues (Barone et al., 2016), and mechanical loads applied to the system (Gordon et al., 2019), which are all strongly related to patient specificity (Mayeur et al., 2016) and pathology (Babayi et al., 2019). To improve the POP's surgical treatment, the numerical simulation may objectively evaluate pelvic system mobilities to evaluate the surgical correction' benefit (Gordon et al., 2019). The pathological mobilities are related to the geometrical aspects (patient anatomy) and tissues' mechanical properties (Mayeur et al., 2016).

This chapter aims to present a methodology to develop a FEM model of the POP system and analyze the different pelvic mobility models. The main objective of this chapter is to allow patient-specific comparison of surgical techniques aiming at a better choice of personalized treatment. To fulfill this aim, we will present the step-by-step methodology required to generate a patient-specific model considering geometrical and mechanical specificities and introduce the pathology and surgical technique models.

17.2 Material and methods

This chapter proposes a method to develop a patient-specific finite element analysis model to evaluate surgical treatment benefits. This construction of the model requires four main steps:

- Analyze medical images to generate a patient-specific model (see Chapter 11).

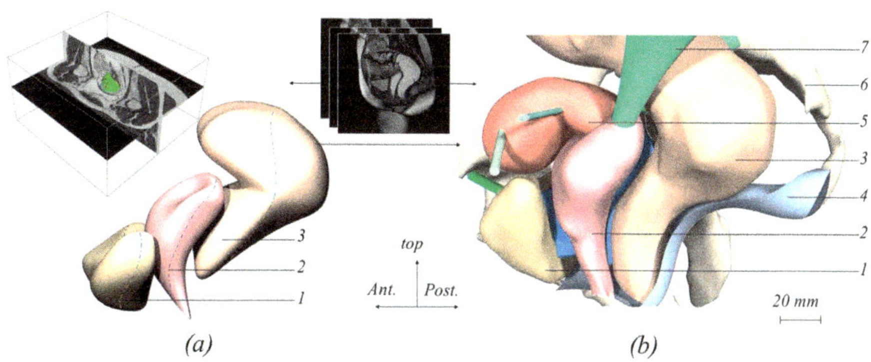

FIGURE 17.1

Patient-specific reconstruction of the pelvic system with 1. Bladder, 2. Vagina and 3. Rectum; (*a*) PS geometry from automatic methodology presented in Chapter 13 and (*b*) PS geometry from manual method to add other anatomical structures, needed for FE Model 4. Pelvic floor, 5. Uterus, 6. Sacrum and 7. Uterosacral ligaments.

- Generate a Finite Element Model (FEM) including mechanical properties (see Chapters 6, and 10), and mechanical loadings.
- Parameterize the FEM to account for the observed prolapse pathology (see Chapter 12).
- Implement surgical techniques on the model (see Chapter 3).

We will not detail the first step related to the patient-specific reconstruction of the geometric model of organs already described in Chapter 11. The sections that follow will focus on the next steps required to develop a FEM to evaluate the benefit of surgical treatment.

17.2.1 Generation of the patient-specific FEM

For instance, the patient-specific geometric models obtained with approaches developed in Chapter 11, in automatic way (Fig. 17.1a), must be compatible with a numerical simulation approach (Mayeur et al., 2016; Jiang et al., 2015). This representative anatomy, enriched with patient-specific mechanical properties (see Chapter 6) and mechanical loadings, will lead to generating a FEM. This will be used in finite element analysis (FEA) simulations, studying the influence of several significant parameters and having quantitative tools to evaluate the accuracy of specific surgery procedures for a particular patient. The different steps of the FEM generation and parameterization are described below.

Before generating a FEM and running FEA simulations, we must ensure that the geometry obtained with image reconstruction is suitable for FEA. The surfaces, being mathematically well defined, help guarantee the mesh's quality to the FEM. This step is performed from the manual protocol already published by Mayeur et al. (Mayeur

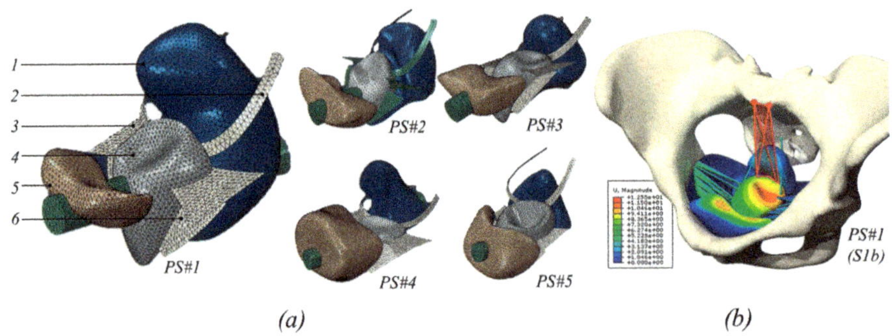

FIGURE 17.2

(*a*) Examples of patient-specific models created by protocol presented on Fig. 17.1 with 1. Rectum, 2. Uterosacral ligaments, 3. Cardinal ligaments, 4. Vagina, 5. Bladder, and 6. Paravaginal ligaments. (*b*) Illustration of the results (ABAQUS, displacement magnitude in mm) obtained for patient-specific surgical corrections PS#1. The surgery presented corresponds to S1b results on Fig. 17.8b.

et al., 2014; Jeanditgautier et al., 2016; Lamblin et al., 2016; Mayeur et al., 2019; Gatellier et al., 2020). This step consists in importing the geometry of organs into computer assisted design (CAD) software to reconstruct a geometrical model compatible with the FE method. Compared to the medical images, the generated geometries are smooth enough without edges and facets. This manual methodology allows us to add anatomical elements, such as suspension structures that are not directly reconstructed with classical thresholding methods (Fig. 17.1b).

17.2.1.1 Integration of non-observable anatomical structures: suspensions system

Once organs and the pelvic floor are generated, with image reconstruction, some anatomical elements of suspension that can barely not be observed on MRI must be added: fascia and ligaments. These structures play a crucial role in balancing pelvic statics (Chen et al., 2009; Vallet et al., 2011; Chen et al., 2013). Due to their complexity, some simplifying assumptions must be made to consider the main structures involved in the pelvic static. Their generation and modeling are performed directly with CAD tools on the geometry of organs and 3D reconstruction generated with image reconstruction methods. Concerning the ligaments, only the most impacted ones in the equilibrium of the pelvic statics are selected, namely the cardinal, paravaginal, and uterosacral ligaments. The generation of such structures is connected with the protocol published by Lamblin et al. and will allow us to understand the pathological mobilities (Lamblin et al., 2016). For the fascia, it is challenging to define their limits geometrically. They are created following this same protocol and allow us to better model the interactions between organs. Each structure is shown in Fig. 17.2.

The geometric model thus defined allows us to import these geometries into a finite element software, ABAQUS, to determine the finite element mesh. The mesh

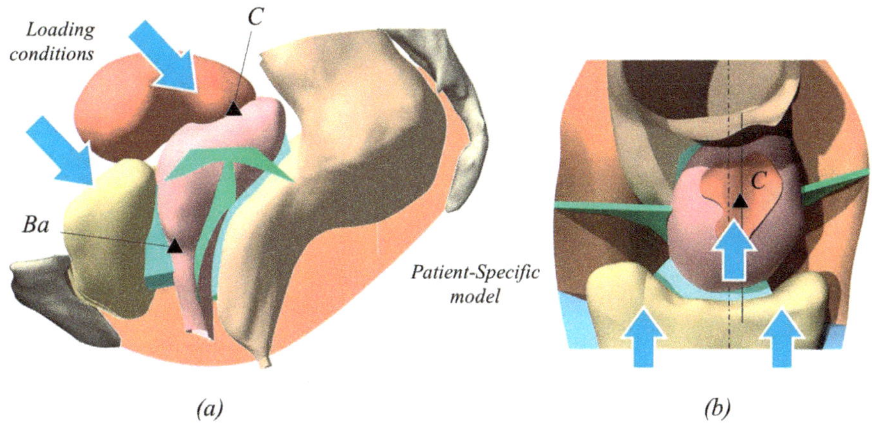

FIGURE 17.3

Patient-specific model showing the boundary conditions with applied pressure in blue (*a*) sagittal view and (*b*) axial view.

chosen is a linear triangular mesh. This last input is refined in the zones that will be highly deformed to prevent the hardening effect due to linear elements. The assembly of all the structures is considered with a tie contact formulation to constrain only translational degrees of freedom. The mesh size and refinement of elements are also carefully designed to ensure the convergence and stability of the calculation.

17.2.1.2 Boundary conditions

The boundary conditions applied to the system consist in on the one hand prescribed loads applied by the surrounding organs and their environment, and on the other hand to the prescribed displacement, fixed or not, imposed by the different anatomical structures surrounding the studied organs. We focus on the most critical case for the induced mobilities of organs: coughing. Zero displacement boundary conditions are applied to the non-displacing parts of the organs in the dynamic MRI analysis (Venugopala Rao et al., 2010; Vallet et al., 2011), also detailed in Chapter 12.

From the loading viewpoint, preliminary studies have shown that the diaphragm pushes on the organs below, with an orientation of 45° to the horizontal axis (Venugopala Rao et al., 2010), from front to back, as illustrated in Fig. 17.3. The intensity of such a surface effort is of the order of 10 kPa (Gordon et al., 2019). Last, the bladder and the rectum are assumed to be closed cavities containing an incompressible fluid.

17.2.1.3 Material properties of the constitutive tissues

The FEM also requires information on the mechanical properties of the connective tissue of organs and ligaments. Unfortunately, obtaining the mechanical properties of the pelvic floor of a specific patient is difficult with qualitative in vivo data (Hollenstein et al., 2013). Even if some method exists, some require already having a

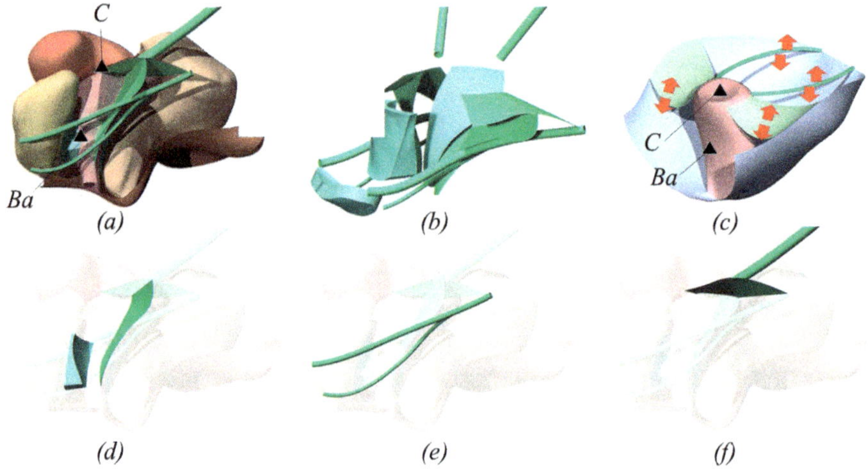

FIGURE 17.4

Parametric models of anatomical structures for pathology: (*a*) Anatomical model, (*b*) Suspension ligaments and fascia, (*c*) Schema of the parametric elongation for apical cystocele, (*d*) Illustration of selected parametric structures to mimic a median cystocele, (*e*) Lateral cystocele, and (*f*) Apical cystocele.

FEM model, such as those presented in Chapter 16 or using some under-development method, such as those shown in Chapter 15.

Other alternatives based on literature are proposed here. Post-mortem samples of pelvic tissues have been studied (Rubod et al., 2008), and some statistical data of the three organs (vagina, bladder, and rectum) show an isotropic hyperelastic behavior (Jean-Charles et al., 2010). This assumption is modeled by second-order Yeoh behavior (Chapter 10), where the free energy depends only on the first invariant I1 of the Cauchy–Green strain tensor. Such strain energy density introduces only two material parameters, C1 and C2, where C1 can be assumed as the initial stiffness, and C2 provides some information about the asymptotic stiffness of the material at large strains. Thus the properties of the FE model structures are based on a statistical study of the existing data available in the literature taking into account the property averages and dispersions (Mayeur et al., 2016). Mechanical properties of the soft tissues are then optimized, thanks to the non-destructive method present in a previous work (Mayeur et al., 2019).

17.2.2 Generation of a patient-specific FEM for pathology

Fig. 17.4a illustrates an anatomical model of a patient with manual creation of the suspension systems (Fig. 17.4b), both constituting a patient-specific physiologic FEM. It is possible to parameterize the patient-specific FEM to make it pathological. Indeed, the occurrence of the pathology is mainly due, according to the literature

(DeLancey, 1992; Petros and Woodman, 2008), to a geometric lengthening of suspension elements (ligaments and fascia). The analysis of the displacements observed with dynamic MRI during a thrust or static MRI at the end of the thrust can also make this assumption and helps to quantify the elongation of each ligament, see Chapter 12.

To transform a patient-specific FEM (PS model) that corresponds to the anatomy of the patient into a FEM (pathologic model) that matches, on the one hand with the anatomical characteristics and, on the other hand, with the patient-specific pathology, is proposed to modify the network of ligaments and fascia to mimic accurately the patient organs' mobilities that can be observed. This parameterization is based on a lengthening of the structures, controlled with B-spline curves and surfaces (Fig. 17.4c). We can modify these elongations simultaneously or independently the structure involved in the different theories (DeLancey, 1992; Petros and Woodman, 2008) to mimic different cases of cystocele (Fig. 17.4d, e, and f). We enriched the model initially created to make it correspond to a pathological situation (PS model). This pathology can also be linked to and characterized by the mobility analysis on dynamic MRI (Courtecuisse et al., 2020), the influence of the mechanical and geometrical parameters (Mayeur et al., 2016), or different criteria of study of the POP-Q score (Mayeur et al., 2014; Lamblin et al., 2016).

Using such a pathologic model initially to control geometrically, with independent parameters for each fascia and ligaments, the FEM model matches the patient's pathology. Several cases of POP can be outlined with variations of the structure's geometric parameters. We can model a median cystocele by modifying the dimensions of paravaginal ligaments and the fascia between bladder and vagina (Fig. 17.4d), a lateral cystocele with arcus tendinous fasciae pelvis (ATFP) and arcus tendinous levator ani structures (ATLA) (Fig. 17.4e), or an apical cystocele with modification of cardinal and uterosacral ligaments (Fig. 17.4f) (DeLancey, 1992; Petros and Woodman, 2008; Mayeur et al., 2014; Lamblin et al., 2016).

17.2.3 Generation of a surgical techniques model

The surgical treatment aims to reduce hyper mobility of organs while maintaining functionality; see Chapter 3 for more information. The different surgical techniques considered are detailed in Table 17.1 and illustrated in Fig. 17.5. Each of the surgical methods studied requires sutures or mesh implants. Both are effectively acting only when they are in tension (Morch et al., 2017). We assumed the constitutive material of sutures and prostheses was made of isotropic elastic material. The sutures are shared nodes between truss elements used to model the sutures or ligament modifications and the shell elements of the vagina. The implementation and description of the attachment areas are summarized in Table 17.1. The location of these sutures and the attachment points of the sacrospinous and other techniques are evaluated by the theory of surgical methods, and the location of the promontories and other distances into the pelvic are measured on MRI.

Such surgical techniques are expected to impact the patient's cervix mobility. It is thus likely to be efficient on patient-specific pathology that can be quantified. It

Table 17.1 Description of principal cure of POP with details of FEM modelization.

Surgical techniques	Type	Implementation	Description for FEM
(S1) Posterior and/or Anterior sacrocolpopexy	Mesh implant (Morch et al., 2017)	Promontory to cervix (posterior and/or anterior vaginal wall)	Creation of a custom prosthesis according to the patient's anatomy. Mesh implant in contact with the anterior vaginal wall (S1a), posterior (S1b), or both (S1c). Sutures on the cervix and wall.
(S2) Bilateral sacrospinous fixation	Mesh implant (Şentürk et al., 2015)	Sacrospinous ligament fixation (Anterior vaginal wall to sacrospinous ligament)	Creation of custom truss elements according to the patient's anatomy. Fixation of the implant on the cervix and the sacrospinous ligament. Distance from the ischial spine (IS) varies from 1 cm (S2a) to 3 cm (S2c).
(S3) Right and/or left post sacrospinous Fixation (Richter)	Implant (Cravello et al., 2008)	Sacrospinous right and/or left to posterior vaginal wall	Changed ligaments truss elements coordinates with locations equivalent to suture points for left (S3a), right (S3b), or both (S3c).
(S4) Anterior vaginal repair	Autologous (Boulanger et al., 2006)	Shortening of the vaginal wall with sutures	Reducing the anterior vaginal wall of 1 cm (S4a) or 0.5 cm (S4b).

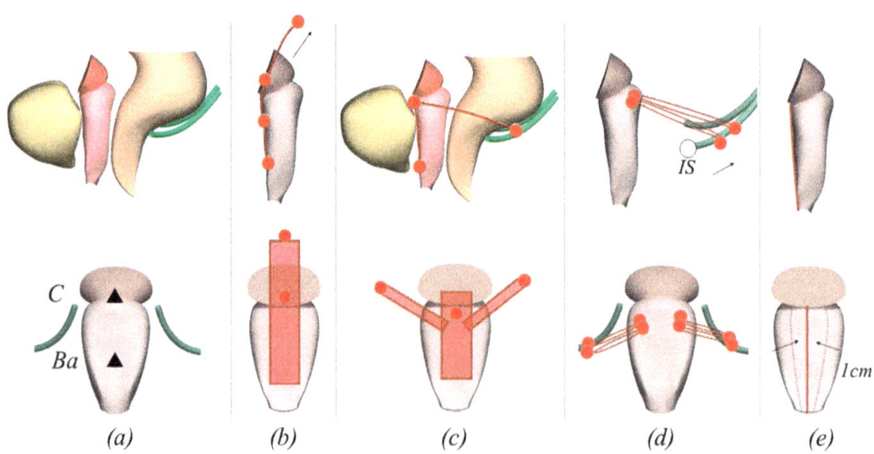

FIGURE 17.5

Numerical model for prolapse surgery (*a*) and example of surgical modelization with the different configuration presented on Table 17.1 such as (*b*) Surgery1 posterior and/or anterior sacrocolpopexy, (*c*) S2 Bilateral sacrospinous fixation, (*d*) S3 Right and/or left post Sacrospinous fixation (Richter), (*e*) S4 anterior vaginal repair.

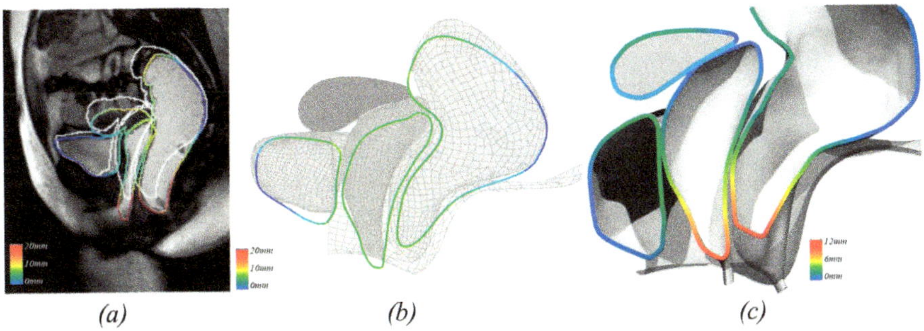

FIGURE 17.6

Validation of the patient specific model thanks to (*a*) evaluation of displacement field on dynamic MRI, (*b*) displacement on the same section into the FE model, and (*c*) gap between MRI and FE evaluations.

can be observed that strain varies in intensity and location concerning the surgical techniques considered.

The numerical twin of pathological model, developed for each specific patient will then be used to test every surgical technique detailed in Table 17.1, and the displacement of the patient's organs will be quantified to estimate the most efficient surgical techniques (Lamblin et al., 2016). The mechanical properties of implants are defined in Morch et al. (2017). To avoid bias in the study, the specific patient studied had similar POP, nature, and severity. The POP is characterized by a POP-Q score (Bump et al., 1996), which quantifies the displacement of a particular location of the pelvic system. However, their anatomy was different: one had a lateralized vagina, the other a symmetric position.

17.3 Results

17.3.1 Validation of the PS model

Pelvic organs' mobility are strongly influenced by the intra-abdominal pressure, the geometry of organs, and the mechanical properties of constitutive soft tissues that we can consider as input parameters of the numerical simulation (Chen et al., 2009; Gordon et al., 2019; Mayeur et al., 2016). To validate the PS model, every following step is based on it; it must compare the numerical results obtained to experimental mobility measured from dynamic MRI. A comparative method, defined by the region of interest, has been proposed (Mayeur et al., 2019). This study is based on acquiring an MRI of a healthy, without prolapse, volunteer. This step allowed us to assess the model's accuracy in predicting pelvic organ mobility.

Fig. 17.6 shows the result of a simulation of a PS model in the sagittal plane to be compared to a dynamic MRI. The left-hand side of Fig. 17.6 shows the results of the analysis of the dynamic MRI, showing the magnitude of the displacement for a

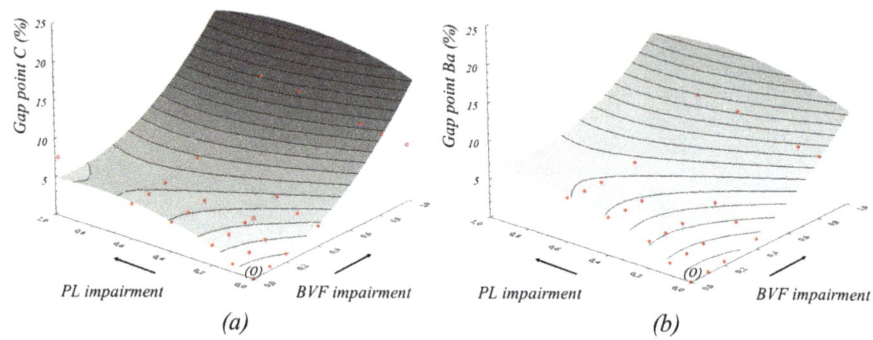

FIGURE 17.7

Surface representation of the gap between physiological and pathological mobility (geometrical impairments), (*a*) on the top of the bladder (POP-Q point C), (*b*) on the bottom of the bladder (POP-Q point Ba). The red dots correspond to the FE results.

specific patient. In contrast, the right-hand side of Fig. 17.6 shows the gap between the measures obtained from the MRI and the results obtained from the PS model. We can observe that difference between and computation is very low.

17.3.2 Validation of the PathoS model

While PS model aimed at predicting the physiological mobility of women, the PathoS model does not aim at predicting the effective pathology of a patient. It rather aims at simulating a realistic representation of the mobility of organs in a pathological situation, thanks to the parametric model of the supporting anatomical structures. We are interested in the mobility of the POP-Q score that allows having a precise localization of the points to be studied to quantify the grade (Persu et al., 2011). We selected the two most specific points for the study of cystocele, namely points C (cervix) and Ba (point B anterior). Although the complete prolapse at advanced grades with externalization of the vagina (POP-Q score at 3 or 4) remains complex with such an FE model of the entire pelvic system, we can reach large mobility of the cervix to correspond to the initiation of the POP occurrence. This allows increasing mobility up to a pathological displacement of organs and considering the simulation of surgical techniques. Such a model brings to study the influence of suspension structures on mobility considering the ligaments or fasciae impairment and elongation. A previous study was based on this approach to understanding pathological mobility better and, in particular, cystocele (Lamblin et al., 2016).

We change the geometric configurations for paravaginal ligaments, corresponding to 10, 20, 40, and 60% of elongation in this context. Equivalent evolution between 10 and 80% when applied to the fasciae between bladder and vagina (shear loading) causes cystocele prolapse (DeLancey, 1992). These results, Fig. 17.7, show a significant and non-negligible impact of ligament elongation and allow quantifying

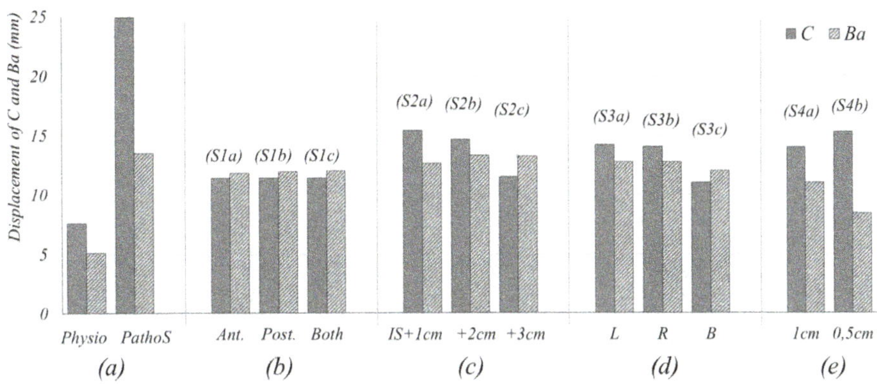

FIGURE 17.8

Mobility analysis on the points C and Ba. (*a*) Comparison between physiological and pathological mobility; (*b*) Comparison between anterior sacrocolpopexy, posterior, and both; (*c*) Bilateral sacrospinous fixation with position of the mesh at +1 cm, +2 cm and +3 cm of ischial spine, (*d*) Right and/or left Post SacroSpinous Fixation (Richter), (*e*) Anterior vaginal repair with a reduction of 0.5 cm and 1 cm.

the differences of mobility to make these models pathological (Mayeur et al., 2014; Lamblin et al., 2016; Jeanditgautier et al., 2016).

Surface representation can thus be considered to represent the influence of each suspension element on the increase of mobility (Fig. 17.7). The use of about twenty simulations on a given geometry, with separate or coupled parameterizations, highlights the importance of taking into account this failure of the suspension system in the analysis of mobility. This study also allowed us to compare models with different mechanical properties between young and old tissues (Chantereau et al., 2014). We thus find that mobility at the cervix is strongly impacted by this coupling between ligament elongation and the size of the fascia between the bladder and the vagina. According to the POP-Q stage, these results allowed us to better model the system's mobility in pathological situations and apply surgical corrections simulation.

Thus we can reach mobility of the cervix from an initial configuration of 7 mm (no failure, physiological configuration) to mobility of 25 mm at the cervix (median cystocele prolapse, pathological configuration), as we see in Fig. 17.8 of the next section. The PathoS model can thus be adapted to every patient-specific pathology.

17.3.3 Evaluation of surgical techniques

A study has been completed on a representative generic model to avoid bias due to patient anatomical specificity to analyze the influence of the surgical techniques. This model can be considered a mean model and has already been the subject of previous publications (Mayeur et al., 2016, 2019). It follows the same approach for modeling. The objective is to underline the differences between surgical treatments,

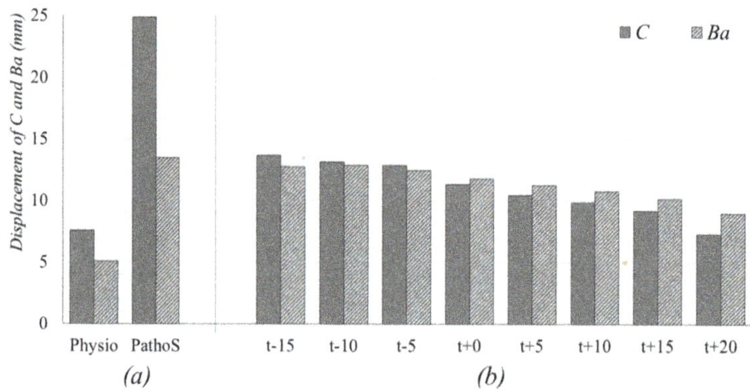

FIGURE 17.9

Mobility analysis on the points C and Ba. (*a*) Comparison between physiological and pathological mobility. (*b*) Influence of the length of the mesh sutured (from −15 mm to +20 mm), zero corresponding to the distance between cervix and promontory into the initial configuration (PS model from MRI when we install the mesh).

avoiding variations due to the geometry and specificity of the patient. The advantage of this model is that it reduces the dispersion of results due to anatomy differences, such as dissymmetry and organ volumes (e.g., bladder and rectum), and it can be used as an alternative to the PS model. The main anatomical structures of suspension underlying pathological parameters are identified and allow us to create mobility for the simulation close to the mobility observed in pathological situations. The different parameters influencing mobility mentioned above have been considered to model physiological and pathological mobility on this generic model.

The obtained mobility of the cervix in physiological conditions is at 7 mm against pathological cases, with an impairment of the sustainable structure, such as ligaments and fascias, at 25 mm. Displacements were analyzed in 2 areas: the cervix (point C) and anterior vaginal wall (point Ba).

The results obtained for the different surgical techniques show a significant decrease in the mobility of the organs compared to the initial displacements of a pathological situation (Fig. 17.8). These results can be evaluated as follows for a selected list of surgery described in Table 17.1.

Three different techniques were studied into this PathoS model:

- The Sacrocolpopexy techniques consist in suturing a mesh on vaginal walls, anterior and/or posterior, on one side and the sacral promontory on the other side. FE simulation reveals that such a technique reduces mobility concerning the generic pathological model (11.4 mm, versus 25 mm in pathological condition for point C). The effect of such techniques depends on the tension applied to the mesh, or in other words, the length of the mesh sutured (loosening or not, Fig. 17.9). A linear correlation is observed between the displacement of the anterior vaginal mesh, and the tension applied (Jeanditgautier et al., 2016). The shorter the anterior mesh at the promontory, the lower the mobility at the cervix and each point of the anterior vaginal wall. For a large shortening of the anterior mesh (20 mm), mobility was small: 7.3 mm at the cervix. These results are similar to those found for the generic physiological model (normal mobility).

- The bilateral sacrospinous fixation consists in suturing the top of the vagina to the sacrospinous ligament on one or both sides. The numerical simulation reveals the mobility decreased from 25 mm, in pathological situations, to 11.5 mm. The further the anchorage zone from the ischial spine (at IS + 3 cm), the lower the mobility at point C and others. This indicates a better apical correction in this configuration but less than the sacrocolpopexy. The simulation also suggests that if we get closer to the ischial spine by 2 cm and then 1 cm, the correction is reduced with mobility of C point at 15.4 and 14.6 mm, respectively.
- The right and/or left sacrospinous fixation (Richter) results showed mobility similar to the bilateral post sacrospinous fixation. Anatomical correction seemed to be better, with less mobility on the cervix (11.1 mm for this correction). There were no significant differences between left or right fixation for the various mobility points (14.1 mm on point C and 12.7 mm on Ba). These results with a unilateral or bilateral installation are to be deepened at the time of more specific simulation by considering the patient's asymmetry.
- The anterior vaginal repair without implant shows a less important surgical correction of the mobilities than for the other techniques. We can underline that the correction is better at a shortening of 1 mm with a mobility of 14 mm at point C against a mobility of 25 mm in pathology.

This study allows observing significant differences induced by different surgical techniques on cervical mobility. It also reveals that each surgical procedure can be optimized for a patient to control its efficiency. It must be mentioned that these results must be cautiously considered since they are model-dependent. It would be interesting to make this analysis considering the patient's specificities, such as her anatomical variations, asymmetry, or the global shape of the organs following the PS models.

17.3.4 Patient-specific model of surgical techniques

Given that the previous results presented allowed us to establish the PS and PathoS models accuracy as our ability to run simulations of surgical techniques on a generic model, we will consider evaluating surgical procedures on patient-specific pathological cases in the sequel. The first study is focused on the posterior and anterior sacropexy with different mesh shapes. This section summarizes the previously published study of Jeanditgautier et al. (2016). The MRI process validated the organ's mobility for this patient (see Fig. 17.6), and a parameterization was achieved to reach a pathological displacement of 22 mm. Based on this mobility, we selected the sacrocolpopexy techniques to study the impact of different parameters on the effectiveness of the surgical correction. Mobility, strains, and stresses results are analyzed according to the mesh size, the number of sutures, Fig. 17.9, or a comparison between the

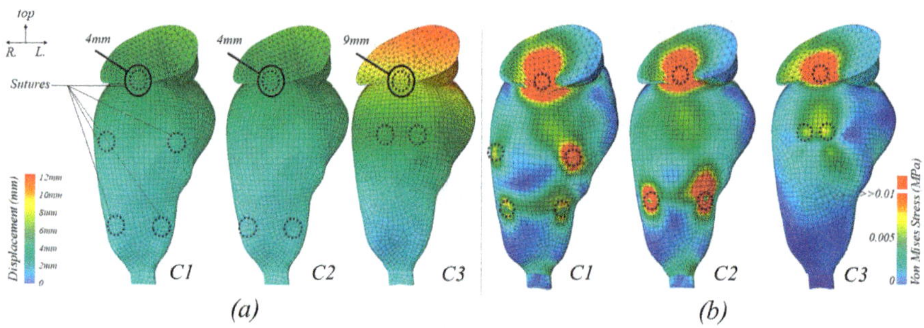

FIGURE 17.10

Patient-specific results for parametric surgery of the Anterior Sacrocolpopexy (S1). (*a*) Presentation of the mobility of the vagina, for different configurations, and (*b*) Corresponding von Mises stress distribution. Configuration C1: Large anterior mesh with four sutures on vaginal wall. Configuration C2: Large anterior mesh with two sutures. Configuration C3: Small anterior mesh with two sutures. The dark circle represents each suture.

anterior and posterior installation. More detailed studies can also be performed on these enriched models, such as the impact of the spacing between two sutures or the tension when shortening the length of the implant/mesh/prosthesis or creating additional tension to maintain the organs. See Fig. 17.10.

Analysis of the mobility of organs, according to the mesh size, revealed that an increase in the mesh size leads to a decrease in organ mobility, both in the anterior and posterior. A 9 mm displacement is observed at the cervix level following the use of a small anterior mesh. This displacement reaches 4 mm using the large mesh. A large anterior mesh and a large posterior one showed a slightly shorter shifting of 2.5 mm. In each case, this analysis of mobility based on the number of sutures on the vaginal wall has not revealed any significant differences in the displacement. A wide prosthesis can significantly reduce mobility but risks creating discomfort and potentially significant risk of failure. It is a compromise to have, and further studies are necessary to confirm these observations. However, it allows us to quantify the differences in results for the same technique.

A study of the von Mises' stresses at the suture level shows an essential concentration of stress in the vicinity of the cervix compared to the one applied on the vaginal wall. Whatever the number of sutures on the mesh and vaginal, this stress zone is 10 times higher than the one used on vaginal sutures. Results of stress levels are by mobility analysis. The lower the mobility, the higher the stresses on the sutures, and therefore the more sutures there are and the better the tear distribution. This conclusion remains the same on the other configuration, small and large mesh, both anterior and posterior. There is no significant impact on structure mobility following the change in the number of sutures. This method also highlights that spaces between sutures impact the stress analysis. With less than 2 cm between sutures, stress values are well distributed. After this 2.3 cm, ±0.3 range, stress increases strongly in a non-linear way (Jeanditgautier et al., 2016).

17.4 Discussion

Optimizing treatment towards a patient-specific treatment remains an open challenge that the numerical simulation may help achieve. Numerical simulation of POP treatment surgery is a field of study with many possibilities. However, many assumptions are made about such simulations using FEA. Such analysis allows us to consider several aspects of the simulation, such as the representative geometry of the patient-specific anatomy, the mechanical properties with realistic behavior, and the different surgical treatment modeling assumptions.

The generation of PS models has already been discussed in Chapter 11. However, the results focus on the visible structures, namely the organs. However, to estimate the mobility after surgical treatment, a more refined model, including a ligament and muscular support system, is required to allow a good representation of the mobilities. In the current chapter, every support system is manually added. It would be interesting to deepen these results to allow the generation of the environment of the organs in a more automated way. The validation of the model and its mobility with the image brings exciting elements to mimic the system's mobilities and help automate this phase.

As with ligaments, we have shown that it is possible to consider fascia and estimate their impact on mobility. From the modeling viewpoint, volumetric fascia models present three main concerns. One is related to the computation time due to local refinement to model thin fascia existing between organs, see Fig. 17.4b. The second one is due to the stiffening effect of the finite element mesh, particularly subject to shear loads. The last is that volumetric fascia meshes are challenging to automate when linking surface elements, from organs to the volumetric fascia elements. An alternative to this mesh approach is to replace the fascia's mesh with contact laws that can be envisioned. It is time-wise computation efficient and does not generate problems while shear motions occur between organs. However, this raises a new difficulty that deals with the characterization of the contract law (Diallo et al., 2021).

The generation of PS or PathoS models is based on the description of the patient-specific anatomy. The generation of such a patient-specific model, described in Chapter 11, induces several difficulties in terms of mesh generation and sometimes in artifact due to very local degeneration of FE meshes. To get rid of this complexity, most of the results of this chapter have been presented on the generic anatomic model, in which geometry was well defined and controlled and generated an FE mesh of quality. These generic models allowed to avoid dispersion of the results to patient specificities only and focus on the impact of a pathology or a surgical treatment. Consequently, the key results presented in this chapter must be considered cautiously. However, using a generic model gives significant numerical advantages, such as having smooth geometry and controlled FE mesh. Such generic models should continue to be considered to understand physiology or pathology better. Moreover, some research aims at morphing generic models, in which geometry and mesh quality are well controlled, to patient-specific anatomy.

The accuracy of the patient-specific simulation and the evaluation of the benefit of surgical treatment is also based on the relevance of the mechanical properties assigned to the numerical model. The results presented in this chapter were based on median mechanical properties for each organ, ligaments, and fascia obtained from destructive tests (Rubod et al., 2008; Jean-Charles et al., 2010; Chantereau et al., 2014; Mayeur et al., 2016). The use of nondestructive, in vivo techniques of characterization would significantly improve the accuracy of numerical simulations. The use of an inverse method, introduced in Chapter 5, or elastography, introduced in Chapter 14, may help to reduce the deviations due to the uncertainty made to define the mechanical properties for each organ or the contact lax between organs. However, non-destructive methods are still under development, and no clinical routine exists.

Out of the FEA and mechanical behavior considerations, the main assumptions made to develop the PathoS and surgical models are how pathology and surgical are modeled. Models of pathology were based on two theories developed in the literature (DeLancey, 1992; Petros and Woodman, 2008). The accuracy of the numerical model thus implicitly depends on the accuracy of the theories cited above. Some of the numerical models of surgery are based on descriptions available in the literature and described by medical doctors, see Chapter 3. In both cases, there are no fundamental objective evaluation tools to assess the accuracy of the model and the way it has been implemented for FEA. The approach proposed in this chapter to validate that the PS model consisted of comparing the numerical results of mobility to those observed with a dynamic MRI. It allows having confidence in the PS and PathoS models. To have the same confidence with the surgery model requires obtaining patients' post-surgery dynamic MRI.

This chapter presented an approach to developing patient-specific pathology and surgical treatment models. It emphasized the different steps to generate such models and the other vital parameters to be considered and their influences. Since physiology and pathology are still not well understood, numerical simulations are a powerful and helpful approach to improve the understanding of the POP genesis, the contribution of ligaments and fascia, and the influence of the quality of the tissues related to aging. And, as a consequence, the quality of treatment. The surgical model aims to study the treatment parameters: the influence of the prosthesis, its size, and the sutures. In each case—physiology, pathology, or surgical treatment—the presented results showed that FEA could provide quantitative pre-operative information to help surgeons to evaluate the pathology with information added to the medical imaging information obtained from MRI, such as the lengthening of the ligaments, and assess the most efficient treatment, such as the selection of the technique, the size of the mesh implanted, or the way it will be fixed. Thanks to FEA, this chapter shows that a quantitative comparison of prolapse surgeries is possible.

17.5 Conclusion

Finite element simulation is more and more used in various fields of surgery to evaluate what might be the consequence of specific surgical procedures to improve the quality of treatment. Women's pelvic surgery is not an exception and will benefit from numerical simulation. Although numerical models of pelvic mobility have existed for several years, the numerical simulation of surgical treatment and procedure is very recent. The development of such numerical tools opens great clinical opportunities and perspectives toward a patient-specific improved treatment.

Numerical simulation is increasingly accessible, and computation time is reducing every day. Such opportunities will allow surgeons to be rapidly informed. However, this requires complete automation of the processes. Such needs will requires new development that remains realistic and opens a future for a numerical planning and optimization of surgical treatment of POP.

References

Babayi, M., Azghani, M.R., Hajebrahimi, S., Berghmans, B., 2019. Three-dimensional finite element analysis of the pelvic organ prolapse: a parametric biomechanical modeling. Neurourol. Urodyn. 38 (2), 591–598. https://doi.org/10.1002/nau.23885. Epub 2018 Nov 29. PMID: 30499117.

Barone, William R., Moalli, Pamela A., Abramowitch, Steven D., 2016. Textile properties of synthetic prolapse mesh in response to uniaxial loading. Am. J. Obstet. Gynecol. (ISSN 0002-9378) 215 (3), 326.e1–326.e9. https://doi.org/10.1016/j.ajog.2016.03.023.

Bland, D.R., Earle, B.B., Vitolins, M.Z., Burke, G., 1999. Use of the Pelvic Organ Prolapse staging system of the International Continence Society, American Urogynecologic Society, and Society of Gynecologic Surgeons in perimenopausal women. Am. J. Obstet. Gynecol. 181 (6), 1324–1327. https://doi.org/10.1016/s0002-9378(99)70371-6. discussion 1327-8. PMID: 10601907.

Boulanger, L., Lucot, J.P., Boukerrou, M., Collinet, P., Cosson, M., 2006. Transvaginal surgery for genital prolapse in women aged over 80 years: results in 48 procedures. J. Gynecol. Obstet. Biol. Reprod. (Paris) 35 (7), 685–690.

Bump, R.C., Mattiasson, A., Bø, K., Brubaker, L.P., DeLancey, J.O., Klarskov, P., Shull, B.L., Smith, A.R., 1996. The standardization of terminology of female pelvic organ prolapse and pelvic floor dysfunction. Am. J. Obstet. Gynecol. 175 (1), 10–17. https://doi.org/10.1016/s0002-9378(96)70243-0. PMID: 8694033.

Chanda, A., Ruchti, T., Upchurch, W., 2018. Biomechanical modeling of prosthetic mesh and human tissue surrogate interaction. Biomimetics (Basel) 3 (3), 27. https://doi.org/10.3390/biomimetics3030027. PMID: 31105249; PMCID, PMC6352698.

Chantereau, P., Brieu, M., Kammal, M., Farthmann, J., Gabriel, B., Cosson, M., 2014. Mechanical properties of pelvic soft tissue of young women and impact of aging. Int. Urogynecol. J. 25 (11), 1547–1553. https://doi.org/10.1007/s00192-014-2439-1. Epub 2014 Jul 10. PMID: 25007897.

Chen, L., Ashton-Miller, J.A., DeLancey, J.O.L., 2009. A 3D finite element model of anterior vaginal wall support to evaluate mechanisms underlying cystocele formation. J. Biomech. 42 (10), 1371–1377. https://doi.org/10.1016/j.jbiomech.2009.04.043. Epub 2009 May 29. PMID: 19481208; PMCID, PMC2744359.

Chen, L., Ramanah, R., Hsu, Y., Ashton-Miller, J.A., Delancey, J.O., 2013. Cardinal and deep uterosacral ligament lines of action: MRI based 3D technique development and preliminary findings in normal women. Int. Urogynecol. J. 24 (1), 37–45. https://doi.org/10.1007/s00192-012-1801-4. Epub 2012 May 23. PMID: 22618207; PMCID, PMC3986864.

Courtecuisse, H., Jiang, Z., Mayeur, O., et al., 2020. Three-dimensional physics-based registration of pelvic system using 2D dynamic magnetic resonance imaging slices. Strain 56, e12339. https://doi.org/10.1111/str.12339.
Cravello, L., Agostini, A., Estrade, J.P., et al., 2008. Techniques, indications et résultats de la sacrospinofixation uni- et bilatérale. Pelv. Perineol. 3, 309–313. https://doi.org/10.1007/s11608-008-0212-y.
DeLancey, J.O.L., 1992. Anatomy aspects of vaginal eversion after hysterectomy. Am. J. Obstet. Gynecol. 166, 1717–1728. https://doi.org/10.1016/0002-9378(92)91562-O.
Diallo, M.N., Mayeur, O., Lecomte-Grosbras, P., Patrouix, L., Witz, J.F., Lesaffre, F., Rubod, C., Cosson, M., Brieu, M., 2021. Simulation of the mobility of the pelvic system: influence of fascia between organs. Comput. Methods Biomech. Biomed. Eng. 16, 1–15. https://doi.org/10.1080/10255842.2021.2001460. Epub ahead of print. PMID: 34783611.
Elahi, S.A., Connesson, N., Chagnon, G., et al., 2019. In-vivo soft tissues mechanical characterization: volume-based aspiration method validated on silicones. Exp. Mech. 59, 251–261. https://doi.org/10.1007/s11340-018-00440-9.
Gatellier, M.A., Dit Gautier, E.J., Mayeur, O., Brieu, M., Cosson, M., Rubod, C., 2020. Complete 3 dimensional reconstruction of parturient pelvic floor. J. Gynecol. Obstet. Hum. Reprod. 49 (1), 101635. https://doi.org/10.1016/j.jogoh.2019.101635. Epub 2019 Sep 6. PMID: 31499277.
Goh, J.T., 2002. Biomechanical properties of prolapsed vaginal tissue in pre- and postmenopausal women. Int. Urogynecol. J. Pelvic Floor Dysfunct. 13 (2), 76–79. https://doi.org/10.1007/s001920200019. discussion 79. PMID: 12054186.
Gordon, M.T., DeLancey, J.O.L., Renfroe, A., Battles, A., Chen, L., 2019. Development of anatomically based customizable three-dimensional finite-element model of pelvic floor support system: POP-SIM1.0. Interface Focus 9 (4), 20190022. https://doi.org/10.1098/rsfs.2019.0022. Epub 2019 Jun 14. PMID: 31263537; PMCID, PMC6597525.
Guenin, M.O., Wolfisberg, L., Kern, B., Stoll, L., Borbély, Y., von Flüe, M., 2011. Érosion rectale d'une prothèse pour cure de prolapsus vaginal (Prolift) (Rectal erosion caused by a prosthesis for treatment of vaginal prolapse (Prolift)). J. Obstet. Gynaecol. Can. 33 (1), 36–37. https://doi.org/10.1016/S1701-2163(16)34770-3 (in French). PMID: 21272434.
Hollenstein, M., Bugnard, G., Joos, R., Kropf, S., Villiger, P., Mazza, E., 2013. Towards laparoscopic tissue aspiration. Med. Image Anal. 17 (8), 1037–1045. https://doi.org/10.1016/j.media.2013.06.001. Epub 2013 Jun 19. PMID: 23876854.
Jean-Charles, C., Rubod, C., Brieu, M., Boukerrou, M., Fasel, J., Cosson, M., 2010. Biomechanical properties of prolapsed or non-prolapsed vaginal tissue: impact on genital prolapse surgery. Int. Urogynecol. J. 21 (12), 1535–1538. https://doi.org/10.1007/s00192-010-1208-z. Epub 2010 Sep 14. PMID: 20838989.
Jeanditgautier, E., Mayeur, O., Brieu, M., et al., 2016. Mobility and stress analysis of different surgical simulations during a sacral colpopexy, using a finite element model of the pelvic system. Int. Urogynecol. J. 27, 951–957. https://doi.org/10.1007/s00192-015-2917-0.
Jelovsek, J.E., Chagin, K., Lukacz, E.S., Nolen, T.L., Shepherd, J.P., Barber, M.D., Sung, V., Brubaker, L., Norton, P.A., Rahn, D.D., Smith, A.L., Ballard, A., Jeppson, P., Meikle, S.F., Kattan, M.W., 2018. NICHD pelvic floor disorders network. Models for predicting recurrence, complications, and health status in women after pelvic organ prolapse surgery. Obstet. Gynecol. 132 (2), 298–309. https://doi.org/10.1097/AOG.0000000000002750. PMID: 29995735; PMCID, PMC6060003.
Jiang, Z., Witz, J.-F., Lecomte-Grosbras, P., Dequidt, J., Duriez, C., Cosson, M., Cotin, S., Brieu, M., 2015. B-spline based multi-organ detection in magnetic resonance imaging. Strain 51, 235–247.

Lamblin, G., Mayeur, O., Giraudet, G., Jean Dit Gautier, E., Chene, G., Brieu, M., Rubod, C., Cosson, M., 2016. Pathophysiological aspects of cystocele with a 3D finite elements model. Arch. Gynecol. Obstet. 294 (5), 983–989. https://doi.org/10.1007/s00404-016-4150-6. Epub 2016 Jul 11. PMID: 27402504.

Long, Cheng-Yu, Lo, Tsia-Shu, Wang, Chiu-Lin, Wu, Chin-Hu, Liu, Cheng-Min, Su, Juin-Huang, 2012. Risk factors of surgical failure following transvaginal mesh repair for the treatment of pelvic organ prolapse. Eur. J. Obstet. Gynecol. Reprod. Biol. (ISSN 0301-2115) 161 (2), 224–227. https://doi.org/10.1016/j.ejogrb.2012.01.009.

Mayeur, O., Lamblin, G., Lecomte-Grosbras, P., Brieu, M., Rubod, C., Cosson, M., 2014. FE simulation for the understanding of the median cystocele prolapse occurrence. In: Bello, F., Cotin, S. (Eds.), Biomedical Simulation. ISBMS 2014. In: Lecture Notes in Computer Science, vol. 8789. Springer, Cham.

Mayeur, O., Witz, J.F., Lecomte, P., Brieu, M., Cosson, M., Miller, K., 2016. Influence of geometry and mechanical properties on the accuracy of patient-specific simulation of women pelvic floor. Ann. Biomed. Eng. 44 (1), 202–212. https://doi.org/10.1007/s10439-015-1401-9. Epub 2015 Jul 28. PMID: 26215307.

Mayeur, O., Witz, J.F., Lecomte-Grosbras, P., Cosson, M., Brieu, M., 2019. Patient-specific simulation: non-destructive identification method for soft tissue under large strain: application to pelvic system. In: Nielsen, P., Wittek, A., Miller, K., Doyle, B., Joldes, G., Nash, M. (Eds.), Computational Biomechanics for Medicine. Springer, Cham.

Morch, A., Doucède, G., Lecomte-Grosbras, P., Brieu, M., Rubod, C., Cosson, M., 2021. Pelvic organ prolapse meshes: can they preserve the physiological behavior? J. Mech. Behav. Biomed. Mater. 120, 104569. https://doi.org/10.1016/j.jmbbm.2021.104569. Epub 2021 May 23. PMID: 34058600.

Morch, A., Pouseele, B., Doucède, G., Witz, J.F., Lesaffre, F., Lecomte-Grosbras, P., Brieu, M., Cosson, M., Rubod, C., 2017. Experimental study of the mechanical behavior of an explanted mesh: the influence of healing. J. Mech. Behav. Biomed. Mater. 65, 190–199. https://doi.org/10.1016/j.jmbbm.2016.07.033. Epub 2016 Aug 10. PMID: 27589261.

Mourtialon, P., Delorme, E., Douvier, S., Marés, P., Sagot, P., Tayrac, R.D., 2010. Proposition d'une classification des techniques chirurgicales de cure de prolapsus (pelvic organ prolapse-surgical technique [POP-ST]) avec mise en place d'une prothèse par voie transvaginale. J. Gynecol. Obstet. Biol. Reprod. 39, 537–548.

Olsen, A.L., Smith, V.J., Bergstrom, J.O., Colling, J.C., Clark, A.L., 1997. Epidemiology of surgically managed pelvic organ prolapse and urinary incontinence. Obstet. Gynecol. 89 (4), 501–506. https://doi.org/10.1016/S0029-7844(97)00058-6. PMID: 9083302.

Peña, E., Calvo, B., Martínez, M.A., Martins, P., Mascarenhas, T., Jorge, R.M., Ferreira, A., Doblaré, M., 2010. Experimental study and constitutive modeling of the viscoelastic mechanical properties of the human prolapsed vaginal tissue. Biomech. Model. Mechanobiol. 9 (1), 35–44. https://doi.org/10.1007/s10237-009-0157-2. Epub 2009 May 6. PMID: 19418081.

Persu, C., Chapple, C.R., Cauni, V., Gutue, S., Geavlete, P., 2011. Pelvic organ prolapse quantification system (POP-Q) - a new era in pelvic prolapse staging. J. Med. Life 4 (1), 75–81. Epub 2011 Feb 25. PMID: 21505577; PMCID, PMC3056425.

Petros, P.E., Woodman, P.J., 2008. The integral theory of continence. Int. Urogynecol. J. Pelvic Floor Dysfunct. 19 (1), 35–40. https://doi.org/10.1007/s00192-007-0475-9. Epub 2007 Oct 30. PMID: 17968480.

Qatawneh, A., Thekrallah, F., Bata, M., et al., 2013. Risk factors of surgical failure following sacrospinous colpopexy for the treatment of uterovaginal prolapse. Arch. Gynecol. Obstet. 287, 1159–1165. https://doi.org/10.1007/s00404-012-2685-8.

Radzimińska, A., Strączyńska, A., Weber-Rajek, M., Styczyńska, H., Strojek, K., Piekorz, Z., 2018. The impact of pelvic floor muscle training on the quality of life of women with urinary incontinence: a systematic literature review. Clin. Interv. Aging 13, 957–965. https://doi.org/10.2147/CIA.S160057. PMID: 29844662; PMCID, PMC5962309.

Rubod, C., Boukerrou, M., Brieu, M., Jean-Charles, C., Dubois, P., Cosson, M., 2008. Biomechanical properties of vaginal tissue: preliminary results. Int. Urogynecol. J. Pelvic Floor Dysfunct. 19 (6), 811–816. https://doi.org/10.1007/s00192-007-0533-3. PMID: 18188492.

Rubod, C., Brieu, M., Cosson, M., Rivaux, G., Clay, J.C., de Landsheere, L., Gabriel, B., 2012. Biomechanical properties of human pelvic organs. Urology 79 (4), 968.e17–968.e22. https://doi.org/10.1016/j.urology.2011.11.010. Epub 2012 Jan 13. PMID: 22245302.

Samuelsson, E.C., Victor, F.T., Tibblin, G., Svärdsudd, K.F., 1999. Signs of genital prolapse in a Swedish population of women 20 to 59 years of age and possible related factors. Am. J. Obstet. Gynecol. 180, 299–305.

Sanches, P.R., Silva Jr, D.P., Müller, A.F., Schmidt, A.P., Ramos, J.G., Nohama, P., 2009. Vaginal probe transducer: characterization and measurement of pelvic-floor strength. J. Biomech. 42 (15), 2466–2471. https://doi.org/10.1016/j.jbiomech.2009.07.021. Epub 2009 Aug 4. PMID: 19656514.

Şentürk, M.B., Güraslan, H., Çakmak, Y., Ekin, M., 2015. Bilateral sacrospinous fixation without hysterectomy: 18-month follow-up. J. Turk. Ger. Gynecol. Assoc. 16 (2), 102–106. https://doi.org/10.5152/jtgga.2015.15220. PMID: 26097393; PMCID, PMC4456967.

Silva, M.E.T., Brandão, S., Parente, M.P.L., Mascarenhas, T., Natal Jorge, R.M., 2017. Biomechanical properties of the pelvic floor muscles of continent and incontinent women using an inverse finite element analysis. Comput. Methods Biomech. Biomed. Eng. 20 (8), 842–852. https://doi.org/10.1080/10255842.2017.1304542. Epub 2017 Mar 17. PMID: 28303730.

Vallet, A., Witz, J.-F., Rubod, C., Brieu, M., Cosson, M., 2011. Simulation of pelvic mobility: topology optimization of ligamentous system. Comput. Methods Biomech. Biomed. Eng. 14, 159–163.

Venugopala Rao, G., Rubod, C., Brieu, M., Bhatnagar, N., Cosson, M., 2010. Experiments and finite element modelling for the study of prolapse in the pelvic floor system. Comput. Methods Biomech. Biomed. Eng. 13 (3), 349–357. https://doi.org/10.1080/10255840903251270. PMID: 20099169.

Weintraub, A.Y., Glinter, H., Marcus-Braun, N., 2020. Narrative review of the epidemiology, diagnosis and pathophysiology of pelvic organ prolapse. Int. Braz. J. Urol. 46 (1), 5–14. https://doi.org/10.1590/S1677-5538.IBJU.2018.0581. PMID: 31851453; PMCID, PMC6968909.

Wu, Y., Hikspoors, J.P., Mommen, G., Dabhoiwala, N.F., Hu, X., Tan, L.-W., Zhang, S.-X., Lamers, W.H., 2020. Interactive three-dimensional teaching models of the female and male pelvic floor. Clin. Anat. 33, 275–285. https://doi.org/10.1002/ca.23508.

CHAPTER

18 Augmented reality biomechanical simulations for pelvic conditions diagnoses

Hadrien Courtecuisse[a] **and Jérémie Dequidt**[b]
[a]*AVR/MLMS ICube, CNRS, University of Strasbourg, Strasbourg, France*
[b]*Univ. Lille, Inria, CNRS, Centrale Lille, UMR 9189 CRIStAL, Lille, France*

18.1 Introduction

Digital tools are increasingly used for intraoperative assistance. Initially designed for training and learning, finite element (FE) simulations are now used and considered essential in the operating room. FE models can be used to display through augmented reality (AR) the internal structures (vessels, tumors, etc.) on top of the intraoperative images. An essential advantage of biomechanical models is to provide physics-based extrapolation, not just geometric, of organs' deformation, even in areas where few or no intraoperative data are available. Nevertheless, a significant difficulty of such simulations concerns the need for real-time computing without sacrificing accuracy. For this purpose, this chapter shows a generic constraint-based biomechanical framework allowing for both the simulations of complex interactions (such as friction, contacts, etc.) and impose displacements coming from medical image to correct modeling errors and maintain low errors with respect to the displacement of real-structures.

In the context of diagnosing women's pelvic pathologies, medical imaging is routinely used by clinicians, especially since modern MRI allows for dynamic scanning without any artifacts due to movement. For instance, the MRI MAGNETOM Aera SIEMENS 1.5 T allows getting 2.5 acquisitions per second of a given slice. This frequency is sufficient to capture the dynamic behavior of tissues during breathing cycles, but it is for now restricted to only one plan of acquisition, and given the absence of any volumetric information, some anatomical structures might be excluded. To overcome this limitation, we propose to describe in this chapter a method that will simulate a bio-mechanical model of the pelvic system, which will be registered on 2D dynamic MR images, allowing to extrapolate the 2D image to estimate off-plane motions and as a result to obtain a 3D motion of the pelvic system in augmented reality. This application will provide better knowledge and understanding of pathologies such as prolapsus or abnormal mobility of tissues by limiting the subjectivity

Biomechanics of the Female Reproductive System: Breast and Pelvic Organs. https://doi.org/10.1016/B978-0-12-823403-7.00032-4

induced by human perception when mentally inferring 3D motion from 2D data in the diagnoses.

To achieve this goal, the obvious requirement is to simulate the mechanical behavior of organs with high accuracy. However, due to patient inter- and intra-variability of the mechanical properties of anatomical structures, having a predictive simulation is not achievable. We propose to mechanically constrain the simulation with MR images by adding boundary conditions that will increase the accuracy of the whole method. With any medical application that involves mechanical simulation, a strong requirement exists for the computation time. Without meeting this requirement, the simulator cannot be used in an interactive context, regardless of the quality and accuracy of models. Therefore the underlying contributions of this chapter are a set of advanced numerical techniques, including modeling, constraints definition, and high parallelization strategies to meet these requirements.

Outline

The remainder of this chapter is organized as follows: Section 18.2 is dedicated to the description of numerical methods for fast simulation of deformable structures with contacts. Section 18.3 concerns the non-rigid registration of soft tissues with intraoperative data. Finally, in Section 18.4, we show the benefits of our approach for the registration of the pelvis organs using dynamic MRI images.

18.2 Simulation of deformable structures

Our methodology is based on the following general background of deformable simulations with contact using an implicit integration time.

18.2.1 FE models and constitutive law

Most of the contributions that will be introduced in this chapter are generic and do not depend on the constitutive model. However, we mainly rely on the linear co-rotational approach to model deformations. This model is extremely fast and not restricted to small deformations, thanks to precomputations.

Real-time implementations of nonlinear models exist. However, although more accurate, hyperelastic models usually raise stability conditions, which are difficult to enforce in a context where the model is subject to user interactions or driven by data extracted from medical images. Since stiffness matrices are precomputed using the rest configuration, the stability of corotational models does not depend on the current positions (at the price of the accuracy in highly deformed states). Since stability is also essential for medical simulation, the corotational model provides a balanced solution between accuracy and computation time.

18.2.1.1 Linear elastic model

Volume organs are meshed with linear tetrahedral elements composed of 4 nodes and 3 degrees of freedom (DOF) per node. The local 12×12 stiffness matrix $\mathbf{K}_e$ for a volume element e can be written with the synthetic formulation:

$$\mathbf{K}_e = \int_{V_e} (\mathbf{C}_e \mathbf{D}_e \mathbf{C}_e \, dV_e), \tag{18.1}$$

where $\mathbf{D}_e$ corresponds to the stress-strain matrix parametrized by the Young's modulus E and the Poisson's ratio ν:

$$\mathbf{D}_e = \frac{E}{(1+\nu)(1-2\nu)} \begin{pmatrix} (1-\nu) & \nu & \nu & 0 & 0 & 0 \\ \nu & (1-\nu) & \nu & 0 & 0 & 0 \\ \nu & \nu & (1-\nu) & 0 & 0 & 0 \\ 0 & 0 & 0 & \frac{1-2\nu}{2} & 0 & 0 \\ 0 & 0 & 0 & 0 & \frac{1-2\nu}{2} & 0 \\ 0 & 0 & 0 & 0 & 0 & \frac{1-2\nu}{2} \end{pmatrix}, \tag{18.2}$$

$\mathbf{C}_e$ is the strain-displacement matrix, which is employed to compute strains at any point inside the element, based on its nodal displacements:

$$\mathbf{C}_e = \frac{1}{6V} \begin{pmatrix} b_1 & 0 & 0 & b_2 & 0 & 0 & b_3 & 0 & 0 & b_4 & 0 & 0 \\ 0 & c_1 & 0 & 0 & c_2 & 0 & 0 & c_3 & 0 & 0 & c_4 & 0 \\ 0 & 0 & d_1 & 0 & 0 & d_2 & 0 & 0 & d_3 & 0 & 0 & d_4 \\ c_1 & b_1 & 0 & c_2 & b_2 & 0 & c_3 & b_3 & 0 & c_4 & b_4 & 0 \\ d_1 & 0 & b_1 & d_2 & 0 & b_2 & d_3 & 0 & b_3 & d_4 & 0 & b_4 \\ 0 & d_1 & c_1 & 0 & d_2 & c_2 & 0 & d_3 & c_3 & 0 & d_4 & c_4 \end{pmatrix}, \tag{18.3}$$

where V is the volume of the tetrahedral element. The method of *cofactors* can be used to invert the following 4×4 matrix and obtain the above coefficients:

$$\begin{pmatrix} 1 & x_1 & y_1 & z_1 \\ 1 & x_2 & y_2 & z_2 \\ 1 & x_3 & y_3 & z_3 \\ 1 & x_4 & y_4 & z_4 \end{pmatrix}^{-1} = \begin{pmatrix} a_1 & a_2 & a_3 & a_4 \\ b_1 & b_2 & b_3 & b_4 \\ c_1 & c_2 & c_3 & c_4 \\ d_1 & d_2 & d_3 & d_4 \end{pmatrix}, \tag{18.4}$$

where x_i, y_i, z_i are the nodal coordinates of the tetrahedra in the global coordinate system.

All the above matrices are constant and can be pre-computed since they depend only on the position of the mesh at the initial state. However, when large rotations occur, the linear approximation of the strain function leads to well-known inflation artifacts. Therefore a co-rotational formulation is mandatory to limit these artifacts.

18.2.1.2 Corotational formulation

The corotational formulation is widely used in computer graphics (Felippa, 2000) for its stability and the fast computation time possible, thanks to precomputations, and the model is not restricted to small displacements. For this purpose, the popular corotational method computes displacements in a rotated local coordinate system.

In each element, the local frame is represented by a rigid rotation matrix $\mathbf{R}_e$, and its transpose is used to align the deformed tetrahedron with its reference, undeformed shape. The forces $\mathbf{f}_e$ are computed in the rotated coordinate system, then transformed back to the world coordinates:

$$\mathbf{f}_e = \mathbf{R}_e \mathbf{K}_e \mathbf{R}_e^T \, \mathbf{u}_e, \tag{18.5}$$

where $\mathbf{u}_e = \mathbf{p} - \bar{\mathbf{p}}$ is the displacement vector between $\mathbf{p}$ and $\bar{\mathbf{p}}$, being, respectively, the current and initial positions.

The rotation matrix $\mathbf{R}_e$ associated to the element e is a 12×12 block-diagonal matrix computed as follows:

$$\mathbf{R}_e = \begin{bmatrix} \bar{\mathbf{R}}_e & & & \\ & \bar{\mathbf{R}}_e & & \\ & & \bar{\mathbf{R}}_e & \\ & & & \bar{\mathbf{R}}_e \end{bmatrix}, \tag{18.6}$$

where $\bar{\mathbf{R}}_e$ is a 3×3 rotation matrix. Several solutions have been proposed to compute the rotation matrix, for instance by computing a polar decomposition of the nodal positions (Nesme et al., 2005).

18.2.2 Implicit time integration

Equations used to model the dynamic behavior of bodies can be written within a synthetic formulation, given by Newton's second law:

$$\mathbf{M}\,\mathbf{a} = \mathcal{G}(t) - \mathcal{F}\,(\mathbf{p}, \mathbf{v}), \tag{18.7}$$

where $\mathbf{p} \in \mathbb{R}^n$ is the vector of generalized degrees of freedom (here, the mesh node positions), $\mathbf{M}$ is the mass matrix assumed constant. The function $\mathcal{G}$ gathers external forces at time t. $\mathcal{F}$ is a nonlinear function providing internal forces of the simulated object, depending on its current position $\mathbf{p}$. The internal forces are derived from the physics-based deformable model described above (i.e., corotational formulation).

Let us consider the time interval $[t_i, t_f]$, whose length is $h = t_f - t_i$. We have

$$\begin{aligned} \mathbf{M}(\mathbf{v}_f - \mathbf{v}_i) &= \int_{t_i}^{t_f} (\mathcal{G}(t) - \mathcal{F}(\mathbf{p}, \mathbf{v}))\, dt, \\ \mathbf{p}_f &= \mathbf{p}_i + \int_{t_i}^{t_f} \mathbf{v}\, dt. \end{aligned} \tag{18.8}$$

To evaluate integrals $\int_{t_i}^{t_f} (\mathcal{G}(t) - \mathcal{F}(\mathbf{p}, \mathbf{v}, t))\, dt$ and $\int_{t_i}^{t_f} \mathbf{v}\, dt$, we chose the following implicit Euler integration scheme:

$$\begin{aligned} \mathbf{M}(\mathbf{v}_f - \mathbf{v}_i) &= h\left(\mathbf{g}_f - \mathcal{F}(\mathbf{p}_f, \mathbf{v}_f)\right), \\ \mathbf{p}_f &= \mathbf{p}_i + h\mathbf{v}_f, \end{aligned} \tag{18.9}$$

where $\mathbf{g}_f$ is the value of function $\mathcal{G}$ at time t_f.

We apply a Taylor series expansion and make a first-order approximation of the nonlinear function $\mathcal{F}$:

$$\mathcal{F}(\mathbf{p}_i + \Delta\mathbf{p}, \mathbf{v}_i + \Delta\mathbf{v}) \simeq \mathbf{f}_i + \frac{\delta\mathcal{F}}{\delta\mathbf{p}}\Delta\mathbf{p} + \frac{\delta\mathcal{F}}{\delta\mathbf{v}}\Delta\mathbf{v}. \tag{18.10}$$

This linearization corresponds to the first iteration of the Newton–Raphson algorithm. Therefore the problem is solved with a single iteration, assuming a temporal coherency of the mechanical behavior. It may lead to minor numerical errors in the dynamic behavior, but these errors tend to decrease at equilibrium or with null velocity.

Replacing (18.10) in (18.9) and using $\Delta\mathbf{p} = \mathbf{p}_f - \mathbf{p}_i = h\mathbf{v}_f$ and $\Delta\mathbf{v} = \mathbf{v}_f - \mathbf{v}_i$, we obtain

$$(\mathbf{M} + h\mathbf{B} + h^2\mathbf{K})\Delta\mathbf{v} = h(\mathbf{g}_f - \mathbf{f}_i) - h^2\mathbf{K}\mathbf{v}_i, \tag{18.11}$$

where $\mathbf{M}$ is the mass matrix, $\mathbf{f}_i$ and $\mathbf{v}_i$ are, respectively, the internal forces and velocities at the beginning of the time step. $\mathbf{K} = \frac{\delta\mathcal{F}}{\delta\mathbf{p}}$ is the global stiffness matrix, and $\mathbf{B} = \frac{\delta\mathcal{F}}{\delta\mathbf{v}}$ is the damping matrix used to dissipate energy.

Following Baraff and Witkin (1998), the damping matrix can be approximated with *Rayleigh damping* $\mathbf{B} \simeq \alpha\mathbf{M} + \beta\mathbf{K}$, where α is the *rayleigh stiffness* and β the *rayleigh damping*. Finally, Eq. (18.11) can be rewritten

$$\underbrace{((1 + h\alpha)\mathbf{M} + h(h + \beta)\mathbf{K})}_{\mathbf{A}}\underbrace{\Delta\mathbf{v}}_{\mathbf{x}} = \underbrace{h(\mathbf{g}_f - \mathbf{f}_i) - h^2\mathbf{K}\mathbf{v}_i}_{\mathbf{b}}. \tag{18.12}$$

Eq. (18.12) provides a linear problem $\mathbf{A}\mathbf{x} = \mathbf{b}$ that must be solved at each time step. Indeed, the left-hand side is a global system matrix $\mathbf{A}$ mainly composed of the mass and stiffness matrices. Using the co-rotational formulation the stiffness matrix $\mathbf{A}$ is obtained with

$$\mathbf{K} = \sum(\mathbf{G}_e\mathbf{R}_e\mathbf{K}_e\mathbf{R}_e^T\mathbf{G}_e{}^T), \tag{18.13}$$

where $\mathbf{G}_e$ is the *globalization matrix* transferring the local stiffness matrix $\mathbf{K}_e$ of an element e to the global stiffness matrix $\mathbf{K}$. The rotation matrices are not constant and must be recomputed at each simulation step. As a result, the linear system must be solved at each step. In an interactive context, Eq. (18.12) is usually solved with the iterative conjugate gradient. A possible hardware-dependent optimization is to implement the full implicit integration solver on the GPU (including CG iterations), where the system matrix is not assembled and the matrix computations are parallelized directly on the original object mesh. It considerably reduces the number of operations required, and more importantly, the consumed bandwidth, enabling the method to be fast enough for highly complex stiff body simulations.

18.2.2.1 Preconditioner

The condition number κ of the matrix $\mathbf{A}$ measures how much the output value of the function can change for a slight change in the input argument. For heterogeneous objects or ill-structured meshes, the condition number κ is often high, which raises convergence issues for the CG algorithm. A common technique is to use a preconditioner to reduce the condition number, ensuring a faster convergence of the algorithm. By definition, a preconditioner approximates the system matrix $\mathbf{A}$, which is less costly to invert. Solving Eq. (18.12) with a preconditioner $\mathbf{P}$ can be written

$$\mathbf{P}^{-1}\mathbf{A}\mathbf{x} = \mathbf{P}^{-1}\mathbf{b}, \quad \text{such that} \quad \kappa(\mathbf{P}^{-1}\mathbf{A}) < \kappa(\mathbf{A}). \tag{18.14}$$

Several preconditioners can be used, ranging from simple diagonal matrices to accurate but costly Cholesky factorization (Baraff and Witkin, 1998). Generally, the performance improvement remains limited since the preconditioner is either too simple to significantly reduce the number of iterations or expensive to compute.

A faster method is to use an asynchronous preconditioning technique. The method relies on the assumption that $\mathbf{A}$ in Eq. (18.12) undergoes small perturbations between two consecutive time steps. Indeed, if $\mathbf{P}_t = \mathbf{A}_t^{-1}$ is available at a specific time t, it remains a "good" approximation for the following time steps. The preconditioner can then be updated at a low frequency on a dedicated CPU thread, and the last preconditioner available can be used to advance the simulation. Therefore the overhead in computing the preconditioner is removed from the simulation loop, which allows using more advanced and computationally costly preconditioners, such as a complete factorization of the system[1]:

$$\mathbf{P} = \mathbf{A} = \mathbf{L}\mathbf{D}\mathbf{L}^{\mathrm{T}}, \tag{18.15}$$

where $\mathbf{D}$ is a diagonal matrix, and $\mathbf{L}$ is a sparse lower-triangular matrix. An essential advantage of this factorization is that the resulting $\mathbf{L}$ matrix remains sparse, making the preconditioner application faster within the CG. Finally, the method allows simulating the deformation of homogeneous and heterogeneous tissues in real-time.

18.2.3 Time-stepping and collision detection

Collision response on mechanical objects leads to discontinuities in the velocities of the colliding points. For such discontinuous events, the acceleration is not defined: the problem belongs to the field of non-smooth mechanics.

To integrate the mechanics and the non-smooth events due to contact over time, we use a *time-stepping* method: The time step is fixed, and there is no limitation on the number of discontinuities that could take place during a time step, but low-order integration schemes should be used. Although it could lead to excessive dissipation if the time step is too significant, it provides stable simulations.

---

[1] Note that even if we compute an exact factorization of $\mathbf{A}_{t-h}$, the preconditioner remains an approximation since its computation is based on a previous configuration of the objects.

18.2.3.1 Collision detection

Before enforcing the contact between anatomical structures or with surgical instruments (rigid or deformable), one needs to detect them (see Teschner et al., 2005, for a survey): a common strategy is to build a collision pipeline, where non-colliding objects will be pruned out of the pipeline, and only parts of meshes that are close enough will remain.

At the end of the pipeline, collision detection returns either penetration depth or a minimal distance between meshes or pairs of primitives (between which distance constraints can be formulated to solve the contacts).

18.2.3.2 Contact mapping

Despite these optimizations, collision detection is a time-consuming operation in the simulation loop. To save computation time, the collision surface of the models is usually simplified compared to the FE mesh used to compute the deformation. In this case, the constraint forces can be transferred from the collision models to the FE model using a *mapping* function $\mathcal{J}$. This relation is used to map the positions $\mathbf{p}_{\text{fe}}$ of the deformable model to the positions $\mathbf{p}_{\text{col}}$ of the collision models:

$$\mathbf{p}_{\text{col}} = \mathcal{J}(\mathbf{p}_{\text{fe}}). \tag{18.16}$$

The most widely used mapping is to couple the collision surface (edges, triangles, or quads) with the tetrahedral mesh supporting the FE computations. In this case, a barycentric mapping is used to link the various representations. At the beginning of the simulation, each vertex of the collision models is associated with the "closest" tetrahedron in the FE mesh. During the simulation, both the collision mesh and the FE model are deformed simultaneously. Therefore this relation remains constant over the simulation:

$$\mathbf{p}_{\text{col}} = \mathbf{J}_{\text{col}}\mathbf{p}_{\text{fe}}, \tag{18.17}$$

where $\mathbf{J}_{\text{col}}$ is the matrix of the *mapping* computed from the barycentric coordinates of collision points with respect to the associated tetrahedron's positions. In addition, the mapping also provides a way to transfer any given force $\mathbf{f}_{\text{col}}$ applied on the collision models to an equivalent force applied on the FE nodes. The transferred forces $\mathbf{f}_{\text{fe}}$ are then weighted with the same barycentric coefficients:

$$\mathbf{f}_{\text{fe}} = \mathbf{J}_{\text{col}}^T \mathbf{f}_{\text{col}} \tag{18.18}$$

Similarly, collision detection may provide contact information at an arbitrary location on the collision surface (for instance, in the middle of a collision triangle). Another mapping relation is, therefore, necessary to transfer contact forces to equivalent nodal forces on the collision mesh (Fig. 18.1a). To simplify the solving process this relation is considered constant during the resolution of each time step. As previously, a mapping matrix $\mathbf{J}_{\text{CD}}$ can be defined to transfer the contact forces $\mathbf{f}_{\text{prox}}$ on the collision mesh:

$$\mathbf{f}_{\text{col}} = \mathbf{J}_{\text{CD}}^T \mathbf{f}_{\text{prox}}. \tag{18.19}$$

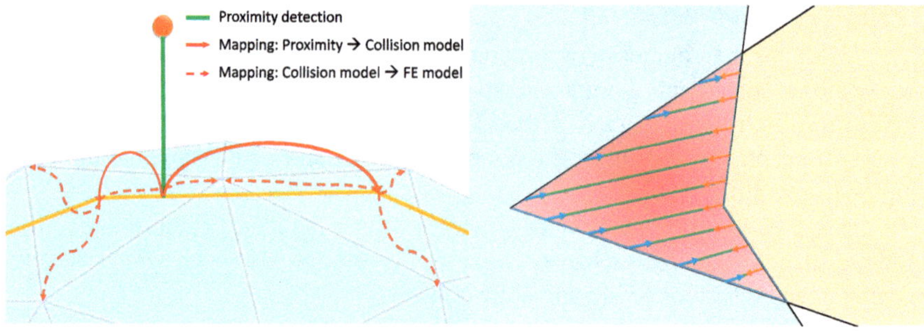

(a) Contact mapping (b) Constraint linearization

FIGURE 18.1

Constraint definition. (Right) The constraint are linearized at the beginning of each time step providing the direction along which contact forces will be applied. (Left) Contact forces are transferred from the collision model to the FE mesh using mapping relations.

Finally, for a given object, an arbitrary number of representation may be used and the mapping relations may be serially applied to finally compute equivalent forces (or displacements) on the DOFs of the FE mesh $\mathbf{f}_{\text{fe}} = \mathbf{J}_{\text{col}}^T \cdots \mathbf{J}_{\text{CD}}^T\, \mathbf{f}_{\text{prox}}$. Since mapping matrices are usually sparse, this operation is inexpensive. Therefore, for simplicity, we now omit this relation considering that the constraints are directly applied to the mechanical model.

18.2.4 Constraint-based simulation

To avoid interpenetrations or to enforce specific behaviors, the simulation of two colliding objects 1 and 2 is now subject to a set of constraints. For this purpose, the simulation of the deformable bodies are coupled with the following equation:

$$\Lambda(\mathbf{p}_{f,1}, \mathbf{p}_{f,2}) = \mathcal{H}_1(\mathbf{p}_{f,1}) - \mathcal{H}_2(\mathbf{p}_{f,2}), \tag{18.20}$$

where Λ is a nonlinear function providing the violation of the constraint with respect to positions $\mathbf{p}_{f,1}$ and $\mathbf{p}_{f,2}$ of the deformable objects at the end of the time steps, and $\mathcal{H}_1$, $\mathcal{H}_2$ provide the value of the constraints. In the case of collisions, Λ corresponds to the shape of the interpenetration (red area in Fig. 18.1b) and must be canceled at the end of each time step. This function is highly nonlinear since constraint forces must be applied to cancel the violation. However, applying such forces will result in displacements, rotations, or deformations of deformable bodies that may generate new collision areas or invalidate the previous ones.

The constraint equations are linearized using a first-order Taylor expansion:

$$\mathcal{H}(\mathbf{p}_f) = \mathcal{H}(\mathbf{p}_i + h\mathbf{v}_f) \approx \mathcal{H}(\mathbf{p}_i) + h\frac{\partial \mathcal{H}(\mathbf{p}_i)}{\partial \mathbf{p}}\mathbf{v}_i + h\frac{\partial \mathcal{H}(\mathbf{p}_i)}{\partial \mathbf{p}}\Delta\mathbf{v}, \tag{18.21}$$

where $\frac{d\mathcal{H}(\mathbf{p}_i)}{d\mathbf{p}}$ is the linearization of constraints at the beginning of the time step (using positions $\mathbf{p}_i$).

In practice, the evaluation and the linearization of the constraints equations are complex. For this purpose, collision detection is performed, providing a discrete set of proximities between both objects (green lines in Fig. 18.1b). The number of proximities usually depends on the resolution of the collision mesh (which may be adapted, thanks to the *mappings*) and the collision detection method itself (for instance, by filtering proximities afterward).

Once the collision information is available, all the constraint equations are evaluated along with the directions of the proximities assumed constant for the time step. This leads to several simplifications: First of all, $\mathbf{H} \approx \frac{d\mathcal{H}(\mathbf{p}_i)}{d\mathbf{p}}$, known as the *Jacobian of the constraints*, can be defined,[2] providing the constraint directions (blue and orange arrows in Fig. 18.1b). The dimension of $\mathbf{H}$ is $c \times n$, where c is the number of proximities, and n is the number of DOFs. Similarly, the violation of the constraints $\mathcal{H}(\mathbf{p}_i) \approx \mathbf{H}\mathbf{p}_i$ is evaluated along the same constraint directions. With these simplification and replacing (18.21) in (18.20), the violation of the constraint at the end of the step can be rewritten as:

$$\boldsymbol{\delta}_f = \underbrace{\left(\mathbf{H}_1\mathbf{p}_{i,1} + h\mathbf{H}_1\mathbf{v}_{i,1}\right) + \left(\mathbf{H}_2\mathbf{p}_{i,2} + h\mathbf{H}_2\mathbf{v}_{i,2}\right)}_{\boldsymbol{\delta}_i} + h\ \mathbf{H}_1\Delta\mathbf{v}_1 + h\ \mathbf{H}_2\Delta\mathbf{v}_2, \quad (18.22)$$

where $\boldsymbol{\delta}_f$ is the violation of the constraints at the end of the time step, and $\mathbf{H}_1 = \frac{d\mathcal{H}_1(\mathbf{p}_{i,1})}{d\mathbf{p}}$ and $\mathbf{H}_2 = -\frac{d\mathcal{H}_2(\mathbf{p}_{i,2})}{d\mathbf{p}}$.

In addition, to enforce nonpenetrating objects, constraint forces $\boldsymbol{\lambda}$ (also called Lagrangian multipliers) are applied along the direction of the constraint. The governing Eq. (18.7) is modified as follows:

$$\mathbf{M}\,\mathbf{a} - \mathcal{G}(t) - \mathcal{F}\,(\mathbf{p}, \mathbf{v}) + h\mathbf{H}^T\,\boldsymbol{\lambda}. \quad (18.23)$$

After the linearization, Eq. (18.12) is modified as follows:

$$\mathbf{A}\mathbf{x} = \mathbf{b} + h\mathbf{H}^T\,\boldsymbol{\lambda}. \quad (18.24)$$

Gathering Eqs. (18.24) and (18.22) for the two interacting objects results in the following Karush–Kuhn–Tucker (KKT) system:

$$\begin{cases} \mathbf{A}_1\mathbf{x}_1 + h\mathbf{H}_1^T\boldsymbol{\lambda} = \mathbf{b}_1 & \text{(a)} \\ \mathbf{A}_2\mathbf{x}_2 + h\mathbf{H}_2^T\boldsymbol{\lambda} = \mathbf{b}_2 & \text{(b)} \\ h\mathbf{H}_1\mathbf{x}_1 + h\mathbf{H}_2\mathbf{x}_2 = \Delta\boldsymbol{\delta}, & \text{(c)} \end{cases} \quad (18.25)$$

where $\Delta\boldsymbol{\delta} = \boldsymbol{\delta}_f - \boldsymbol{\delta}_i$

[2] Note that the matrix may be multiplied by mapping matrices, but this operation is omitted as stated above.

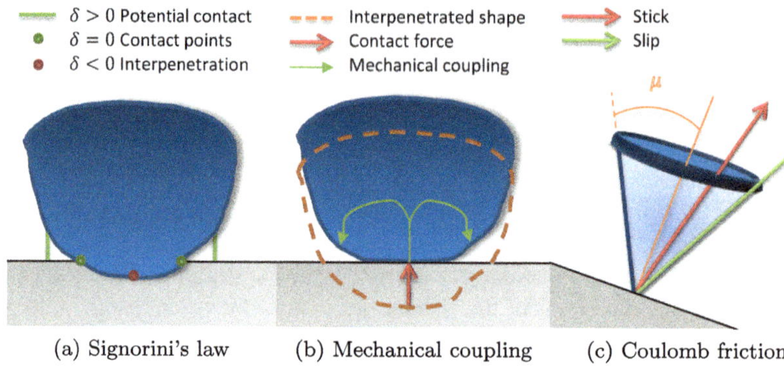

FIGURE 18.2

Illustration of the physical phenomena simulated for the contact's response.

18.2.4.1 Contact and friction models

Contact equations are subject to Signorini's law (Fig. 18.2a). It expresses that there is a complementarity[3] relation between the contact force $\boldsymbol{\lambda}_n$ and the distance $\boldsymbol{\delta}_n$, along the direction $\mathbf{n}$ of the contact:

$$\mathbf{0} \leq \boldsymbol{\delta}_\mathbf{n} \perp \boldsymbol{\lambda}_\mathbf{n} \geq \mathbf{0}. \tag{18.26}$$

This model has several physical justifications, including noninterpenetration and no sticking force. Moreover the contact force vanishes if the points are not strictly in contact. However, using Signorini's law, the contact force creates a frictionless response.

Coulomb's friction law describes the macroscopic behavior of friction in the tangent contact space (Fig. 18.2c). In this law, the reaction force is included in a cone whose height and direction are given by the normal force. If the reaction force is strictly included inside the cone, objects stick together; otherwise, the reaction force is on the cone's border, and objects are slipping along the tangential direction. In this last case, the friction force must be directed along the direction of motion:

$$\begin{aligned} &\dot{\boldsymbol{\delta}}_\mathbf{T} = \mathbf{0} \Rightarrow \|\boldsymbol{\lambda}_\mathbf{T}\| < \mu\ \|\mathbf{f}_\mathbf{n}\|\ \text{(stick)}, \\ &\dot{\boldsymbol{\delta}}_\mathbf{T} \neq \mathbf{0} \Rightarrow \boldsymbol{\lambda}_\mathbf{T} = -\mu\ \|\boldsymbol{\lambda}_\mathbf{n}\| \frac{\dot{\boldsymbol{\delta}}_\mathbf{T}}{\|\dot{\boldsymbol{\delta}}_\mathbf{T}\|} = -\mu\ \|\boldsymbol{\lambda}_\mathbf{n}\|\ \mathbf{T}\ \text{(slip)}, \end{aligned} \tag{18.27}$$

where μ is the friction parameter, and $\mathbf{T}$ is the direction of motion in the tangential plane to the contact normal $\mathbf{n}$.

During 3D slipping motion (also called *dynamic friction*), the tangential direction is unknown. Indeed, the tangential velocity depends on the tangential force that has

[3] Complementarity is noted $\perp$. It states that one of the two values $\boldsymbol{\delta}_n$ or $\boldsymbol{\lambda}_n$ must be null, whereas the other is positive.

to be found during the solving process. It creates a nonlinearity equation in addition to the complementarity state stick/slip.

Signorini's law and Coulomb's law are also valid in a *multi-contact* case. This model is defined for each *potential* contact provided by the contact detection algorithm. However, to solve these laws at every contact point, we have to consider the coupling that exists between these contact points (Fig. 18.2b). This coupling comes from the intrinsic mechanical behavior of deformable objects.

18.2.4.2 Constraint solving and mechanical coupling

When objects are distant, the mechanical matrices $\mathbf{A}_1$ and $\mathbf{A}_2$ are not coupled and can be solved independently. However, coupling terms must be added to take the constraints into account.

Compared to penalty-based methods, a significant advantage of relying on Lagrange multipliers is that collision events never modify the mechanical matrices. It is therefore possible to rely on the *schür complement* method to solve Eq. (18.25a), (18.25b), and (18.25c) with the 5 following steps:

Step 1

interacting objects are solved independently while setting $\boldsymbol{\lambda} = \mathbf{0}$. A set of independent linear systems of equations $\mathbf{Ax} = \mathbf{b}$ must then be solved for each object. The conjugate gradient algorithm is generally proposed to solve this problem since $\mathbf{A}$ is large, sparse, symmetric, and positive definite. It provides $\mathbf{x}_1^{\text{free}} = \mathbf{A}_1{}^{-1}\mathbf{b}_1$ and $\mathbf{x}_2^{\text{free}} = \mathbf{A}_2{}^{-1}\mathbf{b}_2$ called the *free motion* for each object, corresponding to the velocities update without considering any constraint nor collision. After integration, we obtain $\mathbf{p}_1^{\text{free}}$ and $\mathbf{p}_2^{\text{free}}$:

$$\mathbf{p}_1^{\text{free}} = \mathbf{p}_{i,1} + h\mathbf{v}_{i,1} + h\mathbf{x}_1^{\text{free}}, \tag{18.28}$$

$$\mathbf{p}_2^{\text{free}} = \mathbf{p}_{i,2} + h\mathbf{v}_{i,2} + h\mathbf{x}_2^{\text{free}}. \tag{18.29}$$

Step 2

the constraint laws are linearized during the time step. The linearization corresponds to the collision detection that provides the proximities of potential contacts and their associated normals between the position $\tilde{\mathbf{p}}_1$ and $\tilde{\mathbf{p}}_2$ of the collision meshes. Constraint equations are then transferred to the mechanical DOFs using the mapping relation (18.18), providing the Jacobian of the constraint $\mathbf{H}_1$ and $\mathbf{H}_2$, which are assumed constant during the time step. The collision detection is performed at the *beginning* of the time step, whereas the violation $\boldsymbol{\delta}^{\text{free}}$ is instead computed using the *free motion* $\mathbf{p}_1^{\text{free}}$ and $\mathbf{p}_2^{\text{free}}$ obtained at the previous step.

Step 3

The violation of the constraint is evaluated, and constraints are defined according to the collision proximities. With Euler equation one may note that

$$\Delta\boldsymbol{\delta} - \mathbf{H}_1\mathbf{x}_1^{\text{free}} - \mathbf{H}_2\mathbf{x}_2^{\text{free}} = \boldsymbol{\delta}_f - \mathbf{H}_1\mathbf{p}_1^{\text{free}} - \mathbf{H}_2\mathbf{p}_2^{\text{free}}. \tag{18.30}$$

Replacing (18.25a) and (18.25b) in (18.25c) gives

$$\underbrace{\left[\mathbf{H}_1\mathbf{A}_1^{-1}\mathbf{H}_1^T + \mathbf{H}_2\mathbf{A}_2^{-1}\mathbf{H}_2^T\right]}_{\mathbf{W}}\boldsymbol{\lambda} = \underbrace{\mathbf{H}_1\mathbf{p}_1^{\text{free}} - \mathbf{H}_2\mathbf{p}_2^{\text{free}}}_{\boldsymbol{\delta}_f - \boldsymbol{\delta}^{\text{free}}}, \tag{18.31}$$

where $\mathbf{W}$ is known as the *compliance matrix*, or the *Delasus operator*.

In the above equation, both $\boldsymbol{\lambda}$ and $\boldsymbol{\delta}$ are unknown. For Signorini's law (Eq. (18.26)), this equation describes an LCP (linear complementarity problem). If it is combined with Coulomb's law (Eq. (18.27)), we obtain an NLCP (nonlinear complementarity problem). An important difficulty lies in the fact that this equation involves evaluating the inverse of large matrices $\mathbf{A}_1$ and $\mathbf{A}_2$ (same dimension as the number of DOFs) to compute the *compliance matrix* $\mathbf{W}$.

Step 4

We obtain the value of $\boldsymbol{\lambda}$ using a Gauss–Seidel algorithm dedicated to the NLCP created by contact and friction equations. Considering a contact α, among m instantaneous contacts, one can write the behavior of the model in contact space:

$$\underbrace{\boldsymbol{\delta}_\alpha - \mathbf{W}_{\alpha\alpha}\,\boldsymbol{\lambda}_\alpha}_{\text{unknown}} = \underbrace{\sum_{\beta=1}^{\alpha-1}\mathbf{W}_{\alpha\beta}\,\boldsymbol{\lambda}_\beta + \sum_{\beta=\alpha+1}^{m}\mathbf{W}_{\alpha\beta}\,\boldsymbol{\lambda}_\beta +}_{\text{frozen}}\ \boldsymbol{\delta}_\alpha^{\text{free}}, \tag{18.32}$$

where $\mathbf{W}_{\alpha,\beta}$ is a compliance matrix modeling the mechanical coupling between contact points α and β. For each contact α, this method solves the contact equations by considering the other contact points ($\alpha \neq \beta$) as "frozen." The new value of $\boldsymbol{\lambda}_\alpha$ is given by solving Signorini's law and Coulomb's law on this contact (see Duriez et al., 2006, for details of implementation and performance).

Step 5

When the value of $\boldsymbol{\lambda}$ is available for all the contacts, the corrective motion is computed:

$$\begin{aligned} \mathbf{p}_{1,t+h} &= \mathbf{p}_1^{\text{free}} + h\Delta\mathbf{v}_1^{\text{cor}} \quad \text{with} \quad \Delta\mathbf{v}_1^{\text{cor}} = \mathbf{A}_1^{-1}\mathbf{H}_1^T\boldsymbol{\lambda}, \\ \mathbf{p}_{2,t+h} &= \mathbf{p}_2^{\text{free}} + h\Delta\mathbf{v}_2^{\text{cor}} \quad \text{with} \quad \Delta\mathbf{v}_2^{\text{cor}} = \mathbf{A}_2^{-1}\mathbf{H}_2^T\boldsymbol{\lambda}. \end{aligned} \tag{18.33}$$

We finally obtain $\mathbf{p}_{1,t+h}$ and $\mathbf{p}_{2,t+h}$, the positions of objects 1 and 2, which fulfills the contact and friction laws.

18.2.4.3 Compliance and mechanical coupling

In what follow, the term "*mechanical coupling*" describes the coupling between contact constraints applied on two subsets of the boundary of a deformable body. This coupling occurs through the deformation of the body itself. Indeed, even if the contact points are only defined on the boundary of the deformable bodies, they are all influenced by each other through the stiffness of the material, which is represented by **W** in Eq. (18.31).

Formally speaking, the *compliance matrix* **W** requires the computation of $\mathbf{A}^{-1}$, which is a large matrix (same dimension as the number of DOF) and changes at each time step. Although computing this inverse in real-time is possible for coarse models, it raises significant difficulties for large meshes in real-time.

A possible approximation of the *compliance matrix* **W** is to use the asynchronous preconditioner of the previous section:

$$\mathbf{W} = \mathbf{H}\mathbf{A}^{-1}\mathbf{H}^{\mathrm{T}} \approx \mathbf{H}\mathbf{P}^{-1}\mathbf{H}^{\mathrm{T}} = \mathbf{H}\left(\mathbf{R}\,\mathbf{L}\mathbf{D}\mathbf{L}^{\mathrm{T}}\mathbf{R}^{\mathrm{T}}\right)^{-1}\mathbf{H}^{\mathrm{T}}. \tag{18.34}$$

Indeed, since **P** is an efficient preconditioner (i.e., a close approximation of the factorization of **A**), we propose to use it to compute **W** in Eq. (18.31). The above equation requires computing the product of the inverse of the preconditioner with the Jacobian of contacts **H**, which can be achieved by computing columns independently of the following matrix:

$$\mathbf{L}\mathbf{D}\mathbf{L}^{\mathrm{T}}\,\mathbf{X}_i = \mathbf{H}_i \quad \Leftrightarrow \quad \mathbf{X}_i = \left(\mathbf{L}\mathbf{D}\mathbf{L}^{\mathrm{T}}\right)^{-1}\mathbf{H}_i. \tag{18.35}$$

Solving this equation is still expensive since multiple triangular systems must be solved for each right-hand side vector. Again, since multiple right-hand side vectors stored in **H** can be computed independently, GPU optimizations could be considered by assigning the computation of each column to an independent multiprocessor on the GPU.

Then we use another level of parallelism, where each STS is solved with several threads. Indeed, many data can potentially be treated in parallel during the solving process of each STS. This two-level strategy fits the GPU architectures, where local synchronizations within a group of threads are fast, whereas global synchronizations over multiple groups are costly.

18.3 Registration of biomechanical models

Medical imaging is one of the main improvements of modern surgical practice and represents not only a support for diagnostics but also an actual operative instrument during therapeutic procedures. However, from a physiological point of view, organs are not static, and even the most straightforward breathing motion can induce shape deformations, which requires a mental effort of three-dimensional reconstruction to coincide with the pre-operative images. However, as an approximation, FE models cannot be directly used for medical purpose.

Yet, to limit errors, FE models can be registered with intraoperative data extracted from medical images. In this case, biomechanical models can be used to display through augmented reality (AR) the internal structures (vessels, tumors, etc.) on top of the intraoperative images that are often sparse, incomplete, and of poor quality. The advantages of relying on FE models for this task are multiple: FE models can be used to regularize the data extracted from images, being this way less sensitive to noise and tracking errors. In addition, FE models provide accurate solutions to interpolate the whole volume displacement of the organs, including the displacement of internal structures, such as tumors or vessels.

In this section, we introduce a unified constraint-based formalism and solving process combining intraoperative image data with biomechanical models.

Problem definition

The nonrigid registration of a preoperative 3D model S with intra-operative data is an ill-posed problem known to be very difficult to solve. It can be formalized as finding the 3D nonrigid transformation $\mathcal{T}$ mapping S into the image frame. However, since intraoperative data are usually incomplete, errors can only be measured with the information that can practically be extracted from the operating scene (image or sensors). Let $\mathbf{m}$ be the set of data extracted intraoperatively, the nonrigid registration problem can be synthesized with

$$\min_{\mathbf{p}}(||\overline{\mathcal{T}(S)} - \mathbf{m}||), \tag{18.36}$$

where $\mathbf{p}$ are the positions of the model S, and $\overline{\mathcal{T}(S)}$ corresponds to the transformed model observable in the same condition as the intraoperative data.

This process usually admits a unique solution for rigid transformation scenarios (without considering noise or other pairing issues). However, the rigid assumption is not valid for our purpose, and the deformations generate a problem having an infinite number of solutions.

To decrease the number of solutions, we assume an elastic behavior of the organ with preservation of the volume after the deformation. Therefore we propose to rely on a nonlinear biomechanical finite element model coupled with a set of constraints $\mathcal{H}$ to solve the registration problem of Eq. (18.36). This leads to a nonlinear problem, whose solution is given by the positions of the physical model providing the equilibrium between the internal forces and the external forces applied by the constraints

$$\begin{cases} \mathcal{F}(\mathbf{p}) + \mathcal{H}(\mathbf{m}, \mathbf{p}) = \mathbf{0} & \text{(a)} \\ \mathcal{H}(\mathbf{m}, \mathbf{p}) = \boldsymbol{\delta}, & \text{(b)} \end{cases} \tag{18.37}$$

where $\mathcal{F}$ provides the internal forces of the FE model, and $\boldsymbol{\delta}$ is the violation of the constraints, as described in the previous section. Lagrangian multipliers $\boldsymbol{\lambda}$ are used to impose displacements and enforce constraints of Eq. (18.37b). The number and the direction of Lagrangian multipliers will be described below; it depends on the position of the models $\mathbf{p}$ and the data extracted from intraoperative images $\mathbf{m}$.

Note that Eq. (18.37a) and (18.37b) correspond to the static equilibrium of the dynamic equation (18.23). However, in this section, the simulation time t is not considered anymore. Instead, the simulation steps correspond to the iterations of the iterative registration procedure.

Segmentation and meshing and initialization

Preoperatively, a patient-specific model is built from the images acquired before surgery.

The model is usually constructed using either contrast-enhanced CT data or volume MRI sequence. The image is first segmented (using, for instance, the semi-automatic methods available in ITKSnap[4]), then a tetrahedral mesh is obtained from the segmented maps (for instance, using CGAL[5]). Relying on the numerical strategies described above, the target resolution of the FE mesh is around 2000 nodes per organ, to enforce real-time compatible computation times.

18.3.1 Geometrical binding

Vision-based algorithms are traditionally used to extract geometrical primitives (2D/3D positions, partial triangular mesh, etc.) from intraoperative images. These data provide the intraoperative shape (called **target**), on which the preoperative model (called **source**) must be registered.

A significant difficulty is related to the fact that it is complicated to relate the position of the resulting data to the FE model.

Indeed, the correspondences between image data and material points on the model are usually not known. This problem is addressed with an iterative process taking as input the **target** data $\mathbf{m}$ and the **source** positions $\mathbf{p}$ of the *collision* mesh.[6]

As previously, the constraint equations are linearized at each iteration $\mathcal{H}(\mathbf{p},\mathbf{m}) \simeq \frac{\partial \mathcal{H}}{\partial \mathbf{p}} = \mathbf{H}$, providing the directions along which Lagrangian multipliers are computed. The linearization (corresponding to the collision detection in the previous section) corresponds to the pairing operation between the control points $\mathbf{m}$ and material points on the model. The *iterative closest point* (ICP) method (Rusinkiewicz and Levoy, 2001) is employed to associate the target points with their respective closest points on the source mesh. The linearization is performed at the beginning of the simulation step (corresponding to the iteration t) using the current position of the model and assumed constant during the solving process.

18.3.1.1 Outliers and geometrical filtering

The quality of the registration strongly depends on the quality of the pairings performed during the binding process. Indeed, incoherent or antagonistic pairings (called outliers) can appear due to the noise (image artifacts, missing data, etc.) and must be ignored. Depending on the application, the following filters may be used:

---

4 https://www.itksnap.org.

5 https://www.cgal.org.

6 Note that the search is usually performed with positions $\tilde{\mathbf{p}}$ of a dedicated *collision* model, either to save computation time or to have a specific representation of the structures on which constraints will be applied (vessels, contours, etc.). However, as explained in the previous section, a linear relation $\mathbf{J}$ can be precomputed using barycentric coordinates (see Eqs. (18.17) and (18.18)) to transfer the constraint equations to the mechanical DOFs $\mathbf{p}$. For the sake of simplicity, we omit this relation, considering that the constraints are directly applied to the mechanical model.

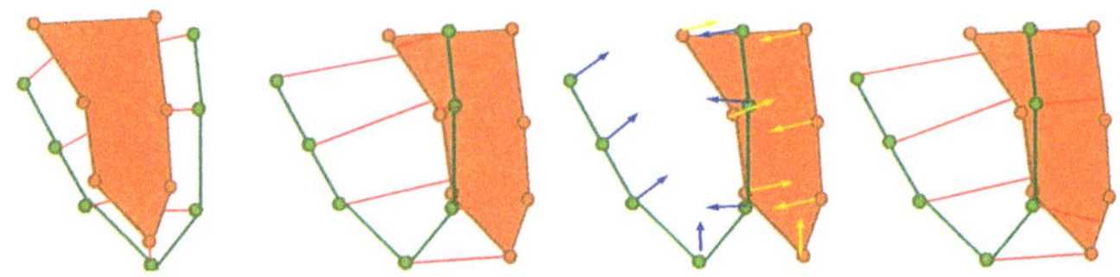

FIGURE 18.3

Binding process of the control points and the FE structure. 18.3a the control points (green) are associated (red lines) with the closer surface of the object (orange). 18.3b the distance criterion is not sufficient since all the control points are associated with the same side of the object and the constraints cannot be satisfied. 18.3c a set of normals aiming inside the surface are computed. 18.3d the control points are associated with the closer triangle, whose normal is oriented in the same direction.

Unique pairing

First, we enforce that control points $\mathbf{m}$ are associated with a unique element on the model; otherwise, it would lead to over-constrained problems. When several vertices of $\mathbf{m}$ are projected onto the same geometrical element of the mesh, only the nearest one is kept.

Filtering with distance

A pairing is considered as an outlier, and then ignored, if it does not satisfy the following condition:

$$\mathbf{d}_j \in [\tilde{\mathbf{d}} - \mathbf{d}_t ; \tilde{\mathbf{d}} + \mathbf{d}_t], \tag{18.38}$$

where $\tilde{\mathbf{d}}$ is the median distance computed over all pairings and $\mathbf{d}_t$, a threshold. In addition, a filtering criterion $\mathbf{d}_{\max}$ can be added to reject pairing above a given distance $\mathbf{d}_{\max}$.

Filtering with normals

When image data are sufficiently dense, the normals of the surface can be estimated using a simple least square plane fitting, as implemented in the Point Cloud Library.[7] In this case, the method can be improved by ensuring that the normals associated with image data have the same orientation as the FE surface (Fig. 18.3).

18.3.2 Constraints definition

This section introduces the various constraints employed for our applications. At this stage, the control points $\mathbf{m}$ of the **target** shape are associated with material positions

[7] https://pointclouds.org/documentation/tutorials/normal_estimation.html.

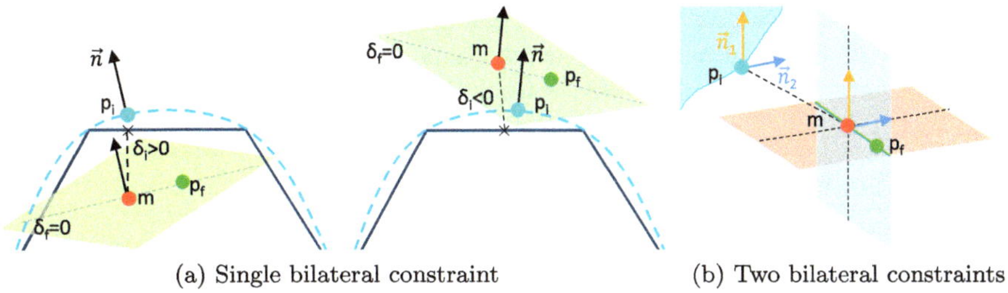

FIGURE 18.4

The broken blue lines represent the Bezier surface defined over the triangles of the source mesh. $\mathbf{m}$ (in red) are paired with positions $\mathbf{p}_i$ (in blue) at the beginning of the step. After the resolution, $\mathbf{p}_f$ (in green) must be located on the tangential plane defined by the constraints. Note that $\mathbf{m}$ does not move, but instead, the model is deformed (assuming linear translations of the solution plane) to enforce the constraint.

$\mathbf{p}$ on the biomechanical model. Lagrange multipliers are used to impose constraint forces on the FE model to cancel the violation of constraints.

18.3.2.1 Bilateral constraints

Bilateral constraints are holonomic constraints, $\boldsymbol{\Omega}(\mathbf{p}) = \mathbf{0}$, used to impose the relative positions of constrained points on the orthogonal plane defined by the constraint normal. For a given control point i, the violation of constraints $\boldsymbol{\delta}_i$ is computed with

$$\boldsymbol{\delta}_i = \mathbf{n}_i \cdot (\mathbf{m}_i - \mathbf{p}_i^{\text{free}}), \tag{18.39}$$

where $\cdot$ is the Euclidean dot product, and $\mathbf{n}_i$ is the normal of the constraint.

Depending on the desired behavior, $\mathbf{n}_i$ may be computed in several ways. It can be either the normal of the source's surface at position $\mathbf{p}_i$, or the direction of the pairing $\mathbf{n}_j = \frac{\mathbf{m}_i - \mathbf{p}_i}{||\mathbf{m}_i - \mathbf{p}_i||}$, or simply the unit vectors $\vec{x}$, $\vec{y}$ or $\vec{z}$.

At the end of each simulation step, the corresponding constraint force $\boldsymbol{\lambda}_i$, applied in the direction of $\mathbf{n}_i$, is computed so that no violation remains ($\bar{\boldsymbol{\delta}}_i = 0$) in that direction. It means that the vertex i, with position $\mathbf{p}_i$, will be located on the tangential plane given by $\mathbf{n}_i$ (Fig. 18.4). The combination of two bilateral constraints forces the points to follow a line at the intersection of the two planes (Fig. 18.4b).

Similarly, three orthogonal bilateral constraints force the point $\mathbf{p}_i$ to be located at the exact 3D location $\mathbf{m}_i$ at the end of the step. As it will be introduced in the next section, the possibility to "slide" on the surface of the FE mesh is essential to stabilize the model around the configuration minimizing the energy and satisfying the constraints.

18.3.2.2 Unilateral constraints

Unilateral constraints are used to simulate collision on the source mesh. It can be either collision with other FE models, the surrounding environment, or even collision with medical instruments segmented in the images. Unilateral constraint satisfies the Signorini conditions $\boldsymbol{\lambda} \perp \boldsymbol{\delta}$, as explained in the previous section (see Section 18.2.4.1). The violation is computed the same way as in Eq. (18.39). At the end of the resolution, the constraint node i is located on the positive side of the constraint.

18.3.3 Constraint solving

The linearization of Eqs. (18.37a) and (18.37b), provides the following KKT system in a quasi-static scenario:

$$\begin{cases} \mathbf{A}\mathbf{x} + \mathbf{H}^T\boldsymbol{\lambda} = \mathbf{0} & \text{(a)} \\ \mathbf{H}\mathbf{x} = \boldsymbol{\delta}. & \text{(b)} \end{cases} \tag{18.40}$$

Note that although we only consider static equations for the registration, the dynamic terms of Eq. (18.12) are usually conserved during the registration process, i.e., $\mathbf{A} = (1 + h\alpha)\mathbf{M} + h(h + \beta)\mathbf{K}$, as in Eq. (18.12). Indeed, dynamic terms increase the stability of the registration procedure, and since we only consider the converged state (i.e., when the velocity update $d\mathbf{v} = \mathbf{0}$) without gravity or external forces, the solution of the dynamic and static equations are identical.

A new pairing procedure is called at each step to associate the control points $\mathbf{m}_i$ with their respective closest point $\mathbf{p}_i$ on the model. For each control point, a unique bilateral constraint is defined to perform the registration. Although for a single constraint, this would move the point $\mathbf{p}_i$ to the exact location $\mathbf{m}_i$, it is not necessarily the case for multiple constraints. Indeed, when several constraints are applied simultaneously on the model, the constraints are coupled through the *compliance matrix* $\mathbf{W}$. Therefore at the end of the simulation step, $\mathbf{p}_i$ and $\mathbf{m}_i$ are located on the tangential plane defined by the normal $\mathbf{n}_i$. However, the cumulative effect of other constraints may move the points in lateral directions, allowing the model to "slide" toward the shape, minimizing the energy necessary to enforce the constraints.

During the Gauss–Seidel iterations, the constraint equations are either activated with a nonzero force or deactivated if the violation is canceled by other constraints. As a result, only the constraints necessary to cancel the violation are activated, and $\boldsymbol{\lambda}$ minimizes the energy to register the model. The section that follows will show that this observation can be used to identify the boundary conditions acting between the preoperative and intraoperative configurations.

Finally, the iterative registration process is stopped when the deformable body is stabilized, i.e., when the variation of velocity between consecutive iterations t and $t + 1$ is lower than a threshold $||\mathbf{x}^{(t+1)} - \mathbf{x}^{(t)}|| < \epsilon$.

18.3.3.1 Outliers and mechanical filtering

The geometrical filters introduced previously may not be sufficient. Additional mechanical filters are employed to stabilize the process and facilitate the convergence.

Force clipping

During the solving phase, the constraint forces are computed as follows:

$$\boldsymbol{\lambda} \in [-\mathbf{f}_{max}; \mathbf{f}_{max}], \tag{18.41}$$

where $\mathbf{f}_{max}$ is a parameter defining the maximal force applied on the constraints. Since the constraint forces are bounded, all the constraints might not be satisfied at the end of the resolution ($\bar{\delta} \neq 0$). However, the cumulative effect of all the constraints will create the general motion/deformation of the model. Strong forces are only necessary if the registration constraints require significant energy to act against the internal elastic forces of the deformable body, which is most of the time attributed to outliers.

Image compliance

In the above formulation, the data extracted from images are never impacted by the biomechanical model, but instead, the model deformed to fit the data. However, as stated in the introduction, the data may be noisy, and their positions are subject to inaccuracy.

A compliance factor is associated with the constraints to improve the robustness of the method:

$$(\mathbf{W} + \mathbf{W}^{\text{soft}})\,\boldsymbol{\lambda} = \boldsymbol{\delta}, \tag{18.42}$$

where $\mathbf{W}^{\text{soft}}$ is a diagonal matrix, whose coefficients w_{soft} are chosen according to the mechanical parameters of the model and the confidence of the tracking method. If $w_{\text{soft}} = 0$, image data $\mathbf{m}$ are not influenced by the constraints, whereas when the higher w_{soft} is chosen, the more the image data will be projected onto the model rather than deforming it. Therefore both the biomechanical model and the data $\mathbf{m}$ can be deformed during the constraints resolution.

18.4 Applications

Having a simulation that can handle the deformations of the anatomical structures, the collisions between tissues and surgical tools, and mechanical constraints that can fit simulation data to input (medical images), many medical applications can be designed (Courtecuisse et al., 2014; Trivisonne et al., 2016; Courtecuisse et al., 2020). In this book chapter, we focus on an application that will ease the diagnoses of pelvic pathologies by adding an augmented reality set-up to 2D MRI images. 2D dynamic MRI sequences are commonly used in clinical routines to evaluate the dynamic of organs. However, due to the limited view, subjectivity related to human perception cannot be avoided in the diagnoses. Consequently, the aim is to provide better knowledge and understanding of pathologies such as prolapsus or abnormal mobility of tissues. Our application will handle the four main structures of the female pelvic floor (bladder, vagina, uterus, and rectum).

18.4.1 Data acquisition

Patient-specific data are generated from two data sets of two volunteers (designed as T005 and T007) presenting no pelvic pathology. The experimental data inputs have been acquired following a close protocol to a standard clinical routine (see Fig. 18.5).

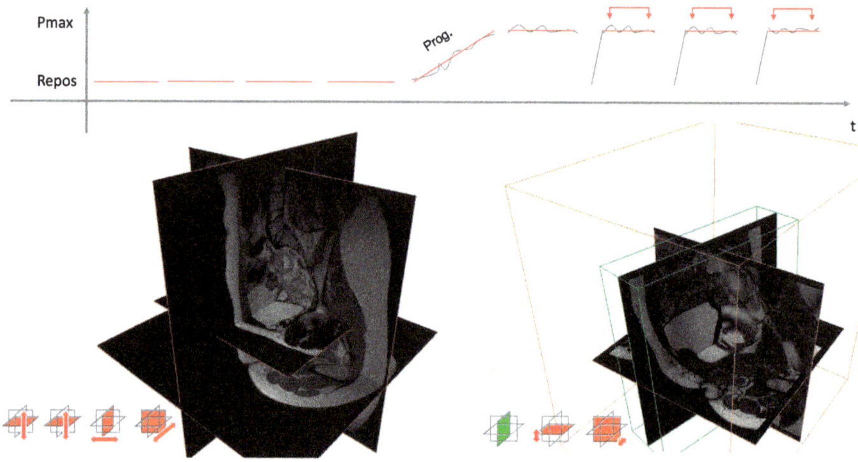

FIGURE 18.5

Protocol for MRI data acquisition composed of three steps: TSE: Volume acquisition is performed providing the geometry of organs. DYN: A dynamic 2D sequence is acquired during breathing and contraction of muscles. FFE: A sparse volume acquisition is performed for the validation, while the volunteer maintains a maximal pressure of organs.

Pelvic MRI requires between 20 and 50 minutes to obtain static volume images with sufficient organ resolution (TSE). The procedure is followed by a dynamic acquisition (DYN) while the patient contracts and relaxes the pelvic muscles. The DYN sequence is performed in the middle plane, representing organ motion the most. Data were obtained with a 3 Tesla MRI (Philips Achieva 3.0T TX), providing up to 2 images per second in dynamic mode. TSE and DYN are already part of standard clinical routines, but a final step (FFE) is added to assess the accuracy and validate our registration method. Volunteers are asked to maintain a maximal contraction of the pelvic muscles without moving or breathing. It significantly limits the time allocated for acquisition, and 15 of sparse sagittal slices uniformly distributed in the volume were acquired. Although the FFE is not required to perform the registration, it provides the only data available to assess the accuracy of our approach.

18.4.2 Contour tracking

The problem of outline organs' tracking in DYN MR images is ill-defined, because it requires finding tangential information that may vary due to off-plane motion. To follow material points over time, especially to avoid drifting along the contour boundaries, we propose to apply additional constraints for the computation of these points to track the material points better and to ensure the plausibility of the computed motion.

Based on the initial B-spline model, each organ is redefined by a composite Bézier curve (a series of segments attached end to end). The beginning and ending points of

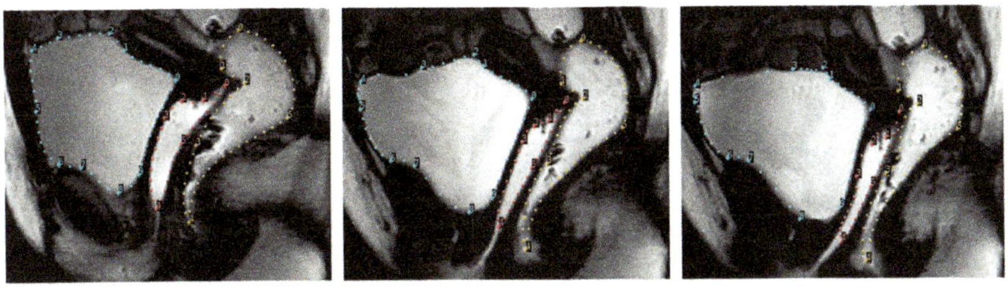

FIGURE 18.6

Outline's tracking with material points in the sequence of dynamic MR images.

Bézier segments are considered material points, selected automatically based on the criteria of high curvature. The "corner" points are identifiable during the deformation over time. Then the conversion from B-spline to composite Bézier is carried out using a least squares minimization.

The whole organ contour is initialized on the first image. Then, the contour is deformed automatically to fit the other images in the DYN sequence. The material points are first registered from image t to $t+1$ using an optical flow method. Then, the energy E_{image} is minimized for each Bézier segment in the neighborhood of the curve. The energy is evaluated from the gray image level, the normal direction of the Bézier curve, and a *level set* function (more details can be found in Jiang et al. (2015)).

Fig. 18.6 shows the semi-automatic tracking of three highlighted organs with material points: bladder (red), vagina (blue), and rectum (green). Concerning the mobility of different areas, the distribution of points matches the organ shape and can provide more physical information for the following 3D registration.

18.4.3 Boundary condition

The original contribution was to combine the 3D FE models, image data, and a priori knowledge of boundary conditions. Boundary conditions include contacts between structures and an approximative definition of ligaments and attached structures (Fig. 18.7).

According to the knowledge of anatomy, anatomical constraints are added to limit organs' off-plane motion (see Fig. 18.7, center). The vagina is attached on each side by ligaments modeled as stiff springs (blue), allowing vertical mobility but preventing lateral motions. The uterus is attached to the vagina through bilateral constraints (red), linking the displacement of both structures. In addition, fasciae constrain lateral motions of the bladder and the rectum on each side. Friction contacts are taken into account to limit the mobility of the tissues. To not generate unsolvable problems (where the outline of the organs overlaps fixed faciae geometries), these structures are considered deformable with fixed constraints on their extremity. Therefore faciae re-

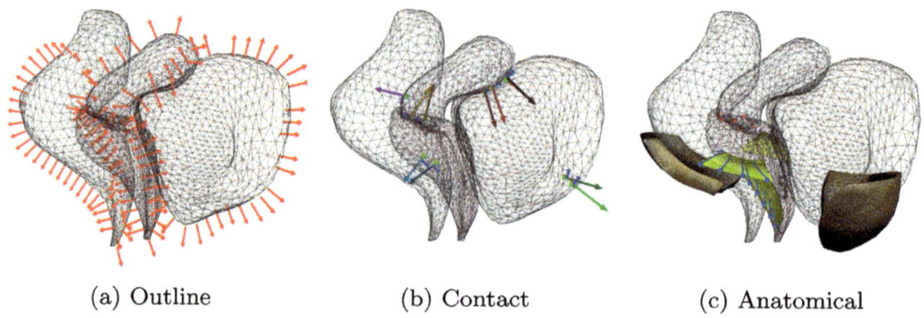

FIGURE 18.7

Constraints set used in the simulation: Outline constraints (left), Contact Constraints (center), anatomical constraints (right).

Table 18.1 Number of degrees of freedom (DoF) and Young's modulus (E) of the FE models used for the registration.

	DoF (T005)	DoF (T007)	E (MPa)
Vagina	928	1789	0.66
Uterus	740	523	0.6
Bladder	806	772	0.24
Rectum	894	2205	0.54
Fascia Pelvis	170	352	0.03
Fascia Sacrum	288	326	0.03
Ligaments	199	233	0.78

mains in contact with tissues, preventing lateral motion, thanks to contact simulation, while deformations are allowed to satisfy image constraints.

The exact patient-specific modeling of these conditions is complex (since boundary conditions are usually not visible in medical images). Therefore the precise location and geometries of fasciae are unknown and may change due to patient motions. However, in the next section, we show that even such approximate information significantly improves the registration accuracy.

18.4.4 Evaluation

Since boundary conditions and mechanical parameters are not accurately known for patient-specific data, all the following simulations have been run 100 times, where the mechanical parameters of each model have been randomized within a range of 10% around the values given in Table 18.1. In addition, the location of springs (used to attach the vagina and model ligaments) have been randomized around the ideal position segmented by an expert. The results that follow include the average error and the standard deviation of this simulations set.

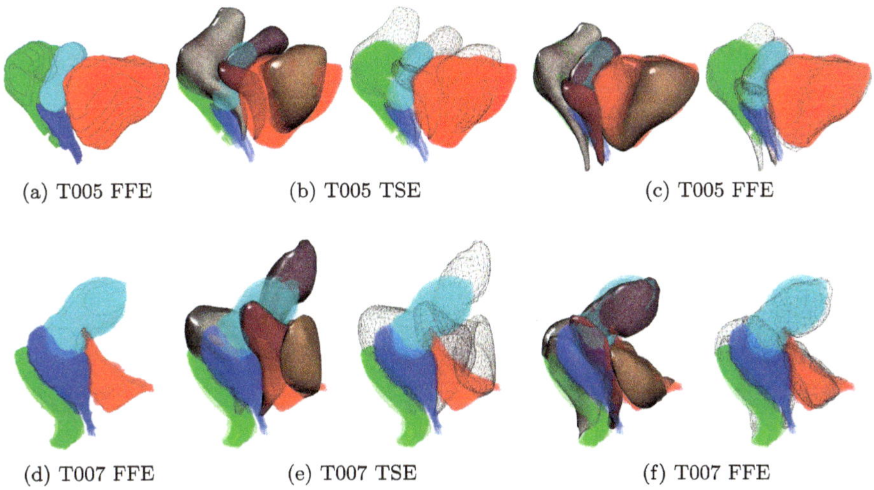

FIGURE 18.8

Overlapping between sparse FFE segmentation 18.8a and organs' geometries for both before 18.8b and after 18.8c the dynamic registration.

Since no off-plane data are available during the DYN sequence, validation can only be performed at the end of the dynamic acquisition using FFE slices. The 15 FFE sagittal images have been segmented, providing a coarse representation of the organs' geometries at maximal pressure state (see Fig. 18.8a).

Overlapping with the organs' geometries is shown in Fig. 18.8. Results show that complex motion and deformations happened, enhancing the need for biomechanical models during the registration: For T005, the vagina has been significantly displaced and contracted, leading to the compression of the rectum and sliding of the uterus along the bladder. For T007, the bladder is compressed by surrounding tissues and the fasciae, which significantly modify the organ's geometry. Our method provides realistic transitions, including far from the image plane.

The average distance error, according to FFE segmentation, is reported in Fig. 18.9. Results show that our biomechanical registration significantly reduces the error in the 2D DYN plane (index 0), where constraints are applied. Still, it can also provide an accurate extrapolation for off-plane motion. The vagina undergoes the most significant displacement. The average error between TSE and FFE surfaces is $15mn$ initially and reduced to $3mm$ using our method. The remaining part of the error is mainly due to a significant amount of contrast gel expelled during the dynamic acquisition (due to compression of muscles), which is not considered in current FE models.

Fig. 18.9 shows a comparison with registration without boundary conditions to enhance the importance of anatomical constraints. Although the method provides similar accuracy results in the dynamical plane, the error increases significantly faster for off-plane motion due to nonrealistic rotations of organs around the constrained

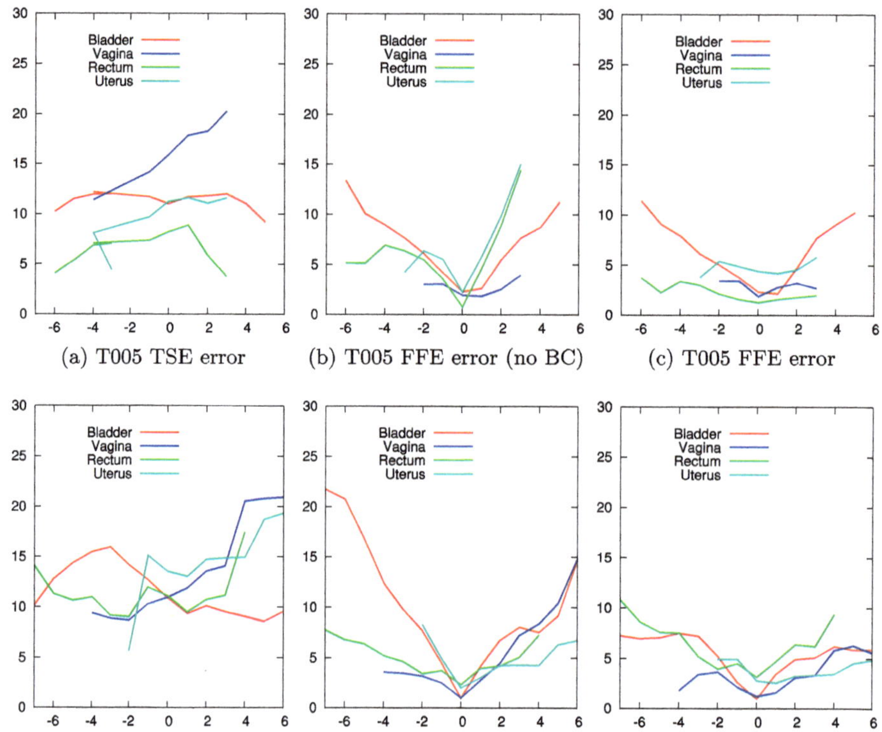

(a) T005 TSE error (b) T005 FFE error (no BC) (c) T005 FFE error

(d) T007 TSE error (e) T007 FFE error (no BC) (f) T007 FFE error

FIGURE 18.9

Average distance error (mm) according to the index of FFE sagittal planes. Index 0 corresponds to the location of DYN slices (i.e., where constraints are applied).

plane. More importantly, without any boundary conditions, the method leads to the interpenetration of organs that are not physically possible.

Finally, the total number of tetrahedral elements is shown in Table 18.1. Each simulation step involves 30 nonlinear friction contact constraints with 70 outline constraints. Apart from the segmentation and meshing of TSE images, the method is compatible with real-time (i.e., 1 second is simulated in 1 second).

18.5 Conclusion

A new constraint-based formalism was introduced for nonrigid registration between a preoperative model and the intraoperative configuration. The method relies on the registration of 3D FE models to 2D dynamic MRI slices, providing a biomechanical extrapolation for the off-plane dynamic motion of organs. We showed that *a priori* knowledge of boundary conditions can be combined with intraoperative image-based data to solve the ill-posed 2D/3D registration problem and provide a 3D extrapolation of the dynamic motion observed in a single 2D MRI slice.

The methodology follows a protocol compatible with current clinical constraints presenting potential short-term medical applications in this way. The registration accuracy was evaluated with two patient-specific data sets of volunteers presenting no pelvic pathology, and a sensitivity study is performed using synthetic data. We showed that the method is fast, robust, and compatible with intraoperative constraints. The resulting simulations provide a 3D extrapolation of the dynamic of the organs observed in a single 2D MRI slice, facilitating diagnosis compared to 2D sequences. As future work, a model of loss of organs' volume related to the expulsion of contrast gel should be added to the biomechanical models.

References

Baraff, D., Witkin, A., 1998. Large steps in cloth simulation. In: Proceedings of the 25th Annual Conference on Computer Graphics and Interactive Techniques. SIGGRAPH 1998. ACM, pp. 43–54.

Courtecuisse, H., Jiang, Z., Mayeur, O., Witz, J.F., Lecomte-Grosbras, P., Cosson, M., Brieu, M., Cotin, S., 2020. Three-dimensional physics-based registration of pelvic system using 2D dynamic magnetic resonance imaging slices. Strain 56 (3), e12339.

Courtecuisse, H., Peterlík, I., Trivisonne, R., Duriez, C., Cotin, S., 2014. Constraint-based simulation for non-rigid real-time registration. In: Medicine Meets Virtual Reality, vol. 196, pp. 76–82.

Duriez, C., Dubois, F., Kheddar, A.A.A., Andriot, C., 2006. Realistic haptic rendering of interactive deformable objects in virtual environments. IEEE Transactions on Visualization and Computer Graphics 12 (1), 36–47.

Felippa, C., 2000. A systematic approach to the element-independent corotational dynamics of finite elements. Tech. Rep. Center for Aerospace Structures.

Jiang, Z., Witz, J.F., Lecomte-Grosbras, P., Dequidt, J., Duriez, C., Cosson, M., Cotin, S., Brieu, M., 2015. B-spline based multi-organ detection in magnetic resonance imaging. Strain 51 (3), 235–247.

Nesme, M., Marchal, M., Promayon, E., Chabanas, M., Payan, Y., Faure, F., 2005. Physically realistic interactive simulation for biological soft tissues. Recent Research Developments in Biomechanics 661 (2), 11–22.

Rusinkiewicz, S., Levoy, M., 2001. Efficient variants of the ICP algorithm. In: Proceedings of International Conference on 3-D Digital Imaging and Modeling. 3DIM. IEEE, pp. 145–152.

Teschner, M., Kimmerle, S., Heidelberger, B., Zachmann, G., Raghupathi, L., Fuhrmann, A., Cani, M.P., Faure, F., Magnenat-Thalmann, N., Strasser, W., Volino, P., 2005. Collision detection for deformable objects. Computer Graphics Forum 24 (1), 61–81.

Trivisonne, R., Peterlik, I., Cotin, S., Courtecuisse, H., Peterlík, I., Courtecuisse, H., Cotin, S., Courtecuisse, H., Peterlik, I., Cotin, S., Courtecuisse, H., 2016. 3D physics-based registration of 2D dynamic MRI data. In: Medicine Meets Virtual Reality, vol. 220. IOS Press, p. 432.

CHAPTER

19 Towards patient-specific treatment in gynecologic surgery: recent development and perspectives

Mathias Brieu[a] and Michel Cosson[b]

[a]*College of Engineering, Computer Science and Technology, Dept. of Mechanical Engineering, California State University, Los Angeles, CA, United States*

[b]*CHU Lille, Service de Chirurgie Gynécologique, Lille, France*

19.1 Introduction

Obstetric and gynecologic surgical procedures represent almost one-fourth of women's surgery in the US (Oliphant et al., 2010). Every year, up to four million gynecologic surgeries are performed in the US (Oliphant et al., 2010; Borahay et al., 2020). These procedures concern the uterus, ovaries, cervix, or vagina. Several pathologies may justify gynecologic surgery: endometriosis, fibroids (benign tumors), ovarian cysts, cancer, chronic pelvic pain, and prolapse. Each of these surgeries will lead surgeons to proceed to resection, anatomical reconstruction, and, or add implants. In every case, the interaction of the surgeon with pelvic tissue and their biomechanical interactions are essential to the success of the treatment. The development of a better understanding of the biomechanics of the women's reproductive system and numerical tools aims to provide the surgeon with tools for training, decision support, and care improvement.

As discussed in Chapter 6 (Mechanical properties of women pelvic soft tissues), the mechanical behavior of pelvic tissues is elastic, nonlinear, and undergoes large deformations, also known as hyperelastic behavior, as explained and modeled in Chapter 10 (Constitutive models of soft connective tissues under large strain: application to pelvic tissue?). The mechanical properties of the soft connective tissue are, in particular, dictated by the mechanical loads they are subjected to Juncosa-Melvin et al. (2006, 2007); Kjaer and Magnussin (2008); Benhardt and Cosgriff-Hernandez (2009). The mechanical properties of the pelvic tissues are thus first due to their physiological functions. Elasticity and stretchability of the pelvic tissue is a requirement for physiological functioning of the pelvic organs: sexual relationship, filling of the bladder by urine, which can exceed 600 mL (Brouwer et al., 2018), or the rectum

Biomechanics of the Female Reproductive System: Breast and Pelvic Organs. https://doi.org/10.1016/B978-0-12-823403-7.00033-6

by stool, which can exceed 150 mL (Onal et al., 2009). Such hyperelastic behavior also helps with the mobility of pelvic organs, which is a physiologic need in various situations: for example, urinary continence allowed by the mobility of the bladder and the fixation of the urethra, bladder emptying with the alignment of the bladder in the axis of the urethra, or defecation allowed by the displacement of the rectum in the axis of the anus.

As discussed in Chapter 8 (Evolution of mechanical properties with pathology & aging: application to pelvic tissues?), the histology and the induced mechanical properties of the pelvic soft tissues evolve with age or pathology. Pathologies or tissue infiltrations, such as cancers but especially endometriosis, lead to changes in the rigidity of pelvic tissues with difficulties in the bladder or rectal emptying and dyspareunia. Other pathologies are associated with an elongation of the tissues, such as genital prolapse or the defect of fixation of the rectum, rectal prolapse. These evolutions lead to or are induced by modification of the observed mobility; see Chapter 1 (Pelvic floor functional anatomy) and Chapter 2 (Epidemiology & pathophysiology of pelvic organ prolapse & urinary incontinence in women) for more details. All those pathologies lead to the development of surgical techniques; for instance, Chapter 3 (Current surgical treatments for women's genital prolapse) sometimes shows moderate success.

Since the seminal work of Fung (1993), the image analysis of the pelvic system, the characterization and the modeling of the pelvic soft connective, and the numerical simulation of pelvic organ mobilities has been extensively developed. Image reconstruction (see Chapter 11—Medical imaging and patient-specific modeling of women pelvic system: application to magnetic resonance images) and image analysis (see Chapter 12—Quantitative assessment of pelvic mobility in women using MRI image analysis) offer opportunities to have an objective assessment of the mobility and pathological situation of specific patients. Coupled to nondestructive methods of characterization of the mechanical properties (see the following chapters: Chapter 5—Inverse problems in the characterization of soft connective tissue: perspective for reproduction system; Chapter 13—Patient-specific biomechanical modeling for applications in breast cancer diagnosis and treatment, or Chapter 14—Ultrasound elastography: in vivo assessment of tissue stiffness), image analysis techniques allow developing numerical approaches of the simulation and the prediction of pelvic organs mobility in various situations (see the following chapters: Chapter 15—Numerical simulation of vaginal delivery; Chapter 17—A numerical model for prolapse surgery, or Chapter 18—Augmented reality biomechanical simulations for pelvic conditions diagnoses).

Those preliminary works have already allowed researchers and clinician to improve their understanding of the mobility of the pelvic organs. They allowed, for instance, to understand that the evolution with the age of the mechanical properties and the lengthening of organs is similar to what is observed on the skin (Chantereau et al., 2014). They helped, among several results, to demonstrate that the standing posture induces more significant mechanical loadings of the uterosacral ligaments, which partly explains the greater rigidity of these ligaments compared to the other ligaments of the pelvic system (Rubod et al., 2013), the significant impact of patient-specific anatomy of the organ mobility and risks of abnormal mobility (Mayeur et al., 2016), and the interest of numerical simulation with potential application in training (Obst et al., 2004; Buttin et al., 2013).

However, most of the pre-cited work is not yet clinical tools used for evaluating the patient specificity, pathology, or evaluation of strategy of treatment. They are based on several assumptions that need to be considered and assessed.

Although we are convinced that engineering and numerical simulation will take an increasingly important part in surgical care, it is appropriate at this stage to define the milestones to be achieved and the clinical perspectives that seem to be within reach of the medical professionals in their quest for continuous improvement of care.

19.2 Research investigation needs

As we have already mentioned, the development of numerical tools for the improvement of treatments relies on a better understanding and modeling of the mechanical behavior of tissues, analysis of information provided by medical imaging and any other non- or minimally invasive technique, description, and modeling of pathologies, their genesis and treatments and in the case of evolution delivery. However, to our knowledge, some key questions remain to be studied before thinking about using numerical simulation to improve care.

19.2.1 Mechanical properties of pelvic tissue

The mechanical properties of pelvic tissue play a significant role in organ mobility and elasticity. As the literature reveals (see Chapter 8), these mechanical properties evolve with age and pathology. However, many questions remain unanswered and sometimes even inaccurately worded. No one has asked:

i. What is the variability of the mechanical properties of the pelvic tissue across the population? While the mechanical properties of pelvic tissue vary with pathology and age, is this variation similar within the population? Is there any overlap between the mechanical properties of healthy and pathological tissues between different parts of the population?
ii. How do the tissues, histology, and mechanical properties evolve with age? What is the impact of pregnancy or vaginal delivery? Are there any cumulative effects or tissue regeneration that can correct all or part of the effects, if they exist?
iii. Suppose pregnancy or vaginal delivery has some impact on some or every pelvic tissue, leading to their mechanical properties' evolution. Can we define some threshold of activation of such remodeling that would help to identify situations of risk due to pregnancy or vaginal delivery?
iv. Can we define some threshold of strain or stress that could occur only in specific cases, such as the vaginal delivery of a tall or big baby, which may lead to damage to the tissues, and thus lead to an evolution of the mechanical properties? Is

the damage to the tissue the same for short-term events, such as vaginal delivery or long-term events, such as aging, physical activity, or being overweight?

v. Could the evolution of the mechanical properties of tissues be an early indicator of the appearance of a pathology? Can we thus define a threshold at which tissue can be considered pathological from a mechanical perspective?

All those questions are fundamental to improving the understanding of pathology and further developing numerical methods that will help to improve treatment. However, answering those questions is almost impossible today. Answering those questions requires the study of a vast population and, ideally, a follow-up during the lifetime of the patients. Therefore such questions will not be answered if reliable nondestructive characterization methods exist.

19.2.2 Nondestructive characterization of the mechanical properties of pelvic tissue

The so-called inverse methods are the main approaches for the nondestructive characterization of the mechanical properties of soft tissue existing in the literature. Nondestructive methods are the keys to a long-term study that would help to answer the question introduced in the previous subsection. Ultrasound elastography methods (see Chapter 14—Ultrasound elastography: in vivo assessment of tissue stiffness) developed are, to date, based on the assumption of the linear elastic behavior of the materials. Consequently, existing ultrasound elastography can provide only Young's moduli. Its use for nonlinear materials, such as pelvic soft tissue, allows for measuring local tangent moduli at prescribed strain to approximate the nonlinear elastic behavior with a multilinear approximation. Some improvements and development are required:

vi. Nonlinear ultrasound elastography, made for material having nonlinear, anisotropic, and heterogeneous elastic properties, should be developed and available for clinical applications.

Other inverse mechanical properties characterization methods (introduced in Chapter 5—Inverse problems in the characterization of soft connective tissue: perspective for reproduction system) allow identifying nonlinear mechanical properties. Such inverse methods compare the measured behavior of a tissue or organ to a prescribed load (displacement or a force/pressure) to the numerical response obtained, while assuming the mechanical properties of the tissue or organ of interest. The gap between numerical simulation and observed behavior will help implement an iterative process of optimization of the mechanical properties of the tissue in the numerical model up to the convergence to the measured in vivo quantities.

Both methods developments, nonlinear ultrasound elastography and numerical inverse methods, require precise measurements of the displacements and forces involved, relying upon validation. This validation should be completed by comparing destructive ex-vivo measures to in vivo measures completed. Such an assessment can only be achieved with animal studies. Such animal studies would also help work on the open questions defined in the previous subsection. However, this raises a new required investigation:

vii. What is an accurate animal model? Which animal model may help answer every question cited in Section 19.2.1 and assess inverse methods?

19.2.3 Medical image analysis

Chapters 11 (Medical imaging and patient-specific modeling of women pelvic system: application to magnetic resonance images), 12 (Quantitative assessment of pelvic mobility in women using MRI image analysis), and 13 (Patient-specific biomechanical modeling for applications in breast cancer diagnosis and treatment) illustrated the developments in medical image analysis to apprehend the patient-specific pelvic or breast anatomy and their mobilities.

Each chapter showed some results of clinical use based on magnetic resonance images or elastography and illustrated the relevance of image analysis for clinical evaluation of pathology. However, the presented results were achieved on specific images obtained for research purposes, and their application to medical routine remains questioned. We may therefore, for instance, raise the following questions:

viii. How can we optimize the sequence of acquisition of medical images to improve the resolution of the presented results?

ix. Would any medical image technics resolution be sufficient for image reconstruction? In other words, are the method presented technology of image acquisition dependent? If yes, can we adapt those methods to every medical image technique?

x. Would artificial intelligence help with image reconstruction, motion analysis, and tumor identification? (To the authors' knowledge, to date, no AI on pelvic system work presenting results of the quality of what has been shown in the pre-cited has been published.)

19.2.4 Numerical simulation of women's reproductive system treatment

Chapters 15 (Numerical simulation of vaginal delivery), 16 (Numerical models for breast surgery and reconstruction), 17 (A numerical model for prolapse surgery), and 18 (Augmented reality biomechanical simulations for pelvic conditions diagnoses) illustrate the state of the art in terms of numerical simulation of women reproductive system. They proved that numerical simulations could simulate the importance of anatomical specificity on the nature of organ motions and load transmission, and that numerical simulation also allows us to simulate in the case of delivery the impact of different situations, such as the configuration of the vaginal delivery or the size of the baby or, in the case of surgery, the difference in the efficiency of surgical technics considering patient specificity. All the results presented allow us to envision fantastic improvement in the quality of the treatment but require some upgrades before being considered for clinical applications:

xi. Every numerical model presented is based on several simplification or approximation of the anatomy. How can the impact of such simplification be evaluated?
xii. In the case of delivery, how can the trajectory of the delivery, in the case of vaginal delivery, and the orientation of the fetal head be anticipated concerning the patient's anatomy or the circumstances of the delivery?
xiii. In the case of surgery. Numerical simulations are performed at the time of surgery. How can healing and remodeling, or even aging several years after surgery, be considered?
xiv. How can substantial deformation be modeled in both cases, delivery and surgery? Can we apprehend them to predict risks of immediate damage or tear of the tissue in the case of delivery or another risk of prolapse or need for reintervention?

19.3 Perspectives: Use of numerical simulation for treatment improvement in gynecology

Though several developments remain required to consider the clinical use of biomechanics numerical tools for improving the treatment quality, all the works presented in this book allow us to easily imagine how all these developments will revolutionize care techniques shortly. We are living in fascinating times, rich in scientific creativity that will make the medicine of tomorrow more and more engineering medicine.

19.3.1 Numerical simulator for training

The learning of care for all medical students is done in two phases, one theoretical and the other practical. The practice phase consists of learning the correct gestures and knowing how to do them at the right time. More and more, this gesture learning is performed by training with manikin or phantoms. This learning is also increasingly associated with using virtual reality tools to facilitate the immersion of learners in an environment conducive to learning. However, to date, in the case of surgical treatment in surgical gynecology, no manikin exists, whereas for vaginal delivery simulation, their realism is limited.

Research on the use of numerical simulation for the control of robotic manikin for the simulation of delivery is in progress. However, the emergence of virtual reality may lead to the development of a generation of training systems made of virtual reality glasses coupled with haptic controllers and accurate real-time numerical models, as described in Chapter 15. The same tools could be developed to train medical students in surgical gynecology.

19.3.2 Numerical simulation for a better understanding of physiology and pathology

Women's pelvic organs are experiencing physiologic large displacements and deformation. The main contribution of every sustaining part of the pelvic system (ligaments, fascia, and muscles) is not well understood. Their physiologic contribution is interpreted by different theories, and none of them is considered the most plausible one. However, understanding the contribution of every ligament, fascia, and muscle in the different phases of the motion and deformation of organs requires understanding physiology and how physiopathology may occur.

Numerical simulation can be used to develop a physiologic model that would, thanks to a step-by-step process, study the contribution of each ligament, fascia, or muscle with care. Once such a model is developed, it would also help to understand the occurrence of the pathology and test the different existing theories. This would also help define and identify stages of progression, or criteria, of pathology occurrence. This would help understand the limits or to define an objective criterion between physiological and pathological situations.

From an engineer's perspective, this would help to define with biomechanical findings what is a pathologic situation, how a situation evolves from a physiologic to a pathologic situation, and how, from a biomechanical point of view, we may act using surgical reconstruction or implants, for instance, to restore the physiological situation to a pathological patient.

19.3.3 Numerical tools for risks or pathology evaluation and prevention

In the case of delivery, the risk evaluation methods are only based on statistical studies of clinical parameters, such as age, weight, parity, size, and weight of the baby, etc. No plans are based on the patient- and baby-specific anatomical, morphological, or mechanical characteristics. Same today, the evaluation of a patient's pathology, such as prolapse, endometriosis, or urinary incontinence, is only based on medical images and clinical exams. In the last analysis, in both case, the evaluation of the patient specificities are only based on the physician evaluation, which is unfortunately subjective and may vary from one physician to another.

Numerical models based on medical images, nondestructive tissue characterization, and accurate biomechanical numerical models will help the clinician to objectively evaluate the pathology's risks or nature. With accurate numerical models, we could run some pre-delivery simulations to estimate, considering the fetus size, the patient's anatomy and the patient's tissue mechanical properties; we could consider every mode of delivery and assess the perineal risks: episiotomy, tear of tissue, etc. We could even estimate deep tissues overstretch and provide recommendations for post-delivery personalized sports or daily activities. The numerical simulation would become a tool to prevent risks and complications.

19.3.4 Numerical tools for treatment planification

Giving birth in a hospital, which is the case in many countries, is not a necessity. It depends, of course, on the mother's wishes and the risks that may occur during the delivery. It is also not desirable to perform the same surgical procedure for every patient with prolapse, urinary incontinence, endometriosis, or cancer. Today, the decision to recommend inpatient delivery or the surgical procedure remains the physician's entire decision. It is based on the information and expertise available to them. Decisions are too frequently based on habits and not on objective patient specificities.

If we had more accurate numerical evaluations, it could help the clinician give better advice and have a more efficient planification of the treatment plan. Women could be duly informed about their future delivery and make their own decision. Clinicians, in the case of prolapse, would have more objective information to think about the different approaches: hormonal treatment, physiotherapy with personalized recommendations, pessaries selection and use, or surgical technique to be used (choice of the technics, use or not of mesh implants, size, and location of the mesh, number of sutures, orientation and tension of the suspensions, etc.). Such numerical tools could also improve the post-surgical and recovery phases with personalized recommendations about physical activities: duration, limitations, etc.

Moreover, such numerical biomechanics tools would also help for a treatment that is more respectful to the patient, allowing her to be accompanied in her choices and the absence of undesirable stress factors.

19.3.5 Numerical tools for treatment assistance

During care, the physician is left to his own choices. They only have limited information about the measured medical constants and what they can observe. This creates high-pressure situations for the doctor and potentially an increased risk in case of complications.

However, the development of digital simulation tools in real-time offers promising prospects for assistance in care. These digital tools will provide the physician with much information on his environment and the potential consequences of his choices. This will give them augmented reality tools in the context of care.

19.3.6 Towards a new generation of treatment

To date, the remodeling of tissue and its aging is not well understood and considered in the treatment, except for the rare medical doctors with expertise built over the years. However, we are aging, and it would be expected to think about therapy, surgery, or delivery, considering that actions on our body today may have unexpected consequences in several years.

Biomechanical numerical simulation allows thinking about a new generation of treatment that considers the treatment's effectiveness over the years. Once the impact of damage, remodeling, and aging is a better model, the new mechanical models will be quickly integrated into numerical models. They will allow predicting the evolution of the patient with years.

This will also open unique perspectives of development towards fully degradable implants that would help, for instance, in the case of prolapse, to think about treatment that would foster tissue regeneration and lead the patient to be healthy without having a permanent implant in their body, a situation of growing concern.

References

Benhardt, H.A., Cosgriff-Hernandez, E.M., 2009. The role of mechanical loading in ligament tissue engineering. Tissue Engineering. Part B: Reviews 15 (4), 467–475.

Borahay, M.A., Wethington, S.L., Wang, K.C., Christianson, M.S., Martin, S., Lawson, S.M., Esguerra, C., Lippitt, M.H., Wu, H., Handa, V.L., Satin, A.J., Fader, A.N., 2020. Patient-centered gynecology-specific prioritization of nonurgent surgeries during the COVID-19 pandemic: proposal of a novel scoring system. Journal of Minimally Invasive Gynecology 27 (6). https://doi.org/10.1016/j.jmig.2020.05.026. Epub 2020 Jun 3. PMID: 32504823; PMCID: PMC7837225.

Brouwer, T.A., van den Boogaard, C., van Roon, E.N., Kalkman, C.J., Veeger, N, 2018. Non-invasive bladder volume measurement for the prevention of postoperative urinary retention: validation of two ultrasound devices in a clinical setting. Journal of Clinical Monitoring and Computing 32, 1117–1126. https://doi.org/10.1007/s10877-018-0123-6.

Buttin, R., Zara, F., Shariat, B., Redarce, T., Grange, G., 2013. Biomechanical simulation of the fetal descent without imposed theoretical trajectory. Computer Methods and Programs in Biomedicine 111 (2), 389–401.

Chantereau, P., Brieu, M., Kammal, M., Farthmann, J., Gabriel, B., Cosson, M., 2014. Mechanical properties of pelvic soft tissue of young women and impact of aging. International Urogynecology Journal 25, 1547–1553.

Fung, Y.C., 1993. Biomechanics - Mechanical Properties of Living Tissues. Springer.

Juncosa-Melvin, N., Matlin, K.S., Holdcraft, R.W., Nirmalanandhan, V.S., Butler, D.L., 2007. Mechanical stimulation increases collagen type I and collagen type III gene expression of stem cell–collagen sponge constructs for patellar tendon repair. Tissue Engineering 13 (6), 1219–1226.

Juncosa-Melvin, N., Shearn, J.T., Boivin, G.P., Gooh, C., Galloway, M.T., West, J.R., Nirmalanandhan, V.S., Bradica, G., Butler, D.L., 2006. Effects of mechanical stimulation on the biomechanics and histology of stem cell–collagen sponge constructs for rabbit patellar tendon repair. Tissue Engineering 12 (8), 2291–2300.

Kjaer, M., Magnussin, S., 2008. Mechanical adaptation and tissue remodeling. In: Fratzl (Ed.), Collagen. Springer.

Mayeur, O., Witz, J.F., Lecomte, P., Brieu, M., Cosson, M., Miller, K., 2016. Influence of geometry and mechanical properties on the accuracy of patient-specific simulation of women pelvic floor. Annals of Biomedical Engineering 44 (1), 202–212. https://doi.org/10.1007/s10439-015-1401-9.

Obst, T., Burgkart, R., Ruckhäberle, E., Riener, R., 2004. The Delivery Simulator: A New Application of Medical VR. Studies in Health Technology and Informatics, vol. 98, pp. 281–287.

Oliphant, S.S., Jones, K.A., Wang, L., Bunker, C.H., Lowder, J.L., 2010. Trends over time with commonly performed obstetric and gynecologic inpatient procedures. Obstetrics and Gynecology 116 (4), 926–931. https://doi.org/10.1097/AOG.0b013e3181f38599.

Onal, C., Topkan, E., Efe, E., Yavuz, M., Sonmez, S., Yavuz, Y., 2009. Comparison of rectal volume definition techniques and their influence on rectal toxicity in patients with prostate cancer treated with 3D conformal radiotherapy: a dose-volume analysis. Radiation Oncology 4, 14. https://doi.org/10.1186/1748-717X-4-14.

Rubod, C., Lecomte-Grosbras, P., Brieu, M., Giraudet, G., Betrouni, N., Cosson, M., 2013. 3D simulation of pelvic system numerical simulation for a better understanding of the contribution of the uterine ligaments. International Urogynecology Journal 20. https://doi.org/10.1007/s00192-013-2135-6.

Index

J

K

L

M

N

O